SIMPLIFIED ENGINEERING FOR ARCHITECTS AND BUILDERS

SIMPLIFIED ENGINEERING FOR ARCHITECTS AND BUILDERS

12th Edition

JAMES AMBROSE AND PATRICK TRIPENY

Cover image: © gyn9037/Shutterstock
Cover design: Wiley

This book is printed on acid-free paper. ♾

Published by John Wiley & Sons, Inc., Hoboken, New Jersey.
Published simultaneously in Canada.

Library of Congress Cataloging-in-Publication Data is available.

ISBN 978-1-118-97504-6 (hardback)
978-1-118-97531-2 (epdf)
978-1-118-97530-5 (epub)

Printed in the United States of America

10 9 8 7 6 5 4 3 2 1

CONTENTS

PREFACE TO THE TWELFTH EDITION

This book treats the topic of design of structures for buildings. As with previous editions, the material in this book has been prepared for persons lacking formal training in engineering. Mathematical work is limited mostly to simple algebra. It is thus well suited for programs in architecture and building construction.

However, as most programs in civil engineering offer little opportunity for study of the general fields of building planning and construction, this book may well be useful as a supplement to engineering texts. The emphasis here is on the development of practical design, which typically involves a relatively small effort in structural investigation and a lot of consideration for circumstantial situations relating to the existence of the building structure.

Changes that occur in reference sources and in design and construction practices make it necessary to revise the material in this book periodically. This edition has indeed received such an updating, although the reader is advised that these changes are continuous, so that it is inevitable that some of the material present here will be outdated in a short time. However, the concentration in this work is on fundamental concepts and processes of investigation and design; thus the use of specific data is of less concern to the learning of the fundamental material. For use in any actual design work, data should be obtained from current references. A list of these references can be found in the reference section of this book.

In addition to updating, each new edition affords an opportunity to reconsider the organization, presentation, and scope of the material contained in the book. This new edition therefore offers some minor alterations of the basic content of previous editions, although just about everything contained in the previous edition is here somewhere. Some trimming has occurred, largely in order to add new material without significantly increasing the size of the book.

This textbook has an accompanying website for students. The purpose of this website is to provide study material that will aid students in learning this complex material. It is available to anyone using this text either as part of the purchase price or at a nominal cost for those obtaining it secondhand. We encourage all students to take advantage of this resource.

In recent editions it has been the practice to provide answers for all of the computational exercise problems. However, this book receives considerable use as a course text, and several teachers have requested that some problems be reserved for use without given answers. To accommodate this request in this edition, additional exercise problems have been provided, with answers given only to alternate problems. These answers are provided at the accompanying website.

For text demonstrations, as well as for the exercise problems, it is desirable to have some data sources contained in this book. We are grateful to various industry organizations for their permission to use excerpts from these data sources, acknowledgment for which is provided where data are provided.

Both personally—as the authors of this edition—and as representatives of the academic and professional communities, we must express our gratitude to John Wiley & Sons for its continued publication of this highly utilized reference source. We are truly grateful for the sympathetic and highly competent support provided by the Wiley editors and production staff.

Finally, we need to express the gratitude we have to our families. Writing work, especially when added to an already full-time occupation, is very time consuming. We thank our spouses and children for their patience, endurance, support, and encouragement in permitting us to achieve this work.

James Ambrose
Patrick Tripeny

PREFACE TO THE FIRST EDITION

(The following is an excerpt from Professor Parker's preface to the first edition.)

To the average young architectural draftsman or builder, the problem of selecting the proper structural member for given conditions appears to be a difficult task. Most of the numerous books on engineering which are available assume that the reader has previously acquired a knowledge of fundamental principles and, thus, are almost useless to the beginner. It is true that some engineering problems are exceedingly difficult, but it is also true that many of the problems that occur so frequently are surprisingly simple in their solution. With this in mind, and with a consciousness of the seeming difficulties in solving structural problems, this book has been written.

In order to understand the discussions of engineering problems, it is essential that the student have a thorough knowledge of the various terms which are employed. In addition, basic principles of forces in equilibrium must be understood. The first section of this book,

"Principles of Mechanics," is presented for those who wish a brief review of the subject. Following this section are structural problems involving the most commonly used building materials, wood, steel, reinforced concrete, and roof trusses. A major portion of the book is devoted to numerous problems and their solution, the purpose of which is to explain practical procedure in the design of structural members. Similar examples are given to be solved by the student. Although handbooks published by the manufacturers are necessities to the more advanced student, a great number of appropriate tables are presented herewith so that sufficient data are directly at hand to those using this book.

Care has been taken to avoid the use of advanced mathematics, a knowledge of arithmetic and high school algebra being all that is required to follow the discussions presented. The usual formulas employed in the solution of structural problems are given with explanations of the terms involved and their application, but only the most elementary of these formulas are derived. These derivations are given to show how simple they are and how the underlying principle involved is used in building up a formula that has practical application.

No attempt has been made to introduce new methods of calculation, nor have all the various methods been included. It has been the desire of the author to present to those having little or no knowledge of the subject simple solutions of everyday problems. Whereas thorough technical training is to be desired, it is hoped that this presentation of fundamentals will provide valuable working knowledge and, perhaps, open the doors to more advanced study.

Harry Parker
Philadelphia, Pennsylvania
March, 1938

INTRODUCTION

The principal purpose of this book is to develop the topic of *structural design*. However, to do the necessary work for design, use must be made of various methods of *structural investigation*. The work of investigation consists of the consideration of the tasks required of a structure and the evaluation of the responses of the structure in performing these tasks. Investigation may be performed in various ways, the principal ones being the use of modeling by either mathematics or the construction of physical models. For the designer, a major first step in any investigation is the visualization of the structure and the force actions to which it must respond. In this book, extensive use is made of graphic illustrations in order to encourage the reader in the development of the habit of first clearly *seeing* what is happening, before proceeding with the essentially abstract procedures of mathematical investigation. When working a problem within the book, the reader is encouraged to begin it by drawing an illustration of the problem.

Structural Mechanics

The branch of physics called *mechanics* concerns the actions of forces on physical bodies. Most of engineering design and investigation is based on applications of the science of mechanics. *Statics* is the branch of mechanics that deals with bodies held in a state of unchanging motion by the balanced nature (called *static equilibrium*) of the forces acting on them. *Dynamics* is the branch of mechanics that concerns bodies in motion or in a process of change of shape due to actions of forces. A static condition is essentially unchanging with regard to time; a dynamic condition implies a time-dependent action and response.

When external forces act on a body, two things happen. First, internal forces that resist the actions of the external forces are set up in the body. These internal forces produce *stresses* in the material of the body. Second, the external forces produce *deformations*, or changes in shape, of the body. *Strength of materials*, or *mechanics of materials*, is the study of the properties of material bodies that enable them to resist the actions of external forces, of the stresses within the bodies, and of the deformations of bodies that result from external forces.

Taken together, the topics of applied mechanics and strength of materials are often given the overall designation of *structural mechanics* or *structural analysis*. This is the fundamental basis for structural investigation, which is essentially an analytical process. On the other hand, *design* is a progressive refining process in which a structure is first visualized; then it is investigated for required force responses and its performance is evaluated. Finally—possibly after several cycles of investigation and modification—an acceptable form is derived for the structure.

Units of Measurement

Early editions of this book used U.S. units (feet, inches, pounds, etc.) with equivalent SI (Standard International—aka metric) units in brackets for the basic presentation. In this edition, the basic work is developed with U.S. units only. While the building industry in the United States is now in the slow process of changing to SI units, our decision for the presentation here is a pragmatic one. Most of the references used for this book are still developed primarily in U.S. units and most readers educated in the United States use U.S. units as their first language, even if they now also use SI units.

Table I.1 lists the standard units of measurement in the U.S. system with the abbreviations used in this work and a description of common usage in structural design work. In similar form, Table I.2 gives the corresponding units in the SI system. Conversion factors to be used for shifting from one unit system to the other are given in Table I.3.

TABLE I.1 Units of Measurement: U.S. System

Name of Unit	Abbreviation	Use in Building Design
Length		
Foot	ft	Large dimensions, building plans, beam spans
Inch	in.	Small dimensions, size of member cross sections
Area		
Square feet	ft^2	Large areas
Square inches	$in.^2$	Small areas, properties of cross sections
Volume		
Cubic yards	yd^3	Large volumes, of soil or concrete (commonly called simply “yards”)
Cubic feet	ft^3	Quantities of materials
Cubic inches	$in.^3$	Small volumes
Force, Mass		
Pound	lb	Specific weight, force, load
Kip	kip, k	1000 pounds
Ton	ton	2000 pounds
Pounds per foot	lb/ft, plf	Linear load (as on a beam)
Kips per foot	kips/ft, klf	Linear load (as on a beam)
Pounds per square foot	lb/ft^2, psf	Distributed load on a surface, pressure
Kips per square foot	k/ft^2, ksf	Distributed load on a surface, pressure
Pounds per cubic foot	lb/ft^3	Relative density, unit weight
Moment		
Foot-pounds	ft-lb	Rotational or bending moment
Inch-pounds	in.-lb	Rotational or bending moment
Kip-feet	kip-ft	Rotational or bending moment
Kip-inches	kip-in.	Rotational or bending moment
Stress		
Pounds per square foot	lb/ft^2, psf	Soil pressure
Pounds per square inch	$lb/in.^2$, psi	Stresses in structures
Kips per square foot	$kips/ft^2$, ksf	Soil pressure
Kips per square inch	$kips/in.^2$, ksi	Stresses in structures
Temperature		
Degree Fahrenheit	°F	Temperature

TABLE I.2 Units of Measurement: SI System

Name of Unit	Abbreviation	Use in Building Design
Length		
Meter	m	Large dimensions, building plans, beam spans
Millimeter	mm	Small dimensions, size of member cross sections
Area		
Square meters	m^2	Large areas
Square millimeters	mm^2	Small areas, properties of member cross sections
Volume		
Cubic meters	m^3	Large volumes
Cubic millimeters	mm^3	Small volumes
Mass		
Kilogram	kg	Mass of material (equivalent to weight in U.S. units)
Kilograms per cubic meter	kg/m^3	Density (unit weight)
Force, Load		
Newton	N	Force or load on structure
Kilonewton	kN	1000 newtons
Stress		
Pascal	Pa	Stress or pressure (1 pascal = 1 N/m^2)
Kilopascal	kPa	1000 pascals
Megapascal	MPa	1,000,000 pascals
Gigapascal	GPa	1,000,000,000 pascals
Temperature		
Degree Celsius	°C	Temperature

Direct use of the conversion factors will produce what is called a *hard conversion* of a reasonably precise form. Even though all of the work done in this book with be in U.S. units, the tables with SI units are given as a handy reference to readers who may be using reference books in SI units or using both systems.

Accuracy of Computations

Structures for buildings are seldom produced with a high degree of dimensional precision. Exact dimensions are difficult to achieve, even for the most diligent of workers and builders. Add this to considerations for the lack of precision in predicting loads for any structure, and the significance of highly precise structural computations

TABLE I.3 Factors for Conversion of Units

To Convert from U.S. Units to SI Units, Multiply by:	U.S. Unit	SI Unit	To Convert from SI Units to U.S. Units, Multiply by:
25.4	in.	mm	0.03937
0.3048	ft	m	3.281
645.2	in.^{2}	mm^2	1.550×10^{-3}
16.39×10^3	in.^{3}	mm^3	61.02×10^{-6}
416.2×10^3	in.^{4}	mm^4	2.403×10^{-6}
0.09290	ft^2	m^2	10.76
0.02832	ft^3	m^3	35.31
0.4536	lb (mass)	kg	2.205
4.448	lb (force)	N	0.2248
4.448	kip (force)	kN	0.2248
1.356	ft-lb (moment)	N-m	0.7376
1.356	kip-ft (moment)	kN-m	0.7376
16.0185	lb/ft^3 (density)	kg/m^3	0.06243
14.59	lb/ft (load)	N/m	0.06853
14.59	kips/ft (load)	kN/m	0.06853
6.895	psi (stress)	kPa	0.1450
6.895	ksi (stress)	MPa	0.1450
0.04788	psf (load or pressure)	kPa	20.93
47.88	ksf (load or pressure)	kPa	0.02093
$0.566 \times (°F - 32)$	°F	°C	$(1.8 \times °C) + 32$

becomes moot. This is not to be used for an argument to justify sloppy mathematical work, overly sloppy construction, or use of vague theories of investigation of behaviors. Nevertheless, it makes a case for not being highly concerned with any numbers beyond three significant digits.

While most professional design work these days is likely to be done with computer support, most of the work illustrated here is quite simple and was actually performed with a hand calculator (the 8-digit, scientific type is adequate). Rounding off of these computations is done with no apologies.

With the use of the computer, accuracy of computational work is a somewhat different matter. Still, it is the designer (a person) who makes judgments based on the computations and who knows how good the input to the computer was and what the real significance of the degree of accuracy of an answer is.

Symbols

The following shorthand symbols are frequently used.

Symbol	Reading
$>$	is greater than
$<$	is less than
$\geq$	equal to or greater than
$\leq$	equal to or less than
Σ	the sum of
ΔL	change in L

Standard Notation

Notation used in this book complies generally with that used in the building design field. A general attempt has been made to conform to usage in the reference standards commonly used by structural designers. The following list includes all of the notation used in this book that is general and is related to the topic of the book. Specialized notation is used by various groups, especially as related to individual materials: wood, steel, masonry, concrete, and so on. The reader is referred to basic references for notation in special fields. Some of this notation is explained in later parts of this book.

Building codes use special notation that is usually carefully defined by the code, and the reader is referred to the source for interpretation of these definitions. When used in demonstrations of computations, such notation is explained in the text of this book.

A_g = gross area of a section, defined by the outer dimensions
A_n = net area
C = compressive force
E = modulus of elasticity
I = moment of inertia
L = length (usually of a span)
M = bending moment
P = concentrated load
S = elastic section modulus
T = tension force
W = (1) total gravity load; (2) weight, or dead load of an object; (3) total wind load force; (4) total of a uniformly distributed load or pressure due to gravity

Z = plastic section modulus
A = unit area
E = eccentricity of a nonaxial load, from point of application of the load to the centroid of the section
F = computed stress
H = effective height (usually meaning unbraced height) of a wall or column
L = length, usually of a span
S = spacing, center to center
W = unit of weight or other uniformly distributed load per unit length of member (note: usually $W = wL$ or wl)

SIMPLIFIED ENGINEERING FOR ARCHITECTS AND BUILDERS

I

FUNDAMENTAL FUNCTIONS OF STRUCTURES

This part presents various considerations regarding the general nature and performance of structures for buildings. A major part of this material consists of basic concepts and applications from the field of applied mechanics as they have evolved in the process of investigation of the behavior of structures. The purpose of studying this material is twofold: first, the general need for a comprehensive understanding of what structures do and how they do it; second, the need for some factual, quantified basis for the exercise of judgment in the process of structural design. If it is accepted that the understanding of a problem is the necessary first step in its solution, this analytical study should be seen as the cornerstone of any successful, informed design process.

A second major concern is for the sources of the tasks that structures must undertake, that is, for what structures are basically needed.

These tasks are defined in terms of the loads that are applied to structures. Considerations in this regard include the load sources, the manner of their application, the combinations in which they occur, and the quantification of their specific values.

Finally, consideration must be given to the possible forms and materials of structures. These concerns affect the determination of the construction of the structures, but they also relate to the general design development of the buildings.

This part provides an introduction; the remaining parts of the book present amplifications of all the issues raised in this part.

1

INVESTIGATION OF FORCES, FORCE SYSTEMS, LOADING, AND REACTIONS

Loads deriving from the tasks of a structure produce forces. The tasks of the structure involve the transmission of the load forces to the supports for the structure. The external loads and support forces produce a resistance from the structure in terms of internal forces that resist changes in the shape of the structure. This chapter treats the basic properties and actions of forces.

1.1 PROPERTIES OF FORCES

Force is a fundamental concept of mechanics but does not yield to simple definition. An accepted concept is that a *force* is an effort that tends to change the form or the state of motion of a physical object. Mechanical force was defined by Isaac Newton as being a product of mass and acceleration; that is, $F = ma$. Gravitational attraction is a form of acceleration, and thus weight—a force we experience—is defined as $W = mg$ with g being the acceleration of gravity. Physical objects have weight, but more precisely they have mass, and will thus have *different* weights when they experience different gravitational effects, for example, on the surface of Earth or on the surface of the moon.

In U.S., aka imperial, units gravity force is quantified as the weight of the body. Gravity forces are thus measured in pounds (lb) or in some other unit such as tons (T) or kips (one kilopound, or 1000 pounds). In the SI (International System, aka metric) system force is measured in a more scientific manner related to the mass of objects, the mass of an object being a constant, whereas weight is proportional to the precise value of the acceleration of gravity, which varies from place to place. Force in metric units is measured in newtons (N) or in kilonewtons (kN) or in meganewtons (mN), whereas weight is measured in grams (g) or in kilograms (kg).

In structural engineering work, forces are described as *loads*. Loads derive from various sources, including gravity, and are dealt with in terms of their application to a given structure. Thus, the gravity load on a beam begins with the weight of the beam itself and goes on to include the weight of everything else supported by the beam.

Vectors

A quantity that involves magnitude, line of action (e.g., vertical), and sense (up, down, etc.) is a vector quantity, whereas a scalar quantity involves only magnitude and sense. Force, velocity, and acceleration are vector quantities, while energy, time, and temperature are scalar quantities. A vector can be represented by a straight line, leading to the possibility of constructed graphical solutions in some cases, a situation that will be demonstrated later. Mathematically, a scalar quantity can be represented completely as +50 or −50, while a vector must somehow have its line of action represented as well (50 vertical, horizontal, etc.).

Properties of Forces

In order to completely identify a force, it is necessary to establish the following:

Magnitude, or the amount of the force, measured in weight units such as pounds or kips.

Line of Action of the force, which refers to the orientation of its path, usually described by the angle that the line of action makes with some reference, such as the horizontal.

Sense of the force, which refers to the manner in which it acts along its line of action (e.g., up or down, right or left). Sense is usually expressed algebraically in terms of the sign of the force, either plus or minus.

Forces can be represented graphically in terms of these three properties by the use of an arrow, as shown in Figure 1.1*a*. Drawn to some scale, the length of the arrow represents the magnitude of the force.

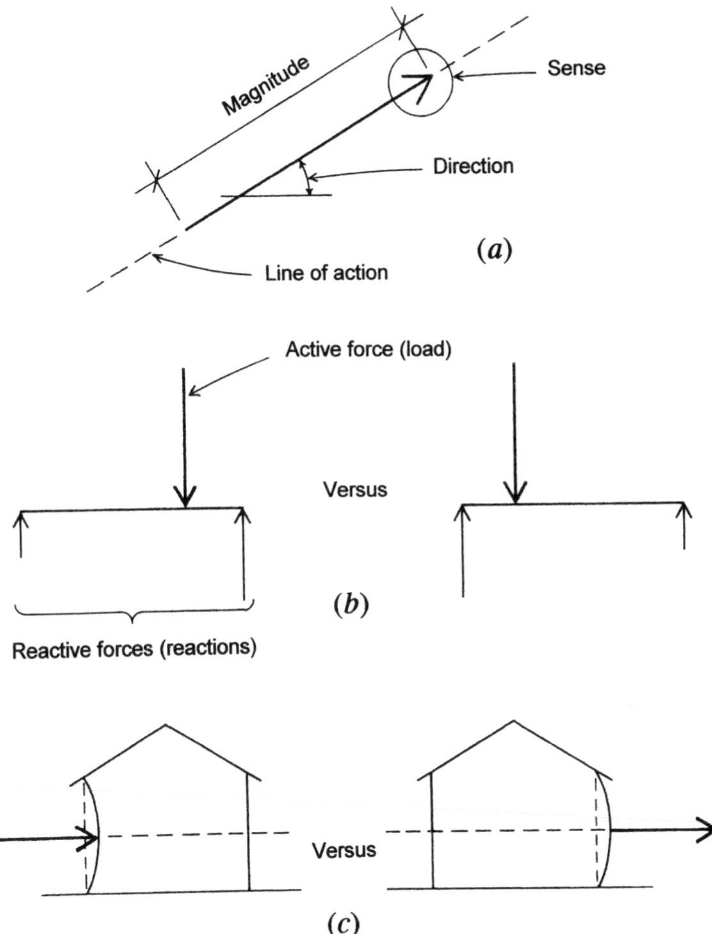

Figure 1.1 Representation of forces and force actions.

The angle of inclination of the arrow represents the direction of the force. The location of the arrowhead represents the sense of the force. This form of representation can be more than merely symbolic, since actual mathematical manipulations may be performed using the vector representation that the force arrows constitute. In the work in this book arrows are used in a symbolic way for visual reference when performing algebraic computations and in a truly representative way when performing graphical analyses.

In addition to the basic properties of magnitude, line of action, and sense, some other concerns that may be significant for certain investigations are:

The *position of the line of action* of the force with respect to the lines of action of other forces or to some object on which the force operates, as shown in Figure 1.1*b*. For the beam, shifting of the location of the load (active force) affects changes in the forces at the supports (reactions).

The *point of application* of the force along its line of action may be of concern in analyzing for the specific effect of the force on a structure, as shown in Figure 1.1*c*.

When forces are not resisted, they tend to produce motion. An inherent aspect of static forces is that they exist in a state of *static equilibrium*, that is, with no motion occurring. In order for static equilibrium to exist, it is necessary to have a balanced system of forces. An important consideration in the analysis of static forces is the nature of the geometric arrangement of forces in a given set of forces that constitute a single system. The usual technique for classifying force systems involves consideration of whether the forces in the system are:

Coplanar. All acting in a single plane, such as the plane of a vertical wall.

Parallel. All having the same direction though not along the same line of action.

Concurrent. All having their lines of action intersect at a common point.

Using these three considerations, the possible variations are given in Table 1.1 and illustrated in Figure 1.2.

TABLE 1.1 Classification of Force Systems[a]

	Qualifications		
System Variation	Coplanar	Parallel	Concurrent
1	Yes	Yes	Yes
2	Yes	Yes	No
3	Yes	No	Yes
4	Yes	No	No
5	No	Yes	No
6	No	No	Yes
7	No	No	No

[a]See Figure 1.2.

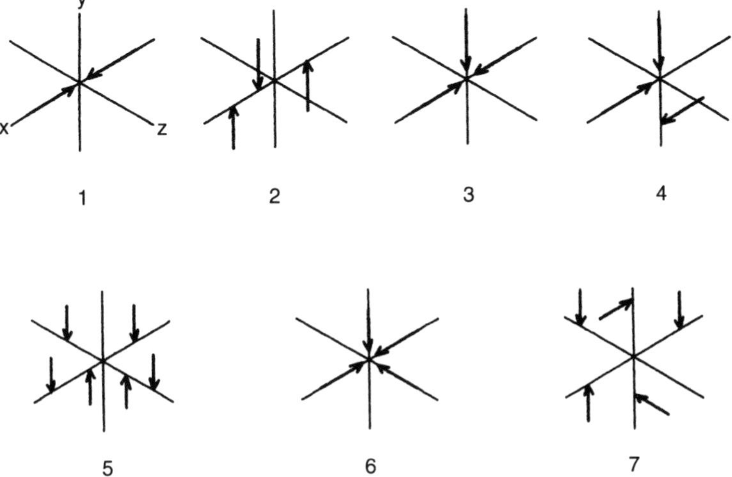

Figure 1.2 Types of force systems.

It is necessary to qualify a set of forces in the manner just illustrated before proceeding with any analysis, whether it is to be performed algebraically or graphically.

1.2 STATIC EQUILIBRIUM

As stated previously, an object is in *equilibrium* when it either is at rest or has uniform motion. When a system of forces acting on an

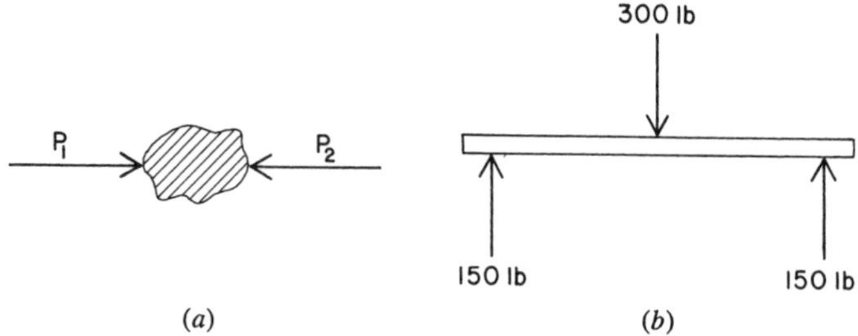

Figure 1.3 Equilibrium of forces.

object produces no motion, the system of forces is said to be in *static equilibrium*.

A simple example of equilibrium is illustrated in Figure 1.3*a*. Two equal, opposite, and parallel forces, having the same line of action, P_1 and P_2, act on a body. If the two forces balance each other, the body does not move and the system of forces is in equilibrium. These two forces are *concurrent*. If the lines of action of a system of forces have a point in common, the forces are concurrent.

Another example of forces in equilibrium is illustrated in Figure 1.3*b*. A vertical downward force of 300 lb acts at the midpoint in the length of a beam. The two upward vertical forces of 150 lb each (the reactions) act at the ends of the beam. The system of three forces is in equilibrium. The forces are parallel and, not having a point in common, are *nonconcurrent*.

1.3 FORCE COMPONENTS AND COMBINATIONS

Individual forces may interact and be combined with other forces in various situations. Conversely, a single force may have more than one effect on an object, such as a vertical action and a horizontal action simultaneously. This section considers both of these issues: adding up of forces (combination) and breaking down of single forces into components (resolution).

Resultant of Forces

The *resultant* of a system of forces is the simplest system (usually a single force) that has the same effect as the various forces in the system acting simultaneously. The lines of action of any system of two coplanar nonparallel forces must have a point in common, and the resultant of the two forces will pass through this common point. The resultant of two coplanar, nonparallel forces may be found graphically by constructing a *parallelogram of forces*.

To construct a parallelogram of two forces, the forces are drawn at any scale (so many pounds to the inch) with both forces pointing toward or both forces pointing away from the point of intersection of their lines of action. A parallelogram is then produced with the two forces as adjacent sides. The diagonal of the parallelogram passing through the common point is the resultant in magnitude, line of action, and sense, the direction of the resultant being similar to that of the given forces, toward or away from the point in common. In Figure 1.4*a*, P_1 and P_2 represent two nonparallel forces whose lines of action intersect at point O. The parallelogram is drawn, and the diagonal R is the resultant of the given system. In this illustration note that the two forces point *away* from the point in common, and hence the resultant also has its sense away from point O. It is a force upward to the right. Notice that the resultant of forces P_1 and P_2 shown in Figure 1.4*b* is R; its sense is toward the point in common.

Forces may be considered to act at any points on their lines of action. In Figure 1.4*c* the lines of action of the two forces P_1 and

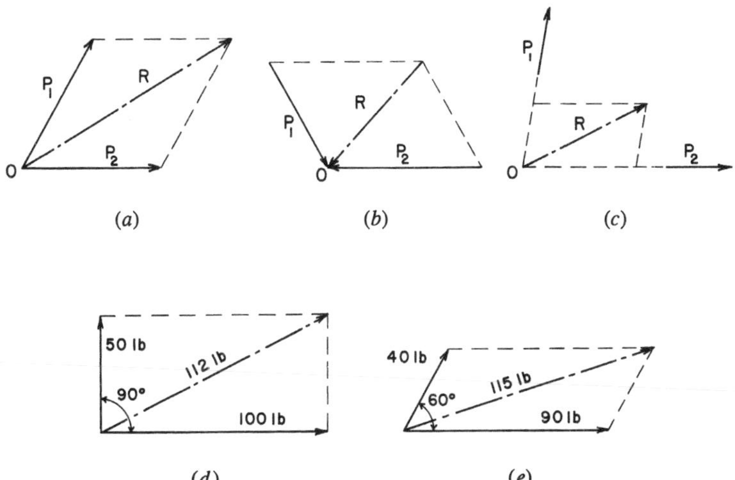

Figure 1.4 Consideration of the resultant of a set of forces.

P_2 are extended until they meet at point O. At this point the parallelogram of forces is constructed, and R, the diagonal, is the resultant of forces P_1 and P_2. In determining the magnitude of the resultant, the scale used is the same scale used in drawing the given system of forces.

Example 1. A vertical force of 50 lb and a horizontal force of 100 lb, as shown in Figure 1.4*d*, have an angle of 90° between their lines of action. Determine the resultant.

Solution: The two forces are laid off at a convenient scale from their point of intersection, the parallelogram is drawn, and the diagonal is the resultant. Its magnitude scales approximately 112 lb, its sense is upward to the right, and its line of action passes through the point of intersection of the lines of action of the two given forces. By use of a protractor it is found that the angle between the resultant and the force of 100 lb is approximately 26.5°.

Example 2. The angle between two forces of 40 and 90 lb, as shown in Figure 1.4*e*, is 60°. Determine the resultant.

Solution: The forces are laid off to scale from their point of intersection, the parallelogram of forces is constructed, and the resultant is found to be a force of approximately 115 lb, its sense is upward to the right, and its line of action passes through the common point of the two given forces. The angle between the resultant and the force of 90 lb is approximately 17.5°.

Attention is called to the fact that these two problems have been solved graphically by the construction of diagrams. Mathematics might have been employed. For many practical problems, carefully constructed graphical solutions give sufficiently accurate answers and frequently require far less time. Do not make diagrams too small as greater accuracy is obtained by using larger parallelograms of forces.

Problems 1.3.A–F By constructing the parallelogram of forces, determine the resultants for the pairs of forces shown in Figures 1.5*a–f*.

Components of a Force

In addition to combining forces to obtain their resultant, it is often helpful to replace a single force by its *components*. The components of

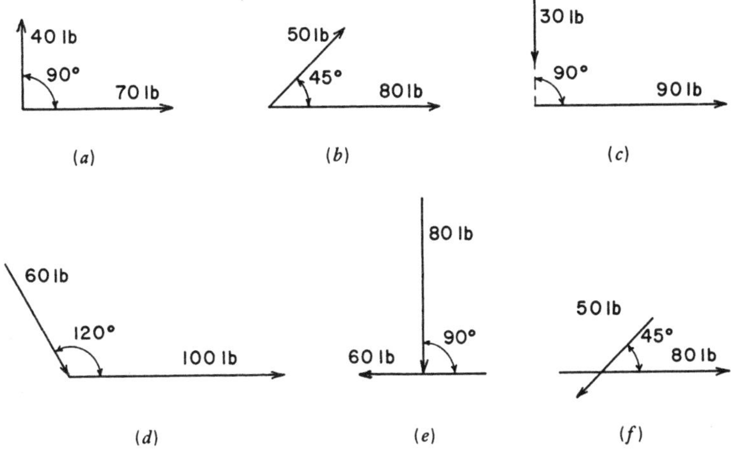

Figure 1.5 Reference for Problem 1.3, Part 1.

a force are the two or more forces that, acting together, have the same effect as the given force. In Figure 1.4*d*, if we are given the force of 112 lb, its vertical component is 50 lb and its horizontal component is 100 lb. That is, the 112-lb force has been *resolved* into its vertical and horizontal components. Any force may be considered as the resultant of its components.

Combined Resultants

The resultant of more than two nonparallel forces may be obtained by finding the resultants of pairs of forces and finally the resultant of the resultants.

Example 3. Find the resultant of the concurrent forces P_1, P_2, P_3, and P_4 shown in Figure 1.6.

Solution: By constructing a parallelogram of forces, the resultant of P_1 and P_2 is found to be R_1. Similarly, the resultant of P_3 and P_4 is R_2. Finally, the resultant of R_1 and R_2 is R, the resultant of the four given forces.

Problems 1.3.G–I Using graphical methods, find the resultant of the systems of concurrent forces shown in Figure 1.7.

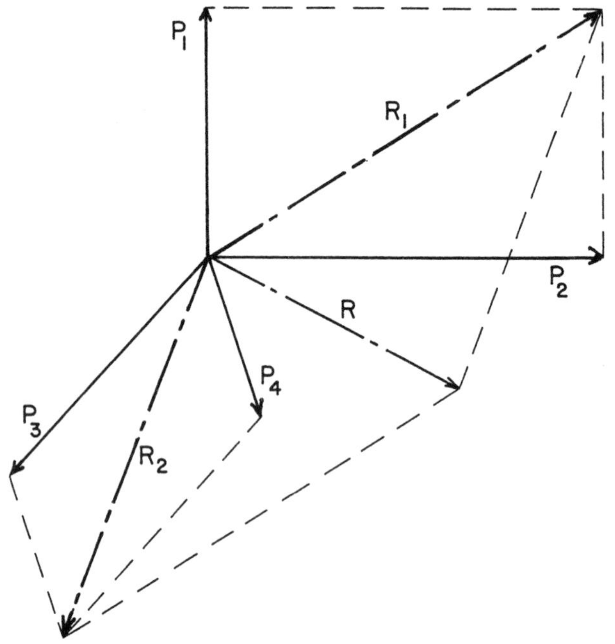

Figure 1.6 Finding a resultant by successive pairs.

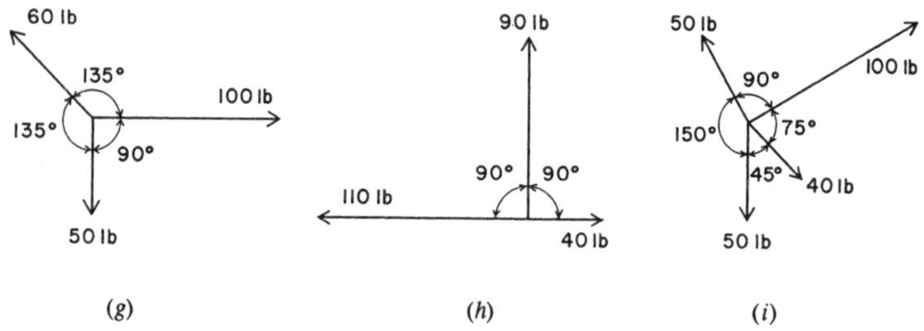

Figure 1.7 Reference for Problem 1.3, Part 2.

Equilibrant

The force, not a part of the system of forces, required to maintain a system of forces in equilibrium is called the *equilibrant* of the system. Suppose that we are required to investigate the system of two forces,

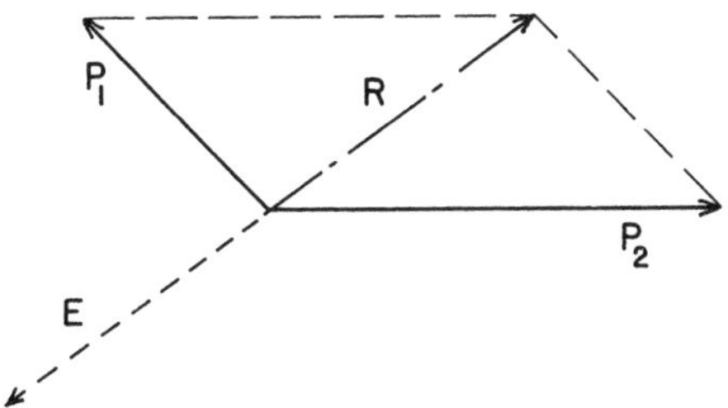

Figure 1.8 Finding an equilibrant.

P_1 and P_2, as shown in Figure 1.8. The parallelogram of forces is constructed, and the resultant is found to be R. The system is not in equilibrium. The force required to maintain equilibrium is force E, shown by the dashed line. The equilibrant, E, is the same as the resultant in magnitude and line of action but is opposite in sense. The three forces, P_1 and P_2 and E, constitute a system in equilibrium.

If two forces are in equilibrium, they must be equal in magnitude and opposite in sense and have the same direction and line of action. Either of the two forces may be said to be the equilibrant of the other. The resultant of a system of forces in equilibrium is zero.

1.4 GRAPHICAL ANALYSIS OF CONCURRENT FORCE SYSTEMS

Force Polygon

The resultant of a system of concurrent forces may be found by constructing a *force polygon*. To draw the force polygon, begin with a point and lay off, at a convenient scale, a line parallel to one of the forces, with its length equal to the force in magnitude, and having the same sense. From the termination of this line draw similarly another line corresponding to one of the remaining forces and continue in the same manner until all the forces in the given system are accounted for. If the polygon does not close, the system of forces is not in equilibrium, and the line required to close the polygon *drawn from the starting point* is the resultant in magnitude and direction. If the forces in the given system are concurrent, the line of action of the resultant passes through the point they have in common.

If the force polygon for a system of concurrent forces closes, the system is in equilibrium and the resultant is zero.

Example 4. Let it be required to find the resultant of the four concurrent forces P_1, P_2, P_3, and P_4 shown in Figure 1.9*a*. This diagram is called the *space diagram*; it shows the relative positions of the forces in a given system.

Solution: Beginning with some point such as O, shown in Figure 1.9*b*, draw the upward force P_1. At the upper extremity of the line representing P_1, draw P_2, continuing in a like manner with P_3 and P_4. The polygon does not close; therefore, the system is not in equilibrium. The resultant R, shown by the dot-and-dash line, is the resultant of the given system. Note that its sense is *from* the starting point O, downward to the right. The line of action of the resultant of the given system shown in Figure 1.9*a* has its line of action passing through the point they have in common, its magnitude and direction having been found in the force polygon.

In drawing the force polygon, the forces may be taken in any sequence. In Figure 1.9*c* a different sequence is taken, but the resultant R is found to have the same magnitude and direction as previously found in Figure 1.9*b*.

Bow's Notation

Thus far, forces have been identified by the symbols P_1, P_2, and so on. A system of identifying forces, known as Bow's notation, affords many advantages. In this system letters are placed in the

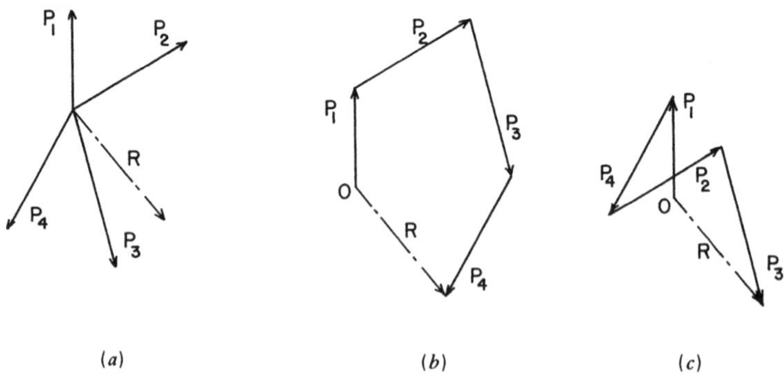

Figure 1.9 Finding a resultant by continuous vector addition of forces.

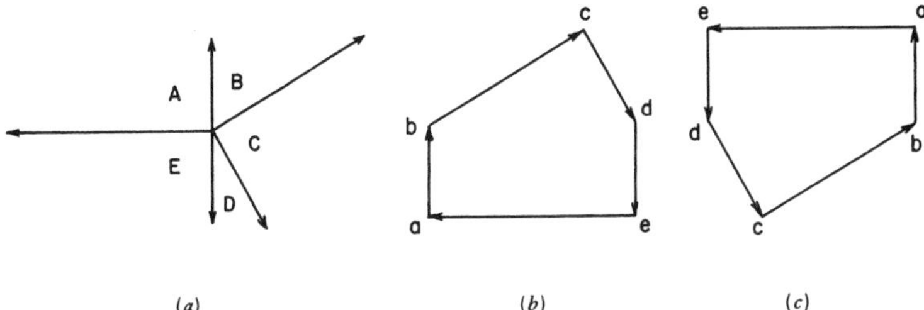

Figure 1.10 Construction of a force polygon.

space diagram on each side of a force and a force is identified by two letters. The sequence in which the letters are read is important. Figure 1.10*a* shows the space diagram of five concurrent forces. Reading about the point in common *in a clockwise manner* the forces are *AB, BC, CD, DE*, and *EA*. When a force in the force polygon is represented by a line, a letter is placed at each end of the line. As an example, the vertical upward force in Figure 1.10*a* is read *AB* (note that this is read clockwise about the common point); in the force polygon (Figure 1.10*b*) the letter *a* is placed at the bottom of the line representing the force *AB* and the letter *b* is at the top. Use capital letters to identify the forces in the space diagrams and lowercase letters in the force polygon. From point *b* in the force polygon draw force *bc*, then *cd*, and continue with *de* and *ea*. Since the force polygon closes, the five concurrent forces are in equilibrium.

In reading forces, a clockwise manner is used in all the following discussions. It is important that this method of identifying forces be thoroughly understood. To make this clear, suppose that a force polygon is drawn for the five forces shown in Figure 1.10*a*, reading the forces in sequence in a counterclockwise manner. This will produce the force polygon shown in Figure 1.10*c*. Either method may be used, but for consistency the method of reading clockwise is used here.

Use of the Force Polygon

Two ropes are attached to a ceiling and their lower ends are connected to a ring, making the arrangement shown in Figure 1.11*a*. A weight of 100 lb is suspended from the ring. Obviously, the force in the rope *AB* is 100 lb, but the magnitudes of the forces in ropes *BC* and *CA* are unknown.

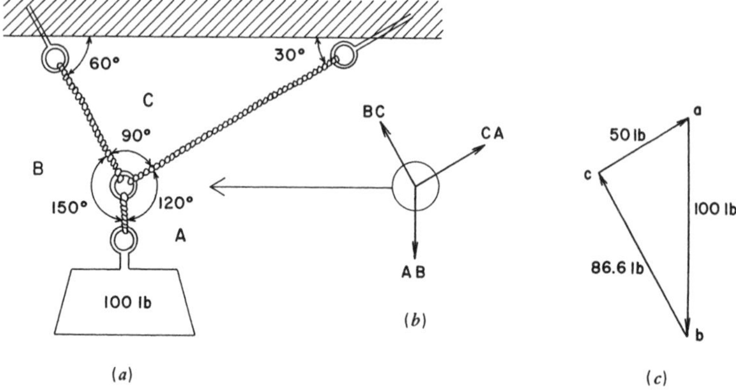

Figure 1.11 Solution of a concentric force system.

The forces in the ropes *AB*, *BC*, and *CA* constitute a concurrent force system in equilibrium (Figure 1.11*b*). The magnitude of only one of the forces is known—it is 100 lb in rope *AB*. Since the three concurrent forces are in equilibrium, their force polygon must close, and this fact makes it possible to find their magnitudes. Now, at a convenient scale, draw the line *ab* (Figure 1.11*c*) representing the downward force *AB*, 100 lb. The line *ab* is one side of the force polygon. From point *b* draw a line parallel to rope *BC*; point *c* will be at some location on this line. Next, draw a line through point *a* parallel to rope *CA*; point *c* will be at some position on this line. Since point *c* is also on the line through *b* parallel to *BC*, the intersection of the two lines determines point *c*. The force polygon for the three forces is now completed; it is *abca*, and the lengths of the sides of the polygon represent the magnitudes of the forces in ropes *BC* and *CA*, 86.6 and 50 lb, respectively.

Particular attention is called to the fact that the lengths of the ropes in Figure 1.11*a* are not an indication of the magnitudes of the forces within the ropes; the magnitudes are determined by the lengths of the corresponding sides of the force polygon (Figure 1.11*c*). Figure 1.11*a* merely determines the geometric layout for the structure.

Problems 1.4.A–D Find the sense (tension or compression) and magnitude of the internal force in the member indicated in Figure 1.12 using graphical methods.

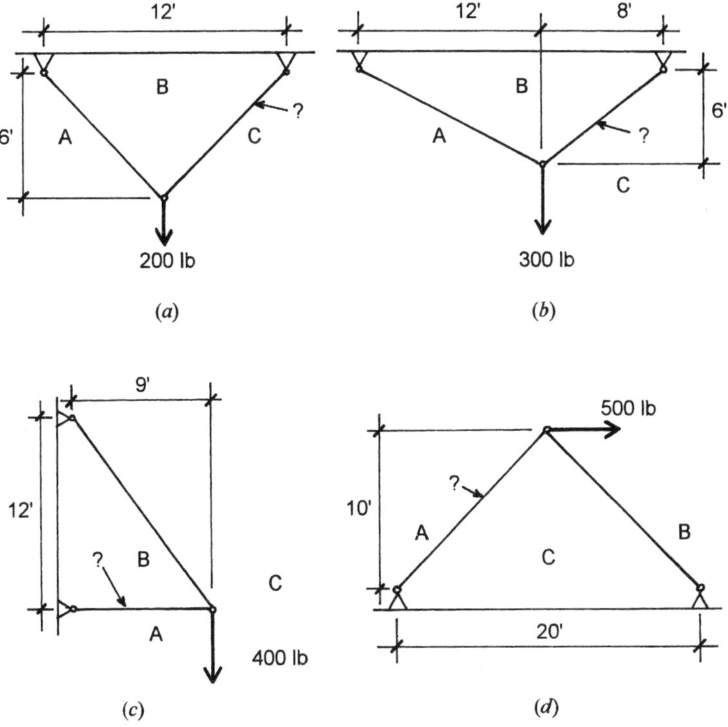

Figure 1.12 Reference for Problem 1.4.

1.5 ALGEBRAIC ANALYSIS OF NONCONCURRENT FORCE SYSTEMS

The simplest analysis for nonconcurrent force systems is not graphical, as it was for concurrent force systems, but rather algebraic. Usually, the forces in the system are resolved into vertical and horizontal components, and the components are added; the result is the components of the resultant, which can be used to determine its actual magnitude, its line of action, and its sense.

Example 5. Find the resultant of the two forces in Figure 1.13*a*.

Solution: The two forces are parallel, so they can simply be added to find the resultant, with a magnitude of 15 lb, a line of action parallel to the forces, and a sense the same as the forces, as shown in

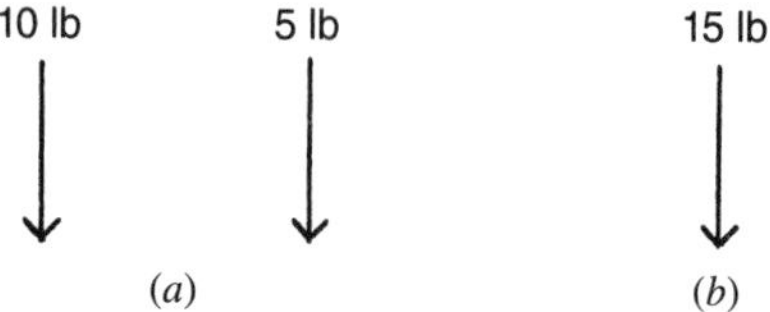

Figure 1.13 Reference for Example 5.

Figure 1.13*b*. Consideration for the location of the lines of action of the forces and their resultant requires something more than simple force addition; this is discussed in the following section.

Moments

The term moment of a force is commonly used in engineering problems. It is fairly easy to visualize a length of 3 ft, an area of 26 in.^{2}, or a force of 100 lb. A moment, however, is less easily understood; it is a force multiplied by a distance. A moment is the *tendency of a force to cause rotation about a given point or axis*. The magnitude of the moment of a force about a given point is the magnitude of the force (pounds, kips, etc.) multiplied by the distance (feet, inches, etc.) from the force to the point of rotation. The point is called the center of moments, and the distance, which is called the *lever arm* or *moment arm*, is measured by a line drawn through the center of moments perpendicular to the line of action of the force. Moments are expressed in compound units such as foot-pounds and inch-pounds or kip-feet and kip-inches. In summary,

$$\text{Moment of force} = \text{magnitude of force} \times \text{moment arm}$$

Consider the horizontal force of 100 lb shown in Figure 1.14*a*. If point A is the center of moments, the lever arm of the force is 5 ft. Then the moment of the 100-lb force with respect to point A is $100 \times 5 = 500$ ft-lb. In this illustration the force tends to cause a *clockwise* rotation (shown by the dashed-line arrow) about point A and is called a positive moment. If point B is the center of moments, the moment arm of the force is 3 ft. Therefore, the moment of the 100-lb force about point B is $100 \times 3 = 300$ ft-lb. With respect to point B, the force tends to cause *counterclockwise* rotation; it is called a negative moment. It is important to remember that you can never consider the moment of a force without having in mind the particular point or axis about which it tends to cause rotation.

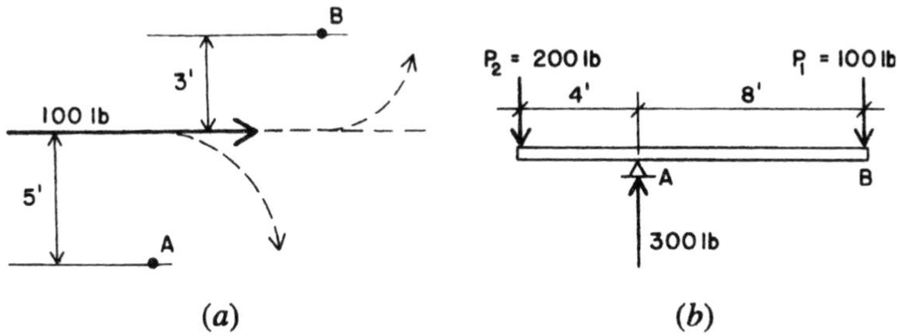

Figure 1.14 Development of moments.

Figure 1.14*b* represents two forces acting on a bar that is supported at point *A*. The moment of force P_1 about point *A* is $100 \times 8 = 800$ ft-lb, and it is clockwise or positive. The moment of force P_2 about point *A* is $200 \times 4 = 800$ ft-lb. The two moment values are the same, but P_2 tends to produce a counterclockwise, or negative, moment about point *A*. The positive and negative moments are equal in magnitude and are in *equilibrium*; that is, there is no motion. Another way of stating this is to say that the sum of the positive and negative moments about point *A* is zero, or

$$\sum M_A = 0$$

Stated more generally, if a system of forces is in equilibrium, the algebraic sum of the moments is zero. This is one of the laws of equilibrium.

In Figure 1.14*b* point *A* was taken as the center of the moments, but the fundamental law holds for any point that might be selected. For example, if point *B* is taken as the center of moments, the moment of the upward supporting force of 300 lb acting at *A* is clockwise (positive) and that of P_2 is counterclockwise (negative). Then

$$(300 \times 8) - (200 \times 12) = 2400 - 2400 = 0$$

Note that the moment of force P_1 about point *B* is $100 \times 0 = 0$; it is therefore omitted in writing the equation. The reader should be satisfied that the sum of the moments is zero also when the center of moments is taken at the left end of the bar under the point of application of P_2.

Example 6. Find the location of the line of action of the resultant for the forces in Figure 1.13 if the two forces are 30 ft apart.

Solution: In Example 5 the magnitude of the resultant was determined to be 15 lb, but the location of its line of action was not determined. Solution of this problem requires the use of moments of the forces. A procedure for this solution is as follows.

Consider the layout of the forces and their resultant as shown in Figure 1.15. A condition for the resultant is that it must be capable of completely replacing the forces; this is true for force summation and also for any moment summation. To use this relationship, we consider an arbitrary reference point (*P* in Figure 1.15). Moments of the forces about this point are

$$\begin{aligned} M_p &= (10 \text{ lb} \times 5 \text{ ft}) + (5 \text{ lb} \times 35 \text{ ft}) \\ &= 50 \text{ ft-lb} + 175 \text{ ft-lb} = 225 \text{ ft-lb} \end{aligned}$$

As shown in Figure 1.15, the distance of the resultant from *P* is defined as x. Its moment about *P* is thus Rx, and equating this to the moment of the forces we determine

$$x = \frac{M_P}{R} = \frac{225}{15} = 15 \text{ ft}$$

As stated, the location of *P* is arbitrary, but the true location of the line of action of the resultant is a constant. The reader should try a different location for *P* to verify this.

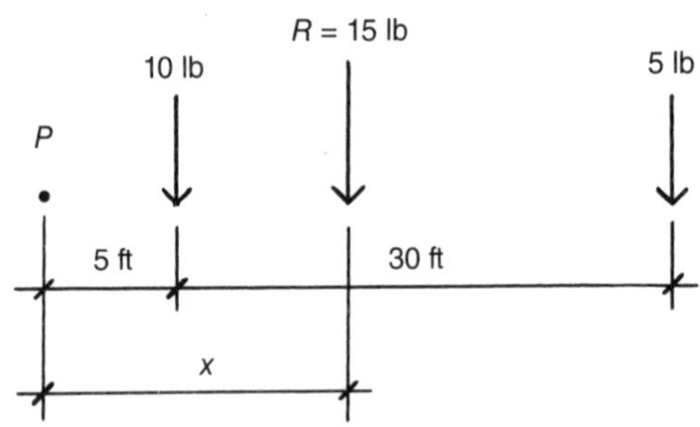

Figure 1.15 Reference for Example 6.

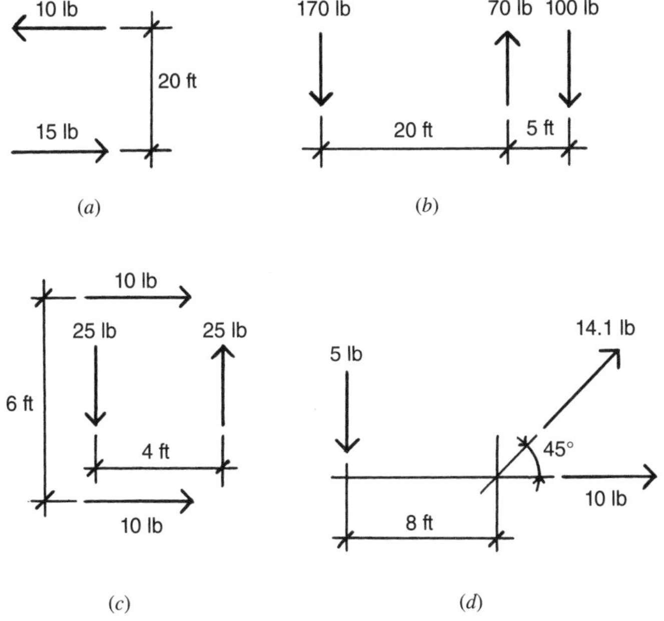

Figure 1.16 Reference for Problem 1.5.

Problems 1.5.A–D Find the resultant for the force systems in Figure 1.16. Find the magnitude, line of action, and sense of the resultant.

1.6 LAWS OF EQUILIBRIUM

When an object is acted on by a number of forces, each force tends to move the object. If the forces are of such magnitude and position that their combined effect produces no motion of the object, the forces are said to be in *equilibrium* (Section 1.2). The three fundamental laws of static equilibrium for a general set of coplanar forces are:

1. The algebraic sum of all the vertical forces equals zero.
2. The algebraic sum of all the horizontal forces equals zero.
3. The algebraic sum of the moments of all the forces about any point equals zero.

These laws, sometimes called the conditions for equilibrium, may be expressed as follows (the symbol $\sum$ indicates a summation, i.e., an algebraic addition of all similar terms involved in the problem):

$$\sum V = 0 \qquad \sum H = 0 \qquad \sum M = 0$$

The law of moments, $\sum M = 0$, was presented in the preceding discussion.

The expression $\sum V = 0$ is another way of saying that *the sum of the downward forces equals the sum of the upward forces*. Thus, the bar of Figure 1.14*b* satisfies $\sum \mathrm{V} = 0$ because the upward force of 300 lb equals the sum of P_1 and P_2.

Moments of Forces

Figure 1.17*a* shows two downward forces of 100 and 200 lb acting on a beam. The beam has a length of 8 ft between the supports; the supporting forces, which are called reactions, are 175 and 125 lb. The four forces are parallel and for equilibrium, therefore, the two laws, $\sum V = 0$ and $\sum M = 0$, apply.

First, because the forces are in equilibrium, the sum of the downward forces must equal the sum of the upward forces. The sum of the downward forces, the loads, is $100 + 200 = 300$ lb; the sum of the upward forces, the reactions, is $175 + 125 = 300$ lb. Thus, the force summation is zero.

Second, because the forces are in equilibrium, the sum of the moments of the forces tending to cause clockwise rotation (positive moments) must equal the sum of the moments of the forces tending to produce counterclockwise rotation (negative moments) about any center of moments. Considering an equation of moments about point A at the right-hand support, the force tending to cause clockwise rotation (shown by the curved arrow) about this point is 175 lb; its moment is $175 \times 8 = 1400$ ft-lb. The forces tending to cause counterclockwise rotation about the same point are 100 and 200 lb, and their moments are 100×6 and 200×4 ft-lb. Thus, if $\sum M_A = 0$, then

$$175 \times 8 = (100 \times 6) + (200 \times 4)$$

$$1400 = 600 + 800$$

$$1400 \text{ ft-lb} = 1400 \text{ ft-lb}$$

which is true.

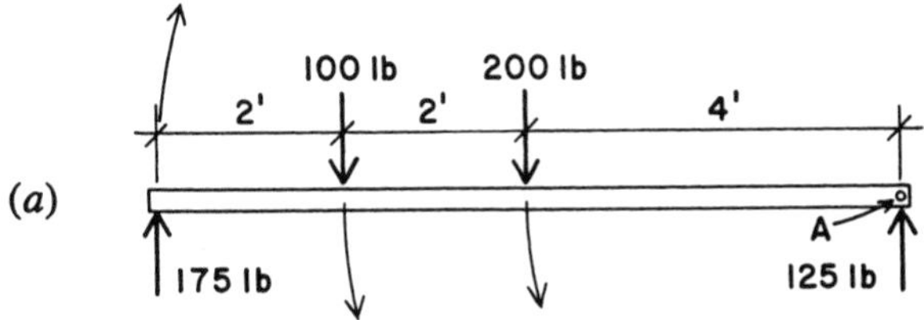

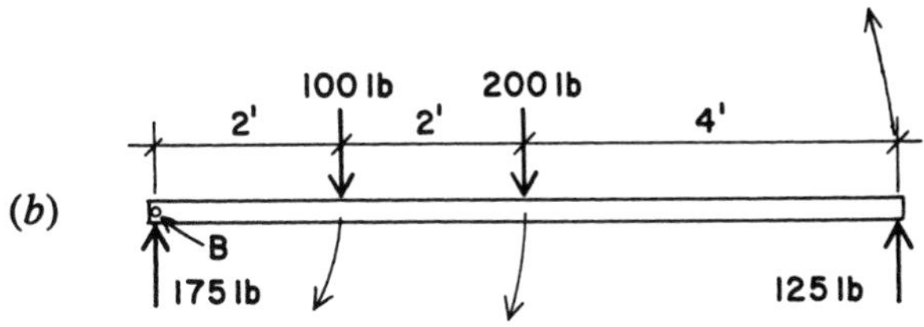

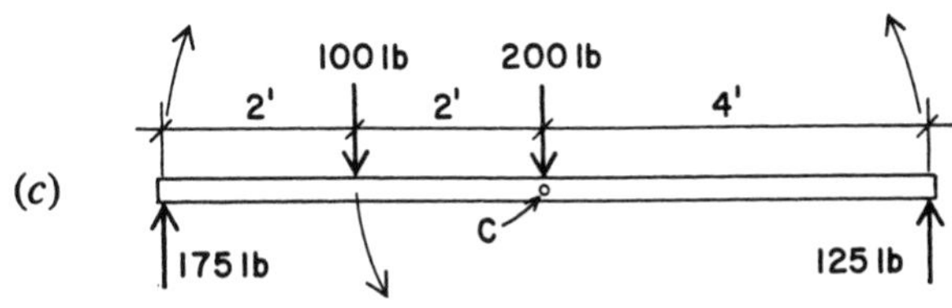

Figure 1.17 Summation of moments about selected points.

The upward force of 125 lb is omitted from the above equation because its lever arm about point *A* is 0 ft, and consequently its moment is zero. A force passing through the center of moments does not cause rotation about that point.

Now select point *B* at the left support as the center of moments (see Figure 1.17*b*). By the same reasoning, if $\sum M_B = 0$, then

$$(100 \times 2) + (200 \times 4) = 125 \times 8$$

$$200 + 800 = 1000$$

$$1000 \text{ ft-lb} = 1000 \text{ ft-lb}$$

Again the law holds. In this case, the force of 175 lb has a lever arm of 0 ft about the center of moments and its moment is zero.

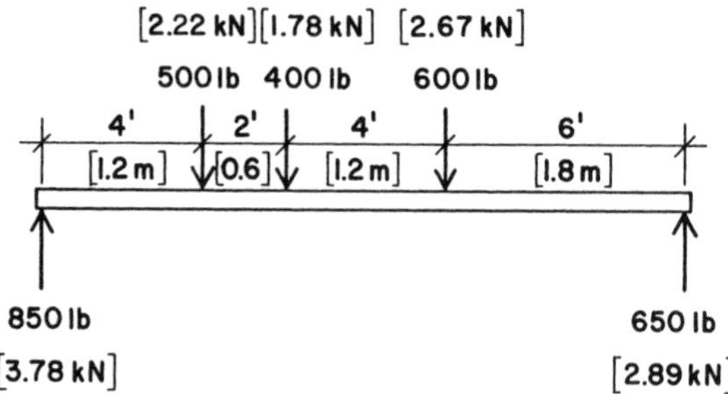

Figure 1.18 Reference for Problem 1.6.A.

The reader should verify this case by selecting any other point, such as point C in Figure 1.17*c* as the center of moments, and confirming that the sum of the moments is zero for this point.

Problem 1.6.A Figure 1.18 represents a beam in equilibrium with three loads and two reactions. Select five different centers of moments and write the equation of moments for each, showing that the sum of the clockwise moments equals the sum of the counterclockwise moments.

1.7 LOADS AND REACTIVE FORCES

Structural members, such as beams, are acted on by external forces that consist of the loads and the reaction forces developed by the beam's supports. The two types of loads that commonly occur on beams are called *concentrated* and *distributed.* A concentrated load is assumed to act at a definite point; such a load is that caused when one beam supports another beam. A distributed load is one that acts over a considerable length of the beam; such a load is caused by a floor deck supported directly by a beam. If the distributed load exerts a force of equal magnitude for each unit of length of the beam, it is known as a *uniformly distributed load.* The weight of a beam is a uniformly distributed load that extends over the entire length of the beam. However, some uniformly distributed loadings supported by the beam may extend over only a portion of the beam length.

Reactive Forces

Reactive forces are the upward forces acting at the supports that hold in equilibrium the downward forces or loads. Reactive forces are often

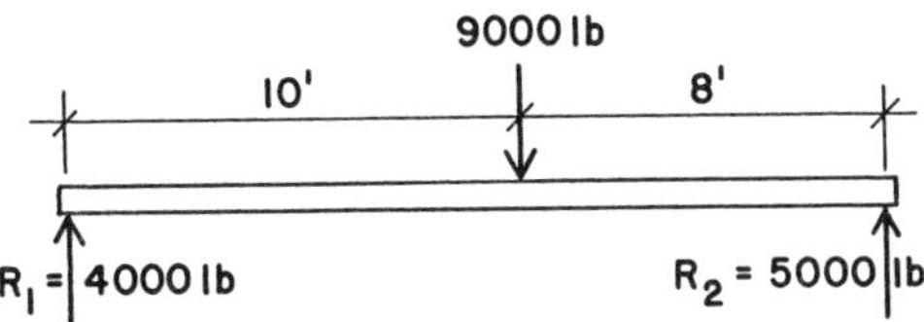

Figure 1.19 Beam reactions for a single load.

refered to as "reactions." The left and right reactions of a simple beam are usually called R_1 and R_2, respectively. Determination of reactions for simple beams is achieved with the use of equilibrium conditions for parallel force systems.

If a beam 18 ft long has a concentrated load of 9000 lb located 9 ft from the supports, it is readily seen that each upward force at the supports will be equal and will be one-half the load in magnitude, or 4500 lb. But consider, for instance, the 9000-lb load placed 10 ft from one end, as shown in Figure 1.19. What will the upward supporting forces be? This is where the principle of moments can be used. Consider a summation of moments about the right-hand support R_2. Thus,

$$\sum M = 0 = +(R_1 \times 18) - (9000 \times 8) + (R_2 \times 0)$$

$$R_1 = \frac{72{,}000}{18} = 4000 \text{ lb}$$

Then, considering the equilibrium of vertical forces,

$$\sum V = 0 = +R_1 + R_2 - 9000$$

$$R_2 = 9000 - 4000 = 5000 \text{ lb}$$

The accuracy of this solution can be verified by taking moments about the left-hand support. Thus,

$$\sum M = 0 = -(R_2 \times 18) + (9000 \times 10) + (R_1 \times 0)$$

$$R_2 = \frac{90{,}000}{18} = 5000 \text{ lb}$$

***Example** 7.* A simple beam 20 ft long has three concentrated loads, as indicated in Figure 1.20. Find the magnitudes of the reactions.

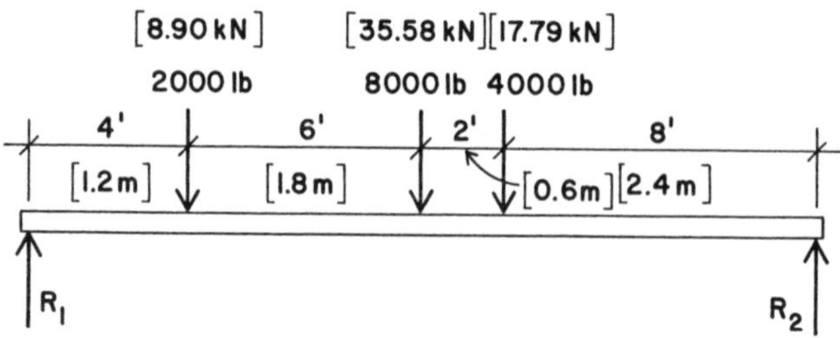

Figure 1.20 Reference for Example 7.

Solution: Using the right-hand support as the center of moments,

$$\sum M = +(R_1 \times 20) - (2000 \times 16) - (8000 \times 10) - (4000 \times 8)$$

from which

$$R_1 = \frac{32{,}000 + 80{,}000 + 32{,}000}{20} = 7200 \text{ lb}$$

From a summation of the vertical forces,

$$\sum V = 0 = +R_2 + 7200 - 2000 - 8000 - 4000$$
$$R_2 = 6800 \text{ lb}$$

With all forces determined, a summation about the left-hand support—or any point except the right-hand support—will verify the accuracy of the work.

The following example demonstrates a solution with uniformly distributed loading on a beam. A convenience in this work is to consider the total uniformly distributed load as a concentrated force placed at the center of the distributed load.

Example 8. A simple beam 16 ft long carries the loading shown in Figure 1.21*a*. Find the reactions.

Solution: The total uniformly distributed load may be considered as a single concentrated load placed at 5 ft from the right-hand support;

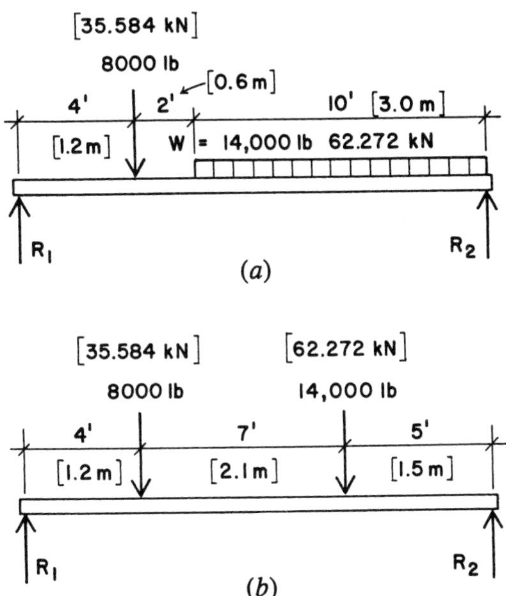

Figure 1.21 Reference for Example 8.

this loading is shown in Figure 1.21*b*. Considering moments about the right-hand support,

$$\sum M = 0 = +(R_1 \times 16) - (8000 \times 12) - (14{,}000 \times 5)$$

$$R_1 = \frac{166{,}000}{16} = 10{,}375 \text{ lb}$$

And, from a summation of vertical forces,

$$R_2 = (8000 + 14{,}000) - 10{,}375 = 11{,}625 \text{ lb}$$

Again, a summation of moments about the left-hand support will verify the accuracy of the work.

In general, any beam with only two supports, for which the supports develop only vertical reaction forces, will be statically determinate. This includes the simple span beams in the preceding examples as well as beams with overhanging ends.

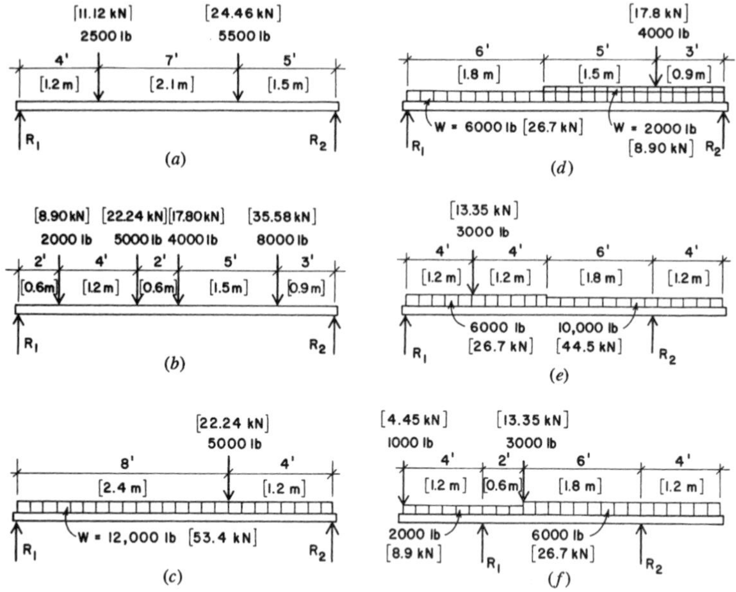

Figure 1.22 Reference for Problem 1.7.

Problems 1.7.A–F Find the reactions for the beams shown in Figure 1.22.

1.8 LOAD SOURCES

Structural tasks are defined primarily in terms of the loading conditions imposed on the structure. There are many potential sources of load for building structures. Designers must consider all the potential sources and the logical combinations with which they may occur. Building codes currently stipulate both the load sources and the form of combinations to be used for design. The following loads are listed in the 2013 edition of *Minimum Design Loads for Buildings and Other Structures* [of the American Society of Civil Engineers (ASCE), Ref. 1], hereinafter referred to as ASCE 2013:

D = Dead load

E = Earthquake-induced force

L = Live load, except roof load

Lr = Roof live load

S = Snow load
W = Load due to wind pressure

Additional special loads are listed, but these are the commonly occurring loads. The following is a description of some of these loads.

Dead Loads

Dead load consists of the weight of the materials of which the building is constructed, such as walls, partitions, columns, framing, floors, roofs, and ceilings. In the design of a beam or column, the dead load used must include an allowance for the weight of the structural member itself. Table 1.2, which lists the weights of many construction materials, may be used in the computation of dead loads. Dead loads are due to gravity and they result in downward vertical forces.

Dead load is generally a permanent load once the building construction is completed unless remodeling or rearrangement of the construction occurs. Because of this permanent, long-time character, the dead load requires certain considerations in design, such as the following:

1. Dead load is always included in design loading combinations, except for investigations of singular effects, such as deflections due to only live load.
2. Its long-time character has some special effects causing permanent sag and requiring reduction of design stresses in wood structures, development of long-term, continuing settlements in some soils, and producing creep effects in concrete structures.
3. Dead load contributes some unique responses, such as the stabilizing effects that resist uplift and overturn due to wind forces.

Although weights of materials can be reasonably accurately determined, the complexity of most building construction makes the computation of dead loads possible only on an approximate basis. This adds to other factors to make design for structural behaviors a very approximate science. As in other cases, this should not be used as an excuse for sloppiness in the computational work, but it should be recognized as a fact to temper concern for high accuracy in design computations.

TABLE 1.2 Weight of Building Construction

	psf[a]
Roofs	
3-ply ready roofing (roll, composition)	1
3-ply felt and gravel	5.5
5-ply felt and gravel	6.5
Shingles: Wood	2
Asphalt	2–3
Clay tile	9–12
Concrete tile	6–10
Slate, 3 in.	10
Insulation: Fiber glass batts	0.5
Foam plastic, rigid panels	1.5
Foamed concrete, mineral aggregate	2.5/in.
Wood rafters: 2 × 6 at 24 in.	1.0
2 × 8 at 24 in.	1.4
2 × 10 at 24 in.	1.7
2 × 12 at 24 in.	2.1
Steel deck, painted: 22 gage	1.6
20 gage	2.0
Skylights: Steel frame with glass	6–10
Aluminum frame with plastic	3–6
Plywood or softwood board sheathing	3.0/in.
Ceilings	
Suspended steel channels	1
Lath: Steel mesh	0.5
Gypsum board, 1/2 in.	2
Fiber tile	1
Drywall, gypsum board, 1/2 in.	2.5
Plaster: Gypsum	5
Cement	8.5
Suspended lighting and heating, ventilation, and air conditioning (HVAC), average	3
Floors	
Hardwood, 1/2 in.	2.5
Vinyl tile	1.5
Ceramic tile: 3/4 in.	10
Thin-set	5
Fiberboard underlay, 0.625 in.	3
Carpet and pad, average	3
Timber deck	2.5/in.
Steel deck, stone concrete fill, average	35–40
Concrete slab deck, stone aggregate	12.5/in.
Lightweight concrete fill	8.0/in.

TABLE 1.2 (*Continued*)

	psf[a]
Wood joists: 2 × 8 at 16 in.	2.1
2 × 10 at 16 in.	2.6
2 × 12 at 16 in.	3.2
Walls	
2 × 4 studs at 16 in., average	2
Steel studs at 16 in., average	4
Lath, plaster—see *Ceilings*	
Drywall, gypsum board, 1/2 in.	2.5
Stucco, on paper and wire backup	10
Windows, average, frame + glazing:	
Small pane, wood or metal frame	5
Large pane, wood or metal frame	8
Increase for double glazing	2–3
Curtain wall, manufactured units	10–15
Brick veneer, 4 in., mortar joints	40
1/2 in., mastic-adhered	10
Concrete block:	
Lightweight, unreinforced, 4 in.	20
6 in.	25
8 in.	30
Heavy, reinforced, grouted, 6 in.	45
8 in.	60
12 in.	85

[a] Average weight per square foot of surface, except as noted. Values given as /in. (per in.) are to be multiplied by actual thickness of material.

Building Code Requirements

Structural design of buildings is most directly controlled by building codes, which are the general basis for the granting of building permits—the legal permission required for construction. Building codes (and the permit-granting process) are administered by some unit of government: city, county, or state. Most building codes, however, are based on some model code.

Model codes are more similar than they are different and are in turn largely derived from the same basic data and standard reference sources, including many industry standards. In the several model codes and many city, county, and state codes, however, there are some items that reflect particular regional concerns. With respect to control of structures, all codes have materials (all essentially the same) that relate to the following issues:

1. *Minimum Required Live Loads*. All building codes have tables that provide required values to be used for live loads. Tables 1.3 and 1.4 contain some loads as specified in ASCE 2013 (Ref. 1).
2. *Wind Loads*. These are highly regional in character with respect to concern for local windstorm conditions. Model codes provide data with variability on the basis of geographic zones.
3. *Seismic (Earthquake) Effects*. These are also regional with predominant concerns in the western states. These data, including recommended investigations, are subject to quite frequent modification, as the area of study responds to ongoing research and experience.
4. *Load Duration*. Loads or design stresses are often modified on the basis of the time span of the load, varying from the life of

TABLE 1.3 Minimum Floor Live Loads

Building Occupancy or Use	Uniformly Distributed Load (psf)	Concentrated Load (lb)
Apartments and Hotels		
Private rooms and corridors serving them	40	
Public rooms and corridors serving them	100	
Dwellings, One and Two Family		
Uninhabitable attics without storage	10	
Uninhabitable attics with storage	20	
Habitable attics and sleeping rooms	30	
All other areas except stairs and balconies	40	
Office Buildings		
Offices	50	2000
Lobbies and first-floor corridors	100	2000
Corridors above first floor	80	2000
Stores		
Retail		
First floor	100	1000
Upper floors	75	1000
Wholesale, all floors	125	1000

Source: ASCE 2013 (Ref 1), used with permission of the publisher, American Society of Civil Engineers.

TABLE 1.4 Live-Load Element Factor, K_{LL}

Element	K_{LL}
Interior columns	4
Exterior columns without cantilever slabs	4
Edge columns with cantilever slabs	3
Corner columns with cantilever slabs	2
Edge beams without cantilever slabs	2
Interior beams	2
All other members not identified above	1

Source: ASCE 2013 (Ref. 1), used with permission of the publisher, American Society of Civil Engineers.

the structure for dead load to a few seconds for a wind gust or a single major seismic shock. Safety factors are frequently adjusted on this basis. Some applications are illustrated in the work in the design examples.

5. *Load Combinations*. These were formerly mostly left to the discretion of designers but are now quite commonly stipulated in codes, mostly because of the increasing use of ultimate strength design and the use of factored loads.
6. *Design Data for Types of Structures*. These deal with basic materials (wood, steel, concrete, masonry, etc.), specific structures (rigid frames, towers, balconies, pole structures, etc.), and special problems (foundations, retaining walls, stairs, etc.). Industrywide standards and common practices are generally recognized, but local codes may reflect particular local experience or attitudes. Minimal structural safety is the general basis, and some specified limits may result in questionably adequate performances (bouncy floors, cracked plaster, etc.).
7. *Fire Resistance*. For the structure, there are two basic concerns, both of which produce limits for the construction. The first concern is for structural collapse or significant structural loss. The second concern is for containment of the fire to control its spread. These concerns produce limits on the choice of materials (e.g., combustible or noncombustible) and some details of the construction (cover on reinforcement in concrete, fire insulation for steel beams, etc.).

The work in the design examples in Chapters 18–20 is based largely on criteria from ASCE 2013 (Ref. 1).

Live Loads

Live loads technically include all the nonpermanent loadings that can occur in addition to the dead loads. However, the term as commonly used usually refers only to the vertical gravity loadings on roof and floor surfaces. These loads occur in combination with the dead loads but are generally random in character and must be dealt with as potential contributors to various loading combinations, as discussed in Section 1.9.

Roof Loads

In addition to the dead loads they support, roofs are designed for a uniformly distributed live load. The minimum specified live load accounts for general loadings that occur during construction and maintenance of the roof. For special conditions, such as heavy snowfalls, additional loadings are specified.

The minimum roof live load in pounds per square foot (psf) is specified in ASCE 2013 (Ref. 1) in the form of an equation, as follows:

$$L_r = 20R_1R_2, \quad \text{in which } 12 \le L_r \le 20$$

In the equation R_1 is a reduction factor based on the tributary area supported by the structural member being designed (designated as A_t and quantified in square feet) and is determined as follows:

$$\begin{aligned} R_1 &= 1 \quad \text{for } A_t \le 200 \text{ ft}^2 \\ &= 1.2 - 0.001A_t \quad \text{for } 200 \text{ ft}^2 < A_t < 600 \text{ ft}^2 \\ &= 0.6 \quad \text{for } A_t \ge 600 \text{ ft}^2 \end{aligned}$$

Reduction factor R_2 accounts for the slope of a pitched roof and is determined as follows:

$$\begin{aligned} R_2 &= 1 \quad \text{for } F \le 4 \\ &= 1.2 - 0.05F \quad \text{for } 4 < F < 12 \\ &= 0.6 \quad \text{for } F \ge 12 \end{aligned}$$

The quantity F in the equations for R_2 is the number of inches of rise per foot for a pitched roof (e.g., $F = 12$ indicates a rise of 12 in. or an angle of 45°).

The design standard also provides data for roof surfaces that are arched or domed and for special loadings for snow or water accumulation. Roof surfaces must also be designed for wind pressures on the roof surface, both upward and downward. A special situation that must be considered is that of a roof with a low dead load and a significant wind load that exceeds the dead load.

Although the term *flat roof* is often used, there is generally no such thing; all roofs must be designed for some water drainage. The minimum required pitch is usually $^1/_4$ in./ft, or a slope of approximately 1 : 50. With roof surfaces that are close to flat, a potential problem is that of *ponding*, a phenomenon in which the weight of the water on the surface causes deflection of the supporting structure, which in turn allows for more water accumulation (in a pond), causing more deflection, and so on, resulting in an accelerated collapse condition.

Floor Live Loads

The live load on a floor represents the probable effects created by the occupancy. It includes the weights of human occupants, furniture, equipment, stored materials, and so on. All building codes provide minimum live loads to be used in the design of buildings for various occupancies. Since there is a lack of uniformity among different codes in specifying live loads, the local code should always be used. Table 1.3 contains a sample of values for floor live loads as given in ASCE 2013 (Ref. 1) and commonly specified by building codes.

Although expressed as uniform loads, code-required values are usually established large enough to account for ordinary concentrations that occur. For offices, parking garages, and some other occupancies, codes often require the consideration of a specified concentrated load as well as the distributed loading. This required concentrated load is listed in Table 1.3 for the appropriate occupancies.

Where buildings are to contain heavy machinery, stored materials, or other contents of unusual weight, these must be provided for individually in the design of the structure.

When structural framing members support large areas, most codes allow some reduction in the total live load to be used for design. These reductions, in the case of roof loads, are incorporated in the formulas for roof loads given previously. The following is the method given in ASCE 2013 (Ref. 1) for determining the reduction permitted for beams, trusses, or columns that support large floor areas.

The design live load on a member may be reduced in accordance with the formula

$$L = L_0 \left(0.25 + \frac{15}{\sqrt{K_{LL} A_T}} \right)$$

where L = reduced live load supported, in psf
L_0 = unreduced live load, in psf
K_{LL} = live-load element factor (see Table 1.4)
A_T = tributary area supported, in ft^2

For members supporting one floor L shall not be less than $0.50L_0$, and L shall not be less than $0.40L_0$ for members supporting two or more floors.

In office buildings and certain other building types, partitions may not be permanently fixed in location but may be erected or moved from one position to another in accordance with the requirements of the occupants. In order to provide for this flexibility, it is customary to require an allowance of 15–20 psf, which is usually added to other dead loads.

Lateral Loads (Wind and Earthquake)

As used in building design, the term *lateral load* is usually applied to the effects of wind and earthquakes, as they induce horizontal forces on stationary structures. From experience and research, design criteria and methods in this area are continuously refined, with recommended practices being presented through the various model building codes.

Space limitations do not permit a complete discussion of the topic of lateral loads and design for their resistance. The following discussion summarizes some of the criteria for design in ASCE 2013 (Ref. 1). Examples of application of these criteria are given in the design examples of building structural design in Chapters 18–20. For a more extensive discussion the reader is referred to *Simplified Building Design for Wind and Earthquake Forces* (Ref. 2).

Wind

Where wind is a regional problem, local codes are often developed in response to local conditions. Complete design for wind effects on buildings includes a large number of both architectural and structural concerns. The following is a discussion of some of the requirements from ASCE 2013 (Ref. 1).

Basic Wind Speed. This is the maximum wind speed (or velocity) to be used for specific locations. It is based on recorded wind histories and adjusted for some statistical likelihood of occurrence. For the United States recommended minimum wind speeds are taken from maps provided in the ASCE standard. As a reference point, the speeds are those recorded at the standard measuring position of 10 m (approximately 33 ft) above the ground surface.

Wind Exposure. This refers to the conditions of the terrain surrounding the building site. The ASCE standard uses three categories, labeled B, C, and D. Qualifications for categories are based on the form and size of wind-shielding objects within specified distances around the building,

Simplified Design Wind Pressure (p_s). This is the basic reference equivalent static pressure based on the critical wind speed and is determined as follows:

$$p_s = \lambda I p_{S30}$$

where λ = adjustment factor for building height and exposure
I = importance factor
p_{S30} = simplified pressure, exposure B, height 30 ft, $I = 1$

The importance factor for ordinary circumstances of building occupancy is 1.0. For other buildings, factors are given for facilities that involve hazard to a large number of people, for facilities considered to be essential during emergencies (such as windstorms), and for buildings with hazardous contents.

The design wind pressure may be positive (inward) or negative (outward, suction) on any given surface. Both the sign and the value for the pressure are given in the design standard. Individual building surfaces, or parts thereof, must be designed for these pressures.

Design Methods. Two methods are described in the code for the application of wind pressures.

Method 1 (Simplified Procedure). This method is permitted to be used for relatively small, low-rise buildings of simple symmetrical shape. It is the method described here and used for the examples in Part V.

Method 2 (Analytical Procedure). This method is much more complex and is prescribed to be used for buildings that do not fit the limitations described for method 1.

Uplift. Uplift may occur as a general effect, involving the entire roof or even the whole building. It may also occur as a local phenomenon such as that generated by the overturning moment on a single shear wall.

Overturning Moment. Most codes require that the ratio of the dead-load resisting moment (called the restoring moment, stabilizing moment, etc.) to the overturning moment be 1.5 or greater. When this is not the case, uplift effects must be resisted by anchorage capable of developing the excess overturning moment. Overturning may be a critical problem for the whole building, as in the case of relatively tall and slender tower structures. For buildings braced by individual shear walls, trussed bents, and rigid-frame bents, overturning is investigated for the individual bracing units.

Drift. Drift refers to the horizontal deflection of the structure due to lateral loads. Code criteria for drift are usually limited to requirements for the drift of a single story (horizontal movement of one level with respect to the next level above or below). As in other situations involving structural deformations, effects on the building construction must be considered; thus, the detailing of curtain walls or interior partitions may affect design limits on drift.

Special Problems. The general design criteria given in most codes are applicable to ordinary buildings. More thorough investigation is recommended (and sometimes required) for special circumstances such as the following:

Tall Buildings. These are critical with regard to their height dimension as well as the overall size and number of occupants inferred. Local wind speeds and unusual wind phenomena at upper elevations must be considered. Tall buildings often require wind tunnel testing (either physical or with computer analysis) to determine appropriate wind loadings.

Flexible Structures. These may be affected in a variety of ways, including vibration or flutter as well as simple magnitude of movements.

Unusual Shapes. Open structures, structures with large overhangs or other projections, and any building with a complex shape should be carefully studied for the special wind effects that may occur. Wind tunnel testing may be advised or even required by some codes.

Earthquakes

During an earthquake, a building is shaken up and down and back and forth. The back-and-forth (horizontal) movements are typically more violent and tend to produce major destabilizing effects on buildings; thus, structural design for earthquakes is mostly done in terms of considerations for horizontal (called lateral) forces. The lateral forces are actually generated by the weight of the building—or, more specifically, by the mass of the building that represents both an inertial resistance to movement and the source for kinetic energy once the building is actually in motion. In the simplified procedures of the equivalent static force method, the building structure is considered to be loaded by a set of horizontal forces consisting of some fraction of the building weight. An analogy would be to visualize the building as being rotated vertically 90° to form a cantilever beam, with the ground as the fixed end and with a load consisting of the building weight.

In general, design for the horizontal force effects of earthquakes is quite similar to design for the horizontal force effects of wind. The same basic types of lateral bracing (shear walls, trussed bents, rigid frames, etc.) are used to resist both force effects. There are indeed some significant differences, but in the main a system of bracing that is developed for wind bracing will most likely serve reasonably well for earthquake resistance as well.

Because of its considerably more complex criteria and procedures, we have chosen not to illustrate the design for earthquake effects in the examples in this book. Nevertheless, the development of elements and systems for the lateral bracing of the building in the design examples here is quite applicable in general to situations where earthquakes are a predominant concern. For structural investigation, the principal difference is in the determination of the loads and their distribution in the building. Another major difference is in the true dynamic effects, critical wind force being usually represented by a single, major, one-direction punch from a gust, while earthquakes represent rapid back-and-forth actions. However, once the dynamic effects are translated into equivalent static forces, design concerns for the bracing systems are very similar, involving considerations for shear, overturning, horizontal sliding, and so on.

For a detailed explanation of earthquake effects and illustrations of the investigation by the equivalent static force method, the reader is referred to *Simplified Building Design for Wind and Earthquake Forces* (Ref. 2).

1.9 LOAD COMBINATIONS

The various types of load sources, as described in the preceding section, must be individually considered for quantification. However, for design work the possible combination of loads must also be considered. Using the appropriate combinations, the design load for individual structural elements must be determined. The first step in finding the design load is to establish the critical combinations of load for the individual element. Using ASCE 2013 (Ref. 1) as a reference, the following combinations are to be considered. As this process is different for the two basic methods of design, they are presented separately.

Allowable Stress Method

For this method the individual loads are used directly for the following possible combinations:

Dead load only

Dead load + live load

Dead load + roof load

Dead load + 0.75(live load) + 0.75(roof load)

Dead load + wind load or 0.7(earthquake load)

Dead load + 0.75(live load) + 0.75(roof load) + 0.75(wind load) or 0.7(earthquake load)

0.6(dead load) + wind load

0.6(dead load) + 0.7(earthquake load)

The combination that produces the critical design situation for individual structural elements depends on the load magnitudes and the loading condition for the elements. Demonstrations of examples of the use of these combinations are given in the building design cases in Part V and isolated problems in Part II.

Strength Design Method

Some adjustment of the percentage of loads (called factoring) is utilized in load combinations within the allowable stress method. However, factoring is done with all the loads for the strength method. The need here is to produce a load higher than the true anticipated load (called the *service load*)—the difference representing a margin of safety. The structural elements will be designed at their failure limits with the design load, and they really should not fail with the actual expected loads. For the strength method the following combinations are considered:

1.4(dead load)
1.2(dead load) + 1.6(live load) + 0.5(roof load)
1.2(dead load) + 1.6(roof load) + live load or 0.8(wind load)
1.2(dead load) + 1.6(wind load) + (live load) + 0.5(roof load)
1.2(dead load) + 1.0(earthquake load) + live load + 0.2(snow load)
0.9(dead load) + 1.0(earthquake load) or 1.6(wind load)

Use of these load combinations is demonstrated in the building design cases in Part V and in isolated problems in Parts III and IV.

1.10 DETERMINATION OF DESIGN LOADS

Figure 1.23 shows the plan layout for the framed structure of a multistory building. The vertical structure consists of columns and the horizontal floor structure of a deck and beam system. The repeating plan unit of 24 × 32 ft is called a column bay. Assuming lateral bracing of the building to be achieved by other structural elements, the columns and beams shown here will be designed for dead load and live load only.

The load to be carried by each element of the structure is defined by the unit loads for dead load and live load and the *load periphery* for the individual elements. The load periphery for an element is established by the layout and dimensions of the framing system. Referring to the labeled elements in Figure 1.23, the load peripheries are as follows:

Beam A: $8 \times 24 = 192\ \text{ft}^2$
Beam B: $4 \times 24 = 96\ \text{ft}^2$

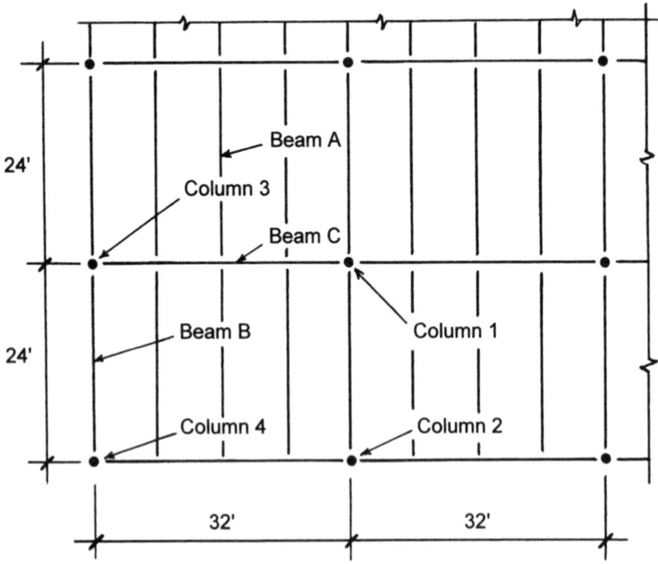

Figure 1.23 Reference for determination of distributed loads.

Beam C: $24 \times 24 = 576\ \text{ft}^2$ (Note that beam C carries only three of the four beams per bay of the system, the fourth being carried directly by the columns.)

Column 1: $24 \times 32 = 768\ \text{ft}^2$

Column 2: $12 \times 32 = 384\ \text{ft}^2$

Column 3: $16 \times 24 = 384\ \text{ft}^2$

Column 4: $12 \times 16 = 192\ \text{ft}^2$

For each of these elements the unit dead load and unit live load from the floor are multiplied by the floor areas computed for the individual elements. Any possible live-load reduction (as described in Section 1.8) is made for the individual elements based on their load periphery area.

Additional dead load for the elements consists of the dead weight of the elements themselves. For the columns and beams at the building edge, another additional dead load consists of the portion of the exterior wall construction supported by the elements. Thus, column 2 carries an area of the exterior wall defined by the multiple of the story height times 32 ft. Column 3 carries 24 ft of wall and column 4 carries 28 ft of wall (12 + 16).

The column loads are determined by the indicated supported floor, to which is added the weight of the columns. For an individual story column this would be added to loads supported above this level—from the roof and any upper levels of floor.

The loads as described are used in the defined combinations described in Section 1.9. If any of these elements are involved in the development of the lateral bracing structure, the appropriate wind or earthquake loads are also added.

Floor live loads may be reduced by the method described in Section 1.8. Reductions are based on the tributary area supported and the number of levels supported by members.

Computations of design loads using the process described here are given for the building design cases in Part V.

1.11 DESIGN METHODS

Use of allowable stress as a design condition relates to the classic method of structural design known as the *working stress method* and now called the *allowable stress design* (ASD) method. The loads used for this method are generally those described as *service loads*; that is, they are related to the service (use) of the structure. Deformation limits are also related to service loads.

Even from the earliest times of use of stress methods, it was known that for most materials and structures the true ultimate capacity was not predictable by use of elastic stress methods. Compensating for this with the working stress method was mostly accomplished by considerations for the establishment of the limiting design stresses. For more accurate predictions of true failure limits, however, it was necessary to abandon elastic methods and to use true ultimate strength behaviors. This led eventually to the so-called *strength method* for design, presently described as the LRFD method, or *load and resistance factor design* method.

The procedures of the stress method are still applicable in many cases—especially for design for deformation limitations. However, the LRFD methods are now very closely related to more accurate use of test data and risk analysis and purport to be more realistic with regard to true structural safety.

The Allowable Stress Design (ASD) Method

The ASD method generally consists of the following:

1. The service (working) load conditions are visualized and quantified as intelligently as possible. Adjustments may be made here by the determination of various statistically likely load combinations (dead load plus live load plus wind load, etc.), by consideration of load duration, and so on.
2. Stress, stability, and deformation limits are set by standards for the various responses of the structure to the loads: in tension, bending, shear, buckling, deflection, uplift, overturning, and so on.
3. The structure is then evaluated (investigated) for its adequacy or is proposed (designed) for an adequate response.

An advantage obtained in working with the stress method is that the real usage condition (or at least an intelligent guess about it) is kept continuously in mind. The principal disadvantage comes from its detached nature regarding real failure conditions, since most structures develop much different forms of stress and strain as they approach their failure limits.

The Strength Design Method (LRFD)

In essence, the ASD method consists of designing a structure to *work* at some established appropriate percentage of its total capacity. The strength method consists of designing a structure to *fail*, but at a load condition well beyond what it should have to experience in use. A major reason for favoring the strength methods is that the failure of a structure is relatively easily demonstrated by physical testing. What is truly appropriate as a working condition, however, is pretty much a theoretical speculation. The strength method is now largely preferred in professional design work. It was first developed mostly for design of concrete structures but has now generally taken over all areas of structural design.

Nevertheless, it is considered necessary to study the classic theories of elastic behavior as a basis for visualization of the general ways that structures work. Ultimate responses are usually some form of variant from the classic responses (because of inelastic materials, secondary effects, multimode responses, etc.). In other words, the usual study procedure is to first consider a classic, elastic response and then to observe (or speculate about) what happens as failure limits are approached.

For the strength method, the process is as follows:

1. The service loads are quantified as in step 1 for the stress method and then are multiplied by an adjustment factor (essentially a safety factor) to produce the *factored load.*
2. The form of response of the structure is visualized and its ultimate (maximum, failure) resistance is quantified in appropriate terms (resistance to compression, to buckling, to bending, etc.). This quantified resistance is also subject to an adjustment factor called the *resistance factor*. Use of resistance factors is discussed in Part II for wood structures, in Part III for steel structures, and in Part IV for concrete structures.
3. The usable resistance of the structure is then compared to the ultimate resistance required (an investigation procedure), or a structure with an appropriate resistance is proposed (a design procedure).

Choice of Design Method

Applications of design procedures in the stress method tend to be simpler and more direct appearing than in the strength methods. For example, the design of a beam may amount to the simple inversion of a few stress or strain equations to derive some required properties (section modulus for bending, area for shear, moment of inertia for deflection, etc.). Applications of strength methods tend to be more obscure, simply because the mathematical formulations for describing failure conditions are more complex than the refined forms of the classic elastic methods.

As strength methods are increasingly used, however, the same kinds of shortcuts, approximations, and round-number rules of thumb will emerge to ease the work of designers. And, of course, use of the computer combined with design experience will permit designers to utilize highly complex formulas and massive databases with ease—all the while hopefully keeping some sense of the reality of it all.

Arguments for use of the stress method or the strength method are essentially academic. An advantage of the stress method may be a closer association with the in-use working conditions of the structure. On the other hand, strength design has a tighter grip on true safety through its focus on failure modes and mechanisms. The successful structural designer, however, needs both forms of consciousness and will develop them through the work of design, whatever methods are employed for the task.

Deformations of structures (such as deflection of beams) that are of concern for design will occur at the working stress level. Visualization and computation of these deformations require the use of basic techniques developed for the ASD method, regardless of whether ASD or LRFD is used for the design work in general.

Work in this book demonstrates the use of both the ASD and LRFD methods. Either method can be used for wood, steel, or concrete structures. However, professional structural design for steel and concrete is now done almost exclusively with the LRFD method. For wood design, work is still done with both methods, although the trend is steadily toward the use of the LRFD method.

2

INVESTIGATION OF AXIAL FORCE ACTIONS

In Chapter 1 the nature of forces, the sources that produce them, and the form of their application to structures were considered. This chapter presents discussion of the force actions of structural members that are subjected to axial forces, including columns and members occurring in trusses, suspension cables, cable-stayed systems, and funicular arches.

2.1 FORCES AND STRESSES

Direct Stress

Figure 2.1*a* represents a block of metal weighing 6400 lb supported on a wooden block having an 8-in. × 8-in. cross section. The wooden block is in turn supported on a base of masonry. The gravity force of the metal block exerted on the wood is 6400 lb, or 6.4 kips. Ignoring its own weight, the wooden block in turn transmits a force of equal magnitude to the masonry base. If there is no motion (a state described

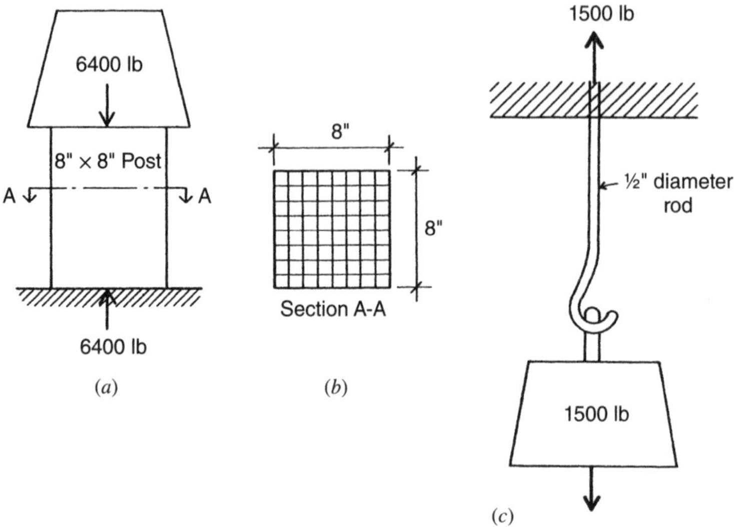

Figure 2.1 Direct force action and stress.

as equilibrium), there must be an equal upward force developed by the supporting masonry. Thus, the wooden block is acted on by a set of balanced forces consisting of the applied (or active) downward load of 6400 lb and the resisting (called reactive) upward force of 6400 lb.

To resist being crushed, the wooden block develops an internal force of compression through stress in the material, stress being defined as internal force per unit area of the block cross section. For the situation shown, each square inch of the block cross section must develop a stress equal to $6400/64 = 100\ \text{lb/in.}^2$ (psi). See Figure 2.1*b*.

Design Use of Direct Stress

The fundamental relationship for simple direct stress may be stated as

$$f = \frac{P}{A} \quad \text{or} \quad P = fA \quad \text{or} \quad A = \frac{P}{f}$$

where f = direct axial stress

P = axial force

A = cross-sectional area of the member

The first form is used for stress determinations with stress defined as *force per unit area* or simply *unit stress*, the second form for finding the total load (total force) capacity of a member, and the third form for determining the required cross-sectional area of a member for a required load with a defined limiting stress condition (called the *allowable stress* or the *working stress*).

In the examples and problems dealing with the direct stress equation, differentiation is made between the unit stress developed in a member sustaining a given load, $f = P/A$, and the *allowable unit stress* used when determining the size of a member required to carry a given load, $A = P/f$. The latter form of the equation is, of course, the one used in design. The procedures for establishing allowable unit stresses in tension, compression, shear, and bending are different for different materials and are prescribed in industry-prepared specifications. A sample of such data is presented in Table 2.1. In actual design work, the building code governing the construction of buildings in the particular locality must be consulted for specific requirements.

The stresses discussed so far have been direct or axial stresses. This means they are assumed to be uniformly distributed over the cross section. The examples and problems presented fall under three general types: first, the design of structural members ($A = P/f$); second, the determination of safe loads ($P = fA$); and third, the investigation

TABLE 2.1 Selected Values for Common Structural Materials

Material and Property	Common Values (psi)
Structural Steel	
Yield strength	50,000
Allowable tension	30,000
Modulus of elasticity, E	29,000,000
Concrete	
f'_c (specified compressive strength)	3,000
Usable compression in bearing	900
Modulus of elasticity, E	3,000,000
Structural Lumber (Douglas fir–larch, select structural grade, posts and timbers)	
Compression, parallel to grain	1,150
Modulus of elasticity, E	1,600,000

of members for safety ($f = P/A$). The following examples will serve to fix in mind each of these types.

Example 1. Design (determine the size of) a short, square post of select structural grade Douglas fir to carry a compressive load of 30,000 lb.

Solution: Referring to Table 2.1, the allowable unit compressive stress for this wood parallel to the grain is 1150 psi. The required area of the post is

$$A = \frac{P}{F} = \frac{30{,}000}{1150} = 26.1 \text{ in.}^2$$

From Table A.8 in Appendix A an area of 30.25 in.^{2} is provided by a 6 × 6-in. post with a dressed size of 5.5 × 5.5 in.

Example 2. Determine the safe axial compressive load for a short, square concrete pier with a side dimension of 2 ft.

Solution: The area of the pier is 4 ft^2, or 576 in.^{2} Table 2.1 gives the allowable unit compressive stress for concrete as 900 psi. Therefore, the safe load on the pier is

$$P = (F)(A) = (900)(576) = 518{,}400 \text{ lb}$$

Example 3. A running track in a gymnasium is hung from the roof trusses by steel rods, each of which supports a tensile load of 11,200 lb. The round rods have a diameter of $^7/_8$ in. with the ends upset, that is, made larger by forging. This upset allows the full cross-sectional area of the rod (0.601 in.^{2}) to be utilized; otherwise, the cutting of the threads will reduce the cross section of the rod. Investigate this design to determine whether it is safe.

Solution: Since the gross area of the hanger rod is effective, the unit stress developed is

$$f = \frac{P}{A} = \frac{11{,}200}{0.601} = 18{,}600 \text{ psi}$$

Table 2.1 gives the allowable unit tensile stress for steel as 30,000 psi, which is greater than that developed by the loading. Therefore, the design is safe.

Problem 2.1.A What axial compression load may be placed on a short timber post whose cross-sectional dimensions are 9.5 × 9.5 in. if the allowable unit compressive stress is 1100 psi?

Problem 2.1.B The allowable compressive bearing capacity of a soil is 8000 psf. What should be the length of the side of a square footing if the total load (including the weight of the footing) is 240 kips?

Problem 2.1.C Determine the minimum cross-sectional area of a steel bar required to support a tensile force of 50 kips if the allowable unit tensile stress is 20 ksi.

Problem 2.1.D A short, square timber post supports a load of 115 kips. If the allowable unit compressive stress is 1000 psi, what nominal size square timber should be used. (See Table A.8.)

2.2 DEFORMATION

Whenever a force acts on a body, there is an accompanying change in shape or size of the body. In structural mechanics this is called *deformation*. Regardless of the magnitude of the force, some deformation is always present, although often it is so small that it is difficult to measure. In the design of structures it is often necessary to know what the deformation in certain members will be. A floor joist, for instance, may be large enough to support a given load safely but may *deflect* (the term for deformation that occurs with bending) to such an extent that the plaster ceiling below will crack or the floor may feel excessively springy to persons walking on it. For the usual cases we can readily determine what the deformation will be.

Stress is a major issue, primarily for determination of the strength of structures. However, deformation due to stress is often of concern, and the relation of stress to strain is one that must be quantitatively established. These relations and the issues they raise are discussed in this section.

Hooke's Law

As a result of experiments with clock springs, Robert Hooke, a mathematician and physicist working in the seventeenth century, developed the theory that "deformations are directly proportional to stresses." In other words, if a force produces a certain deformation, twice the force will produce twice the amount of deformation. This law of physics is of utmost importance in structural engineering although, as we shall find, Hooke's law holds true only up to a certain limit.

Elastic Limit and Yield Point

Suppose that a bar of structural steel with a cross-sectional area of 1 in.^{2} is placed into a machine for making tension tests. Its length is accurately measured and then a tensile force of 5000 lb is applied, which, of course, produces a unit tensile stress of 5000 psi in the bar. Measuring the length again, it is found that the bar has lengthened a definite amount, call it x inches. On applying 5000 lb more, the amount of lengthening is now $2(x)$, or twice the amount noted after the first 5000 lb. If the test is continued, it will be found that for each 5000-lb increment of additional load, the length of the bar will increase the same amount as noted when the initial 5000 lb was applied; that is, the deformations (length changes) are directly proportional to the stresses. So far Hooke's law has held true, but when a unit stress of about 50,000 psi is reached, the length increases more than x for each additional 5000 lb of load. This unit stress is called the *elastic limit*, or the *yield stress*. Beyond this stress limit, Hooke's law will no longer apply.

Figure 2.2 is a graph that displays the relationship between stress and strain for a ductile steel member subjected to tension stress. Strain (ε—epsilon) is a form of measurement of deformation expressed as a percentage of length change; thus,

$$\varepsilon = \frac{\text{change in length}}{\text{original length}} = \frac{\Delta L}{L}$$

Note on the graph that within the elastic stress range the straight line of the graph indicates a constant proportionality between the stress and the strain.

Another phenomenon may be noted in this connection. In the test just described, it will be observed that when any applied load that produces a unit stress *less* than the elastic limit is removed, the bar returns to its original length. If the load producing a unit stress *greater* than the elastic limit is removed, it will be found that the bar has permanently increased its length. This permanent deformation is called the *permanent set*. This fact permits another way of defining the elastic limit: It is that unit stress beyond which the material does not return to its original length when the load is removed.

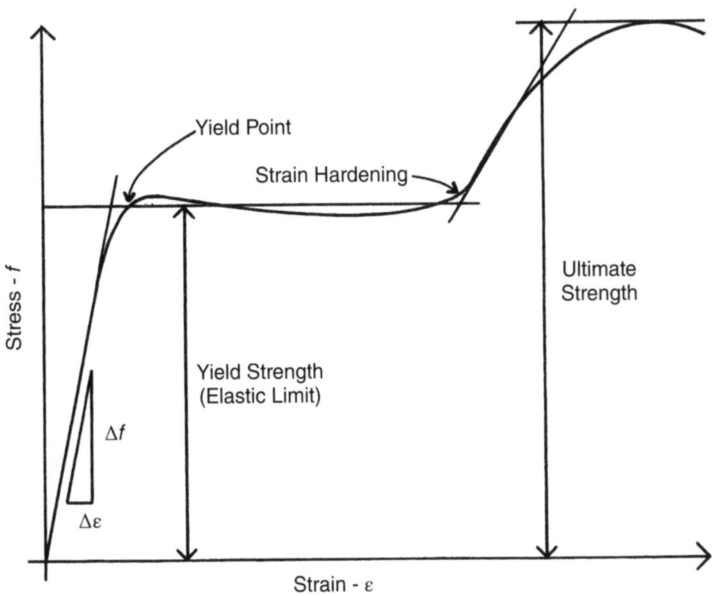

Figure 2.2 Stress–strain graph for ductile structural steel.

If this test is continued beyond the elastic limit, a point is reached where the deformation increases without any increase in the load. The unit stress at which this deformation occurs is called the *yield point*; it has a value only slightly higher than the elastic limit. Since the yield point, or yield stress, as it is sometimes called, can be determined more accurately by test than the elastic limit, it is a particularly important unit stress. Nonductile materials such as wood and cast iron have poorly defined elastic limits and no yield point.

Ultimate Strength

After passing the yield point, the steel bar of the test described in the preceding discussion again develops resistance to the increasing load. When the load reaches a sufficient magnitude, rupture occurs. The unit stress in the bar just before it breaks is called the ultimate strength. For the grade of steel assumed in the test, the ultimate strength may occur at a stress as high as about 65,000 psi.

Steel members are designed so that stresses under normal service conditions will not exceed the elastic limit, even though there is considerable reserve strength between this value and the ultimate strength. This procedure is followed because deformations produced by stresses above the elastic limit are permanent and hence change the shape of the structure in a permanent manner.

Factor of Safety

The degree of uncertainty that exists, with respect to both actual loading of a structure and uniformity in the quality of materials, requires that some reserve strength be built into the design. This degree of reserve strength is the *factor of safety*. Although there is no general agreement on the definition of this term, the following discussion will serve to fix the concept in mind.

Consider a structural steel that has an ultimate tensile unit stress of 65,000 psi, a yield point stress of 50,000 psi, and an allowable stress of 30,000 psi. If the factor of safety is defined as the ratio of the ultimate stress to the allowable stress, its value is 65,000/30,000, or 2.16. On the other hand, if it is defined as the ratio of the yield point stress to the allowable stress, its value is 50,000/30,000, or 1.67. This is a considerable variation, and since deformation failure of a structural member begins when it is stressed beyond the elastic limit, the higher value may be misleading. Consequently, the term *factor of safety* is not employed extensively today. Building codes generally specify the allowable unit stresses that are to be used in design for the grades of structural steel to be employed.

If one should be required to pass judgment on the safety of a structure, the problem resolves itself into considering each structural element, finding its actual unit stress under the existing loading conditions, and comparing this stress with the allowable stress prescribed by the local building regulations. This procedure is called *structural investigation*.

Modulus of Elasticity

Within the elastic limit of a material, deformations are directly proportional to the stresses. The magnitude of these deformations can be computed by use of a number (ratio), called the *modulus of elasticity*, that indicates the degree of *stiffness* of a material.

A material is said to be stiff if its deformation is relatively small when the unit stress is high. As an example, a steel rod 1 in.^{2} in cross-sectional area and 10 ft long will elongate about 0.008 in.

under a tensile load of 2000 lb. However, a piece of wood of the same dimensions will stretch about 0.24 in. with the same tensile load. The steel is said to be stiffer than the wood because, for the same unit stress, the deformation is not so great.

The modulus of elasticity of a structural member is defined as the unit stress divided by the unit deformation. Unit deformation refers to the percent of deformation and is usually called strain. It is dimensionless since it is expressed as a ratio, as follows:

$$\varepsilon = \frac{\Delta L}{L}$$

where ε = strain, or unit deformation
ΔL = actual dimensional change of length
L = original length of the member

The modulus of elasticity for direct stress is represented by the letter E, expressed in pounds per square inch, and has the same value in compression and tension for most structural materials. It is defined as

$$E = \frac{f}{\varepsilon}$$

From Section 2.1, $f = P/A$; then

$$E = \frac{P/A}{\Delta L/L} = \frac{PL}{A(\Delta L)}$$

This can also be written in the form

$$\Delta L = \frac{PL}{AE}$$

where ΔL = total deformation, in.
P = force, lb
L = length, in.
A = cross-sectional area, in.^{2}
E = modulus of elasticity, psi

Note that E is expressed in the same units as f (pounds per square inch) because in the equation $E = f/\varepsilon$, ε is a dimensionless number. For steel E = 29,000,000 psi, and for wood, depending on the species and grade, it varies from something less than 1,000,000 psi to about 1,900,000 psi. For concrete E ranges from about 2,000,000 psi to about 5,000,000 psi for common structural grades. It should be noted that the slope in the elastic zone for concrete is not linear but curved. The modulus of elasticity is an approximation of an "average" slope as opposed to an actual slope.

Example 4. A 2-in.-diameter round steel rod 10 ft long is subjected to a tensile force of 60 kips. How much will it elongate under the load?

Solution: The area of the 2-in. rod is 3.1416 in.^{2} Checking to determine whether the stress in the bar is within the elastic limit, we find that

$$f = \frac{P}{A} = \frac{60}{3.142} = 19.1 \text{ ksi}$$

This is within the elastic limit of ordinary structural steel (50 ksi), so the formula for finding the deformation is applicable. From data, $P = 60$ kips, $L = 120$ (length in inches), $A = 3.1416$, and $E = 29{,}000{,}000$. Substituting these values, we calculate the total lengthening of the rod as

$$\Delta L = \frac{PL}{AE} = \frac{(60{,}000)(120)}{(3.142)(29{,}000{,}000)} = 0.079 \text{ in.}$$

Problem 2.2.A What force must be applied to a steel bar 1 in. square and 2 ft long to produce an elongation of 0.016 in.?

Problem 2.2.B How much will a nominal 8 × 8-in. Douglas fir post 12 ft long shorten under an axial load of 45 kips?

Problem 2.2.C A routine quality control test is made on a structural steel bar that is 1 in. square and 16 in. long. The data developed during the test show that the bar elongated 0.0111 in. when subjected to a tensile force of 20.5 kips. Compute the modulus of elasticity of the steel.

Problem 2.2.D A $\frac{1}{2}$-in.-diameter round steel rod 40 ft long supports a load of 4 kips. How much will it elongate?

2.3 SUSPENSION CABLES

Suspension cables are structures that transmit applied loads to the supports by tension in a hanging cable or rope. A famous example of a suspension cable is the Golden Gate Bridge over San Francisco Bay. A more modest example is that of a clothes line. Suspension cables for buildings fall somewhere between these two examples.

Since cables have little resistance to compression or bending, the profile shape of a suspension cable changes when the applied load changes. Structures whose profile shape is derived from their applied loads are known as *funicular structures*.

Graphical Investigation Method

Since the shape of a cable changes with the applied load, the simplest way to determine the cable shape and the tension forces within the cable is by use of a graphical method of investigation. This method employs Bow's notation, as explained in Section 1.4.

Example 5. Determine the cable profile shape and the internal tension forces for the cable in Figure 2.3*a*, with three loads of 10 kips each placed at the quarter points of the span. (Note that the profile and tension are related not to the actual cable span but to the relative positions of the loads and supports.)

Solution: Using the algebraic method described in Section 1.7, the vertical components of the reactions are determined to be 15 kips each. The Maxwell diagram for this cable, illustrated in Figure 2.3*b*, is begun by drawing the graphical plot of the loads and vertical reactions: *a* to *b* (10 kips down), *b* to *c* (10 kips down), *c* to *d* (10 kips down), *d* to *e* (15 kips up), and *f* to *a* (15 kips up). The location of the focus (point *o*) on a line horizontal from points *e* and *f* will determine both the cable shape and the horizontal components of the reactions. For this example, two foci are investigated, one whose horizontal reaction is 10 kips (designated o_{10}) and one whose horizontal reaction is 20 kips (designated o_{20}).

Figure 2.3*c* illustrates the cable for the system with the horizontal reaction force of 10 kips. The cable shape is determined by beginning at the left support and drawing a line parallel to a–o_{10} from the Maxwell diagram until it intersects the line of action of the first 10-kip load. Next, a line parallel to b–o_{10} is drawn to intersect with the second 10-kip load. This procedure continues until the complete cable—from support to support—is drawn. The magnitudes of the cable's internal tension forces are determined by measuring the corresponding line of the Maxwell diagram.

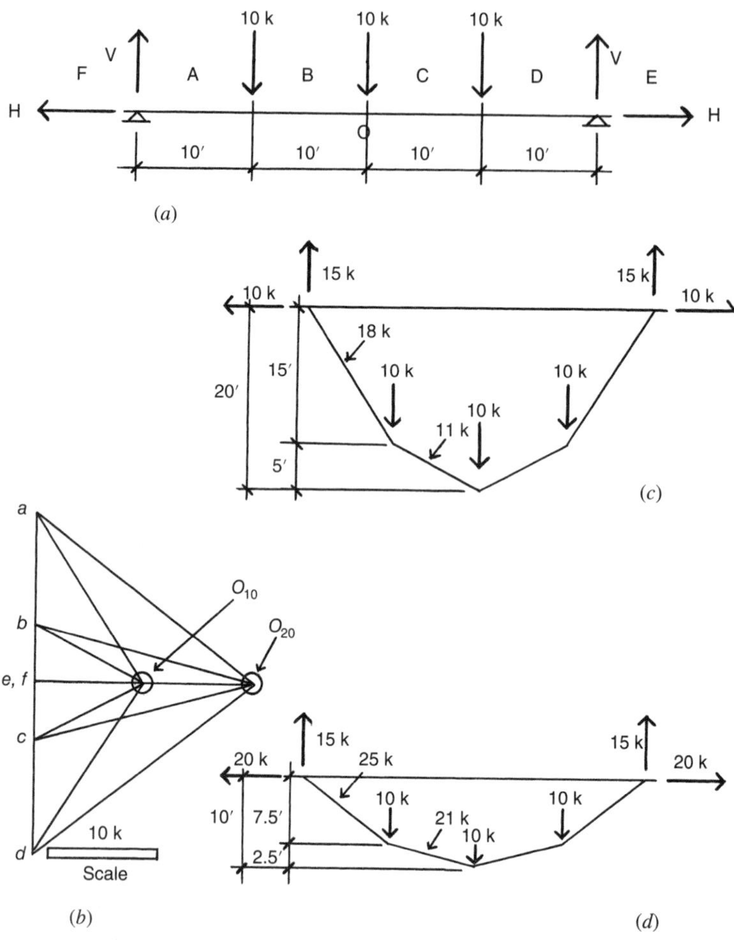

Figure 2.3 Reference for Example 5.

Figure 2.3*d* illustrates the same procedure for the cable with 20-kip horizontal reactions. It should be noted that the result of the larger horizontal force is a cable with less sag and greater internal tension forces.

Problems 2.3.A–D Determine the cable shape and internal forces for the applied forces illustrated in Figure 2.4 using a horizontal reaction force of 20 kips.

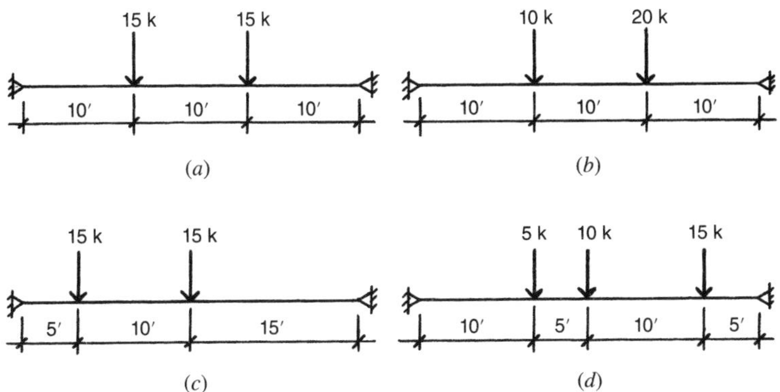

Figure 2.4 Reference for Problems 2.3 and 2.4.

Graphical Form Finding Method

The graphical method for suspension cable can also be used to determine a funicular shape within geometric constraints. In Example 5, the geometric constraints were that the ends of the cables were located horizontally level to each other. These constraints were used to determine the reactive forces (F_{DE} and F_{FE}). A force constraint was also added during the solution of how large the horizontal reaction would be (10 and 20 kips). Instead of this force constraint, an additional geometric constraint could have been added, such as the location of the center of the cable.

Example 6. Determine the cable profile shape for the loading and support conditions shown in Figure 2.5*a* with the right support being located 10 ft above the left support and the center of the cable located 10 ft below the left support.

Solution: Form finding graphical methods include a trail form based on placing the focus point *o* either arbitrarily or based on some assumption such as having the supports level to each other as it was in Example 5. Once the trail focus point *o* is set, which will be labeled *o*′, the trail shape is drawn. For this example *o*′ will be set at the location it was in Example 5 when the horizontal reactions were equal to 10 kips. The Maxwell diagram for this trail is illustrated in Figure 2.5*b*, which produces the trail funicular shape in Figure 2.5*c*.

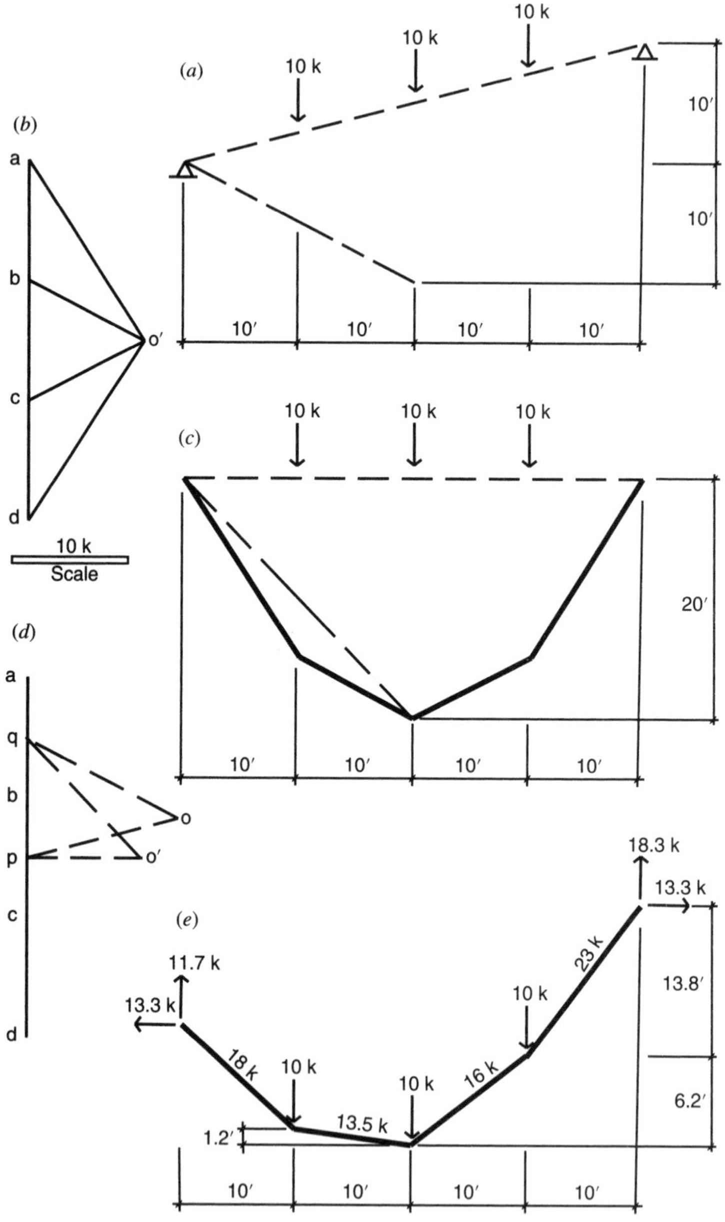

Figure 2.5 Reference for Example 6.

In the trial funicular shape the left-hand support is in the correct location but neither the right-hand support nor the center of the cable are in the correct location. This discrepancy is how we know that o' is not the correct location for the focus point and how the correct location will be located. On the trial shape, Figure 2.5*c*, the relationship between the two supports is horizontal. The load line on the Maxwell diagram and the location of o' have been duplicated in Figure 2.5*d*. The lines that determine the trial shape have been removed for clarity. On this diagram, a horizontal line replicating the horizontal relationship between the two supports is drawn from the trial focus o' to the load line and labeled p. The actual relationship between the two supports, shown in Figure 2.5*a*, is an upward-sloping line with a slope of 1–4. In Figure 2.5*d* we draw this line starting at point p on the load line. Any focus point chosen along this line will give a funicular shape that has the desired relationship between the two supports. Similarly, the relationship between the left-hand support and the center of the cable in the trial shape, Figure 2.5*c*, is transferred from the trial focus, o', back to the load line and labeled q. The desired relationship, Figure 2.5*a*, is transferred to the Maxwell diagram starting at point q. Any focus point chosen along this line will give the desired relationship between the left-hand support and the center of the cable. The intersection of these two lines will give us the final location of o, which produces a funicular shape that has the center of the cable 10 ft below the left-hand support and the right-hand support 10 ft above the left-hand support. When using geometric constraints, no more than three points on the shape can be predetermined.

Problems 2.3.E–H Determine the cable shape and internal forces for the applied forces illustrated in Figure 2.4 using geometric constraints of the center of the cable being 10 ft below the left-hand support and the right-hand support being 10 ft above it.

2.4 FUNICULAR ARCHES

Funicular arches are arches whose shape is determined by the loading (usually the dead loads) applied to the structure. Unlike the shape of a suspension cable, which changes with the loading, the shape of an arch is fixed, and therefore an arch is only truly funicular when the shape matches the anticipated loading.

A noteworthy funicular arch is the Gateway Arch in St. Louis, Missouri. The loading that was used to determine its shape was the dead weight of the total arch construction.

Graphical Investigation Method

The graphical method for arches is similar to that used for suspension cables, as discussed in Section 2.3. The major difference is that the focus point *o* of the Maxwell diagram is placed on the opposite side of the load line; this reflects the situation of internal compression versus internal tension.

Example 7. Determine a funicular arch shape for the loading and support conditions shown in Figure 2.6*a*. The maximum compression force in the arch is to be limited to 50 kips.

Solution: As with the suspension cable, the vertical reaction components may be found with an equilibrium analysis similar to that for a beam: for example, a summation of moments about one support plus a summation of vertical forces. For this example, such an analysis will determine vertical reactions of 15 kips at the left support and 30 kips at the right support.

Construction of the Maxwell diagram begins with the layout of the load line *a–b–c*, as shown in Figure 2.6*b*. Point *d* is located horizontally from point *c* and point *e* is located horizontally from point *a*. Line *a–e* thus represents the horizontal component of the left reaction, and line *c–d* represents the horizontal component of the right reaction. The focus point *o* is located on a horizontal line that is 30 kips above point *d* and 15 kips below point *e*, as established by the known values of the vertical components of the reactions. The horizontal position of the focus on this line is determined by measuring a distance of 50 kips from point *c*. This establishes the maximum limit of compression in member *c–o* and—by measurement on the diagram—a compression force of approximately 42.7 kips in member *a–o*.

The funicular arch shape in Figure 2.6*c* is derived by using lines parallel to those in the Maxwell diagram, as was done for the cable in Example 5.

Problem 2.4.A Determine the funicular arch shape for the loading shown in Figure 2.4*a* if the horizontal reaction is equal to 20 kips.

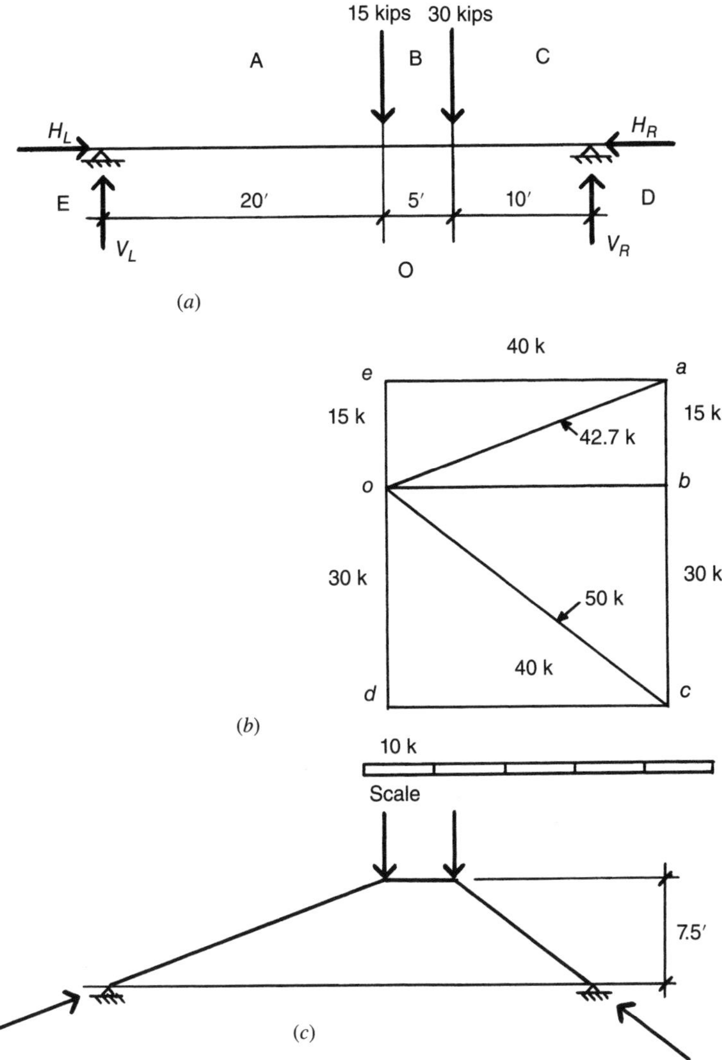

Figure 2.6 Reference for Example 7.

Problem 2.4.B Determine the funicular arch shape for the loading in Figure 2.4*b* if the greatest compression force in the arch is limited to 26 kips.

Problem 2.4.C Determine the funicular arch shape for the loading in Figure 2.4*c* if the greatest compression force in the arch is limited to 25 kips.

Problem 2.4.D Determine the funicular arch shape for the loading in Figure 2.4*d* if the greatest compression force in the arch is limited to 30 kips.

Graphical Form Finding Method

Like suspension cables, the funicular arch shape can be determined through three predetermined points. The difference between arches and cables is that the location of the focus point, o or o', is left of the load line on the Maxwell diagram. The same procedure is used in setting a trial focus point o', which will result in a trial funicular shape. From the trial shape, the location of the final focus point, o, is determined by drawing lines back onto the load line from o' and then finding the intersection of the lines out from p and q.

Problems 2.4.E–H Determine the cable shape and internal forces for the applied forces illustrated in Figure 2.4 using the geometric constraints of the center of the arch being 20 ft above the left-hand support and the right-hand support being 10 ft below the left-hand support.

2.5 GRAPHICAL ANALYSIS OF PLANAR TRUSSES

Planar trusses, comprised of linear elements assembled in triangulated frameworks, have been used for spanning structures in buildings for many centuries. Investigation for internal forces in trusses is typically performed by the basic methods illustrated in the preceding sections. In this section these procedures are demonstrated using both algebraic and graphical methods of solution.

When the so-called *method of joints* is used, finding the internal forces in the members of a planar truss consists of solving a series of concurrent force systems. Figure 2.7 shows a truss with the truss form, the loads, and the reactions displayed in a space diagram. Below the space diagram is a figure consisting of the free-body diagrams of the individual joints of the truss. These are arranged in the same manner as they are in the truss in order to show their interrelationships. However, each joint constitutes a complete concurrent planar force

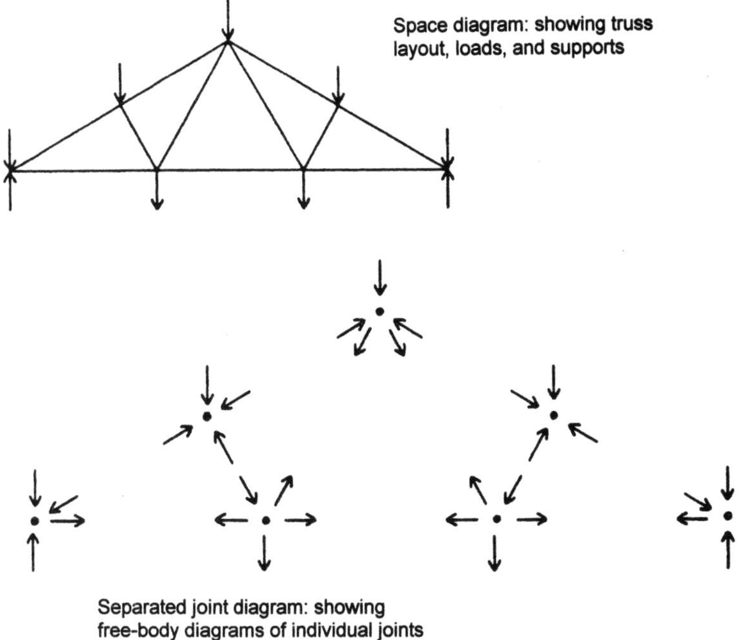

Figure 2.7 Examples of diagrams used to represent trusses and their actions.

system that must have its independent equilibrium. Solving the problem consists of determining the equilibrium conditions for all of the joints. The procedures used for this solution will now be illustrated.

Example 8. Figure 2.8 shows a single-span, planar truss subjected to vertical gravity loads. Find the magnitude and sense of the internal forces in the truss members.

Solution: The space diagram in the figure shows the truss form and dimensions, the support conditions, and the loads. The letters on the space diagram identify individual forces at the truss joints, as discussed in Section 1.4. The sequence of placement of the letters is arbitrary, the only necessary consideration being to place a letter in each space between the loads and the individual truss members so that each force at a joint can be identified by a two-letter symbol.

The separated joint diagram in the figure provides a useful means for visualization of the complete force system at each joint as well as

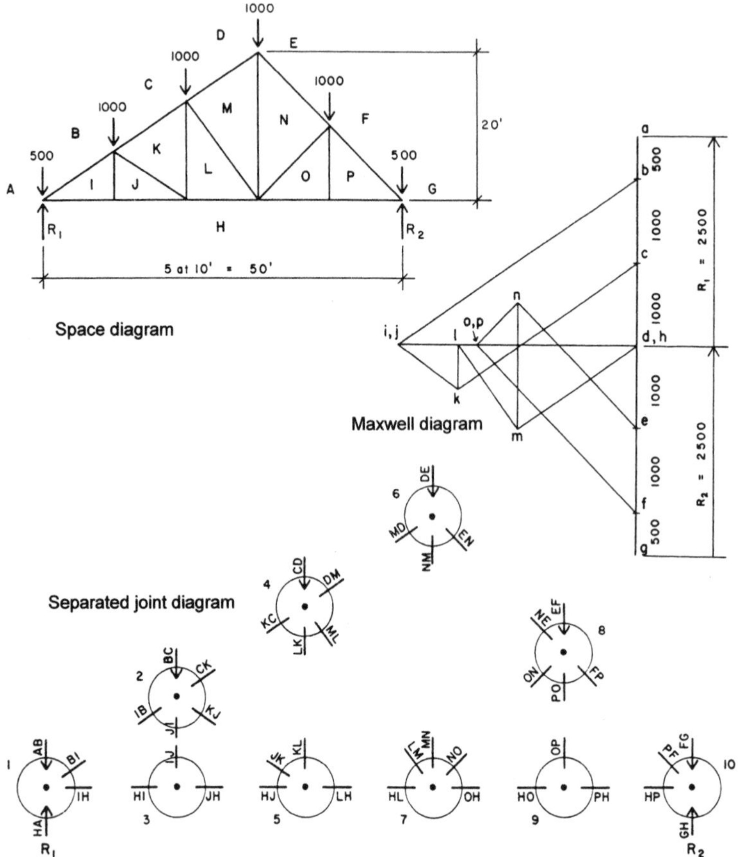

Figure 2.8 Graphical diagrams for the sample problem.

the interrelation of the joints through the truss members. The individual forces at each joint are designated by two-letter symbols that are obtained by simply reading around the joint in the space diagram in a clockwise direction. Note that the two-letter symbols are reversed at the opposite ends of each of the truss members. Thus, the top chord member at the left end of the truss is designated as *BI* when shown in the joint at the left support (joint 1) and is designated as *IB* when shown in the first interior upper chord joint (joint 2). The purpose of this procedure will be demonstrated in the following explanation of the graphical analysis.

The third diagram in Figure 2.8 is a composite force polygon for the external and internal forces in the truss. It is called a Maxwell diagram after one of its early promoters, James Maxwell, a British engineer. The construction of this diagram constitutes a complete solution for the magnitudes and senses of the internal forces in the truss. The procedure for this construction is as follows:

1. *Construct the force polygon for the external forces* (also described as the load line). Before this can be done, the values for the reactions must be found. There are graphical techniques for finding the reactions, but it is usually much simpler and faster to find them with an algebraic solution. In this example, although the truss is not symmetrical, the loading is, and it may simply be observed that the reactions are each equal to one-half of the total load on the truss, or 5000 ÷ 2 = 2500 lb. Since the external forces in this case are all in a single direction, the force polygon for the external forces is actually a straight line. Using the two-letter symbols for the forces and starting with the letter *A* at the left end, we read the force sequence by moving in a clockwise direction around the outside of the truss. The loads are thus read as *AB, BC, CD, DE, EF*, and *FG*, and the two reactions are read as *GH* and *HA*. Beginning at *a* on the Maxwell diagram, the force vector sequence for the external forces is read from *a* to *b*, *b* to *c*, *c* to *d*, and so on, ending back at *a*, which shows that the force polygon closes and the external forces are in the necessary state of static equilibrium. Note that we have pulled the vectors for the reactions off to the side in the diagram to indicate them more clearly. Note also that we have used lowercase letters for the vector ends in the Maxwell diagram, whereas uppercase letters are used on the space diagram. The alphabetic correlation is thus retained (*A* to *a*), while any possible confusion between the two diagrams is prevented. The letters on the space diagram designate open spaces, while the letters on the Maxwell diagram designate points of intersection of lines.
2. *Construct the force polygons for the individual joints.* The graphical procedure for this consists of locating the points on the Maxwell diagram that correspond to the remaining letters, *I* through *P*, on the space diagram. When all the lettered points on the diagram are located, the complete force polygon for

each joint may be read on the diagram. In order to locate these points, we use two relationships. The first is that the truss members can resist only forces that are parallel to the members' positioned directions. Thus, we know the directions of all the internal forces. The second relationship is a simple one from plane geometry: A point may be located at the intersection of two lines. Consider the forces at joint 1, as shown in the separated joint diagram in Figure 2.8. Note that there are four forces and that two of them are known (the load and the reaction) and two are unknown (the internal forces in the truss members). The force polygon for this joint, as shown on the Maxwell diagram, is read as *ABIHA*, where *AB* represents the load, *BI* the force in the upper chord member, *IH* the force in the lower chord member, and *HA* the reaction. Thus, the location of point *i* on the Maxwell diagram is determined by noting that *i* must be in a horizontal direction from *h* (corresponding to the horizontal position of the lower chord) and in a direction from *b* that is parallel to the position of the upper chord.

The remaining points on the Maxwell diagram are found by the same process, using two known points on the diagram to project lines of known direction whose intersection will determine the location of an unknown point. Once all the points are located, the diagram is complete and can be used to find the magnitude and sense of each internal force. The process for construction of the Maxwell diagram typically consists of moving from joint to joint along the truss. Once one of the letters for an internal space is determined on the Maxwell diagram, it may be used as a known point for finding the letter for an adjacent space on the space diagram. The only limitation of the process is that it is not possible to find more than one unknown point on the Maxwell diagram for any single joint. Consider joint 7 on the separated joint diagram in Figure 2.8. To solve this joint first, knowing only the locations of letters *a* through *h* on the Maxwell diagram, it is necessary to locate four unknown points: *l, m, n*, and *o*. This is three more unknowns than can be determined in a single step, so three of the unknowns must be found by using other joints.

Solving for a single unknown point on the Maxwell diagram corresponds to finding two unknown forces at a joint, since each letter on the space diagram is used twice in the force identification for the internal forces. Thus, for joint 1 in the previous example, the letter *I* is part of the identity of forces *BI* and *IH*, as shown on the separated joint diagram. The graphical determination of single points on the Maxwell diagram, therefore, is analogous to finding two unknown quantities in an algebraic solution. As discussed previously, two unknowns are the maximum that can be solved for the equilibrium of a coplanar, concurrent force system, which is the condition of the individual joints in the truss.

When the Maxwell diagram is completed, the internal forces can be read from the diagram as follows:

1. The magnitude is determined by measuring the length of the line in the diagram using the scale that was used to plot the vectors for the external forces.
2. The sense of individual forces is determined by reading the forces in clockwise sequence around a single joint in the space diagram and tracing the same letter sequences on the Maxwell diagram.

Figure 2.9*a* shows the force system at joint 1 and the force polygon for these forces as taken from the Maxwell diagram. The forces known initially are shown as solid lines on the force polygon, and the unknown forces are shown as dashed lines. Starting with letter *A* on the force system, we read the forces in a clockwise sequence as *AB, BI, IH*, and *HA*. Note that on the Maxwell diagram moving from *a* to *b* is moving in the order of the sense of the force, that is, from the tail to the end of the force vector that represents the external load on the joint. Using this sequence on the Maxwell diagram, this flow, moving in the sense of the force, will be a continuous one. Thus, reading from *b* to *i* on the Maxwell diagram is reading from the tail to the head of the force vector, which indicates that force *BI* has its head at the left end. Transferring this sense indication from the Maxwell diagram to the joint diagram indicates that force *BI* is in compression; that is, it is pushing, rather than pulling, on the joint. Reading from *i* to *h* on the Maxwell diagram shows that the arrowhead for this vector is on the right, which translates to a tension effect on the joint diagram.

Having solved for the forces at joint 1 as described, the fact that the forces in truss members *BI* and *IH* are known can be used to consider the adjacent joints, 2 and 3. However, it should be noted that

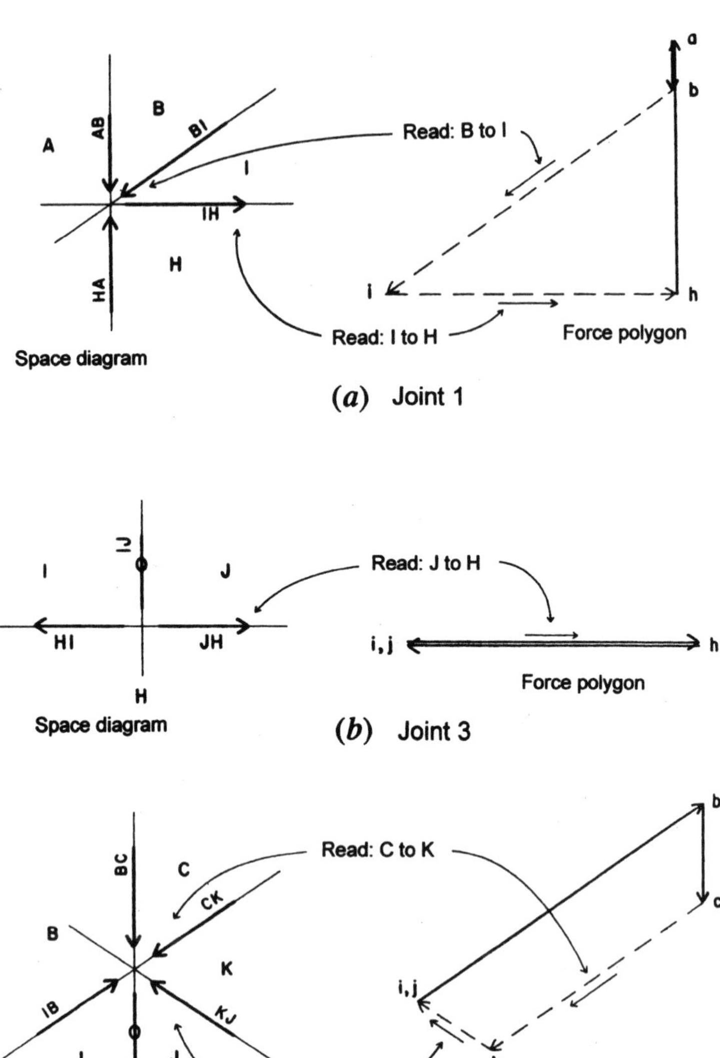

Figure 2.9 Graphical solutions for joints 1, 2, and 3.

the sense reverses at the opposite ends of the members in the joint diagrams. Referring to the separated joint diagram in Figure 2.8, if the upper chord member shown as force *BI* in joint 1 is in compression, its arrowhead is at the lower left end in the diagram for joint 1, as shown in Figure 2.9*a*. However, when the same force is shown as *IB* at joint 2, its pushing effect on the joint will be indicated by having the arrowhead at the upper right end in the diagram for joint 2. Similarly, the tension effect of the lower chord is shown in joint 1 by placing the arrowhead on the right end of the force *IH*, but the same tension force will be indicated in joint 3 by placing the arrowhead on the left end of the vector for force *HI*.

If the solution sequence of solving joint 1 and then joint 2 is chosen, it is now possible to transfer the known force in the upper chord to joint 2. Thus, the solution for the five forces at joint 2 is reduced to finding three unknowns since the load *BC* and the chord force *IB* are now known. However, it is still not possible to solve joint 2 since there are two unknown points on the Maxwell diagram (*k* and *j*) corresponding to the three unknown forces. An option, therefore, is to proceed from joint 1 to joint 3, where there are presently only two unknown forces. On the Maxwell diagram the single unknown point *j* can be found by projecting vector *IJ* vertically from *i* and projecting vector *JH* horizontally from point *h*. Since point *i* is also located horizontally from point *h*, this shows that the vector *IJ* has zero magnitude since both *i* and *j* must be on a horizontal line from *h* in the Maxwell diagram. This indicates that there is actually no stress in this truss member for this loading condition and that points *i* and *j* are coincident on the Maxwell diagram. The joint force diagram and the force polygon for joint 3 are as shown in Figure 2.9*b*. In the joint force diagram place a zero, rather than an arrowhead, on the vector line for *IJ* to indicate the zero-stress condition. In the force polygon in Figure 2.9*b*, the two force vectors are slightly separated for clarity, although they are actually coincident on the same line.

Having solved for the forces at joint 3, proceed to joint 2 since there remain only two unknown forces at this joint. The forces at the joint and the force polygon for joint 2 are shown in Figure 2.9*c*. As for joint 1, read the force polygon in a sequence determined by reading clockwise around the joint: *BCKJIB*. Following the continuous direction of the force arrows on the force polygon in this sequence, it is possible to establish the sense for the two forces *CK* and *KJ*.

It is possible to proceed from one end and to work continuously across the truss from joint to joint to construct the Maxwell diagram in this example. The sequence in terms of locating points on the Maxwell diagram would be *i–j–k–l–m–n–o–p*, which would be accomplished by solving the joints in the following sequence: 1,3,2,5,4,6,7,9,8. However, it is advisable to minimize the error in graphical construction by working from both ends of the truss. Thus, a better procedure would be to find points *i–j–k–l–m*, working from the left end of the truss, and then to find points *p–o–n–m*, working from the right end. This would result in finding two locations for *m*, whose separation constitutes an error in drafting accuracy.

Problems 2.5.A, B Using a Maxwell diagram, find the internal forces in the truss in Figure 2.10.

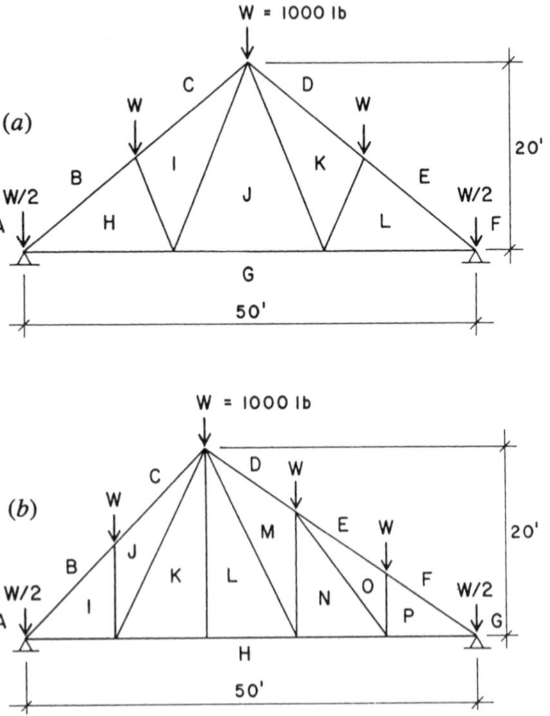

Figure 2.10 Reference for Problems 2.5 and 2.6.

2.6 ALGEBRAIC ANALYSIS OF PLANAR TRUSSES

Graphical solution for the internal forces in a truss using the Maxwell diagram corresponds essentially to an algebraic solution by the *method of joints*. This method consists of solving the concentric force systems at the individual joints using simple force equilibrium equations. The process will be illustrated using the previous example.

As with the graphical solution, first determine the external forces, consisting of the loads and the reactions. Then proceed to consider the equilibrium of the individual joints, following a sequence as in the graphical solution. The limitation of this sequence, corresponding to the limit of finding only one unknown point in the Maxwell diagram, is that only two unknown forces at any single joint can be found in a single step. (Two conditions of equilibrium produce two equations.) Referring to Figure 2.11, the solution for joint 1 is as follows.

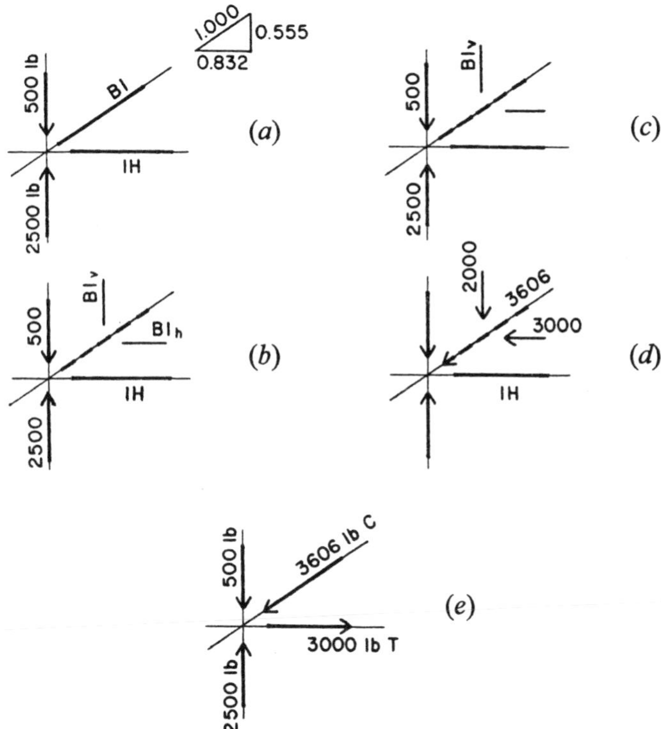

Figure 2.11 Algebraic solution for joint 1.

The force system for the joint is drawn with the sense and magnitude of the known forces shown, but with the unknown internal forces represented by lines without arrowheads, since their senses and magnitudes initially are unknown. For forces that are not vertical or horizontal, replace the forces with their horizontal and vertical components. Then consider the two conditions necessary for the equilibrium of the system: The sum of the vertical forces is zero and the sum of the horizontal forces is zero.

If the algebraic solution is performed carefully, the sense of the forces will be determined automatically. However, it is recommended that whenever possible the sense be predetermined by simple observations of the joint conditions, as will be illustrated in the solutions.

The problem to be solved at joint 1 is as shown in Figure 2.11*a*. In Figure 2.11*b* the system is shown with all forces expressed as vertical and horizontal components. Note that although this now increases the number of unknowns to three (*IH*, *BIv*, and *BIh*), there is a numeric relationship between the two components of *BI*. When this condition is added to the two algebraic conditions for equilibrium, the number of usable relationships totals three, so that the necessary conditions to solve for the three unknowns are present.

The condition for vertical equilibrium is shown in Figure 2.11*c*. Since the horizontal forces do not affect the vertical equilibrium, the balance is between the load, the reaction, and the vertical component of the force in the upper chord. Simple observation of the forces and the known magnitudes makes it obvious that force BI_v must act downward, indicating that *BI* is a compression force. Thus, the sense of *BI* is established by simple visual inspection of the joint, and the algebraic equation for vertical equilibrium (with upward force considered positive) is

$$\sum F_v = 0 = +2500 - 500 - BI_v$$

From this equation BI_v is determined to have a magnitude of 2000 lb. Using the known relationships among BI, BI_v, and BI_h, the values of these three quantities can be determined if any one of them is known. Thus,

$$\frac{BI}{1.000} = \frac{BI_v}{0.555} = \frac{BI_h}{0.832}$$

from which

$$BI_h = \left(\frac{0.832}{0.555}\right)(2000) = 3000 \text{ lb}$$

and

$$BI = \left(\frac{1.000}{0.555}\right)(2000) = 3606 \text{ lb}$$

The results of the analysis to this point are shown in Figure 2.11*d*, from which it may be observed that the conditions for equilibrium of the horizontal forces can be expressed. Stated algebraically (with force sense toward the right considered positive), the condition is

$$\sum F_h = 0 = IH - 3000$$

from which it is established that the force in *IH* is 3000 lb.

The final solution for the joint is then as shown in Figure 2.11*e*. In this diagram the internal forces are identified as to sense by using *C* to indicate compression and *T* to indicate tension.

As with the graphical solution, proceed to consider the forces at joint 3. The initial condition at this joint is as shown in Figure 2.12*a*, with the single known force in member *HI* and the two unknown forces in *IJ* and *JH*. Since the forces at this joint are all vertical and horizontal, there is no need to use components. Consideration of vertical equilibrium makes it obvious that it is not possible to have a force in member *IJ*. Stated algebraically, the condition for vertical equilibrium is

$$\sum F_v = 0 = IJ(\text{since IJ is the only force})$$

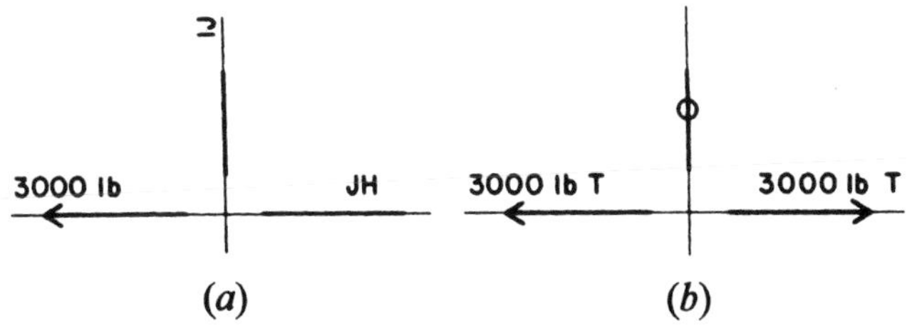

Figure 2.12 Algebraic solution for joint 3.

It is equally obvious that the force in JH must be equal and opposite to that in HI since they are the only two horizontal forces. That is, stated algebraically,

$$\sum F_h = 0 = JH - 3000$$

The final answer for the forces at joint 3 is as shown in Figure 2.12*b*. Note the convention for indicating a truss member with no internal force.

Now proceed to consider joint 2; the initial condition is as shown in Figure 2.13*a*. Of the five forces at the joint only two remain unknown. Following the procedure for joint 1, first resolve the forces into their vertical and horizontal components, as shown in Figure 2.13*b*.

Since the sense of forces CK and KJ is unknown, use the procedure of considering them to be positive until proven otherwise. That is, if they are entered into the algebraic equations with an assumed sense and the solution produces a negative answer, then the assumption was wrong. However, be careful to be consistent with the sense of the force vectors, as the following solution will illustrate.

Arbitrarily assume that force CK is in compression and force KJ is in tension. If this is so, the forces and their components will be as shown in Figure 2.13*c*. Then consider the conditions for vertical equilibrium; the forces involved will be those shown in Figure 2.13*d*, and the equation for vertical equilibrium will be

$$\sum F_v = 0 = -1000 + 2000 - CK_v - KJ_v$$

or

$$0 = +1000 - 0.555CK - 0.555KJ \qquad (2.6.1)$$

Now consider the conditions for horizontal equilibrium; the forces will be as shown in Figure 2.12*e*, and the equation will be

$$\sum F_h = 0 = +3000 - CK_h + KJ_h$$

or

$$0 = +3000 - 0.832CK + 0.832KJ \qquad (2.6.2)$$

Note the consistency of the algebraic signs and the sense of the force vectors, with positive forces considered as upward and toward

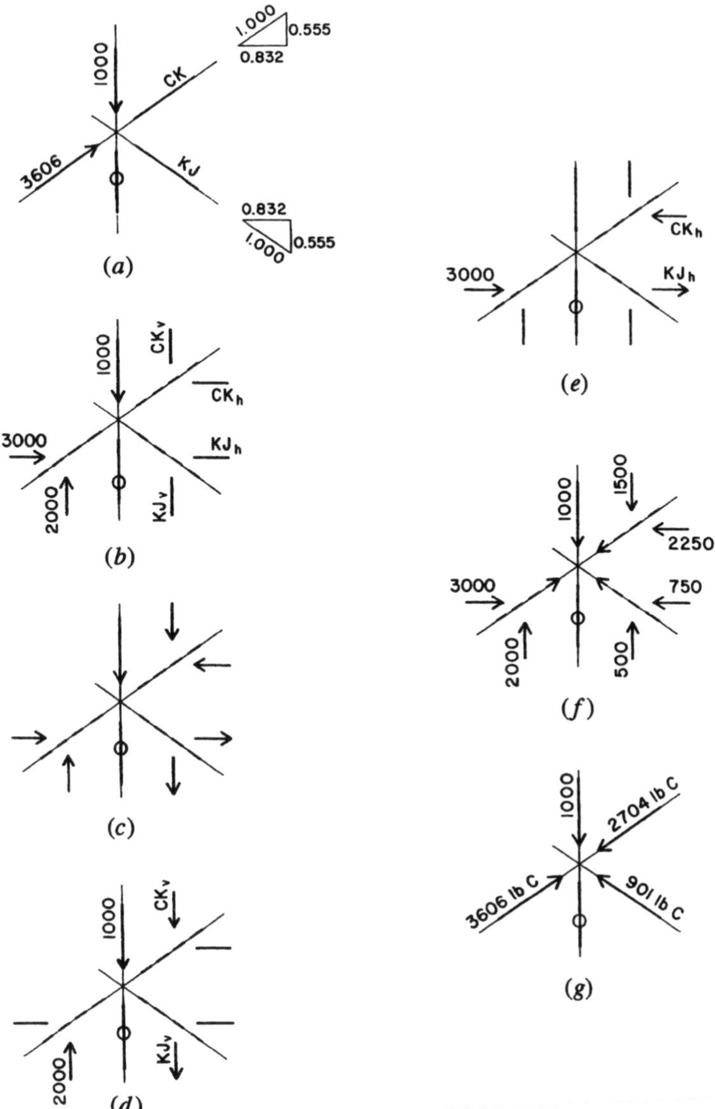

Figure 2.13 Algebraic solution for joint 2.

the right. Now solve these two equations simultaneously for the two unknown forces as follows:

1. Multiply equation (2.6.1) by $\frac{0.832}{0.555}$

$$0 = \left(\frac{0.832}{0.555}\right)(+1000) + \left(\frac{0.832}{0.555}\right)(-0.555\ CK) + \left(\frac{0.832}{0.555}\right)(-0.555\ KJ)$$

or

$$0 = +1500 - 0.832\ CK - 0.832\ KJ$$

2. Add this equation to equation (2.6.2) and solve for CK:

$$0 = +4500 - 1.664\ CK$$

$$CK = \frac{4500}{1.664} = 2704\ \text{lb}$$

Note that the assumed sense of compression in CK is correct since the algebraic solution produces a positive answer. Substituting this value for CK in equation (2.6.1),

$$0 = +1000 - 0.555(2704) - 0.555(KJ)$$

and

$$KJ = \frac{-500}{0.555} = -901\ \text{lb}$$

Since the algebraic solution produces a negative quantity for KJ, the assumed sense for KJ is wrong and the member is actually in compression.

The final answers for the forces at joint 2 are as shown in Figure 2.13*g*. In order to verify that equilibrium exists, however, the forces are shown in the form of their vertical and horizontal components at (*f*) in the illustration.

When all of the internal forces have been determined for the truss, the results may be recorded or displayed in a number of ways. The most direct way is to display them on a scaled diagram of the truss, as shown in Figure 2.14*a*. The force magnitudes are recorded next

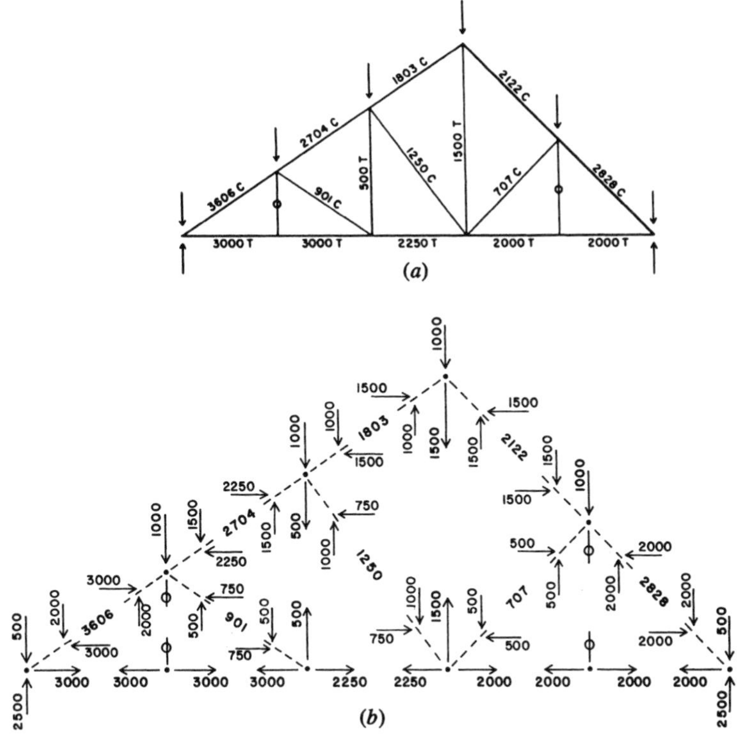

Figure 2.14 Presentation of the internal forces in the truss.

to each member with the sense shown as *T* for tension or *C* for compression. Zero-stress members are indicated by the conventional symbol consisting of a zero placed directly on the member.

When solving by the algebraic method of joints, the results may be recorded on a separated joint diagram as shown in Figure 2.14*b*. If the values for the vertical and horizontal components of force in sloping members are shown, it is a simple matter to verify the equilibrium of the individual joints.

Problems 2.6.A, B Using the algebraic method of joints, find the internal forces in the trusses in Figure 2.10.

2.7 CABLE-STAYED STRUCTURES

Cable-stayed structures are special truss systems in which the members in tension are cables. Many bridges have been built in recent times with this system, along with a few roof structures for buildings. Since these structures are a subset of truss structures, the methods of analysis described in Sections 2.5 and 2.6 can be applied to them.

Example 9. Determine the internal forces in the members of the cable-stayed structure shown in Figure 2.15*a*.

Solution: The two reaction forces are determined as follows. The reaction at the bottom support will be aligned with and have the same magnitude as the compression members at the bottom of the structure. For this example, that means that the reaction force will be a horizontal force; thus, the vertical component of the reaction at the upper support will be equal to the total of the vertical loads. Although it is actually not necessary to find the reactions for this structure before finding the internal forces, the task is quite easily performed algebraically, as follows:

The bottom reaction is determined by a summation of moments about the upper support to be 45 kips.

The horizontal component of the upper reaction is determined by equilibrium of horizontal forces to be 45 kips.

The vertical component of the upper reaction is determined by equilibrium of vertical forces to be equal to 50 kips.

The first step in determining the Maxwell diagram (Figure 2.15*b*) is to plot the load line *a–b–c–d*, which establishes the location of these four points. For the rest of the points, begin by finding point *h*, which is accomplished by drawing lines horizontally from point *b* and in a direction parallel to member *AH* from point *a*. Next, find point *g* by drawing lines horizontally from point *c* and in a direction parallel to member *GH* from point *h*. Finally, find point *e* by drawing lines horizontally from point *d* and in a direction parallel to member *EG* from point *g*. The member forces and reactions are displayed in Figure 2.15*c*.

Problems 2.7.A, B Determine the internal forces and the support reactions for the cable-stayed structures in Figure 2.16.

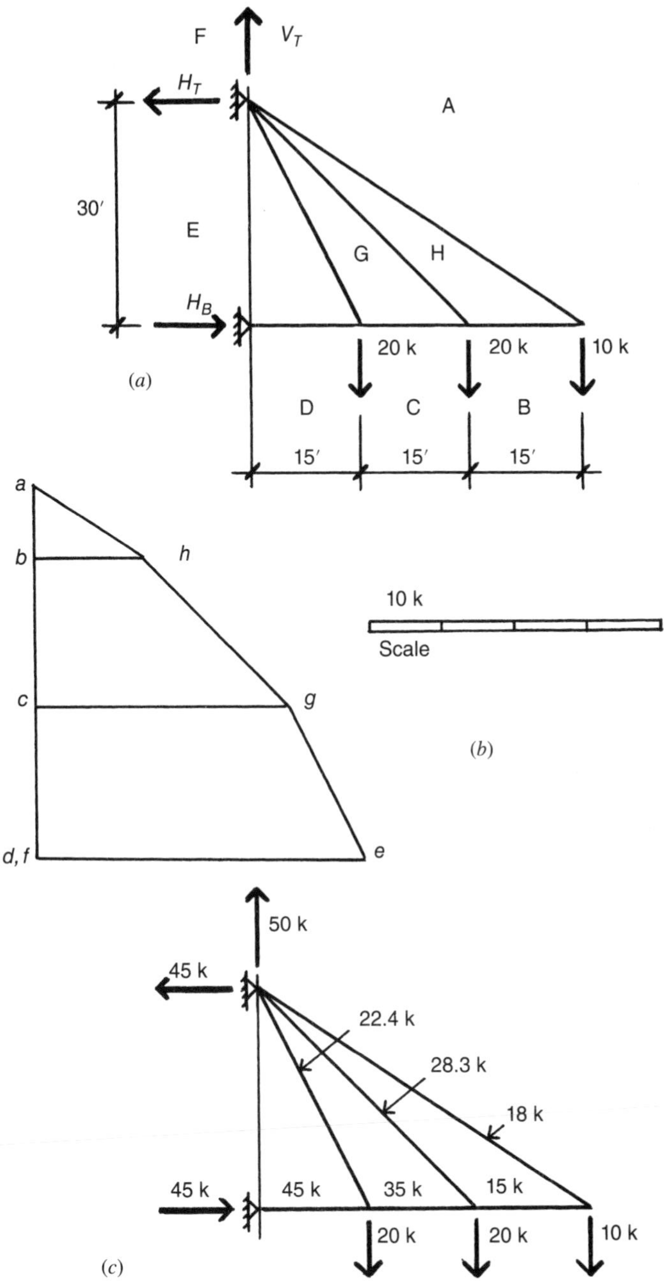

Figure 2.15 Reference for Example 9.

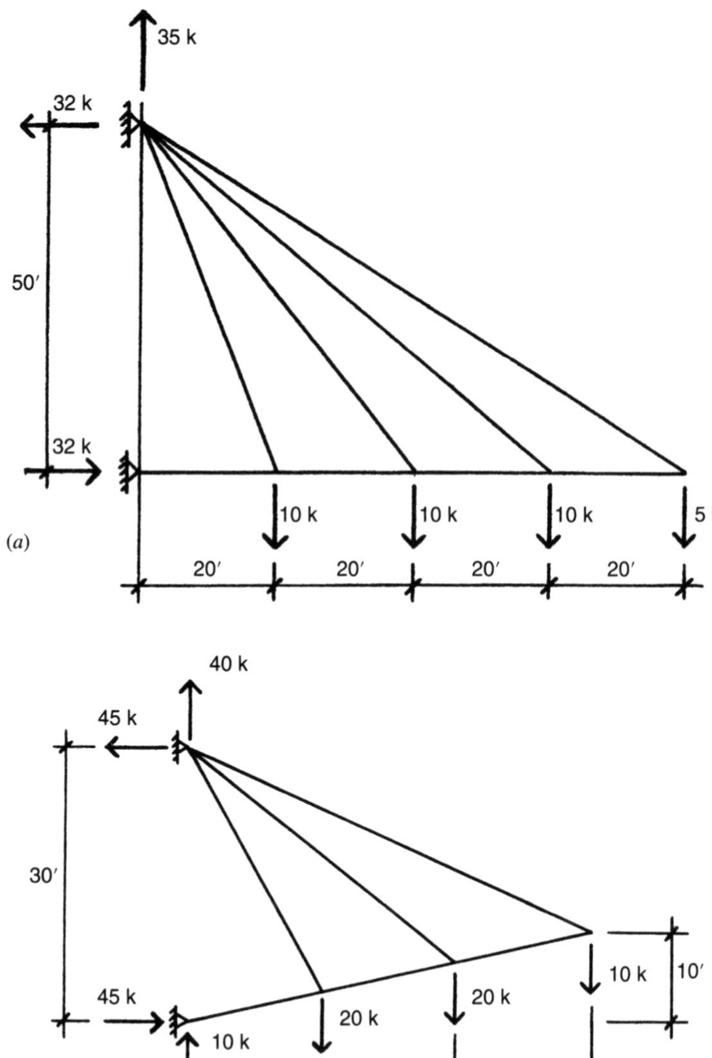

Figure 2.16 Reference for Problem 2.7.

2.8 COMPRESSION MEMBERS

Compression is developed in a number of ways in structures, including the compression component that accompanies the development of internal bending. In this section consideration is given to elements whose primary purpose is resistance of compression. This includes truss members, piers, bearing walls, and bearing footings, although major treatment here is given to columns, which are linear compression members. Building columns may be free-standing architectural elements, with the structural column itself exposed to view. However, for fire or weather protection the structural column must often be incorporated into other construction and may in some cases be fully concealed from view.

Slenderness Effects

Structural columns are often quite slender, and the aspect of slenderness (called *relative slenderness*) must be considered (see Figure 2.17). At the extremes the limiting situations are those of the very stout or short column that fails by crushing and the very slender or tall column that fails by lateral buckling.

The two basic limiting response mechanisms—crushing and buckling—are entirely different in nature. Crushing is a stress resistance phenomenon, and its limit is represented on the graph in Figure 2.17 as a horizontal line, basically established by the compression resistance of the material and the amount of material (area of the cross section) in the compression member. This behavior is limited to the range labeled zone 1 in Figure 2.17.

Buckling actually consists of lateral deflection in bending, and its extreme limit is affected by the bending stiffness of the member, as related to the stiffness of the material (modulus of elasticity) and to the geometric property of the cross section directly related to deflection—the moment of inertia of the cross-sectional area. The classic expression for elastic buckling is stated in the form of the equation developed by Euler:

$$P = \frac{\pi^2 EI}{(KL)^2}$$

The curve produced by this equation is of the form shown in Figure 2.17. It closely predicts the failure of quite slender compression members in the range labeled zone 3 in Figure 2.17.

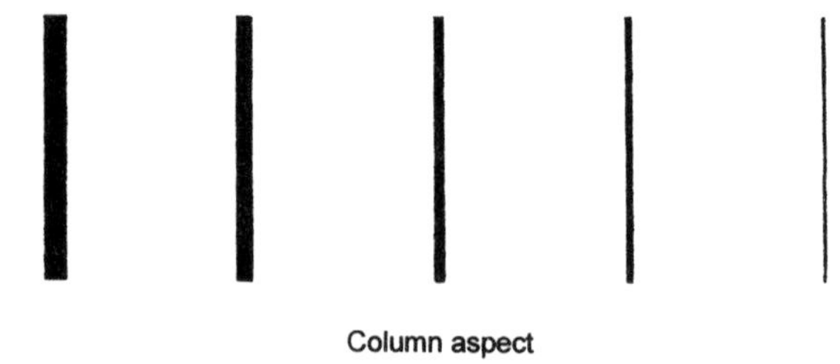

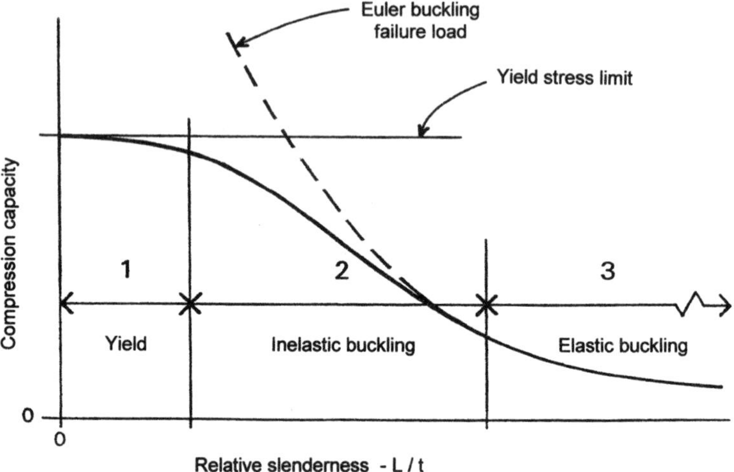

Figure 2.17 Effect of column slenderness on axial compression capacity.

Most building columns fall somewhere between very stout and very slender, in other words in the range labeled zone 2 in Figure 2.17. Their behavior therefore is one of an intermediate form, somewhere between pure stress response and pure elastic buckling. Predictions of structural response in this range must be established by empirical equations that somehow make the transition from the horizontal line to the Euler curve. Equations currently used are explained in Chapter 6 for wood columns and in Chapter 10 for steel columns.

Buckling may be affected by constraints, such as lateral bracing that prevents sideways movement or support conditions that restrain the rotation of the member's ends. Figure 2.18*a* shows the case for the

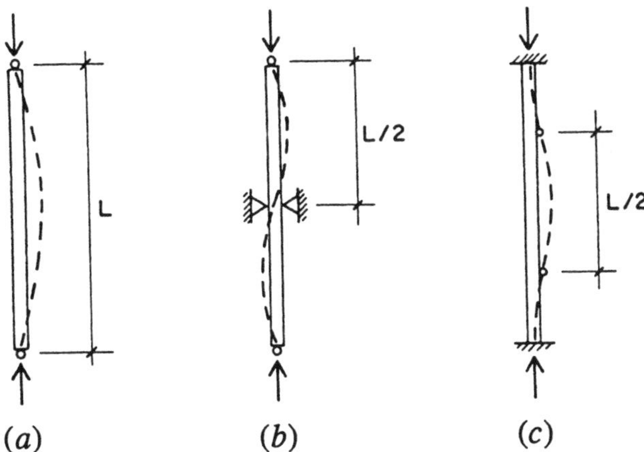

Figure 2.18 Form of buckling of a column as affected by various end conditions and lateral restraint.

member that is the general basis for response as indicated by the Euler formula. This form of response can be altered by lateral constraints, as shown in Figure 2.18*b*, that result in a multimode deflected shape. The member in Figure 2.18*c* has its ends restrained against rotation (described as a fixed end). This also modifies the deflected shape and thus the value produced from the buckling formula. One method used for adjustment is to modify the column length used in the buckling formula to that occurring between inflection points; thus, the *effective buckling length* for the columns in both Figures 2.18*b* and *c* would be one-half that of the true column total length. Inspection of the Euler formula will indicate the impact of this modified length on buckling resistance.

3

INVESTIGATION OF STRUCTURES FOR SHEAR AND BENDING

This chapter investigates the behavior of beams, columns, and simple frames. For a basic explanation of relationships, the units used for forces and dimensions are of less significance than their numeric values.

3.1 DIRECT SHEAR STRESS

Consider the two steel bars held together by a 0.75-in.-diameter bolt, as shown in Figure 3.1*a*, and subjected to a tension force of 5000 lb. The tension force in the bars becomes a shear force on the bolt, described as a direct shear force. There are many results created by the force in Figure 3.1*a*, including tensile stress in the bars and bearing on the sides of the hole by the bolt. For now we are concerned with the slicing action on the bolt, described as direct shear stress. The bolt cross section has an area of $3.1416(0.375)^2 = 0.4418$ in.^{2} and

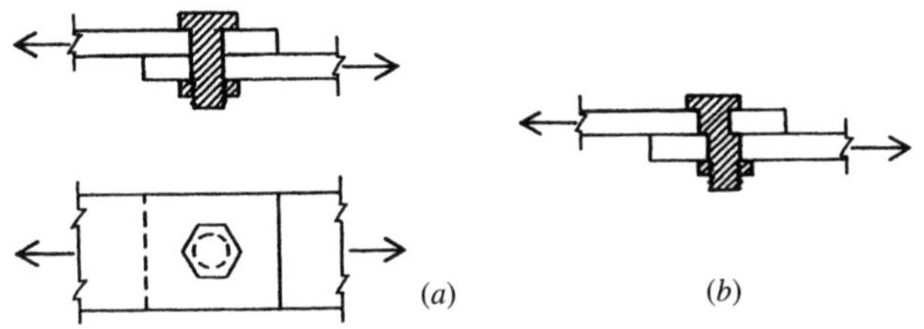

Figure 3.1 Direct shear stress.

the shear stress in the bolt is thus equal to 5000/0.4418 = 11,317 psi. Note that this type of stress is visualized as acting in the plane of the bolt cross section, as a slicing or sliding effect, while both compressive and tensile stresses are visualized as acting perpendicular to a stressed cross section.

The foregoing manipulations of the direct stress formula can, of course, be carried out also with the shearing stress formula, $f_v = P/A$. However, it must be borne in mind that the shearing stress acts transversely to the cross section—not at right angles to it. Furthermore, while the shearing stress equation applies directly to the situation illustrated by Figures 3.1*a* and *b*, it requires modification for application to beams.

3.2 SHEAR IN BEAMS

Figure 3.2*a* represents a simple beam with a uniformly distributed load over its entire length. Examination of an actual beam so loaded

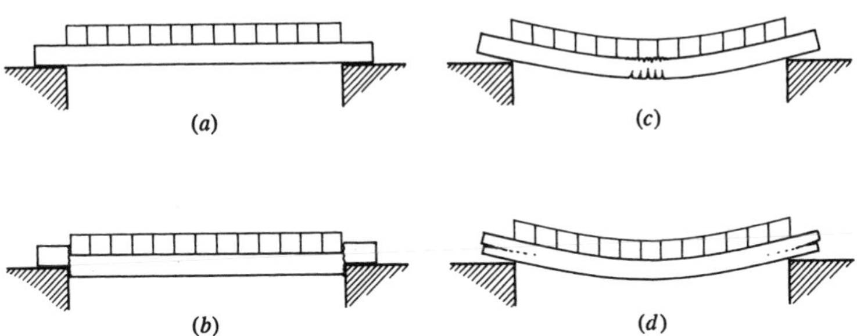

Figure 3.2 Stress failure in beams: (*a*) simple beam; (*b*) bending, (*c*) vertical shear, (*d*) horizontal shear.

probably would not reveal any effects of the loading on the beam. However, there are three distinct major tendencies for the beam to fail. Figures 3.2*b*–*d* illustrate the three phenomena.

First, there is a tendency for the beam to fail by dropping between the supports (Figure 3.2*b*). This is called *vertical shear.* Second, the beam may fail by bending (Figure 3.2*c*). Third, there is a tendency in wood beams for the fibers of the beam to slide past each other in a horizontal direction (Figure 3.2*d*), an action described as horizontal shear. Naturally, a beam properly designed does not fail in any of the ways just mentioned, but these tendencies to fail are always present and must be considered in structural design.

Vertical Shear

Vertical shear is the tendency for one part of a beam to move vertically with respect to an adjacent part. The magnitude of the shear force at any section in the length of a beam is equal to the algebraic sum of the vertical forces on either side of the section. Vertical shear force is usually represented by the capital letter V. In computing its values in the examples and problems, consider the forces to the left of the section but keep in mind that the same resulting force magnitude will be obtained with the forces on the right. To find the magnitude of the vertical shear at any section in the length of a beam, simply add up the forces to the right or the left of the section. It follows that the maximum value of the shear for simple beams is equal to the greater reaction.

Example 1. Figure 3.3*a* illustrates a simple beam with concentrated loads of 600 and 1000 lb. The problem is to find the value of the vertical shear at various points along the length of the beam. Although

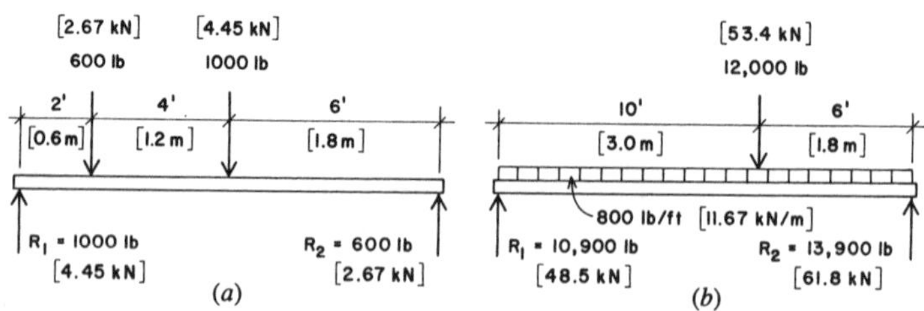

Figure 3.3 Reference for Examples 1 and 2.

the weight of the beam constitutes a uniformly distributed load, it is neglected in this example.

Solution: The reactions are computed as previously described in Section 1.7 and are found to be $R_1 = 1000$ lb and $R_2 = 600$ lb.

Consider next the value of the vertical shear V at an infinitely short distance to the right of R_1. Applying the rule that the shear is equal to the reaction minus the loads to the left of the section, we write

$$V = R_1 - 0 \quad \text{or} \quad V = 1000 \text{ lb}$$

The zero represents the value of the loads to the left of the section, which of course is zero. Now take a section 1 ft to the right of R_1; again

$$V_{x=1} = R_1 - 0 \quad \text{or} \quad V_{x=1} = 1000 \text{ lb}$$

The subscript $x = 1$ indicates the position of the section at which the shear is taken, the distance in feet of the section from R_1. At this section the shear is still 1000 lb and has the same magnitude up to the 600-lb load.

The next section to consider is a very short distance to the right of the 600-lb load. At this section,

$$V_{x=2+} = 1000 - 600 = 400 \text{ lb}$$

Because there are no loads intervening, the shear continues to be the same magnitude up to the 1000-lb load. At a section a short distance to the right of the 1000-lb load,

$$V_{x=6+} = 1000 - (600 + 1000) = -600 \text{ lb}$$

This magnitude continues up to the right-hand reaction R_2.

Example 2. The beam shown in Figure 3.3*b* supports a concentrated load of 12,000 lb located 6 ft from R_2 and a uniformly distributed load of 800 pounds per linear foot (lb/ft) over its entire length. Compute the value of vertical shear at various sections along the span.

Solution: By use of the equations of equilibrium, the reactions are determined to be $R_1 = 10{,}900$ lb and $R_2 = 13{,}900$ lb. Note that the total distributed load is $800 \times 16 = 12{,}800$ lb. Now consider the vertical shear force at the following sections at a distance measured from the left support:

$$V_{x=0} = 10{,}900 - 0 = 10{,}900 \text{ lb}$$

$$V_{x=1} = 10{,}900 - (800 \times 1) = 10{,}100 \text{ lb}$$

$$V_{x=5} = 10{,}900 - (800 \times 5) = 6900 \text{ lb}$$

$$V_{x=10-} = 10{,}900 - (800 \times 10) = 2900 \text{ lb}$$

$$V_{x=10+} = 10{,}900 - [(800 \times 10) + 12{,}000] = -9100 \text{ lb}$$

$$V_{x=16} = 10{,}900 - [(800 \times 16) + 12{,}000] = -13{,}900 \text{ lb}$$

Shear Diagrams

In the two preceding examples the value of the shear at several sections along the length of the beams was computed. In order to visualize the results, it is common practice to plot these values on a diagram, called the *shear diagram*, which is constructed as explained below.

To make such a diagram, first draw the beam to scale and locate the loads. This has been done in Figures 3.4*a* and *b* by repeating the load diagrams of Figures 3.3*a* and *b*, respectively. Beneath the beam draw a horizontal baseline representing zero shear. Above and below

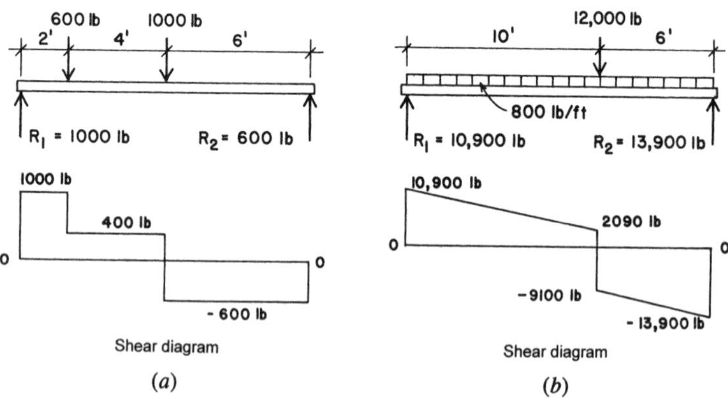

Figure 3.4 Construction of shear diagrams.

this line, plot at any convenient scale the values of the shear at the various sections; the positive, or plus, values are placed above the line and the negative, or minus, values below. In Figure 3.4*a*, for instance, the value of the shear at R_1 is +1000 lb. The shear continues to have the same value up to the load of 600 lb, at which point it drops to 400 lb. The same value continues up to the next load, 1000 lb, where it drops to −600 lb and continues to the right-hand reaction.

Obviously, to draw a shear diagram it is necessary to compute the values at significant points only. Having made the diagram, we may readily find the value of the shear at any section of the beam by scaling the vertical distance in the diagram. The shear diagram for the beam in Figure 3.4*b* is made in the same manner.

There are two important facts to note concerning the vertical shear. The first is the maximum value. The diagrams in each case confirm the earlier observation that the maximum shear is at the reaction having the greater value, and its magnitude is equal to that of the greater reaction. In Figure 3.4*a* the maximum shear is 1000 lb, and in Figure 3.4*b* it is 13,900 lb. We disregard the positive or negative signs in reading the maximum values of the shear, for the diagrams are merely conventional methods of representing the absolute numerical values.

Another important fact to note is the point at which the shear changes from a plus to a minus quantity. We call this the point at which the shear passes through zero. In Figure 3.4*a* it is under the 1000-lb load, 6 ft from R_1. In Figure 3.4*b* it is under the 12,000-lb load, 10 ft from R_1. A major concern for noting this point is that it indicates the location of the maximum value of bending moment in the beam, as discussed in the next section.

Problems 3.2.A–F For the beams shown in Figure 3.5, draw the shear diagrams and note all critical values for shear. Note particularly the maximum value for shear and the point at which the shear passes through zero.

3.3 BENDING MOMENTS IN BEAMS

The forces that tend to cause bending in a beam are the reactions and the loads. Consider section *X–X*, 6 ft from R_1 (Figure 3.6). The force R_1, or 2000 lb, tends to cause a clockwise rotation about this point. Because the force is 2000 lb and the lever arm is 6 ft, the moment of the force is $2000 \times 6 = 12{,}000$ ft-lb. This same value may be found by considering the forces to the right of section *X–X*: R_2, which is

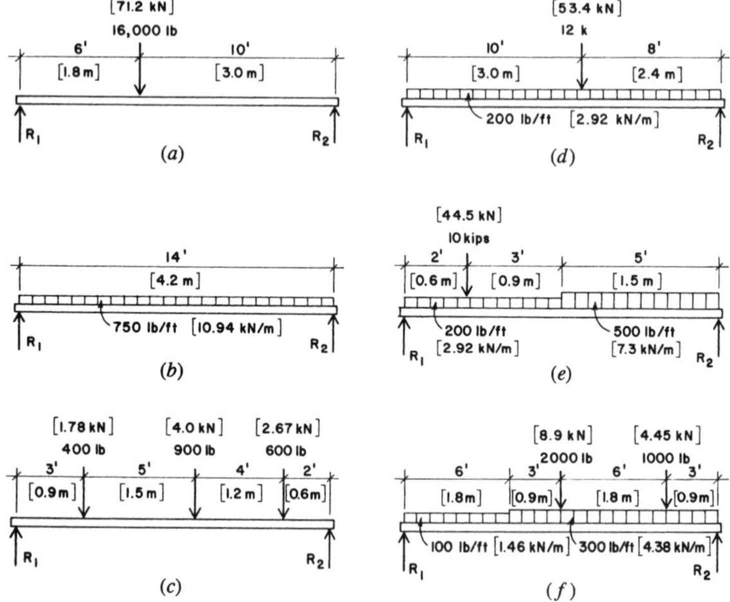

Figure 3.5 Reference for Problems 3.2 and 3.3.

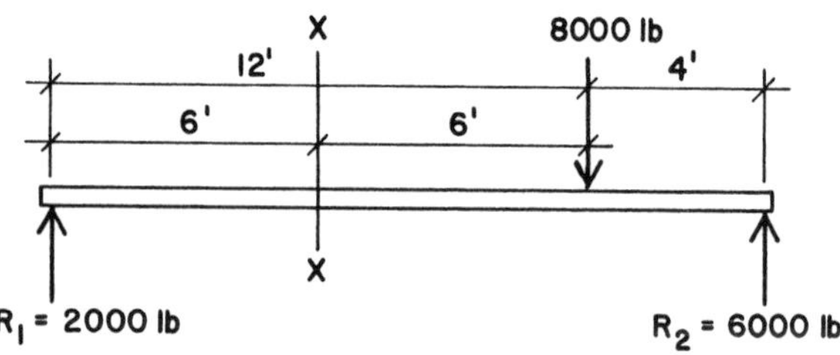

Figure 3.6 Internal bending at a selected beam cross section.

6000 lb, and the load 8000 lb, with lever arms of 10 and 6 ft, respectively. The moment of the reaction is $6000 \times 10 = 60{,}000$ ft-lb, and its direction is counterclockwise with respect to section *X–X*. The moment of the 8000 lb force is $8000 \times 6 = 48{,}000$ ft-lb, and its direction is clockwise. Then $60{,}000 - 48{,}000 = 12{,}000$ ft-lb, which is the resultant moment tending to cause counterclockwise rotation about section *X–X*. This is the same magnitude as the moment of the forces on the left, which tend to cause a clockwise rotation.

Thus, it makes no difference whether use is made of the forces to the right of the section or the left, as the magnitude of the moment obtained is the same. It is called the *bending moment* (or the *internal bending moment*) because it is the moment of the forces that causes bending stresses in the beam. Its magnitude varies throughout the length of the beam. For instance, at 4 ft from R_1 it is only $2000 \times 4 = 8000$ ft-lb. The bending moment is the algebraic sum of the moments of the forces on either side of the section. For simplicity, take the forces on the left; then the bending moment at any section of a beam is equal to the moments of the reactions minus the moments of the loads to the left of the section. Because the bending moment is the result of multiplying forces by distances, the denominations are foot-pounds or kip-feet.

Bending Moment Diagrams

The construction of bending moment diagrams follows the procedure used for shear diagrams. The beam span is drawn to scale, showing the locations of the loads. Below this, and usually below the shear diagram, a horizontal baseline is drawn representing zero bending moment. Then the bending moments are computed at various sections along the beam span, and the values are plotted vertically to any convenient scale. In simple beams all bending moments are positive and therefore are plotted above the baseline. In overhanging or continuous beams there are also negative moments, and these are plotted below the baseline.

Example 3. The load diagram in Figure 3.7 shows a simple beam with two concentrated loads. Draw the shear and bending moment diagrams.

Solution: First R_1 and R_2 are computed and are found to be 16,000 and 14,000 lb, respectively. These values are recorded on the load diagram.

The shear diagram is drawn as described in Section 3.2. Note that in this instance it is necessary to compute the shear at only one section (between the concentrated loads) because there is no distributed load, and we know that the shear at the supports is equal to the reactions.

Because the value of the bending moment at any section of the beam is equal to the moments of the reactions minus the moments of the loads to the left of the section, the moment at R_1 must be zero, for

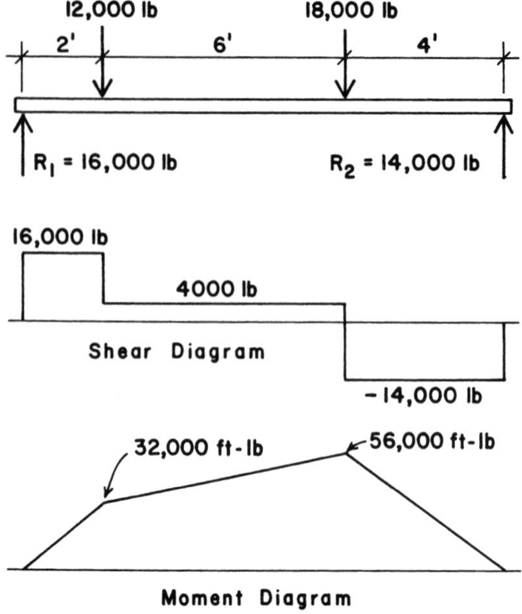

Figure 3.7 Reference for Example 3.

there are no forces to the left. Other values in the beam are computed as follows. The subscripts ($x = 1$, etc.) show the distance in feet from R_1 at which the bending moment is computed.

$$M_{x=1} = (16{,}000 \times 1) = 16{,}000 \text{ ft-lb}$$

$$M_{x=2} = (16{,}000 \times 2) = 32{,}000 \text{ ft-lb}$$

$$M_{x=5} = (16{,}000 \times 5) - (12{,}000 \times 3) = 44{,}000 \text{ ft-lb}$$

$$M_{x=8} = (16{,}000 \times 8) - (12{,}000 \times 6) = 56{,}000 \text{ ft-lb}$$

$$M_{x=10} = (16{,}000 \times 10) - [(12{,}000 \times 8) + (18{,}000 \times 2)] = 28{,}000 \text{ ft-lb}$$

$$M_{x=12} = (16{,}000 \times 12) - [(12{,}000 \times 10) + (18{,}000 \times 4)] = 0$$

The result of plotting these values is shown in the bending moment diagram of Figure 3.7. More moments were computed than were necessary. We know that the bending moments at the supports of simple beams are zero, and in this example only the bending moments directly under the loads were needed for the determination of the moment diagram.

Relations between Shear and Bending Moment

In simple beams the shear diagram passes through zero at some point between the supports. As stated earlier, an important principle in this respect is that the bending moment has a maximum magnitude wherever the shear passes through zero. In Figure 3.7 the shear passes through zero under the 18,000 lb load, that is, at $x = 8$ ft. Note that the bending moment has its greatest value at this same point, 56,000 ft-lb.

Example 4. Draw the shear and bending moment diagrams for the beam shown in Figure 3.8.

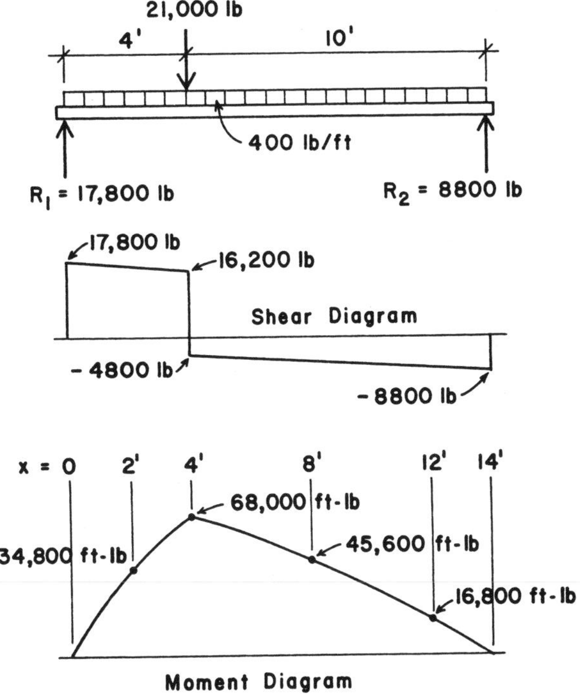

Figure 3.8 Reference for Example 4.

Solution: Computing the reactions, we find $R_1 = 17{,}800$ lb and $R_2 = 8800$ lb. By use of the process described in Section 3.2, the critical shear values are determined and the shear diagram is drawn as shown in the figure.

Although the only value of bending moment that must be computed is that where the shear passes through zero, some additional values are determined in order to plot the true form of the moment diagram. Thus,

$$M_{x=2} = (17{,}800 \times 2) - (400 \times 2 \times 1) = 34{,}800 \text{ ft-lb}$$

$$M_{x=4} = (17{,}800 \times 4) - (400 \times 4 \times 2) = 68{,}000 \text{ ft-lb}$$

$$M_{x=8} = (17{,}800 \times 8) - [(400 \times 8 \times 4) + (21{,}000 \times 4)] = 45{,}600 \text{ ft-lb}$$

$$M_{x=12} = (17{,}800 \times 12) - [400 \times 12 \times 6) + (21{,}000 \times 8)] = 16{,}800 \text{ ft-lb}$$

From the two preceding examples (Figures 3.7 and 3.8), it will be observed that the shear diagram for the parts of the beam on which no loads occur is represented by horizontal lines. For the parts of the beam on which a uniformly distributed load occurs, the shear diagram consists of straight inclined lines. The bending moment diagram is represented by straight inclined lines when only concentrated loads occur and by a curved line if the load is distributed.

Occasionally, when a beam has both concentrated and uniformly distributed loads, the shear does not pass through zero under one of the concentrated loads. This frequently occurs when the distributed load is relatively large compared with the concentrated loads. Since it is necessary in designing beams to find the maximum bending moment, we must know the point at which it occurs. This, of course, is the point where the shear passes through zero, and its location is readily determined by the procedure illustrated in the following example.

Example 5. The load diagram in Figure 3.9 shows a beam with a concentrated load of 7000 lb applied 4 ft from the left reaction and a uniformly distributed load of 800 lb/ft extending over the full span. Compute the maximum bending moment on the beam.

Solution: The values of the reactions are found to be $R_1 = 10{,}600$ lb and $R_2 = 7600$ lb and are recorded on the load diagram.

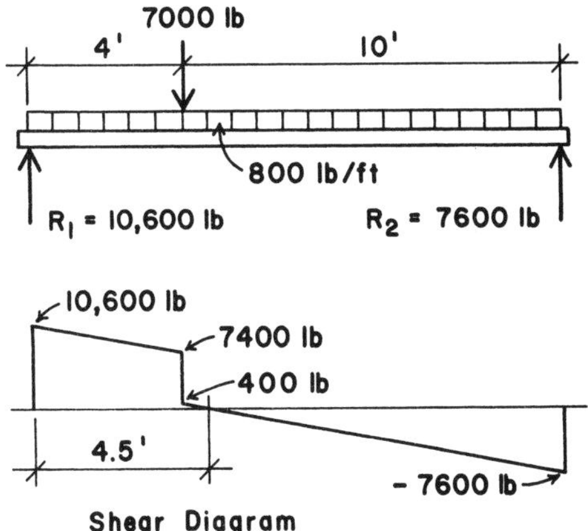

Figure 3.9 Reference for Example 5.

The shear diagram is constructed, and it is observed that the shear passes through zero at some point between the concentrated load of 7000 lb and the right reaction. Call this distance x feet from R_2. The value of the shear at this section is zero; therefore, an expression for the shear for this point, using the reaction and loads, is equal to zero. This equation contains the distance x:

$$V_{\text{at } x} = -7600 + 800x = 0 \qquad x = \frac{7600}{800} = 9.5 \text{ ft}$$

The zero-shear point is thus at 9.5 ft from the right support and (as shown in the diagram) at 4.5 ft from the left support. This location can also be determined by writing an equation for the summation of shear from the left of the point, which should produce the answer of 4.5 ft.

Following the convention of summing up the moments from the left of the section, the maximum moment is determined as

$$\begin{aligned} M_{x=4.5} &= +(10{,}600 \times 4.5) - (7000 \times 0.5) \\ &\quad - \left[800 \times 4.5 \times \left(\frac{4.5}{2}\right)\right] \\ &= 36{,}100 \text{ ft-lb} \end{aligned}$$

Problems 3.3.A–F Draw the shear and bending moment diagrams for the beams in Figure 3.5, indicating all critical values for shear and moment and all significant dimensions. (*Note:* These are the beams for Problem 3.2 for which the shear diagrams were constructed.)

3.4 SENSE OF BENDING IN BEAMS

When a simple beam bends, it has a tendency to assume the shape shown in Figure 3.10*a*. In this case the fibers in the upper part of the beam are in compression. For this condition the bending moment is considered as positive. Another way to describe a positive bending moment is to say that it is positive when the curve assumed by the bent beam is concave upward. When a beam projects beyond a support (Figure 3.10*b*), this portion of the beam has tensile stresses in the upper part. The bending moment for this condition is called negative; the beam is bent concave downward. When constructing moment diagrams, following the method previously described, the positive and negative moments are shown graphically.

Example 6. Draw the shear and bending moment diagrams for the overhanging beam shown in Figure 3.11.

Solution: Computing the following reactions:

$$\text{From } \sum M \text{ about } R_1\text{: } R_2 \times 12 = 600 \times 16 \times 8 \quad R_2 = 6400 \text{ lb}$$

$$\text{From } \sum M \text{ about } R_2\text{: } R_1 \times 12 = 600 \times 16 \times 4 \quad R_1 = 3200 \text{ lb}$$

With the reactions determined, the construction of the shear diagram is quite evident. For the location of the point of zero shear, considering its distance from the left support as x:

$$3200 - 600x = 0 \qquad x = 5.33 \text{ ft}$$

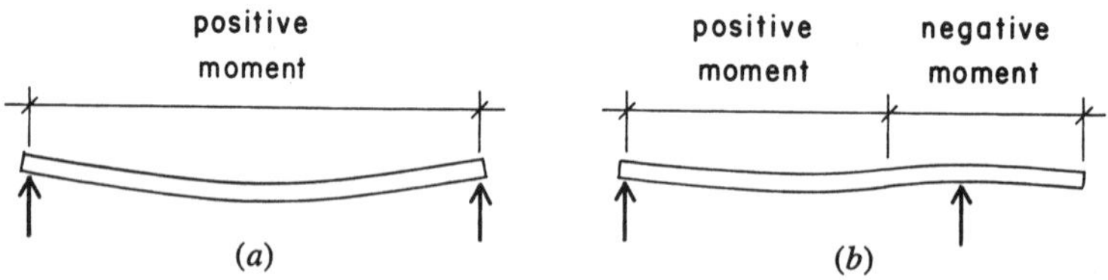

Figure 3.10 Sense of bending moment in beams.

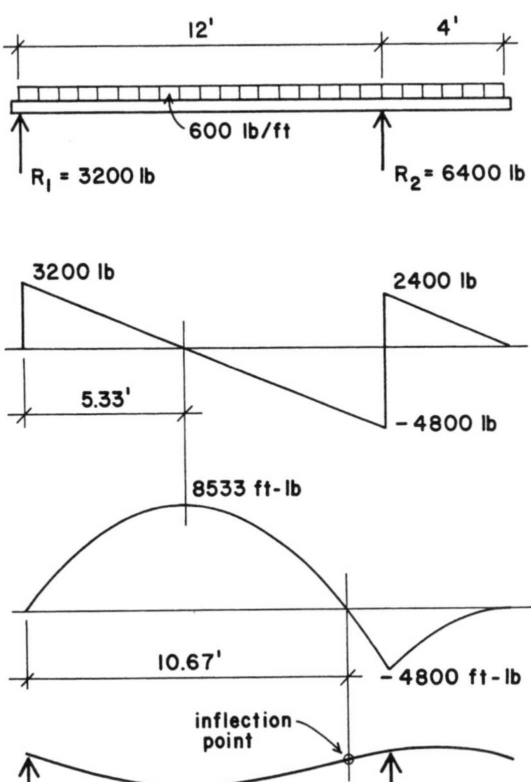

Figure 3.11 Reference for Example 6.

For the critical moment values at $x = 5.33$ ft and $x = 12$ ft needed to plot the moment diagram:

$$M_{x=5.33} = +(3200 \times 5.33) - \left(600 \times 5.33 \times \frac{5.33}{2}\right) = 8533 \text{ ft-lb}$$

$$M_{x=12} = +(3200 \times 12) - (600 \times 12 \times 6) = -4800 \text{ ft-lb}$$

The form of the moment diagram for the distributed loading is a curve (parabolic), which may be verified by plotting some additional points on the graph.

For this case the shear diagram passes through zero twice, with both points indicating peaks of the moment diagram—one positive and one negative. As the peak in the positive portion of the moment diagram is actually the apex of the parabola, the location of the zero-moment value is simply twice the value previously determined as x. This point corresponds to the change in the form of curvature on the elastic curve (deflected shape) of the beam and is described as the *inflection point* for the deflected shape. The location of the point of zero moment can also be determined by writing an equation for the sum of moments at the unknown location. In this case, call the new unknown point x:

$$M = 0 = +(3200 \times x) - \left(600 \times x \times \frac{x}{2}\right)$$

The solution of this quadratic equation should produce the value of $x = 10.67$ ft.

Example 7. Compute the maximum bending moment for the overhanging beam shown in Figure 3.12.

Solution: Computing the reactions, $R_1 = 3200$ lb and $R_2 = 2800$ lb. As usual, the shear diagram can now be plotted as the graph of the loads and reactions, proceeding from left to right. Note that the shear passes through zero at the location of the 4000 lb load and at both supports. As usual, these are clues to the form of the moment diagram.

With the usual moment summations, values for the moment diagram can now be found at the locations of the supports and all of the concentrated loads. From this plot it will be noted that there are two inflection points (locations of zero moment). As the moment diagram is composed of straight-line segments in this case, the locations of these points may be found by writing simple linear equations for their locations. However, use can also be made of some relationships between the shear and moment graphs. One of these has already been used, relating to the correlation of zero shear and maximum moment. Another relationship is that the change of the value of moment between any two points along the beam is equal to the total area of the shear diagram between the points. If the value of the moment is known at some point, it is thus a simple matter to find values at other points. For example, starting from the left

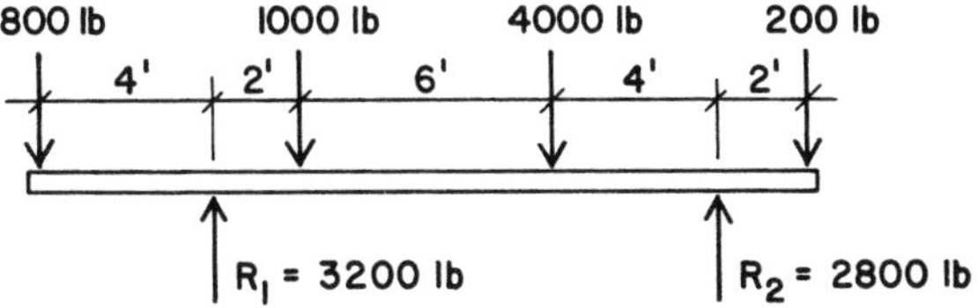

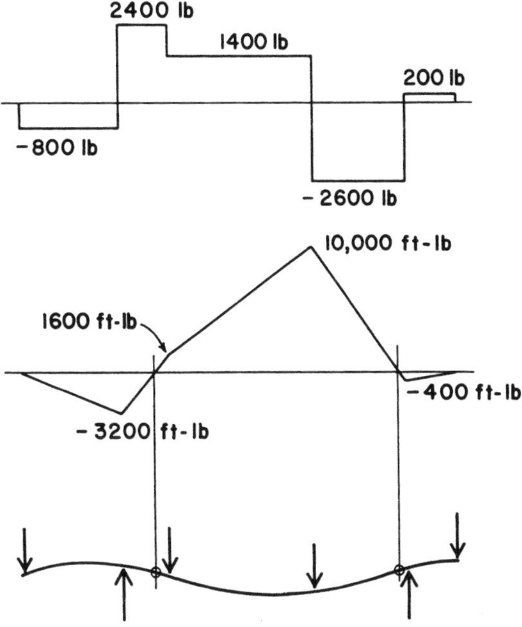

Figure 3.12 Reference for Example 7.

end, the value of the moment is known to be zero at the left end of the beam; then the value of the moment at the support is the area of the rectangle on the shear diagram with a base of 4 ft and a height of 800 lb—the area being $4 \times 800 = 3200$ ft-lb.

Now, proceeding along the beam to the point of zero moment (call it x distance from the support), the change is again 3200, which relates to an area of the shear diagram that is $x \times 2400$. Thus,

$$2400x = 3200 \qquad x = \frac{3200}{2400} = 1.33 \text{ ft}$$

and, calling the distance from the right support to the point of zero moment x,

$$2600x = 400 \qquad x = \frac{400}{2600} = 0.154 \text{ ft}$$

Problems 3.4.A–D Draw the shear and bending moment diagrams for the beams in Figure 3.13, indicating all critical values for shear and moment and all significant dimensions.

Cantilever Beams

In order to keep the signs for shear and moment consistent with those for other beams, it is convenient to draw a cantilever beam with its fixed end to the right, as shown in Figure 3.14. We then plot the values for the shear and moment on the diagrams as before, proceeding from the left end.

Example 8. The cantilever beam shown in Figure 3.14*a* projects 12 ft from the face of the wall and has a concentrated load of 800 lb at the unsupported end. Draw the shear and moment diagrams. What are the values of the maximum shear and maximum bending moment?

Solution: The value of the shear is −800 lb throughout the entire length of the beam. The bending moment is maximum at the wall; its value is 800 × 12 = −9600 ft-lb. The shear and moment diagrams

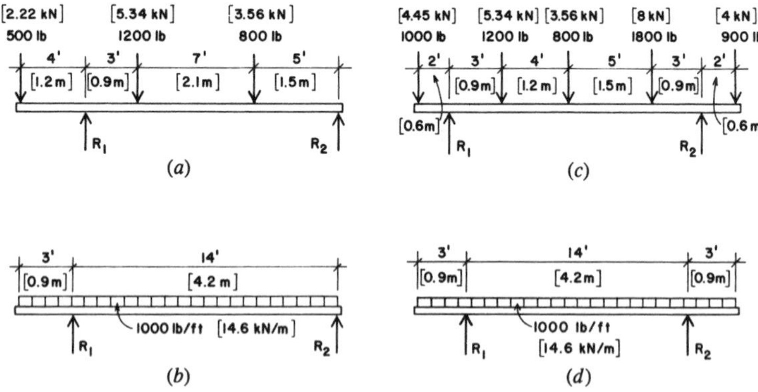

Figure 3.13 Reference for Problem 3.5.A–D.

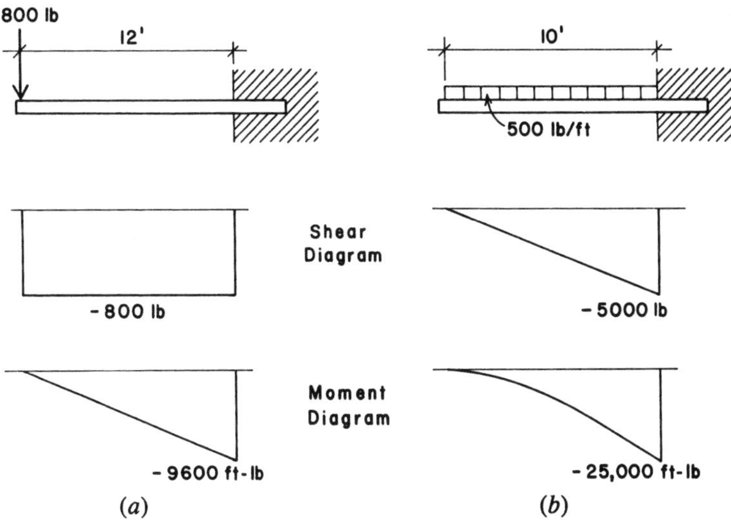

Figure 3.14 Reference for Examples 8 and 9.

are as shown in Figure 3.14*a*. Note that the moment is negative for the entire length of the cantilever beam, corresponding to its concave downward shape throughout its length.

Although not shown in the figure, the reactions in this case are a combination of an upward force of 800 lb and a clockwise resisting moment of 9600 ft-lb.

Example 9. Draw shear and bending moment diagrams for the beam in Figure 3.14*b*, which carries a uniformly distributed load of 500 lb/ft over its full length.

Solution: The total load is $500 \times 10 = 5000$ lb. The reactions are an upward force of 5000 lb and a moment determined as

$$M = -500 \times 10 \times \left(\frac{10}{2}\right) = -25{,}000 \text{ ft-lb}$$

which—it may be noted—is also the total area of the shear diagram between the outer end and the support.

Example 10. The cantilever beam indicated in Figure 3.15 has a concentrated load of 2000 lb and a uniformly distributed load of 600 lb/ft

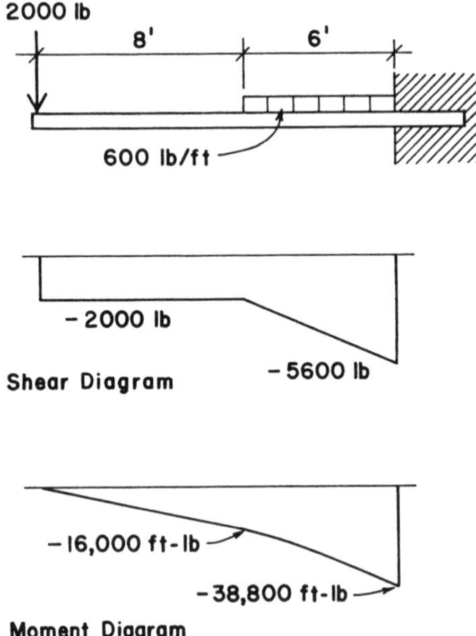

Figure 3.15 Reference for Example 10.

at the positions shown. Draw the shear and bending moment diagrams. What are the magnitudes of the maximum shear and maximum bending moment?

Solution: The reactions are actually *equal* to the maximum shear and bending moment. Determined directly from the forces, they are

$$V = 2000 + (600 \times 6) = 5600 \text{ lb}$$

$$M = -(2000 \times 14) - \left[600 \times 6 \times \left(\frac{6}{2}\right)\right] = -38{,}800 \text{ ft-lb}$$

The diagrams are quite easily determined. The other moment value needed for the moment diagram can be obtained from the moment of the concentrated load or from the simple rectangle of the shear diagram: $2000 \times 8 = 16{,}000$ ft-lb.

Note that the moment diagram has a straight-line shape from the outer end to the beginning of the distributed load and becomes a curve from this point to the support.

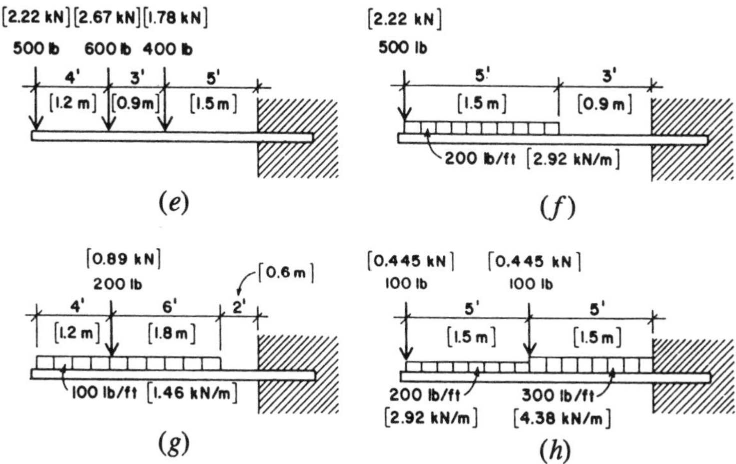

Figure 3.16 Reference for Problem 3.4.E–H.

It is suggested that Example 10 be reworked with Figure 3.22 reversed, left for right. All numerical results will be the same, but the shear diagram will be positive over its full length.

Problems 3.4.E–H Draw the shear and bending moment diagrams for the beams in Figure 3.16, indicating all critical values for the shear and moment and all significant dimensions.

3.5 TABULATED VALUES FOR BEAM BEHAVIOR

Bending Moment Formulas

The methods of computing beam reactions, shears, and bending moments presented thus far in this chapter make it possible to find critical values for design under a wide variety of loading conditions. However, certain conditions occur so frequently that it is convenient to use formulas that give the maximum values directly. Structural design handbooks contain many such formulas; two of the most commonly used formulas are derived in the following examples.

Simple Beam, Concentrated Load at Center of Span

A simple beam with a concentrated load at the center of the span occurs very frequently in practice. Call the load P and the span length between supports L, as indicated in the load diagram of Figure 3.17*a*.

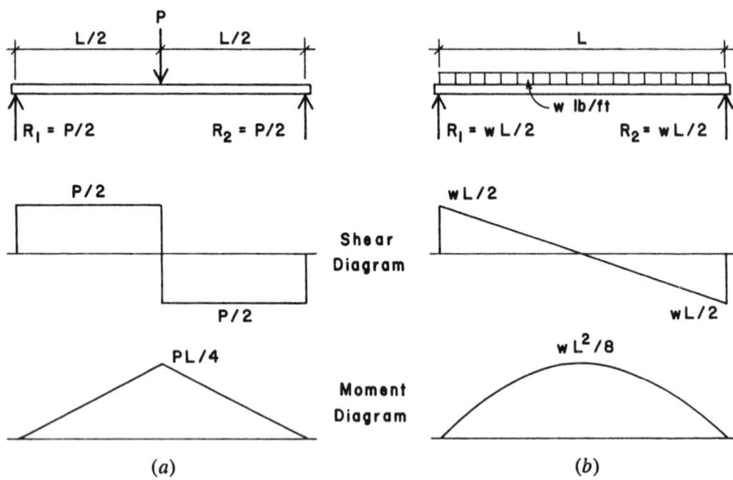

Figure 3.17 Values for simple-beam loadings.

For this symmetrical loading each reaction is $P/2$, and it is readily apparent that the shear will pass through zero at distance $x = L/2$ from R_1. Therefore, the maximum bending moment occurs at the center of the span, under the load. Computing the value of the bending moment at this section,

$$M = \frac{P}{2} \times \frac{L}{2} = \frac{PL}{4}$$

Example 11. A simple beam 20 ft in length has a concentrated load of 8000 lb at the center of the span. Compute the maximum bending moment.

Solution: As just derived, the formula giving the value of the maximum bending moment for this condition is $M = PL/4$. Therefore,

$$M = \frac{PL}{4} = \frac{8000 \times 20}{4} = 40{,}000 \text{ ft-lb}$$

Simple Beam, Uniformly Distributed Load

This is probably the most common beam loading; it occurs time and again. For any beam, its own dead weight as a load to be carried is usually of this form. Calling the span L and the unit load w, as indicated in Figure 3.17*b*, the total load on the beam is $W = wL$; hence each reaction is $W/2$ or $wL/2$. The maximum bending moment occurs at the center of the span at distance $L/2$ from R_1. Writing the value of M for this section,

$$M = +\left(\frac{wL}{2} \times \frac{L}{2}\right) - \left[w \times \left(\frac{L}{2}\right) \times \left(\frac{L}{4}\right)\right] = \frac{wL^2}{8} \quad \text{or} \quad \frac{WL}{8}$$

Note the alternative use of the unit load w or the total load W, where $W = wl$, in this formula. Both forms will be seen in various references. It is important to carefully identify the use of one or the other.

Example 12. A simple beam 14 ft long has a uniformly distributed load of 800 lb/ft. Compute the maximum bending moment.

Solution: As just derived, the formula that gives the maximum bending moment for a simple beam with uniformly distributed load is $M = wL^2/8$. Substituting these values,

$$M = \frac{wL^2}{8} = \frac{800 \times 14^2}{8} = 19{,}600 \text{ ft-lb}$$

or, using the total load W of $800 \times 14 = 11{,}200$ lb,

$$M = \frac{WL}{8} = \frac{11{,}200 \times 14}{8} = 19{,}600 \text{ ft-lb}$$

Use of Tabulated Values for Beams

Some of the most common beam loadings are shown in Figure 3.18. In addition to the formulas for the reactions R, for maximum shear V, and for maximum bending moment M, expressions for maximum deflection D are given also. Discussion of deflections formulas will be deferred for the time being but will be considered in Sections 5.5, 5.7, 9.7 and 13.8.

In Figure 3.18 if the loads P and W are in pounds or kips, the vertical shear V will also be in units of pounds or kips. When the loads are given in pounds or kips and the span is given in feet, the bending moment M will be in units of foot-pounds or kip-feet.

Also given in Figure 3.18 are values designated ETL, which stands for *equivalent tabular load*. These may be used to derive a hypothetical

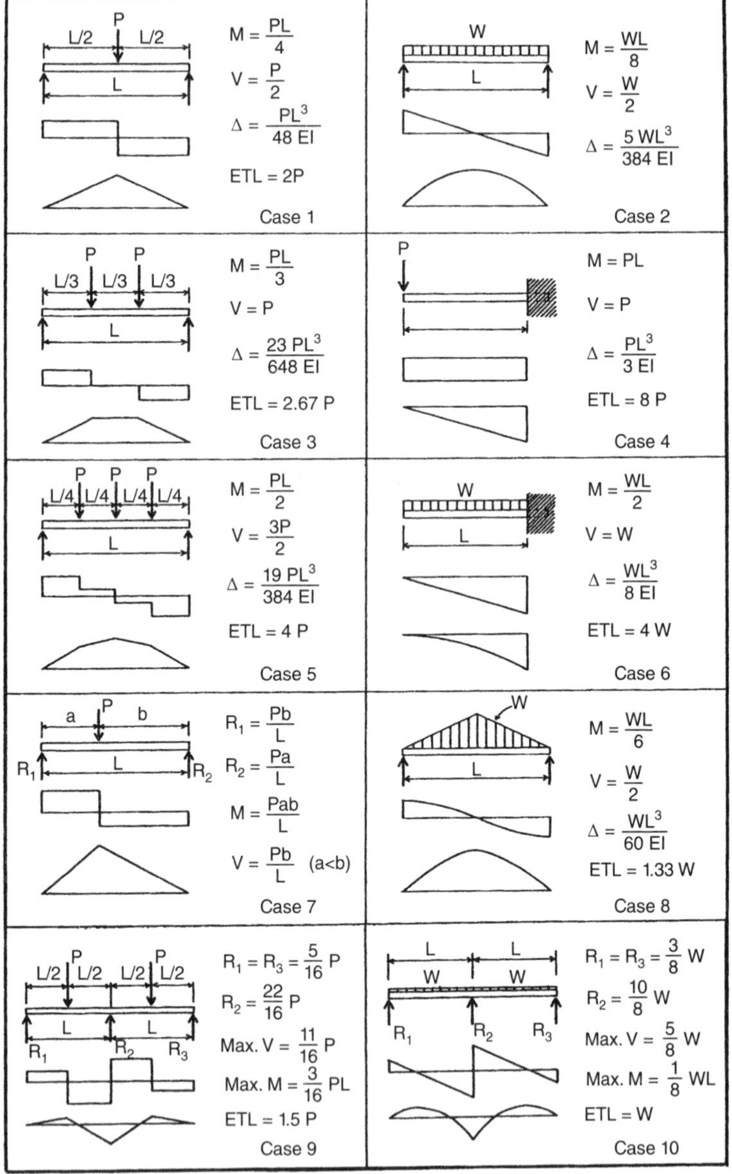

Figure 3.18 Values for typical beam loadings and support conditions.

total uniformly distributed load W, which when applied to the beam will produce the same magnitude of maximum bending moment as that for the given case of loading. Use of these factors is illustrated in later parts of the book.

Problem 3.5.A A simple span beam has two concentrated loads of 4 kips each placed at the third points of the 24-ft span. Find the value for the maximum bending moment in the beam.

Problem 3.5.B A simple span beam has a uniformly distributed load of 2.5 kips/ft on a span of 18 ft. Find the value for the maximum bending moment in the beam.

Problem 3.5.C A simple beam with a span of 32 ft has a concentrated load of 12 kips at 12 ft from one end. Find the value for the maximum bending moment in the beam.

Problem 3.5.D A simple beam with a span of 36 ft has a distributed load that varies from a value of 0 at its ends to a maximum of 1000 lb/ft at its center (case 8 in Figure 3.18). Find the value for the maximum bending moment in the beam.

3.6 DEVELOPMENT OF BENDING RESISTANCE

As developed in the preceding sections, the bending moment is a measure of the tendency of the external forces on a beam to deform it by bending. The purpose of this section is to consider the action within the beam that resists bending and is called the *resisting moment*.

Figure 3.19*a* shows a simple beam, rectangular in cross section, supporting a single concentrated load P. Figure 3.19*b* is an enlarged sketch of the left-handed portion of the beam between the reaction and section X–X. It is observed that the reaction R_1 tends to cause a clockwise rotation about point A in the section under consideration; this is defined as the bending moment at the section. In this type of beam the fibers in the upper part are in compression, and those in the lower part are in tension. There is a horizontal plane separating the compressive and tensile stresses; it is called the *neutral surface*, and at this plane there are neither compressive nor tensile stresses with respect to bending. The line in which the neutral surface intersects the beam cross section (Figure 3.19*c*) is called the *neutral axis* (NA).

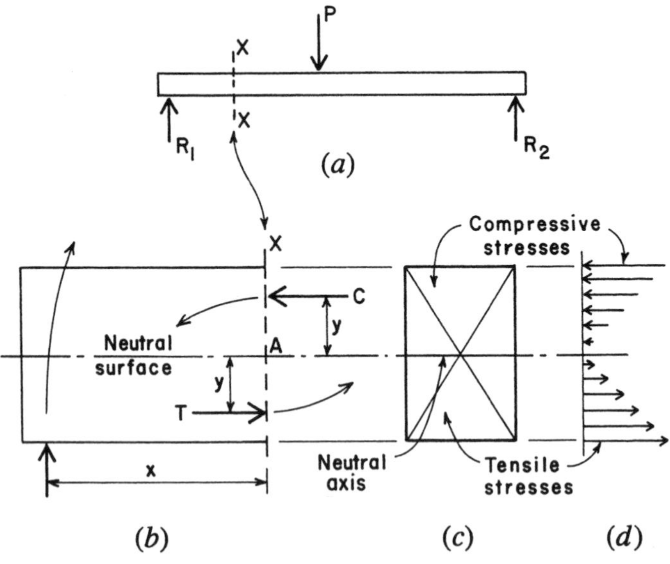

Figure 3.19 Development of bending stress in beams.

Call C the sum of all the compressive stresses acting on the upper part of the cross section, and call T the sum of all the tensile stresses acting on the lower part. It is the sum of the moments of those stresses at the section that holds the beam in equilibrium; this is called the *resisting moment* and is equal to the bending moment in magnitude. The bending moment about A is $R_1 \times x$, and the resisting moment about the same point is $(C \times y) + (T \times y)$. The bending moment tends to cause a clockwise rotation, and the resisting moment tends to cause a counterclockwise rotation. If the beam is in equilibrium, these moments are equal, or

$$R_1 \times x = (C \times y) + (T \times y)$$

that is, the bending moment equals the resisting moment. This is the theory of flexure (bending) in beams. For any type of beam, it is possible to compute the bending moment and to design a beam to withstand this tendency to bend; this requires the selection of a member with a cross section of such shape, area, and material that it is capable of developing a resisting moment equal to the bending moment.

The Flexure Formula

The flexure formula, $M = fS$, is an expression for resisting moment that involves the size and shape of the beam cross section (represented by S in the formula) and the material of which the beam is made (represented by f). It is used in the design of all homogeneous beams, that is, beams made of one material only, such as steel or wood. The following brief derivation is presented to show the principles on which the formula is based.

Figure 3.20 represents a partial side elevation and the cross section of a homogeneous beam subjected to bending stresses. The cross section shown is unsymmetrical about the neutral axis, but this discussion applies to a cross section of any shape. In Figure 3.19*a* let c be the distance of the fiber farthest from the neutral axis, and let f be the unit stress on the fiber at distance c. If f, the extreme fiber stress, does not exceed the elastic limit of the material, the stresses in the other fibers are directly proportional to their distances from the neutral axis. That is to say, if one fiber is twice the distance from the neutral axis than another fiber, the fiber at the greater distance will have twice the stress.

The stresses are indicated in the figure by the small lines with arrows, which represent the compressive and tensile stresses acting toward and away from the section, respectively. If c is in inches, the unit stress on a fiber at 1 in. distance is f/c. Now imagine an infinitely small area a at z distance from the neutral axis. The unit stress on this fiber is $(f/c) \times z$, and because this small area contains a square inches, the total stress on fiber a is $(f/c) \times z \times a$. The moment of the

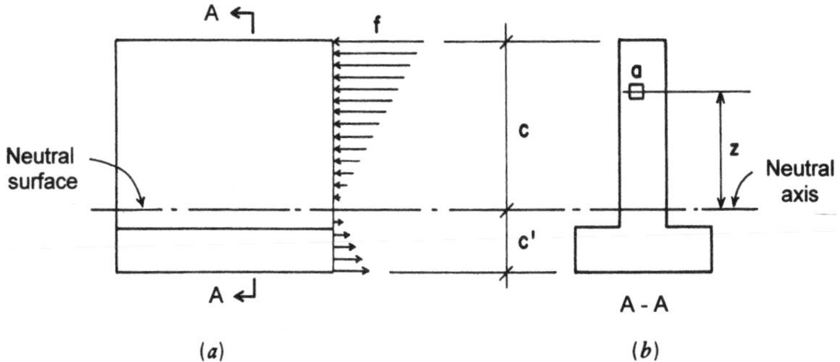

Figure 3.20 Distribution of bending stress on a beam cross section.

stress on fiber a at z distance is

$$\frac{f}{c} \times z \times a \times z \quad \text{or} \quad \frac{f}{c} az^2$$

There are an extremely large number of these minute areas. Using the symbol $\sum$ to represent the sum of this very large number,

$$\sum \left(\frac{f}{c} az^2 \right)$$

which means the sum of the moments of all the stresses in the cross section with respect to the neutral axis. This is the *resisting moment*, and it is equal to the bending moment. Therefore,

$$M_R = \frac{f}{c} \sum az^2$$

The quantity $\sum az^2$ may be read "the sum of the products of all the elementary areas times the square of their distances from the neutral axis." This is called the *moment of inertia* and is represented by the letter I (see Section A.2). Therefore, substituting in the above,

$$M_R = \frac{f}{c} I \quad \text{or} \quad M_R = \frac{fI}{c}$$

This is known as the *flexure formula* or *beam formula*, and by its use it is possible to design any beam that is composed of a single material. The expression may be simplified further by substituting S for I/c, called the *section modulus*, a term that is described more fully in Section A.4. Making this substitution, the formula becomes

$$M = fS$$

Use of the flexural formula is discussed in Chapter 5 for wood beams, in Chapter 9 for steel beams, and in Chapter 13 for reinforced concrete beams.

Inelastic Stress Conditions

At the limits of bending resistance the preceding descriptions of stress and deformation do not generally represent true conditions. The limit states are described as the *ultimate resistance*, and design based on the limit states is called *ultimate strength design*. These conditions and the procedures for their use in design are described in Part III for steel members and in Part IV for reinforced concrete members.

3.7 SHEAR STRESS IN BEAMS

Shear is developed in beams in direct resistance to the vertical force at a beam cross section. Because of the interaction of shear and bending in the beam, the exact nature of stress resistance within the beam depends on the form and materials of the beam. For example, in wood beams the wood grain is normally oriented in the direction of the span and the wood material has a very low resistance to horizontal splitting along the grain. An analogy to this is represented in Figure 3.21, which shows a stack of loose boards subjected to a beam loading. With nothing but minor friction between the boards, the individual boards will slide over each other to produce the loaded form indicated in the bottom figure. This is the failure tendency in the wood beam, and the shear phenomenon for wood beams is usually described as one of *horizontal shear*.

Shear stresses in beams are not distributed evenly over the cross section of the beam, as was assumed for the case of simple direct shear (see Section 2.1). From observations of tested beams and derivations considering the equilibrium of beam segments under combined actions of shear and bending, the following expression has been obtained for shear stress in a beam:

$$f_v = \frac{VQ}{Ib}$$

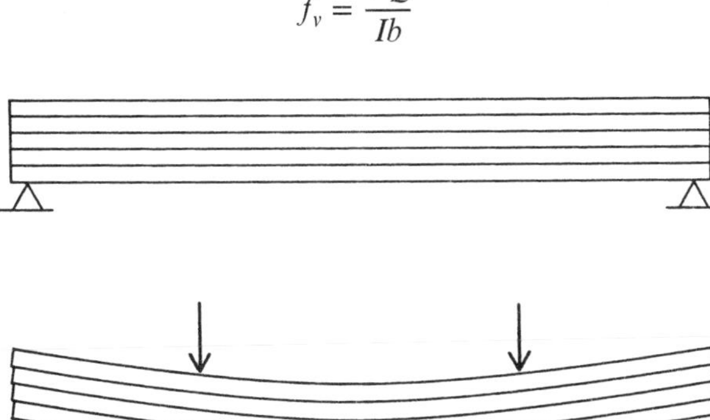

Figure 3.21 Nature of horizontal shear in beams.

where V = shear force at the beam section
Q = moment about the neutral axis of the portion of the cross-sectional area between the edge of the section and the point where stress is being computed
I = moment of inertia of the section
b = width of the section where stress is being computed

It may be observed that the highest value for Q, and thus usually for shear stress, depending on variable widths, will occur at the neutral axis and that shear stress will be zero at the top and bottom edges of the section. This is essentially opposite to the form of distribution of bending stress on a section. The form of shear distribution for various geometric shapes of beam sections is shown in Figure 3.22.

The following examples illustrate the use of the general shear stress formula.

Example 13. A rectangular beam section with a depth of 8 in. and width of 4 in. sustains a shear force of 4 kips. Find the maximum shear stress. (See Figure 3.23*a*.)

Solution: For the rectangular section the moment of inertia about the centroidal axis is (see Figure A.11)

$$I = \frac{bd^3}{12} = \frac{4 \times 8^3}{12} = 171 \text{ in.}^4$$

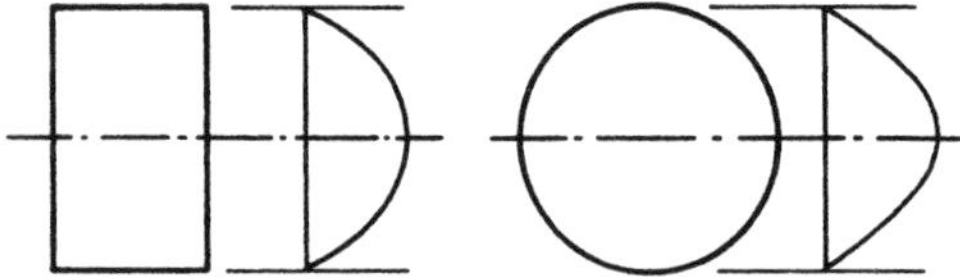

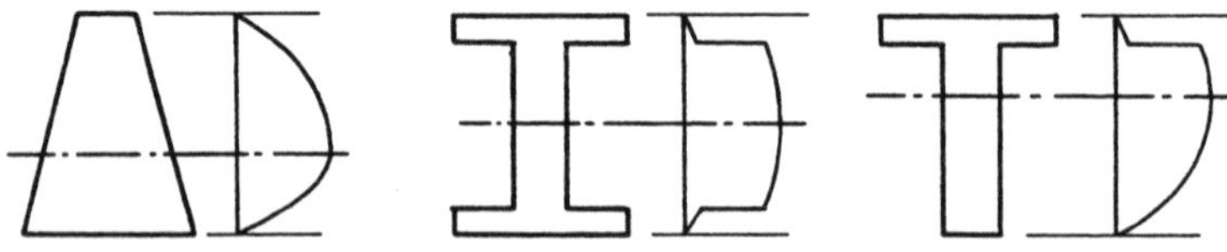

Figure 3.22 Distribution of shear stress in beams with various shapes of cross sections.

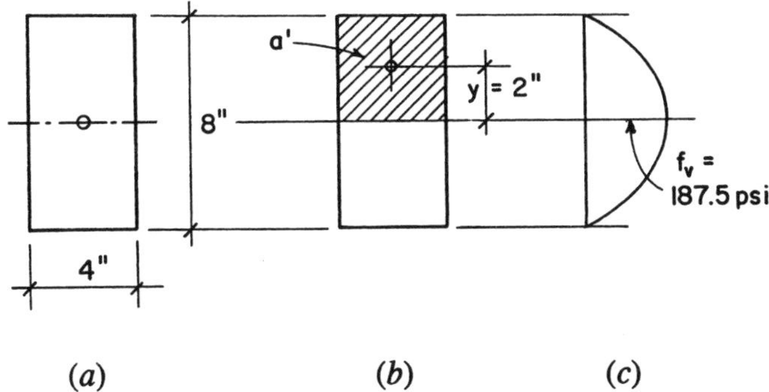

Figure 3.23 Reference for Example 13.

The static moment (Q) is the product of the area a' and its centroidal distance from the neutral axis of the section (y as shown in Figure 3.23*b*). This is the greatest value that can be obtained for Q and will produce the highest shear stress for the section. Thus,

$$Q = a'y = (4 \times 4)(2) = 32 \text{ in.}^3$$

and

$$f_v = \frac{VQ}{Ib} = \frac{4000 \times 32}{171 \times 4} = 187 \text{ psi}$$

The distribution of shear stress on the beam cross section is as shown in Figure 3.23*c*.

Example 14. A beam with the T section shown in Figure 3.24*a* is subjected to a shear force of 8 kips. Find the maximum shear stress and the value of shear stress at the location of the juncture of the web and the flange of the T.

Solution: Since this section is not symmetrical with respect to its horizontal centroidal axis, the first steps for this problem consist of locating the neutral axis and determining the moment of inertia for the section with respect to the neutral axis. To save space, this work is not shown here, although it is performed as Examples 1 and 8 in Appendix A. From that work it is found that the centroidal neutral axis is located at 6.5 in. from the bottom of the T and the moment of inertia about the neutral axis is 1046.7 in.^{4}

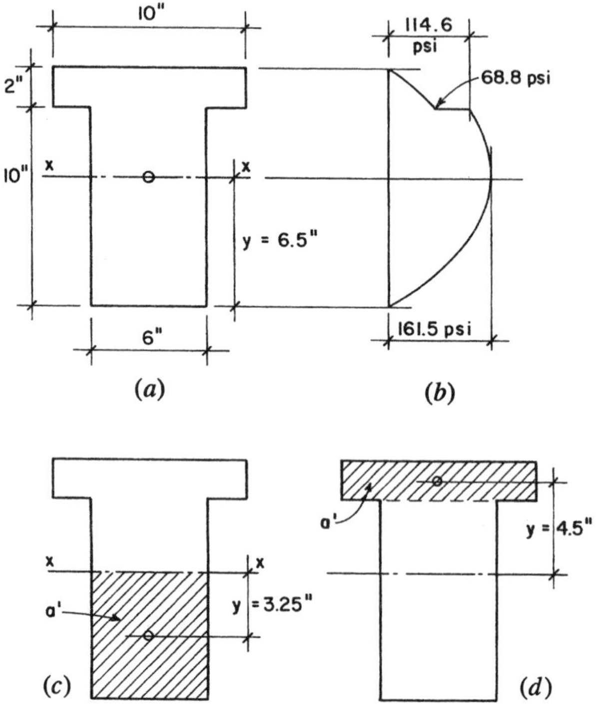

Figure 3.24 Reference for Example 14.

For computation of the maximum shear stress at the neutral axis, the value of Q is found by using the portion of the web below the neutral axis, as shown in Figure 3.24*c*. Thus,

$$Q = a'y = (6.5 \times 6) \times \left(\frac{6.5}{2}\right) = 127 \text{ in.}^3$$

and the maximum stress at the neutral axis is, thus,

$$f_v = \frac{VQ}{Ib} = \frac{8000 \times 127}{1046.7 \times 6} = 161 \text{ psi}$$

For the stress at the juncture of the web and flange, Q is determined using the area shown in Figure 3.24*d*. Thus,

$$Q = (2 \times 10)(4.5) = 90 \text{ in.}^3$$

And the two values for shear stress at this location, as displayed in Figure 3.24*b*, are

$$f_v = \frac{8000 \times 90}{1046.7 \times 6} = 114 \text{ psi} \quad \text{(in the web)}$$

and

$$f_v = \frac{8000 \times 90}{1046.7 \times 10} = 68.8 \text{ psi} \quad \text{(in the flange)}$$

In many situations it is not necessary to use the complex form of the general expression for shear stress in a beam. For wood beams, the sections are mostly simple rectangles, for which the following simplification can be made.

For the simple rectangle, from Figure 1, $I = bd^3/12$, and

$$Q = \left(b \times \frac{d}{2}\right)\frac{d}{4} = \frac{bd^2}{8}$$

Thus,

$$f_v = \frac{VQ}{Ib} = \frac{V(bd^2 8)}{(bd^3 12)b} = 1.5\frac{V}{bd}$$

This is the formula specified by design codes for investigation of shear in wood beams. The somewhat more complex investigations for shear stress are discussed in Chapter 9 for steel beams and in Chapter 13 for reinforced concrete beams.

Problem 3.7.A A beam has an I-shaped cross section with an overall depth of 16 in., web thickness of 2 in., and flanges that are 8 in. wide and 3 in. thick. Compute the critical shear stresses and plot the distribution of shear stress on the cross section if the beam sustains a shear force of 20 kips.

Problem 3.7.B A T-shaped beam cross section has an overall depth of 18 in., web thickness of 4 in., flange width of 8 in., and flange thickness of 3 in. Compute the critical shear stresses and plot the distribution of shear stress on the cross section if the beam sustains a shear force of 12 kips.

3.8 CONTINUOUS AND RESTRAINED BEAMS

Continuous Beams

A *continuous beam* is a beam that rests on more than two supports. Continuous beams are characteristic of site-cast concrete construction but occur less often in wood and steel construction.

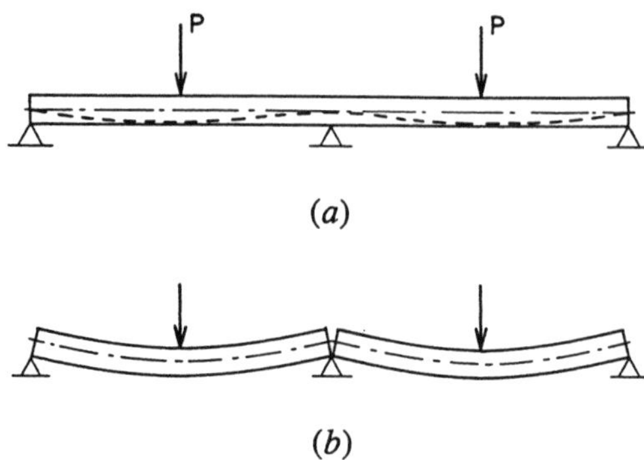

Figure 3.25 Continuous versus simple beams.

The concepts underlying continuity and bending under restraint are illustrated in Figure 3.25. Figure 3.25*a* represents a single beam resting on three supports and carrying equal loads at the centers of the two spans. If the beam is cut over the middle support as shown in Figure 3.25*b*, the result will be two simple beams. Each of these simple beams will deflect as shown. However, when the beam is made continuous over the middle support, its deflected form is as indicated by the dashed line in Figure 3.25*a*.

It is evident that there is no bending moment developed over the middle support in Figure 3.25*b*, while there must be a moment over the support in Figure 3.25*a*. In both cases there is a positive moment at the midspan; that is, there is tension in the bottom and compression in the top of the beam at these locations. In the continuous beam, however, there is a negative moment over the middle support; that is, there is tension in the top and compression in the bottom of the beam. The effect of the negative moment over the support is to reduce the magnitudes of both maximum bending moment and deflection at midspan, which is a principal advantage of continuity.

Values for reaction forces and bending moments cannot be found for continuous beams by use of the equations for static equilibrium alone. For example, the beam in Figure 3.25*a* has three unknown reaction forces, which constitute a parallel force system with the loads. For this condition there are only two conditions of equilibrium and thus only two available equations for solving for the three unknowns. This presents a situation in algebra that is qualified as *indeterminate*, and the structure so qualified is said to be *statically indeterminate*.

Solutions for investigation of indeterminate structures require additional conditions to supplement those available from simple statics. These additional conditions are derived from the deformation and the stress mechanisms of the structure. Various methods for investigation of indeterminate structures have been developed. Of particular interest now are those that yield to application to computer-aided processes. Just about any structure, with any degree of indeterminacy, can now be investigated with readily available programs.

A procedural problem with highly indeterminate structures is that something about the structure must be determined before an investigation can be performed. Useful for this purpose are shortcut methods that give reasonably approximate answers without an extensive investigation. One of these approximation methods is demonstrated in the investigation of a rigid frame structure in Chapter 20.

Theorem of Three Moments

One method for determining reactions and constructing the shear and bending moment diagrams for continuous beams is based on the *theorem of three moments*. This theorem deals with the relation among the bending moments at any three consecutive supports of a continuous beam. Application of the theorem produces an equation, called the *three-moment equation*. The three-moment equation for a continuous beam of two spans with uniformly distributed loading and constant moment of inertia is

$$M_1L_1 + 2M_2(L_1 + L_2) + M_3L_2 = -\frac{w_1{L_1}^3}{4} - \frac{w_2{L_2}^3}{4}$$

in which the various terms are as shown in Figure 3.26. The following examples demonstrate the use of this equation.

Continuous Beam with Two Equal Spans

This is the simplest case with the formula reduced by the symmetry plus the elimination of M_1 and M_3 due to the discontinuity of the beam at its outer ends. The equation is reduced to

$$4M_2 = -\frac{wL^2}{2}$$

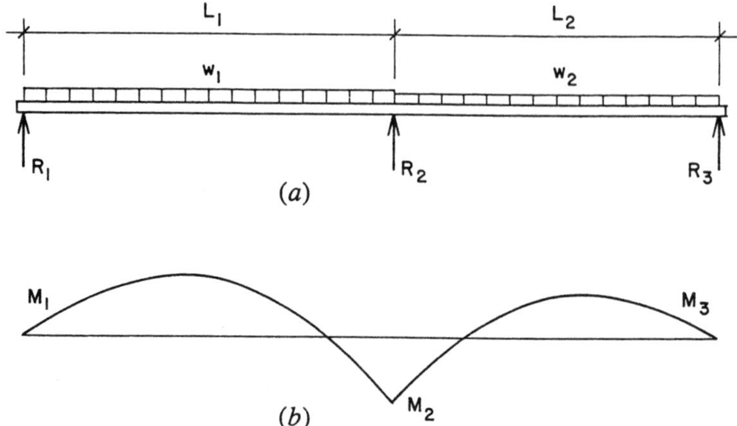

Figure 3.26 Reference for the three-moment equation.

With the loads and spans as given data, a solution for this case is reduced to solving for M_2, the negative moment at the center support. Transforming the equation produces a form for direct solution of the unknown moment; thus

$$M_2 = -\frac{wL^2}{8}$$

With this moment determined, it is possible to now use the available conditions of statics to solve the rest of the data for the beam. The following example demonstrates the process.

Example 15. Compute the values for the reactions and construct the shear and moment diagrams for the beam shown in Figure 3.27*a*.

Solution: With only two conditions of statics for the parallel force system, it is not possible to solve directly for the three unknown reactions. However, use of the equation for the moment at the middle support yields a condition that can be used as shown in the following work:

$$M_2 = -\frac{wL^2}{8} = -\frac{100 \times (10)^2}{8} = -1250 \text{ ft-lb}$$

Next, an equation for the bending moment at 10 ft to the right of the left support is written in the usual manner and is equated to the

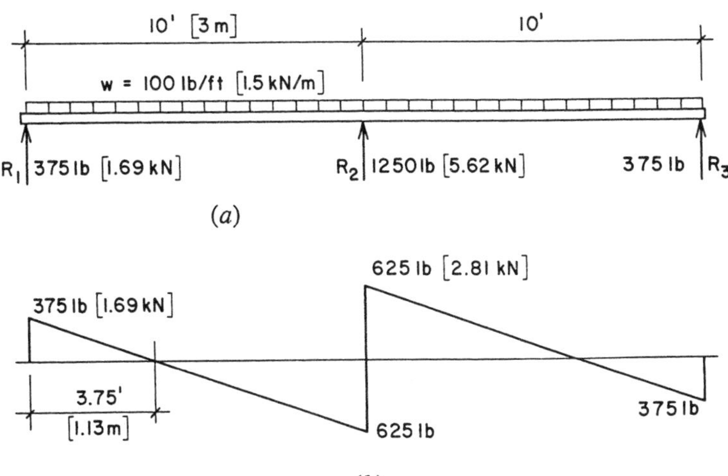

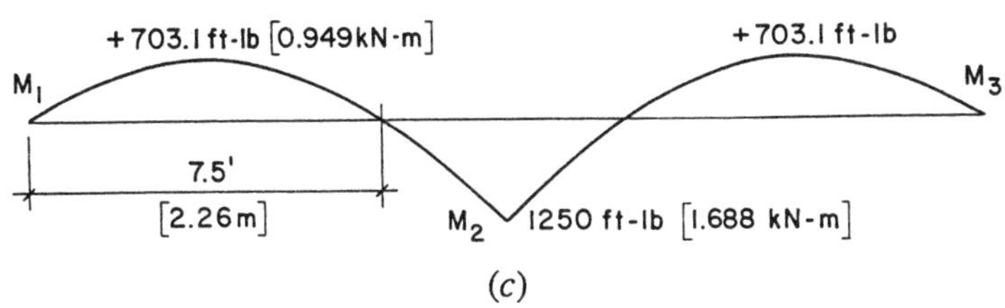

Figure 3.27 Reference for Example 15.

now known value of 1250 ft-lb:

$$M_{x=10} = (R_1 \times 10) - (100 \times 10 \times 5) = -1250 \text{ ft-lb}$$

from which

$$10R_1 = 3750 \qquad R_1 = 375 \text{ lb}$$

By symmetry this is also the value for R_3. The value for R_2 can then be found by a summation of vertical forces. Thus,

$$\sum F_V = 0 = (375 + 375 + R_2) - (100 \times 20) \quad R_2 = 1250 \text{ lb}$$

Sufficient data have now been determined to permit the complete construction of the shear diagram, as shown in Figure 3.27*b*.

The location of zero shear is determined by the equation for shear at the unknown distance x from the left support:

$$375 - (100 \times x) = 0 \qquad x = 3.75 \text{ ft}$$

The maximum value for positive moment at this location can be determined with a moment summation or by finding the area of the shear diagram between the end and the zero-shear location:

$$M = \frac{375 \times 3.75}{2} = 703 \text{ ft-lb}$$

Because of symmetry, the location of zero moment is determined as twice the distance of the zero-shear point from the left support. Sufficient data are now available to plot the moment diagram as shown in Figure 3.27c.

Problems 3.8.A, B Using the three-moment equation, find the bending moments and reactions and draw the complete shear and moment diagrams for the following beams that are continuous over two equal spans and carry uniformly distributed loadings.

	Span Length (ft)	Load (lb/ft)
A	16	200
B	24	350

Continuous Beam with Unequal Spans

The following example shows the slightly more complex problem of dealing with unequal spans.

Example 16. Construct the shear and moment diagrams for the beam in Figure 3.28a.

Solution: In this case the moments at the outer supports are again zero, which reduces the task to solving for only one unknown. Apply the given values to the equation:

$$2M_2(14 + 10) = -\frac{1000 \times (14)^3}{4} - \frac{1000 \times (10)^3}{4}$$

$$M_2 = -19{,}500 \text{ ft-lb}$$

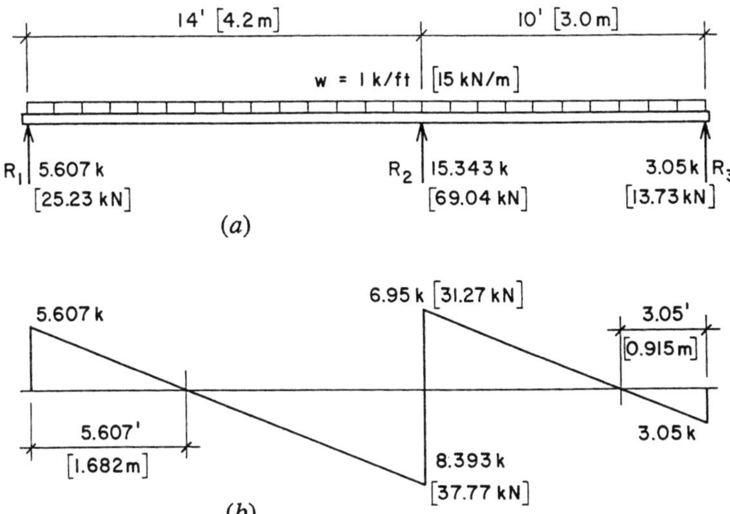

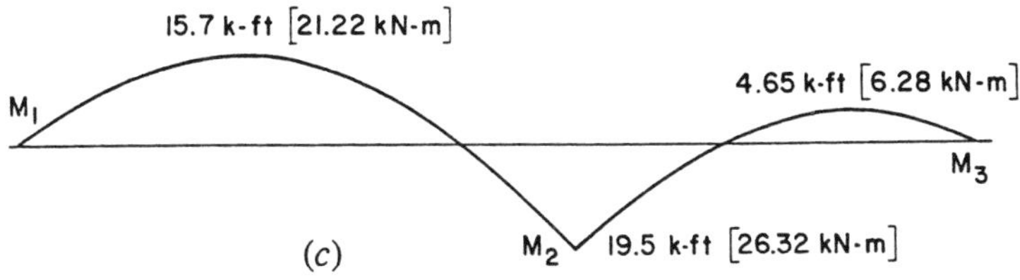

Figure 3.28 Reference for Example 16.

Write a moment summation about a point 14 ft to the right of the left end support using the forces to the left of the point:

$$14R_1 - (1000 \times 14 \times 7) = -19{,}500 \qquad R_1 = 5607 \text{ lb}$$

Then write a moment summation about a point 10 ft to the left of the right end using the forces to the right of the point:

$$10R_3 - (1000 \times 10 \times 5) = -19{,}500 \qquad R_3 = 3050 \text{ lb}$$

A vertical force summation will yield the value of $R_2 = 15{,}343$ lb. With the three reactions determined, the shear values for completing the shear diagram are known. Determination of the points of zero shear and zero moment and the values for positive moment in the two spans can be done as demonstrated in Example 15. The completed diagrams are shown in Figure 3.28.

Problems 3.8.C, D Find the reactions and draw the complete shear and moment diagrams for the following continuous beams with two unequal spans and uniformly distributed loading.

	First Span (ft)	Second Span (ft)	Load (lb/ft)
C	12	16	2000
D	16	20	1200

Continuous Beam with Concentrated Loads

In the previous examples the loads were uniformly distributed. Figure 3.29*a* shows a two-span beam with a single concentrated load in each span. The shape for the moment diagram for this beam is shown in Figure 3.29*b*. For these conditions, the form of the three-moment equation is

$$M_1L_1 + 2M_2(L_1 + L_2) + M_3L_2$$
$$= -P_1L_1^2[n_1(1 - n_1)(1 + n_1)] - P_2L_2^2[n_2(1 - n_2)(2 - n_2)]$$

in which the various terms are as shown in Figure 3.29.

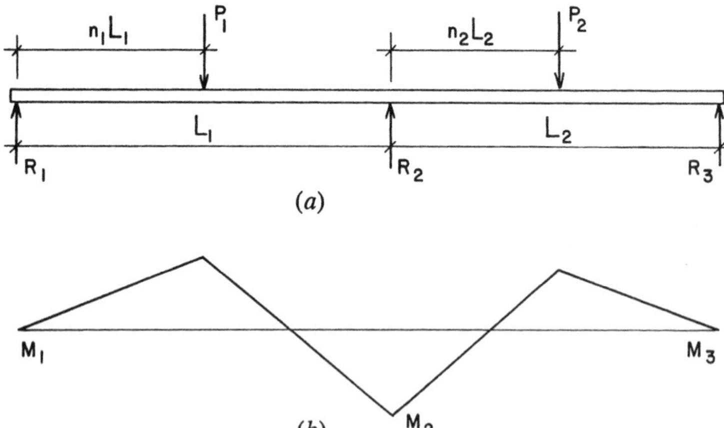

Figure 3.29 Two-span beam with concentrated loads.

Example 17. Compute the reactions and construct the shear and moment diagrams for the beam in Figure 3.30*a*.

Solution: For this case note that $L_1 = L_2$, $P_1 = P_2$, $M_1 = M_3 = 0$, and both n_1 and n_2 equal 0.5. Substituting these conditions and given data into the equation

$$2M_2(20 + 20)$$
$$= -4000(20)^2\,(0.5 \times 0.5 \times 1.5) - 4000(20)^2\,(0.5 \times 0.5 \times 1.5)$$

from which $M_2 = 15{,}000$ ft-lb.

The value of the moment at the middle support can now be used as in Examples 15 and 16 to find the end reaction, from which it is determined that the value is 1250 lb. Then a summation of vertical

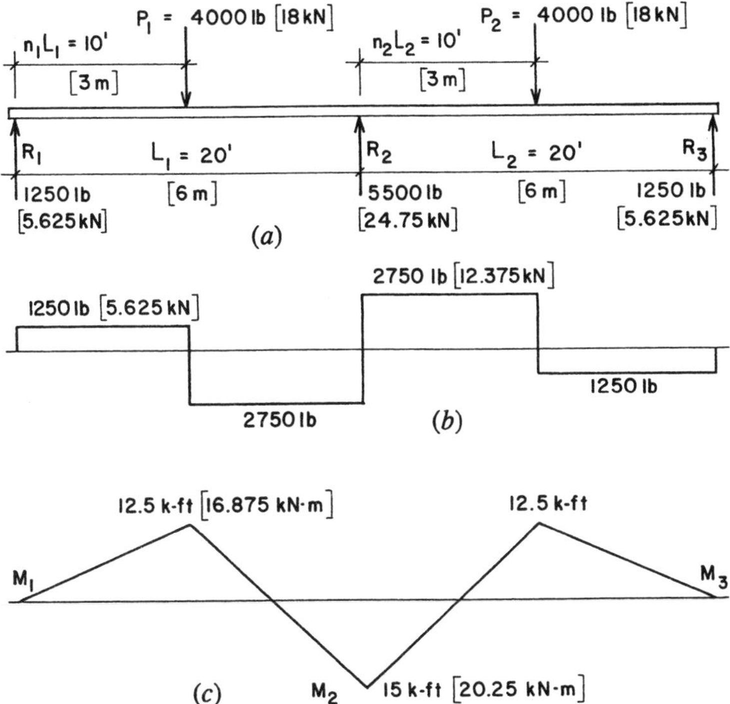

Figure 3.30 Reference for Example 17.

forces will determine the value of R_2 to be 5500 lb. This is sufficient data for construction of the shear diagram. Note that points of zero shear are evident on the diagram.

The values for maximum positive moment can be determined from moment summations at the sections or simply from the areas of the rectangles in the shear diagrams. The locations of points of zero moment can be found by simple proportion, since the moment diagram is composed of straight lines.

Problems 3.8.E, F Find the reactions and draw the complete shear and moment diagrams for the following continuous beams with two equal spans and a single concentrated load at the center of each span.

	Span Length (ft)	Load (kips)
E	24	3
F	32	2.4

Continuous Beam with Three Spans

The preceding examples demonstrate that the key operation in investigation of continuous beams is the determination of negative moment values at the supports. Use of the three-moment equation has been demonstrated for a two-span beam, but the method may be applied to any two adjacent spans of a beam with multiple spans. For example, when applied to the three-span beam shown in Figure 3.31*a*, it would first be applied to the left span and the middle span and next to the middle span and right span. This would produce two equations involving the two unknowns: the negative moments at the two interior supports. In this example the process would be simplified by the symmetry of the beam, but the application is a general one, applicable to any arrangement of spans and loads.

As with simple beams and cantilevers, common situations of spans and loading may be investigated and formulas for beam behavior values derived for subsequent application in simpler investigation processes. Thus, the values of reactions, shears, and moments displayed for the beam in Figure 3.31 may be used for any such support and loading conditions. Tabulations for many ordinary situations are available from various references.

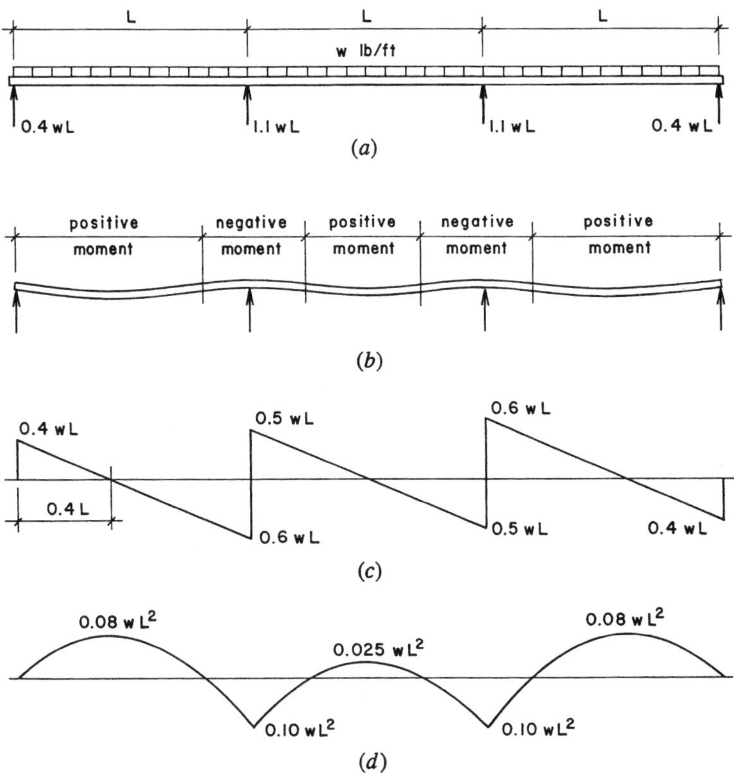

Figure 3.31 Three-span beam with distributed loading.

Example 18. A continuous beam has three equal spans of 20 ft each and a uniformly distributed load of 800 lb/ft extending over the entire length of the beam. Compute the maximum bending moment and the maximum shear.

Solution: Referring to Figure 3.31*d*, the maximum positive moment ($0.08wL^2$) occurs near the middle of each end span, and the maximum negative moment ($0.10wL^2$) occurs over each of the interior supports. Using the larger value, the maximum bending moment on the beam is

$$M = -0.1wL^2 = -(0.1 \times 800 \times 20^2) = 32{,}000 \text{ ft-lb}$$

Figure 3.30*c* shows that the maximum shear occurs at the face of the first interior support and is

$$V = 0.6wL = 0.6 \times 800 \times 20 = 9600 \text{ lb}$$

Using this process it is possible to find the values of the reactions and then to construct the complete shear and moment diagrams if the work at hand warrants it.

Problems 3.8.G, H For the following continuous beams with three equal spans and uniformly distributed loading, find the reactions and draw the complete shear and moment diagrams.

	Span Length (ft)	Load (lb/ft)
G	24	1000
H	32	1600

Restrained Beams

A simple beam was previously defined as a beam that rests on a support at each end, there being no restraint against bending at the supports; the ends are *simply supported.* The shape a simple beam tends to assume under load is shown in Figure 3.32*a*. Figure 3.32*b* shows a beam whose left end is *restrained* or *fixed*, meaning that free rotation of the beam end is prevented. Figure 3.32*c* shows a beam with both ends restrained. End restraint has an effect similar to that caused by the continuity of a beam at an interior support: A negative bending moment is induced in the beam. The beam in Figure 3.32*b* has a profile with an inflection point, indicating a change of sign of the moment within the span. This span behaves in a manner similar to one of the spans in the two-span beam.

The beam with both ends restrained has two inflection points, with a switch of sign to negative bending moment near each end. Although values are slightly different for this beam, the general form of the

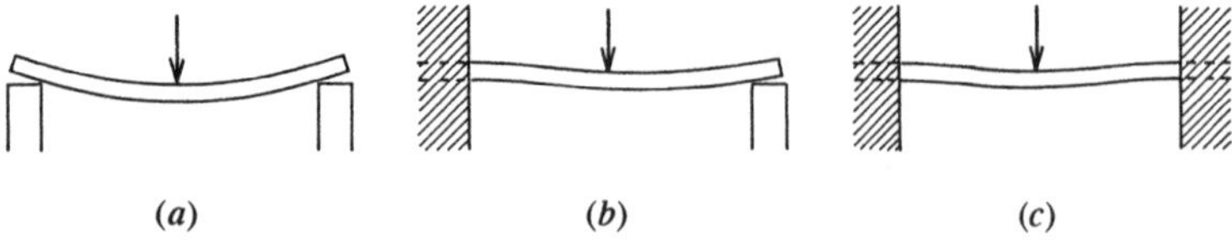

Figure 3.32 Behavior of beams with various forms of rotational restraint at supports: (*a*) no restraint, (*b*) one end restrained, and (*c*) both ends restrained.

deflected shape is similar to that for the middle span in the three-span beam (see Figure 3.31).

Although they have only one span, the beams in Figures 3.32*b* and *c* are both indeterminate. Investigation of the beam with one restrained end involves finding three unknowns: the two reactions plus the restraining moment at the fixed end. For the beam in Figure 3.32*c*, there are four unknowns. There are, however, only a few ordinary cases that cover most common situations, and tabulations of formulas for these ordinary cases are readily available from references. Figure 3.33 gives values for the beams with one and two fixed ends under both uniformly distributed load and a single

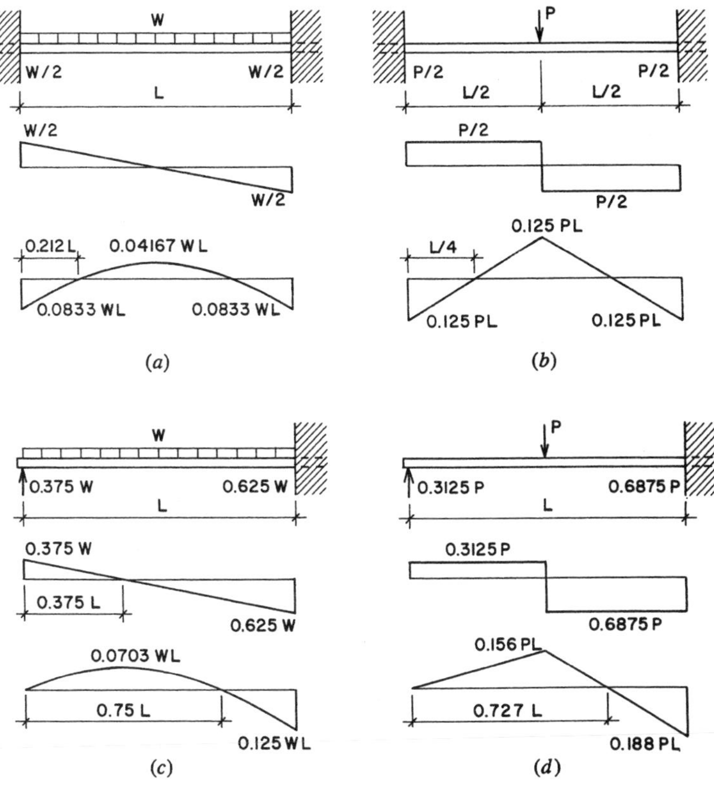

Figure 3.33 Values for restrained beams.

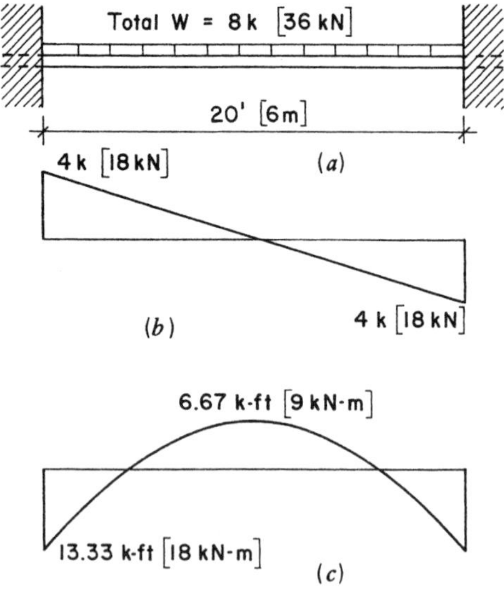

Figure 3.34 Reference for Example 19.

concentrated load at center span. Values for other loadings are also available from references.

Example 19. Figure 3.34*a* represents a 20-ft-span beam with both ends fixed and a total uniformly distributed load of 8 kips. Find the reactions and construct the complete shear and moment diagrams.

Solution: Despite the fact that this beam is indeterminate to the second degree (four unknowns; only two equations of static equilibrium), its symmetry makes some investigation data self-evident. It can be observed that the two vertical reaction forces, and thus the two end shear values, are each equal to one-half of the total load, or 4000 lb. Symmetry also indicates that the location of the point of zero moment and thus the point of maximum positive bending moment are at the center of the span. Also, the end moments, although indeterminate, are equal to each other, leaving only a single value to be determined.

From data in Figure 3.33*a*, the negative end moment is $0.0833WL$ (actually $^1/_{12}\ WL$). Thus,

$$M = \frac{WL}{12} = \frac{8000 \times 20}{12} = 13{,}333 \text{ ft-lb}$$

The maximum positive moment at midspan is $0.04167WL$ (actually $^1/_{24}\ WL$). Thus,

$$M = \frac{WL}{24} = \frac{8000 \times 20}{24} = 6667 \text{ ft-lb}$$

The distance from the beam end to the point of zero moment is

$$x = 0.212L = 0.212(20) = 4.24 \text{ ft}$$

The complete shear and moment diagrams are as shown in Figure 3.34.

Example 20. A beam fixed at one end and simply supported at the other end has a span of 20 ft and a total uniformly distributed load of 8000 lb (Figure 3.35*a*). Find the reactions and construct the shear and moment diagrams.

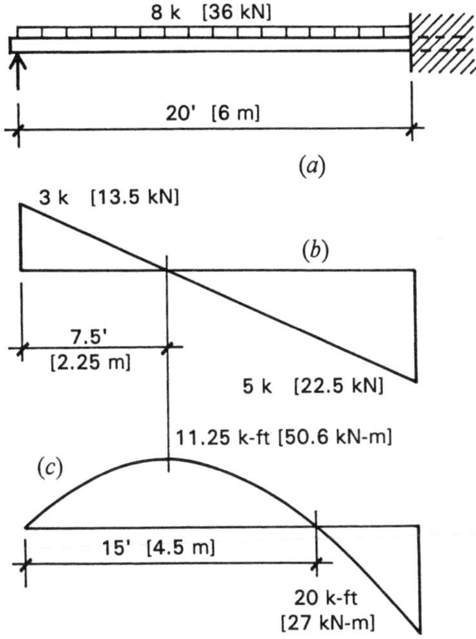

Figure 3.35 Reference for Example 20.

Solution: This is the same span and loading as in the preceding example. Here, however, one end is fixed and the other simply supported (loading case in Figure 3.33*c*). The beam vertical reactions are equal to the end shears; thus, from the data in Figure 3.33*c*:

$$R_1 = V_1 = 0.375(8000) = 3000 \text{ lb}$$

$$R_2 = V_2 = 0.625(8000) = 5000 \text{ lb}$$

and, for the maximum moments,

$$+M = 0.0703(8000 \times 20) = 11{,}248 \text{ ft-lb}$$

$$-M = 0.125(8000 \times 20) = 20{,}000 \text{ ft-lb}$$

The point of zero shear is at 0.375(20) = 7.5 ft from the left end, and the point of zero moment is at twice this distance, 15 ft, from the left end. The complete shear and moment diagrams are shown in Figure 3.35.

Problem 3.8.I A 22-ft span beam is fixed at both ends and carries a single concentrated load of 16 kips at midspan. Find the reactions and construct the complete shear and moment diagrams.

Problem 3.8.J A 16-ft span beam is fixed at one end and simply supported at the other end. A single concentrated load of 9600 lb is placed at the center of the span. Find the vertical reactions and construct the complete shear and moment diagrams.

3.9 MEMBERS EXPERIENCING COMPRESSION PLUS BENDING

Development of Bending in Columns

Bending moments can be developed in structural members in a number of ways. When a member is subjected to an axial compression force, there are various ways in which the compression effect and any bending present can relate to each other.

Figure 3.36*a* shows a very common situation that occurs in building structures when an exterior wall functions as a bearing wall or contains a column. The combination of vertical gravity load and lateral load due to wind or seismic action can result in the

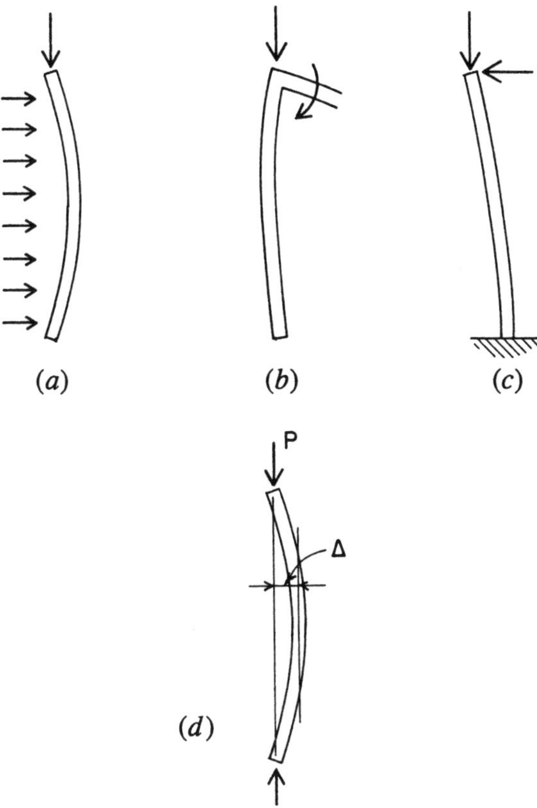

Figure 3.36 Development of bending in columns.

loading shown. If the member is quite flexible, an additional bending is developed as the axis of the member deviates from the action line of the vertical compression load. This added bending is the product of the load and the member deflection; that is, P times Δ, as shown in Figure 3.36*d*. It is thus referred to as the *P–delta* effect.

There are various other situations that can result in the *P*–delta effect. Figure 3.36*b* shows an end column in a rigid frame structure, where the moment is induced at the top of the column by the moment-resistive connection to the beam. Although slightly different in its profile, the column response is similar to that in Figure 3.36*a*. Figure 3.36*c* shows the effect of a combination of gravity and lateral loads on a vertically cantilevered structure that supports a sign or a tank at its top.

In any of these situations, the *P*–delta effect may or may not be critical. The major factor that determines its seriousness is the relative stiffness of the structure as it relates to the magnitude of deflection produced. However, even a significant deflection may not be of concern if the vertical load (*P*) is quite small. In a worst-case scenario, a major *P*–delta effect may produce an accelerating failure, with the added bending producing more deflection, which in turn produces more bending, and so on.

Interaction of Bending and Axial Compression

There are a number of situations in which structural members are subjected to the combined effects that result in development of axial compression and internal bending. Stresses developed by these two actions are both of the direct stress type (tension and compression) and can be combined for consideration of a net stress condition. This is useful for some cases—as is considered following this discussion—but for columns with bending the situation involves two essentially different actions: column action in compression and beam behavior. For column investigation, therefore, it is the usual practice to consider the combination by what is called *interaction.*

The classic form of interaction is represented by the graph in Figure 3.37*a*. Referring to the notation on the graph:

1. The maximum axial load capacity of the member (with no bending) is P_o.
2. The maximum bending capacity of the member (without compression) is M_o.
3. At some compression load below P_o (indicated as P_n) the member is assumed to have some tolerance for a bending moment (indicated as M_n) in combination with the axial load.
4. Combinations of P_n and M_n are assumed to fall on a line connecting points P_o and M_o. The equation of this line has the form expressed as

$$\frac{P_n}{P_o} + \frac{M_n}{M_o} = 1$$

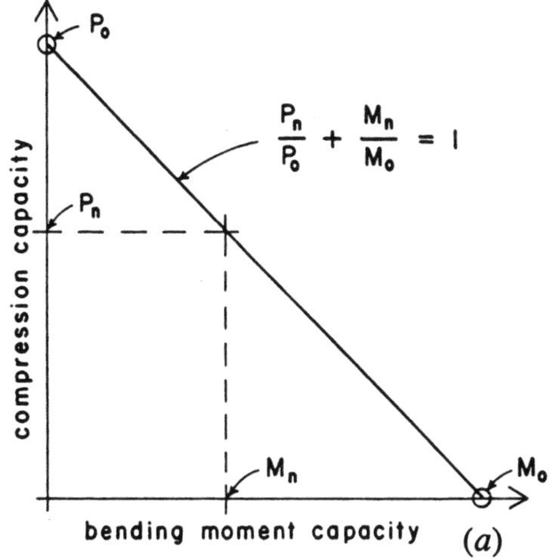

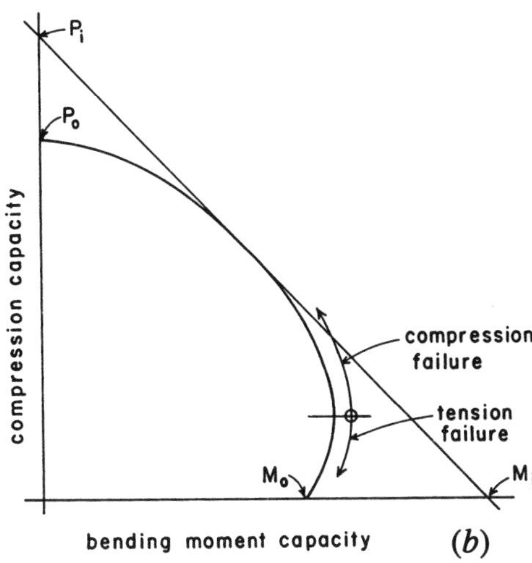

Figure 3.37 Interaction of axial compression and bending in a column: (*a*) classic form of the interaction formula and (*b*) form of interaction response in a reinforced concrete column.

A graph similar to that in Figure 3.37*a* can be constructed using stresses rather than loads and moments. This is the procedure used for wood and steel members; the graph taking the form is expressed as

$$\frac{f_a}{F_a} + \frac{f_b}{F_b} \leq 1$$

where f_a = computed stress due to axial load
F_a = allowable column action stress in compression
f_b = computed stress due to bending
F_b = allowable beam action stress in flexure

For various reasons, real structures do not adhere strictly to the classic straight-line form of response shown in Figure 3.37*a*. Figure 3.37*b* shows a form of response characteristic of reinforced concrete columns. In the midrange, there is some approximation of the theoretical straight-line behavior, but at the terminal ends of the response graph there is considerable variation. This has to do with the nature of the ultimate failure of the reinforced concrete materials in both compression (upper end) and tension (lower end), as explained in Chapter 15.

Steel and wood members also have various deviations from the straight-line interaction response. Special problems include inelastic behavior, effects of lateral stability, geometry of member cross sections, and lack of initial straightness of members. Wood columns are discussed in Chapter 6 and steel columns in Chapter 10.

Combined Stress: Compression Plus Bending

Combined actions of compression plus bending produce various effects on structures. In some situations the actual stress combinations may of themselves be critical, one such case being the development of bearing stress on soils. At the contact face of a bearing footing and its supporting soil, the "section" for stress investigation is the contact face, that is, the bottom surface of the footing. The following discussion presents an approach to this investigation.

Figure 3.38 illustrates the situation of combined direct force and bending moment at a cross section. In this case the "cross section" is the contact face of the footing bottom with the soil. Whatever produces the combined load and moment, a transformation is made to an equivalent eccentric force that produces the same effect. The value for

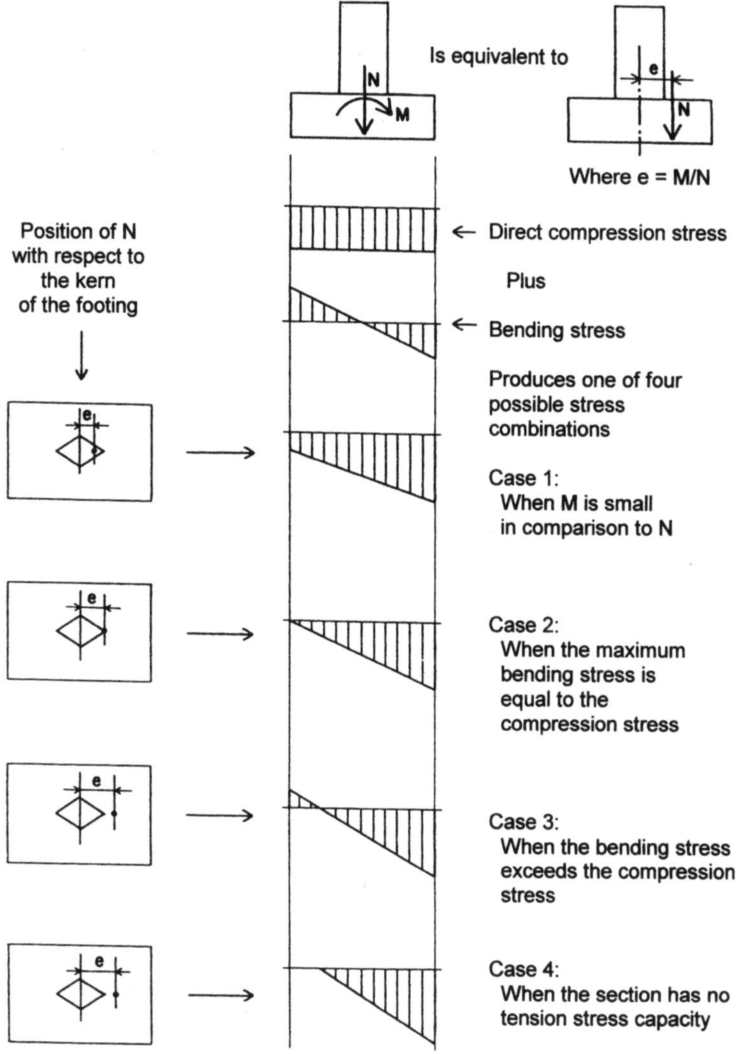

Figure 3.38 Combinations of compression and bending stress considered as generated by an eccentric compression force.

the hypothetical eccentricity e is established by dividing the moment by the load, as shown in the figure. The net, or combined, stress distribution at the section is visualized as the sum of separate stresses caused by the load and the bending moment. For the limiting stresses at the edges of the section the general equation for the combined stress is

$$P = (\text{direct stress}) \pm (\text{bending stress})$$

or

$$p = \frac{N}{A} \pm \frac{Nec}{I}$$

Four cases for this combined stress are shown in Figure 3.38. The first case occurs when e is small, resulting in very little bending stress. The section is thus subjected to all compressive stress that varies from a maximum value at one edge to a minimum value at the opposite edge.

The second case occurs when the two stress components are equal, so that the minimum stress is zero. This is the boundary condition between the first and third cases since an increase in e will produce some reversal stress—in this situation some tension stress.

The second stress case is significant for a footing since tension stress is not possible between the footing and soil. Case 3 is only possible for a beam or column or some other continuously solid element. The value for e that produces case 2 can be derived by equating the two stress components. Thus,

$$\frac{N}{A} = \frac{Nec}{I} \quad \text{and thus} \quad e = \frac{I}{Ac}$$

This value for e establishes what is known as the *kern limit* of the section, which defines a zone around the centroid of the section within which an eccentric force will not cause reversal stress on the section. The form and dimensions for this zone can be established for any geometric shape by use of the derived equation for e. The kern limit zones for three common geometric shapes are shown in Figure 3.39.

When tension stress is not possible, larger eccentricities of the compression normal force will produce a so-called *cracked section*, as shown for case 4 in Figure 3.38. In this situation some portion of the cross section becomes unstressed, or cracked, and the compressive stress on the remainder of the section must develop the entire resistance to the loading effects of the combined force and moment.

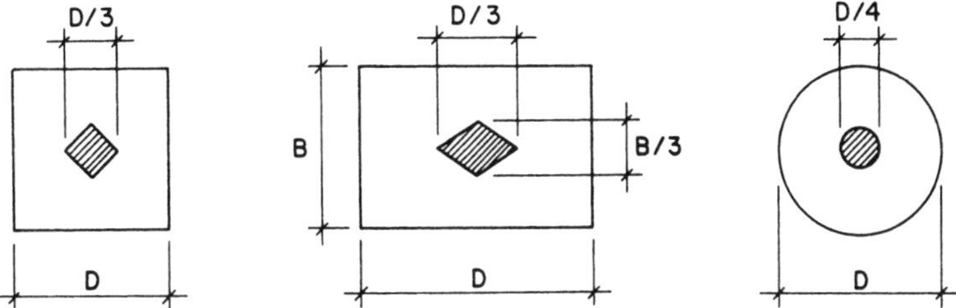

Figure 3.39 Form of the kern for common shapes of cross sections.

Figure 3.40 shows a technique for the analysis of a cracked section, called the *pressure wedge method.* The wedge is a volume that represents the total compressive force as developed by the soil pressure; its volumetric unit is in force units produced by multiplying stress times area. Analysis of the static equilibrium of this wedge produces two relationships that can be used to establish the dimensions of the wedge. These relationships are:

1. The volume of the wedge is equal to the vertical force.
2. The centroid of the wedge is located on a vertical line that coincides with the location of the hypothetical eccentric force.

Referring to Figure 3.40, the three dimensions of the wedge are w (width of the footing), p (maximum soil pressure), and x (limiting dimension of the stressed portion of the cracked section). In this situation the footing width is known so the definition of the wedge requires only the determination of p and x.

For the rectangular section the centroid is at the third point of the triangle. Defining this distance from the edge as a, as shown in Figure 3.40, x is equal to three times a. It may be observed that a is equal to half the footing width minus the eccentricity e. Thus, once the eccentricity is computed, the values of a and x can be determined.

The volume of the stress wedge can be expressed in terms of its three dimensions as

$$V = \frac{wpx}{2}$$

With w and x determined, the remaining dimension of the wedge can be established by transforming the equation of the volume to

$$p = \frac{2V}{wx}$$

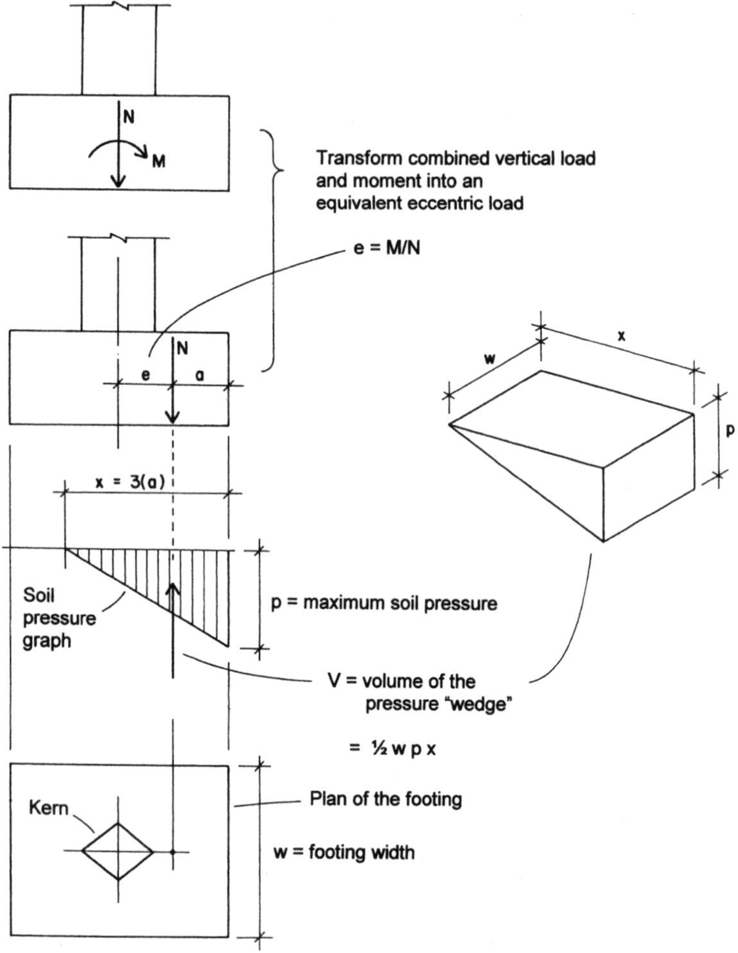

Figure 3.40 Investigation of combined stress on a cracked section by the pressure wedge method.

or, since the volume is equal to the force N,

$$p = \frac{2N}{wx}$$

All four cases of combined stress in Figure 3.38 will cause some tilting of the footing due to deformation of the compressible soil. The extent of this tilting and its effect on the structure supported by the footing must be carefully considered in the design of the footing. It is usually desired that the soil pressure be evenly distributed for anything other than very short time loadings such as those caused by wind or seismic effects.

Example 21. Find the value of maximum soil pressure for a square footing subjected to a load of 100 kips and a moment of 100 kip-ft. Find values for footing widths of (a) 8 ft, (b) 6 ft, and (c) 5 ft.

Solution: The first step is to find the equivalent eccentricity and compare it to the kern limit for the footing to establish which of the cases shown in Figure 3.38 applies. For all parts

$$e = \frac{M}{N} = \frac{100}{100} = 1 \text{ ft}$$

For (a) the 8-ft-wide footing has a kern limit of $8/6 = 1.33$ ft, and thus case 1 applies. For computations the properties of the 8-ft^2 footing are

$$A = 8 \times 8 = 64 \text{ ft}^2$$

$$I = \frac{bd^3}{12} = \frac{(8)(8)^3}{12} = 341 \text{ ft}^4$$

and the maximum soil pressure is

$$p = \frac{N}{A} + \frac{Mc}{I} = \frac{100}{64} + \frac{(100)(4)}{341} = 1.56 + 1.17 = 2.73 \text{ ksf}$$

For (b) the 6-ft-wide footing has a kern limit of 1 ft, the same as the eccentricity. Thus, the situation is case 2 in Figure 3.37 with $N/A = Mc/I$ and

$$p = 2\left(\frac{N}{A}\right) = 2\left(\frac{100}{36}\right) = 5.56 \text{ ksf}$$

For (c) the eccentricity exceeds the kern limit and the investigation must be done as illustrated in Figure 3.40. Thus,

$$a = \frac{5}{2} - e = 2.5 - 1 = 1.5 \text{ ft}$$

$$x = 3a = 3(1.5) = 4.5 \text{ f}$$

$$tp = \frac{2N}{wx} = \frac{2(100)}{5(4.5)} = 8.89 \text{ ksf}$$

Problem 3.9.A A square footing sustains a force of 40 kips and a bending moment of 30 kip-ft. Find the maximum soil pressure for widths of (a) 5 ft and (b) 4 ft.

Problem 3.9.B A square footing sustains a force of 60 kips and a bending moment of 60 kip-ft. Find the maximum soil pressure for widths of (a) 7 ft and (b) 5 ft.

3.10 RIGID FRAMES

Frames in which two or more of the members are attached to each other with connections that are capable of transmitting bending between the ends of the members are called *rigid frames*. The connections used to achieve such a frame are called *moment connections* or *moment-resisting connections*. Most rigid frame structures are statically indeterminate and do not yield to investigation by consideration of static equilibrium alone. The rigid frame structure occurs quite frequently as a multiple-level, multiple-span bent, constituting part of the structure for a multistory building. In most cases, such a bent is used as a lateral bracing element, although once it is formed as a moment-resistive framework, it will respond as such for all types of loads. The computational examples presented in this section are all rigid frames that have conditions that make them statically determinate and thus capable of being fully investigated by methods developed in this book.

Cantilever Frames

Consider the frame shown in Figure 3.41*a*, consisting of two members rigidly joined at their intersection. The vertical member is fixed at its base, providing the necessary support condition for stability of the frame. The horizontal member is loaded with a uniformly distributed loading and functions as a simple cantilever beam. The frame is described as a cantilever frame because of the single fixed support. The five sets of figures shown in Figure 3.41*b*–*f* are useful elements for the investigation of the behavior of the frame. They consist of the following:

1. The free-body diagram of the entire frame, showing the loads and the components of the reactions (Figure 3.41*b*). Study of this figure will help in establishing the nature of the reactions

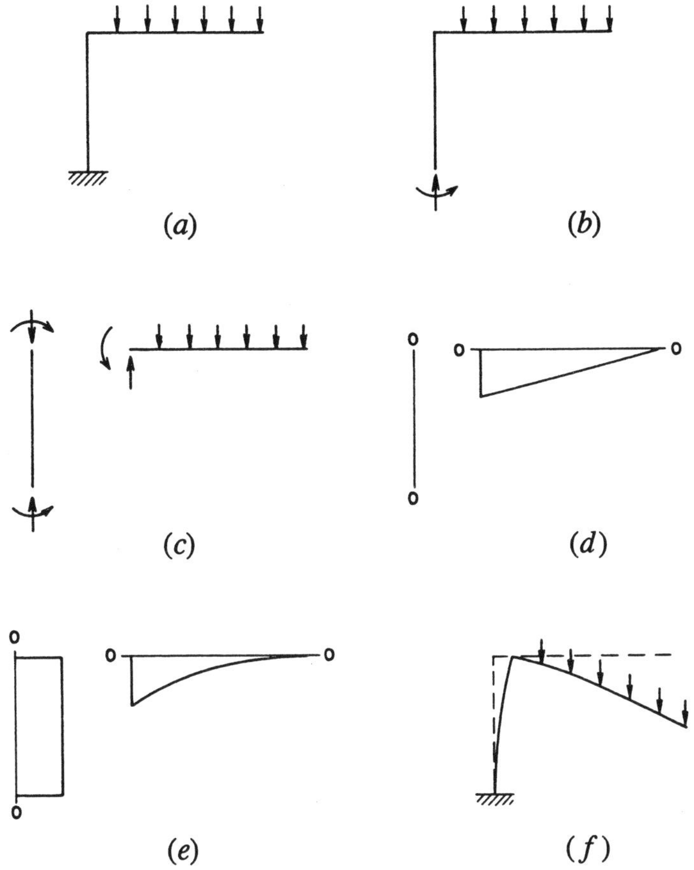

Figure 3.41 Diagrams for investigation of the rigid frame.

and in the determination of the conditions necessary for stability of the frame as a whole.

2. The free-body diagrams of the individual elements (Figure 3.41*c*). These are of great value in visualizing the interaction of the parts of the frame. They are also useful in the computations for the internal forces in the frame.
3. The shear diagrams of the individual elements (Figure 3.41*d*). These are sometimes useful for visualizing or actually computing the variations of the moment in the individual elements.

No particular sign convention is necessary unless in conformity with the sign used for the moment.

4. The moment diagrams for the individual elements (Figure 3.41*e*). These are very useful, especially in determination of the deformation of the frame. The sign convention used is that of plotting the moment on the compression (concave) side of the flexed element.
5. The deformed shape of the loaded frame (Figure 3.41*f*). This is the exaggerated profile of the bent frame, usually superimposed on an outline of the unloaded frame for reference. This is very useful for the general visualization of the frame behavior. It is particularly useful for determination of the character of the external reactions and the form of interaction between the parts of the frame. Correlation between the deformed shape and the form of the moment diagram is a useful check.

When performing investigations, these elements are not usually produced in the sequence just described. In fact, it is generally recommended that the deformed shape be sketched first so that its correlation with other factors in the investigation may be used as a check on the work. The following examples illustrate the process of investigation for simple cantilever frames.

***Example* 22.** Find the components of the reactions and draw the free-body diagram, the shear and moment diagrams, and the deformed shape of the frame shown in Figure 3.42*a*.

Solution: The first step is the determination of the reactions. Considering the free-body diagram of the whole frame (Figure 3.42*b*),

$$\sum F = 0 = +8 - R_v \qquad R_v = 8 \text{ kips (up)}$$

and with respect to the support,

$$\sum M = 0 = M_R - (8 \times 4) \qquad M_R = 32 \text{ kip-ft (clockwise)}$$

Note that the sense, or sign, of the reaction components is visualized from the logical development of the free-body diagram.

Consideration of the free-body diagrams of the individual members will yield the actions required to be transmitted by the moment connection. These may be computed by application of the conditions

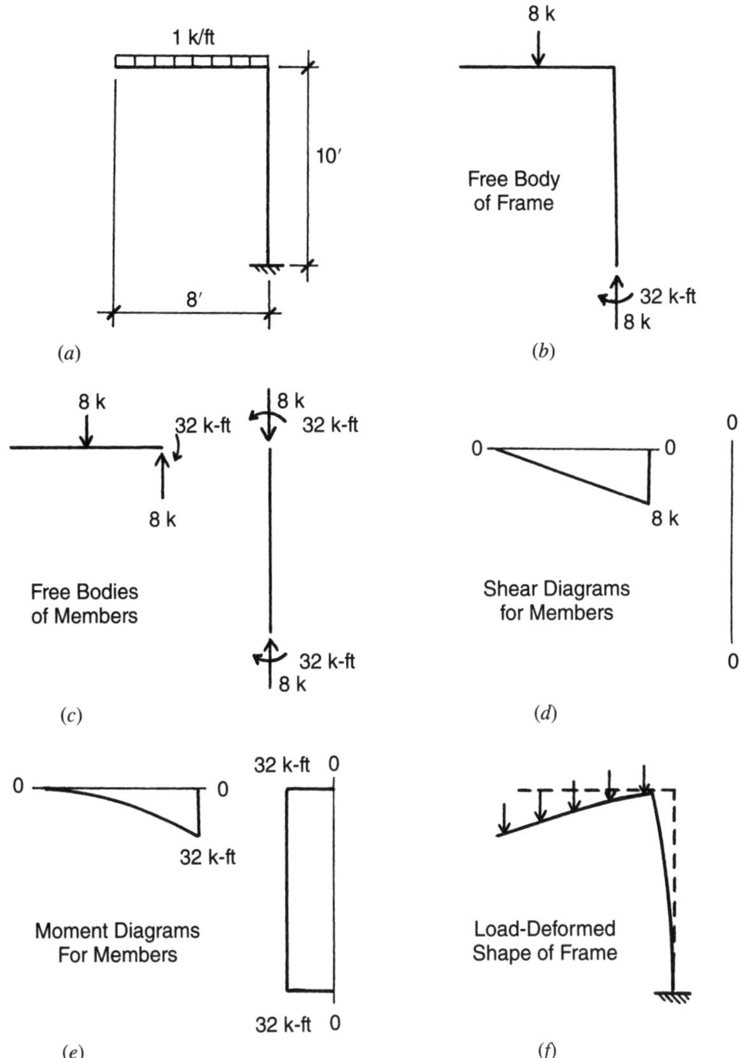

Figure 3.42 Reference for Example 22.

for equilibrium for either of the members of the frame. Note that the sense of the force and moment is opposite for the two members, simply indicating that what one does to the other is the opposite of what is done to it.

In this example there is no shear in the vertical member. As a result, there is no variation in the moment from the top to the bottom of the member. The free-body diagram of the member, the shear and moment diagrams, and the deformed shape should all corroborate this fact. The shear and moment diagrams for the horizontal member are simply those for a cantilever beam.

It is possible with this example, as with many simple frames, to visualize the nature of the deformed shape without recourse to any mathematical computations. It is advisable to attempt to do so as a first step in the investigation and to check continually during the work that individual computations are logical with regard to the nature of the deformed structure.

Example 23. Find the components of the reactions and draw the shear and moment diagrams and the deformed shape of the frame in Figure 3.43*a*.

Solution: In this frame there are three reaction components required for stability since the loads and reactions constitute a general coplanar force system. Using the free-body diagram of the whole frame (Figure 3.43*b*), the three conditions for equilibrium for a coplanar system are used to find the horizontal and vertical reaction components and the moment component. If necessary, the reaction force components could be combined into a single-force vector, although this is seldom required for design purposes.

Note that the inflection occurs in the larger vertical member because the moment of the horizontal load about the support is greater than that of the vertical load. In this case, this computation must be done before the deformed shape can be accurately drawn.

The reader should verify that the free-body diagrams of the individual members are truly in equilibrium and that there is the required correlation between all the diagrams.

Problems 3.10.A–C For the frames shown in Figures 3.44*a–c*, find the components of the reactions, draw the free-body diagrams of the whole frame and the individual members, draw the shear and moment diagrams for the individual members, and sketch the deformed shape of the loaded structure.

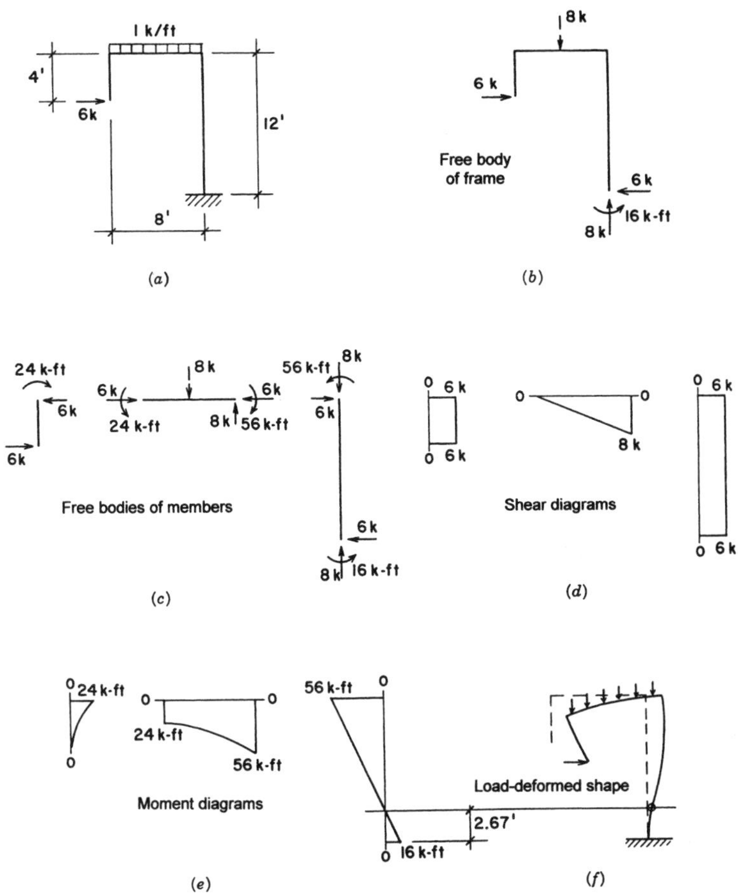

Figure 3.43 Reference for Example 23.

Single-Span Frames

Single-span rigid frames with two supports are ordinarily statically indeterminate. The following example illustrates the case of a statically determinate, single-span frame, made so by the particular conditions of its support and internal construction. In fact, these conditions are technically achievable but a little unusual for practical use. The example is offered here as an exercise for readers that is within the scope of the work in this section.

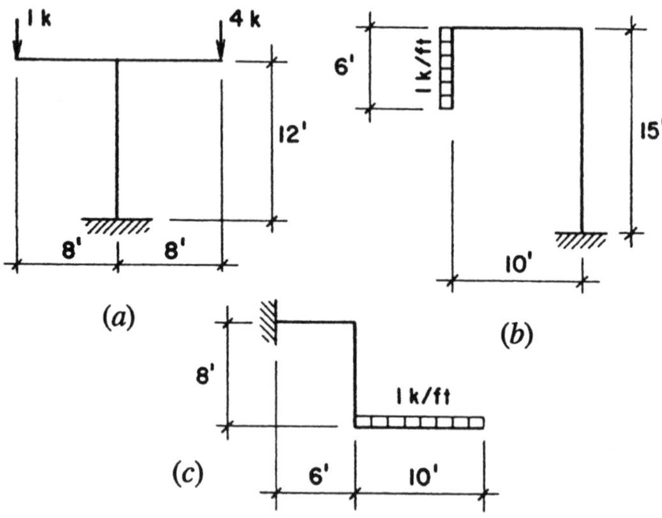

Figure 3.44 Reference for Problem 3.10.A-C.

Example 24. Investigate the frame shown in Figure 3.45 for the reactions and internal conditions. Note that the right-hand support allows for an upward vertical reaction only, whereas the left-hand support allows for both vertical and horizontal components. Neither support provides moment resistance.

Solution: The typical elements of investigation, as illustrated for the preceding examples, are shown in Figure 3.45. The suggested procedure for the work is as follows:

1. Sketch the deflected shape (a little tricky in this case, but a good exercise).
2. Consider the equilibrium of the free-body diagram for the whole frame to find the reactions.
3. Consider the equilibrium of the left-hand vertical member to find the internal actions at its top.
4. Proceed to the equilibrium of the horizontal member.
5. Finally, consider the equilibrium of the right-hand vertical member.
6. Draw the shear and moment diagrams and check for correlation of all work.

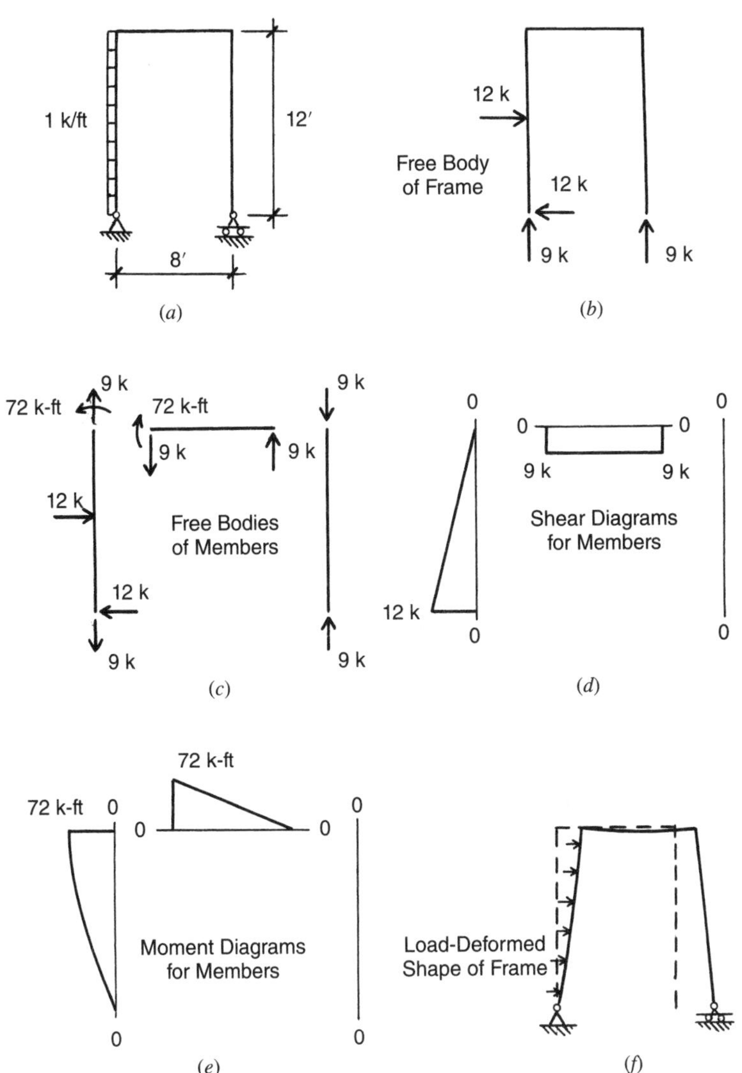

Figure 3.45 Reference for Example 24.

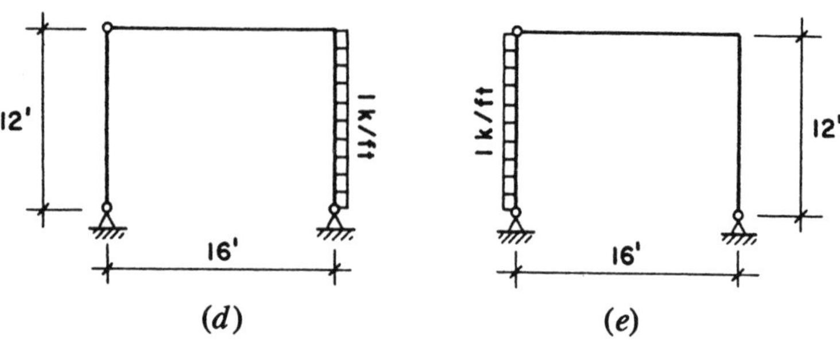

Figure 3.46 Reference for Problem 3.10.D and E.

Before attempting the exercise problems, the reader is advised to attempt to produce the results shown in Figure 3.45 independently.

Problems 3.10.D, E Investigate the frames shown in Figures 3.46*d* and *e* for reactions and internal conditions using the procedure shown for the preceding examples.

3.11 BUCKLING OF BEAMS

Buckling of beams—in one form or another—is mostly a problem with beams that are relatively weak on their transverse axes, that is, the axis of the beam cross section at right angles to the axis of bending. This is not a frequent condition in concrete beams, but it is a common one with beams of wood or steel or with trusses that perform beam functions. The cross sections shown in Figure 3.47 illustrate members that are relatively susceptible to buckling in beam action.

When buckling is a problem, one solution is to redesign the beam for more resistance to lateral movement. Another possibility is to analyze for the lateral buckling effect and reduce the usable bending capacity as appropriate. However, the solution most often used is to brace the beam against the movement developed by the buckling effect. To visualize where and how such bracing should be done, it is first necessary to consider the various possibilities for buckling. The three main forms of beam buckling are shown in Figure 3.48.

Figure 3.48*b* shows the response described as *lateral* (i.e., *sideways*) *buckling*. This action is caused by the compressive stresses in the top of the beam that make it act like a long column, which is thus subject to a sideways movement as with any slender column. Bracing

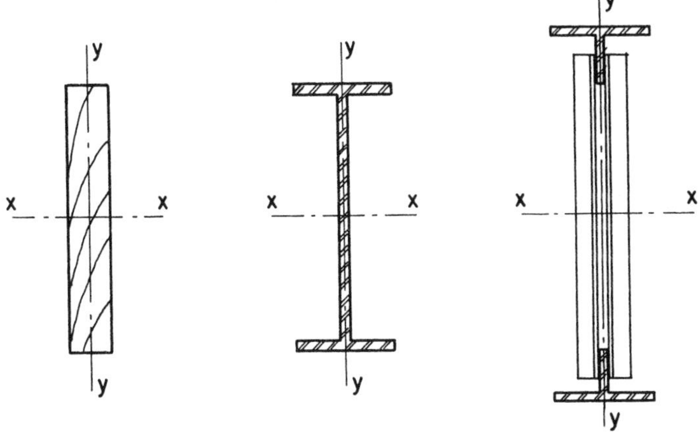

Figure 3.47 Beam shapes with low resistance to lateral bending and buckling.

the beam for this action means simply preventing its sideways movement at the beam edge where compression exists. For simple span beams this edge is the top of the beam. For beams, joists, rafters, or trusses that directly support roof or floor decks, the supported deck may provide this bracing if it is adequately attached to the supporting members. For beams that support other beams in a framing system, the supported beams at right angles to the supporting member may provide lateral bracing. In the latter case, the unsupported length of the buckling member becomes the distance between the supported beams, rather than its entire span length.

Another form of buckling for beams is that described as *torsional buckling*, as shown in Figure 3.48*d*. This action may be caused by tension stress, resulting in a rotational, or twisting, effect. This action can occur even when the top of the beam is braced against lateral movement and is often due to a lack of alignment of the plane of the loading and the vertical axis of the beam. Thus, a beam that is slightly tilted is predisposed to a torsional response. An analogy for this is shown in Figure 3.48*e*, which shows a trussed beam with a vertical post at the center of the span. Unless this post is perfectly vertical, a sideways motion at the bottom end of the post is highly likely.

To prevent both lateral and torsional buckling, it is necessary to brace the beam sideways at both its top and bottom. If the roof or floor deck is capable of bracing the top of the beam, the only extra

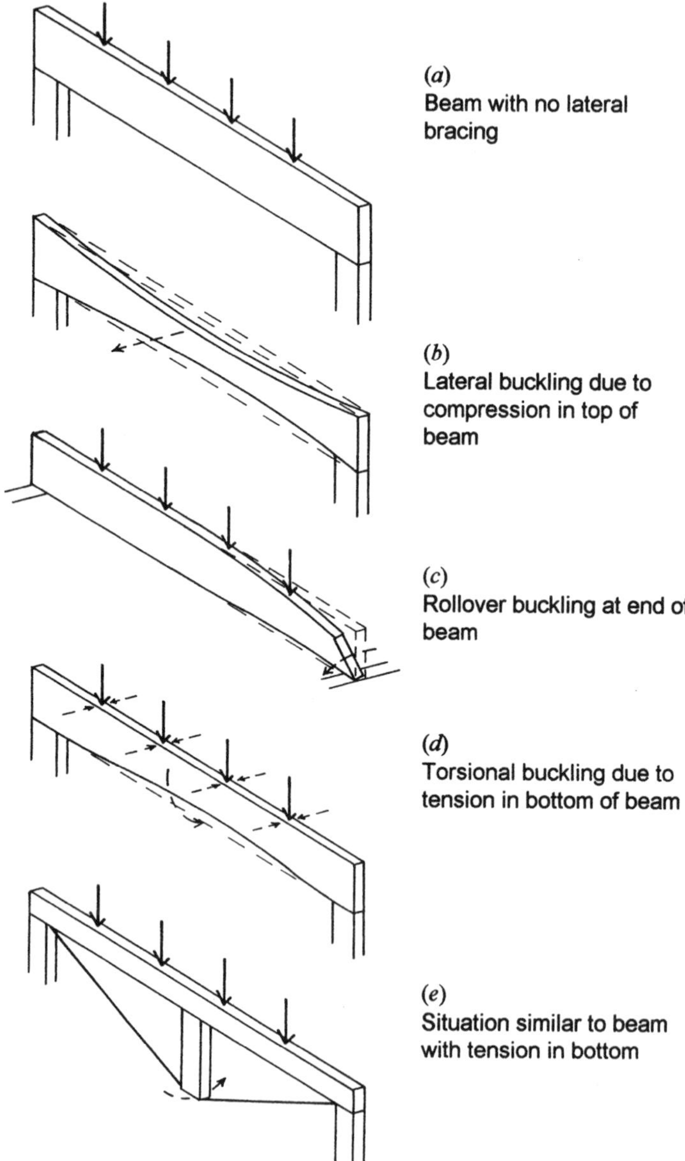

Figure 3.48 Forms of buckling of beams.

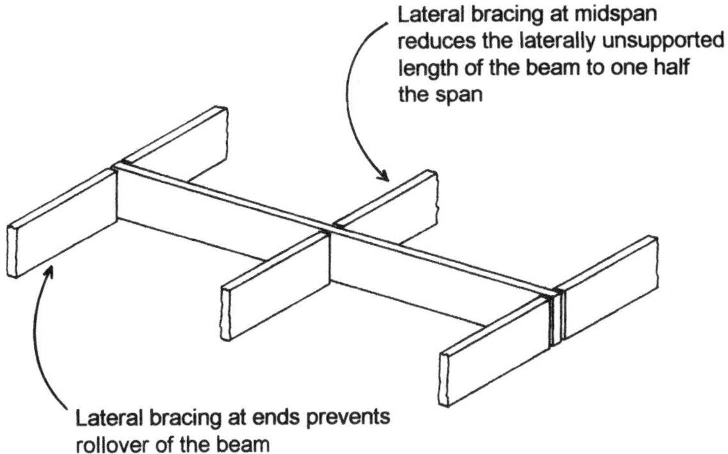

Figure 3.49 Lateral bracing for beams.

bracing required is that for the bottom. For closely spaced trusses this bracing is usually provided by simple horizontal ties between adjacent trusses. For beams in wood or steel framing systems, lateral bracing may be provided as shown in Figure 3.49. The beam shown is braced for both lateral and torsional buckling. Other forms of bracing are described in Part II for wood structures and in Part III for steel structures.

3.12 SECOND-ORDER ANALYSIS

Sections 3.9 and 3.10 introduced the concept of complex loading and complex structural systems, i.e., frames. Figures 3.36*a–c* show the deflected shapes of columns that experience either lateral loading or a bending moment as well as the axial loading. Using traditional methods of analysis, also known as first-order analysis, the internal bending moment of the column is determined from the lateral loading and the internal axial force is determined from the gravity loading. In Figure 3.36*d*, the deflected shape due to the lateral loading is illustrated for the loading given in Figure 3.36*a* along with its maximum deflection, Δ, being noted. The gravity load, P, now produces both an internal axial force and an internal bending moment at the maximum deflection due to the lateral load equal to the applied gravity load, P, times the deflection ($M = P \times \Delta$). Second-order analysis requires that this internal bending moment be added to the internal bending moment caused by the lateral loading and determined during the first-order analysis. Second-order analysis is required by building codes for complex structural systems.

Example 25. Investigate a 13-ft-tall W 8 × 24 steel column that is illustrated in Figure 3.36*c* that has a gravitational load of 10 kips and a lateral load of 5 kips to determine if it is acceptable to carry such loading using the classic form of the interaction formula, Figure 3.37*a*. Both first- and second-order analysis should be investigated.

Solution: The first step to solve this problem is to determine the capacity of the member for bending, M_o, and axial compression, P_o. This is a fairly complex procedure covered in Part II and will be summarized here:

$$P_o = 40 \text{ kips}$$

$$M_o = 87{,}000 \text{ K-ft}$$

$$E = 29{,}000 \text{ ksi}$$

$$I = 82.7 \text{ in.}^4$$

For the first-order analysis, one can use the tabular formulas for a cantilevered beam given in Figure 3.18, case 4:

$$M_n = PL = (5)(13) = 65{,}000 \text{ K-ft}$$

$$\Delta = \frac{PL^3}{3EI} = \frac{(5)(13 \times 12)^3}{3(29{,}000)(82.7)} = 2.64 \text{ in.}$$

Using the classic form of the interaction formula,

$$\frac{P_n}{P_0} + \frac{M_n}{M_0} \leq 1$$

$$\frac{10}{40} + \frac{65}{87} = 0.99 \leq 1$$

Therefore, based on first-order analysis, the column is acceptable for this loading.

For second-order analysis, the internal bending moment needs to include the moment due to the deflection (2.64 in.) caused by the 5-kip horizontal force:

$$M_n = 65 + (10)\left(\frac{2.64}{12}\right) = 67{,}200 \text{ K-ft}$$

Again using the interaction formula,

$$\frac{P_n}{P_0} + \frac{M_n}{M_0} \leq 1$$

$$\frac{10}{40} + \frac{67.2}{87} = 1.02$$

Therefore, based on second-order analysis, this column is not capable of handling the given loading.

3.13 COMPUTER SOFTWARE FOR STRUCTURAL ANALYSIS

Computer software is prevalent in the engineering industry to aid in the determination of internal forces within a building's structural system. The sophistication varies from simple spreadsheet programs for specific structural types and materials to sophisticated finite element analysis packages that are capable of determining stress concentrations within individual members and their connections. All computer engineering software needs to be used as a tool by experienced persons who are capable of verifying the results.

Building structural systems are rarely statically determinate as assumed in most of the examples of this book. Computer software is usually the most efficient way of analyzing complex indeterminate structural systems. With indeterminate structures, each member's stiffness determines how the applied forces work their way through a structure to the support systems. The stiffer a member is relative to other members, the more force is directed through that member. Member stiffness is determined by the member's modulus of elasticity, moment of inertia, and length. In most software packages, the person imputing the structure must know the structure's geometry, material, and an estimate of its size before the analysis is able to be accomplished by the computer. Estimation of structural members' sizes is illustrated in Chapter 4 and Parts II, III, and IV of this book.

Once the first analysis is completed, more accurate member sizing is completed and analysis is done again. This iterative process continues until the analysis reveals that the member sizing is appropriate for the given structural system and loading.

Besides analyzing indeterminate structures, software packages can also be used to illustrate how a structure will deform under different loading conditions. The accuracy of such deformation is dependent on many factors. It can be very insightful for students to use such software to understand how buildings react to various loading such as wind and seismic. Students should understand that the actual wind and seismic forces are very fluid and dynamic and the results that the computer demonstrates are usually static approximations.

4

STRUCTURAL SYSTEMS AND PLANNING

This chapter presents the major structural systems used in architectural structures and then discusses how to plan for the structural system in early phases of building design. The work includes the presentation of approximate dimensions of structural members in order to plan for the structural members with regards to other building systems such as those for mechanical, plumbing, and electrical services. These rules of thumb are derived in later chapters and should not be considered replacements for determining the actual size of members using the later work. The material presented here should be considered the beginning of the structural design process, not the end.

4.1 GENERAL CONSIDERATIONS FOR STRUCTURAL SYSTEMS

Structural systems are complex, usually highly redundant systems that must be able to resist any load imposed on them. Most examples up to this point have dealt with gravitational loading. Individual structural elements have been analyzed for dead and live gravitational loading. When looking at a system as a whole, lateral loading becomes another key loading that must be dealt with by the structure.

Structural system names are largely based on how the system deals with lateral loading. *Shear wall systems* use walls to transfer any lateral load to the foundations. The walls may also carry the gravitational loading, but they may also be for lateral loading alone. *Moment frame systems* transfer the lateral loading by allowing the transfer of internal bending moments from one member to the next. *Braced frame systems* use trussed braces to transfer the lateral loading to the foundations. There are other structural systems that could be included in this discussion, but the majority of buildings use one of the three systems.

Lateral loads in architectural structures are complex and dynamic in nature. Different regions are affected by different lateral loading. Seismic loading is considered in regions with active seismic faults such as the Pacific Rim. Wind affects all buildings but strong destructive winds are regional. Hurricanes cause damage on the western edge of the Atlantic Ocean, islands within large bodies of water, and on the Gulf Coasts. Tornados are dominant in the interior regions of continents, and high-velocity winds affect mountainous regions or areas along large bodies of water. Other dynamic loadings include those due to impact or explosions. Individual buildings are rarely designed to withstand all of these loads. A designer should be familiar with the lateral loads prevalent in the region where the building is being built and should be aware that when working in a different region the critical lateral loading could be very different.

Building behavior (how a building moves) under lateral loading is of concern in the design of structural systems. In general, different structural systems have different stiffness. A shear wall system is stiffer than a braced frame system, which is stiffer than a moment

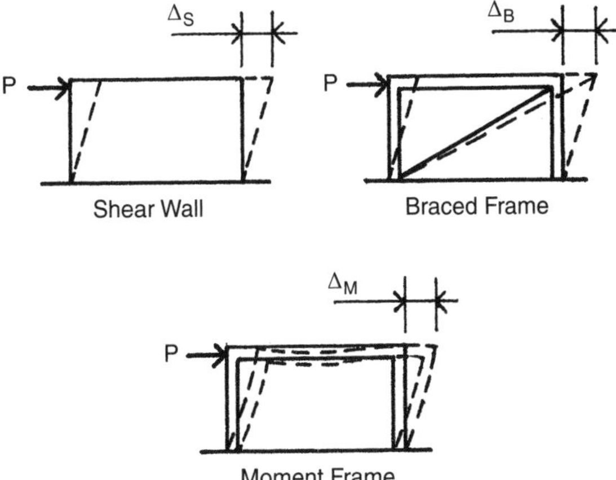

Figure 4.1 Relative deflection of a shear wall, a braced frame, and a moment frame.

frame system. This means that under the same lateral load a shear wall will deflect less than a braced frame, which will deflect less than a moment frame (Figure 4.1).

Care must be taken when laying out a structural system to balance stiffness throughout the system. Examples of common unbalanced lateral stiffness in buildings are stiffer elements on one side of the building than on the other (Figure 4.2*a* for shear wall and Figure 4.2*b* for moment frame), mixing of two structural systems (Figure 4.2*c*), geometric irregularities (Figure 4.2*d*), and combinations of these irregularities (Figure 4.2*e*).

An unbalanced stiffness can lead to the development of torsional forces in the structure when a lateral load is applied to it. This is caused by the center of the lateral force not being collinear to the reactive force from the building, which produces a twisting moment or torsion (Figure 4.3).

The three basic structural systems are presented as if they are totally distinct. The reality is that they are often combined together in a single building. A good example is that most frame systems have horizontal diaphragms (roof and floor decks) to help transfer the lateral loads to the ground.

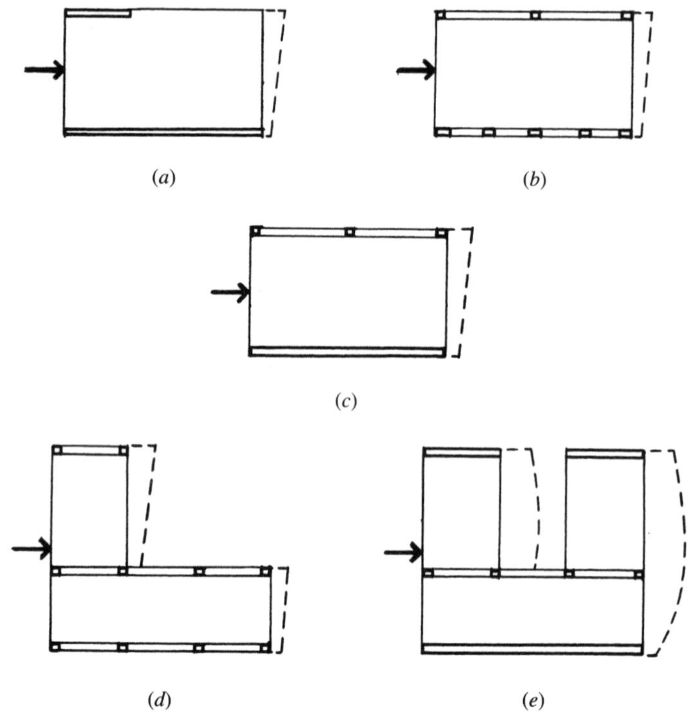

Figure 4.2 Common situations of unbalanced stiffness.

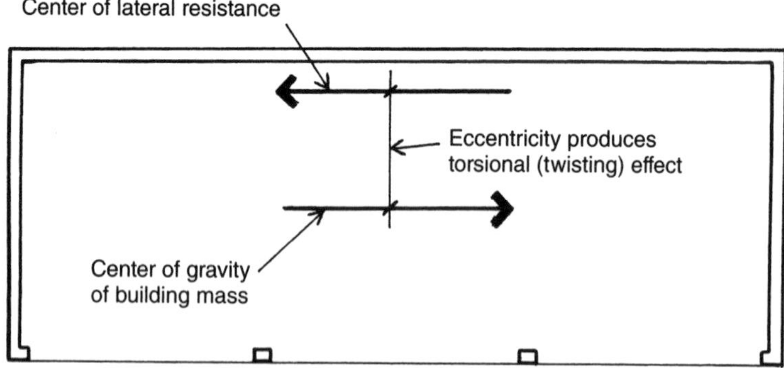

Figure 4.3 Torsional effect of unbalanced stiffness.

4.2 SHEAR WALL AND DIAPHRAGM STRUCTURAL SYSTEM

The shear wall and diaphragm system is the stiffest of the three major structural systems. This system transmits the lateral loads to the ground by acting as large, deep beams. If the lateral force is caused by wind, the wind is collected by the exterior wall and transferred to the horizontal diaphragms. The diaphragms act as deep beams supported by shear walls and transfer the loads to the shear walls (Figure 4.4). The shear walls act as short deep cantilevered beams (Figure 4.5) and transfer the loads to the foundations through the development of shearing actions and overturn resistance in the walls. If the lateral load is a seismic load, the difference is that the lateral force is collected directly into the diaphragm by its own mass and the mass of the objects resting on it. Earthquakes produce accelerations

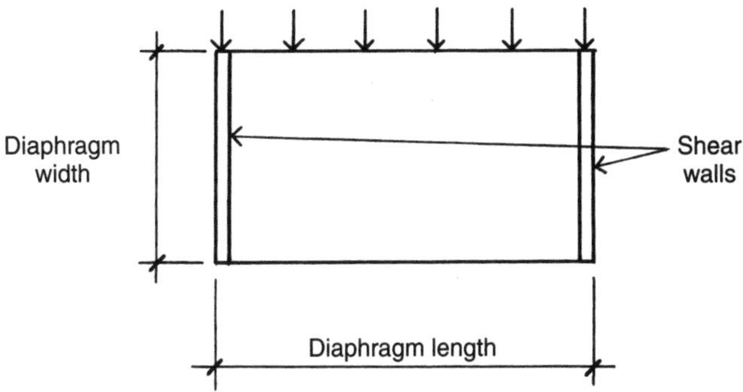

Figure 4.4 Plan of a horizontal diaphragm.

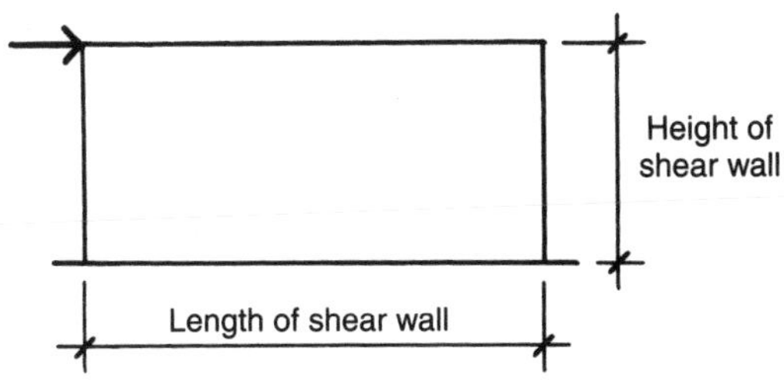

Figure 4.5 Elevation of a shear wall.

that act on the mass of the building and its contents to produce the dynamic seismic force: force = mass × acceleration.

The proportions of the dimensions of the diaphragms and the shear walls are important to make sure that the load transfer happens through shearing action. In the diaphragm, the proportion of width to length should be less than 1 : 3. If the length becomes greater than this, a new intermediate shear wall should be introduced to place the proportions back into an acceptable range. Similarly, the optimum shear wall proportion of length to height is less than 1 : 3. If the shear wall proportions are considerably greater than 1 : 3 and there is no way to bring them back within it, then it is an indication that a shear wall system may not be the most appropriate for the building and a new system should be selected.

Shear walls are commonly made of reinforced concrete, concrete masonry units (CMUs), and plywood or oriented strand board (OSB) over wood or metal studs. Brick or CMUs may also be used as infill shear walls within a frame, which handles the gravitational loading. Diaphragms are commonly made of concrete, formed sheet steel, or plywood/OSB over wood joists.

4.3 BRACED FRAME SYSTEMS

Braced frame systems consist of several subsets, including X bracing, K bracing, cross-bracing, and eccentric bracing (Figures 4.6*a–d*, respectively). The braced frame systems are designed so that the braces transfer only lateral loading. Under lateral loading, the systems act like trusses to transmit the load through a series of compression and tension axial members (Figure 4.7). The very small deformation of individual members within the systems, consisting of elongation or shortening along the member's major axis, makes the system relatively stiff. Careful consideration needs

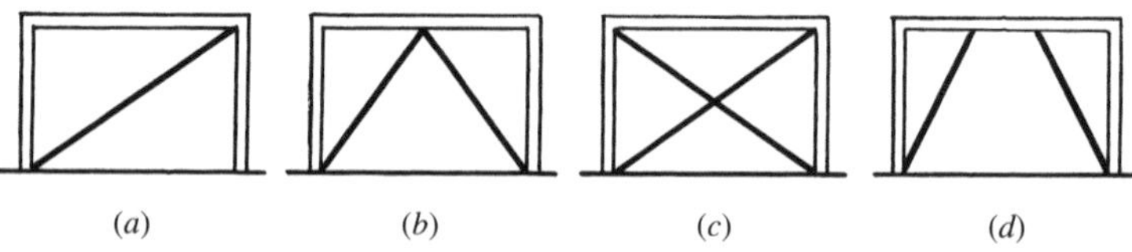

Figure 4.6 Forms of bracing: (*a*) diagonal bracing, (*b*) K bracing, (*c*) cross-bracing, and (*d*) eccentric bracing.

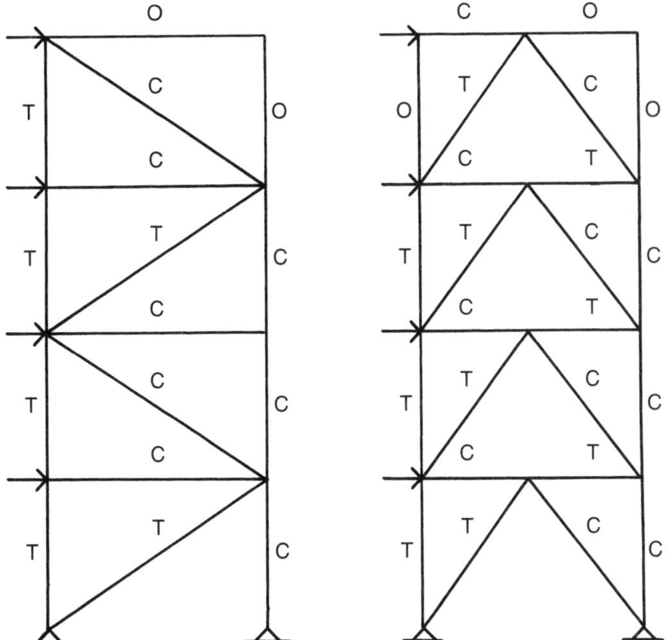

Figure 4.7 Trusslike actions of bracing under lateral loading. T indicates tension and C indicates compression. Members with no notation experience no truss action.

to be given to members that experience compressive forces in order to prevent the buckling of the member and thus the failure of the system.

Horizontal or sloped frames can also be braced similar to the vertical frames or they can utilize a diaphragm to transmit the lateral forces to the vertical bracing. Braced horizontal frames are commonly used in long-span roof structures while diaphragms are commonly used in multistory buildings where floors are commonly made of concrete.

Materials commonly used in braced frame systems are steel and timber. Diagonal bracing was commonly used for lateral resistance in wood stud construction prior to the availability of plywood but is no longer common.

In recent years a few new braced frame systems have been developed in steel for use in seismic regions. The point of these new systems is to have a system that is more ductile under lateral loading than traditional systems and uses movement to dissipate the seismic energy. The first of these systems is the *eccentric braced frame system* (Figure 4.6*d*). By moving the braces so that they do not intersect with the beam at a single point as they do with a K-braced system (Figure 4.6*b*), the beam experiences bending under lateral loading and therefore has more movement in it than a similar K-braced system. A second system is an *unbonded brace system*, which is also a modification to the K-braced system in which the braces are made of thin steel members encased in concrete to prevent buckling under compression. The concrete and steel are developed so that they are not bonded together, and the steel member is allowed to elongate and shorten more than a K-braced system where the steel brace is sized to handle buckling by itself. In the unbonded brace system, the braces act similar to a shock absorber in a car, allowing for more movement under large seismic accelerations.

4.4 MOMENT FRAME SYSTEMS

Moment frame systems are the most ductile of the three major systems. In a moment frame system, some or all of the connections between the system's beams and columns are made rigid so that one is not able to rotate without rotating the adjacent members (Figure 4.8). In other words, the angle between a column and a beam will be the same after loading as it was before (usually 90°). Moment frame systems are sometimes referred to as rigid frame systems or rigid connection frame systems.

There are several variations of moment frames, which typically have to do with which connections are made rigid and which are allowed to freely rotate. An example of this is a frame in which only the connections between the exterior columns and beams are made rigid, producing a moment frame called a *tube structure*. These variations are important to the design of the structure, but, unlike variations in braced frame systems, variations in moment frame systems have little effect on the appearance of the building or the layout of the frame members.

A major advantage of moment frame structures is the simplicity of the arrangements of the frame members. For architectural planning, there is no need to work around braces or to figure out how much you can penetrate a shear wall with a window or door without compromising the integrity of the system. The major disadvantage

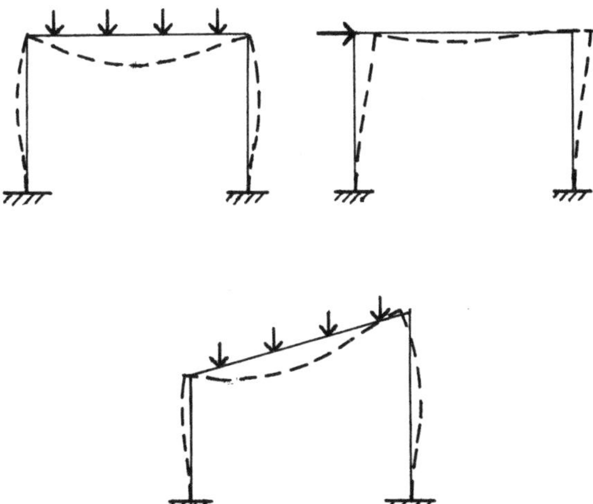

Figure 4.8 Deflected forms of moment frames.

is the introduction of bending into columns, which requires larger columns.

Steel and reinforced concrete are the materials most used for moment frame structures. The choice of which material to use is usually based on economics and regional preferences, not on structural advantages of one material over another.

Similar to braced frame systems, special moment frame systems have been developed for seismic regions. These systems are designed to release the rigidity of one or more beam–column connections during large seismic activity. This can be thought of as a fuse in the structural system. This fuse dissipates large seismic energy through the breaking of the fuse and subsequently through the additional movement of the building. The fuse is usually located near the beam–column connection on the beam. The fuse is designed to release seismic energy without compromising the integrity of the structure itself. These fuses come in many forms, from expensive proprietary systems to relatively inexpensive techniques to weaken beams in order to create a plastic hinge during major seismic events. After a seismic event, any fuses that have been used must be replaced. In general terms, the more expensive the up-front cost of the system, the less expensive the replacement of the fuse. If the fuse is a plastic hinge in a beam, the entire beam will need to be replaced.

4.5 WOOD CONSTRUCTION

Wood by its very nature is quite variable in its structural properties. Wood species are regional and the grades or quality of wood may be unavailable at certain times or in certain places. The assumptions taken in this section are that higher-quality dimensional lumber is being used for the structure and that the rules of thumb presented here are not to replace actual engineering, which is presented in Part II.

Structural Layout

Wood construction is typically made up of repetitive wood joists held up by either wood stud bearing walls or beams, which are in turn held up by columns or bear directly on the foundation (Figure 4.9). Lateral resistance is usually provided by a shear wall and diaphragm system composed of plywood/OSB sheathing over wood studs and joists. All shear walls in wood construction are bearing walls, although not all bearing walls are shear walls. Interior bearing walls not covered in plywood/OSB are usually not used for shear walls, even though they have rated capacity for shear resistance.

When planning a structural layout, the floor and roof joists generally span in the short direction of the floor plan. Joist lengths are limited by both availability and acceptable deflection criteria to approximately 18–22 ft; therefore, one should plan for either interior bearing walls or beams no more than 22 ft apart. If one is using wood beams, they are generally limited to about 20 ft in length or will need columns at intervals of 20 ft or less. If you need a beam span greater

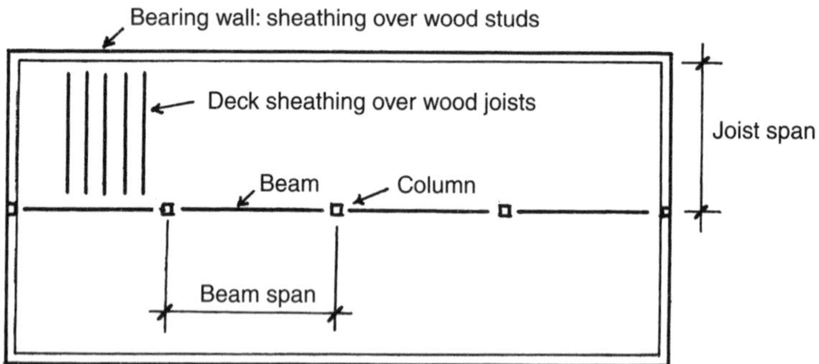

Figure 4.9 Construction form of a typical horizontal wood roof or floor structure.

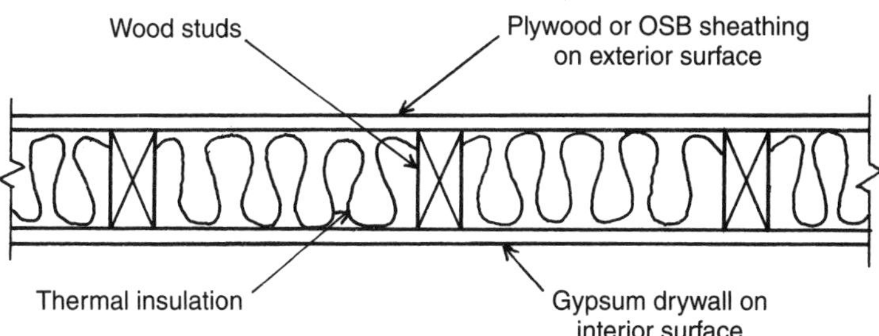

Figure 4.10 Plan section of a typical wood stud wall.

than 20 ft, one should use steel beams and steel columns, which will allow a longer span (see Section 4.6). If a steel beam is used, it should never bear on a wood column or stud wall.

Stud Bearing Wall Sizing

Wood stud bearing walls (Figure 4.10) are incredibly strong when properly braced by plywood, OSB, or wood blocking. For generations, the 2 × 4 wood stud wall was the standard for wood construction. More recently, the 2 × 6 wood stud wall has become more popular and has replaced the 2 × 4 stud wall in many regions. This is not because of its superior strength, which it has, but for its ability to contain more thermal insulation, thus making it a more energy-efficient system. In 2 × 4 walls, the studs are usually spaced 16 in. on center. In 2 × 6 walls, the studs are also usually spaced 16 in. on center, but in some regions they are spaced 24 in. on center. If the project that you are working on is in a year-round temperate climate, the bearing walls will most likely be made 2 × 4. If the project will experience cold winters, hot summers, or both, then the exterior walls will probably be made of 2 × 6's either at 16 or 24 in. on center.

Floor and Roof Joist Sizing

Roofs and floors are generally both designed with 2-in. thick dimensional lumber with plywood or OSB sheathing on top of it, creating a diaphragm (Figure 4.11). With joists, it is important to get high-quality wood. Floor joists are generally sized based on allowable deflection criteria, while sloped roof members are generally sized based on allowable strength criteria. Flat (or near-flat) roof

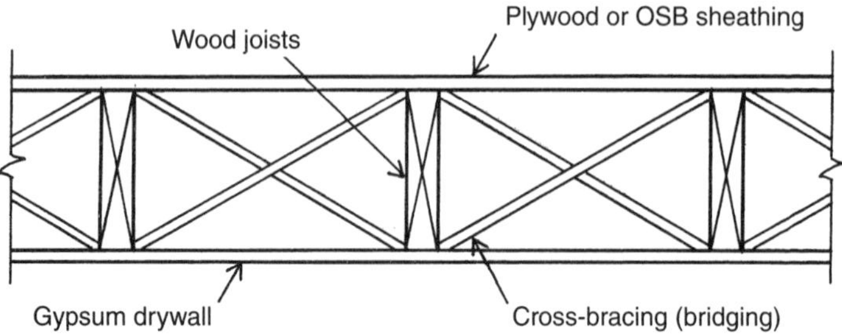

Figure 4.11 Typical form of construction of wood floor structure.

TABLE 4.1 Approximate Span for Floor and Roof Joists

	Maximum Span (ft)	
Lumber Size (in.)	Floor Joists	Roof Joists
2 × 6	9	10
2 × 8	13	14
2 × 10	17	18
2 × 12	21	22

joists should be sized as a floor joist during preliminary design. The spacing of floor and roof joist is typically 16 in. on center but can also be 12, 19.2, or 24 in. on center depending on the particular project.

Table 4.1 gives the approximate span limits for floor and sloped roof joists using dimensional lumber. It should be noted that the span of a sloped roof joist is measured horizontally, not along the length of the member. If you need spans greater than those listed in Table 4.1, then engineered wood joists may be more appropriate for the project. Roof joists can be deeper than that recommended in Table 4.1 if more thermal insulation is required.

Beam Sizing

Wood beams should always be made of high-grade lumber. Beams are generally limited in length to approximately 20 ft. The approximate beam depths are listed in Table 4.2. The beam width varies with actual loading calculations. Headers over window and door openings in bearing walls are considered beams.

TABLE 4.2 Approximate Depths for Wood Beams

Beam Span (ft)	Nominal Beam Depth (in.)
Span ≤ 9	6
9 < Span ≤ 13	8
13 < Span ≤ 17	10
17 < Span ≤ 20	12

Column Sizing

Wood columns are practically limited to 4 × 4's and 6 × 6's. The unbraced length of the column is often the controlling factor in sizing wood columns: 4 × 4's are limited to approximately 12 ft and 6 × 6's are limited to about 20 ft; 8 × 8's are available but are becoming harder to find and more expensive with the reduction of timber still coming from old growth forests. If a column is needed that is larger than a 6 × 6, then a steel column is usually used. Steel columns are also commonly used in areas where a wood column is vulnerable to a large lateral load or impact such as in a garage or basement.

Example 1. Determine the preliminary design and layout for the house illustrated in Figure 4.12. The project is located in a mountainous region that experiences both cold winters and hot summers. The system will be designed to be a shear wall and diaphragm system.

Solution: First, we need to check if the diaphragm is in an acceptable proportion or if interior shear walls need to be added. The overall plan dimensions are 50 and 28 ft, which give a ratio of 1 : 1.79; as this is less than 1 : 3, interior shear walls will not be required.

The exterior walls will be shear walls and their proportions need to be checked. The shear walls between the first and second floors will be of most concern. The height of this shear wall will be 10 ft floor-to-floor height. For one of the north–south shear walls, the ratio will be 1 : 5, which is greater than the required 1 : 3. On the east–west shear walls the ratio will be 1 : 2.8, which is close to the optimal range.

Since the project is not in a temperate climate, we will make the exterior shear walls of 2 × 6 construction to accommodate additional insulation. For the north–south central wall we will use a 2 × 4 bearing wall between the first and second floor with beams and columns

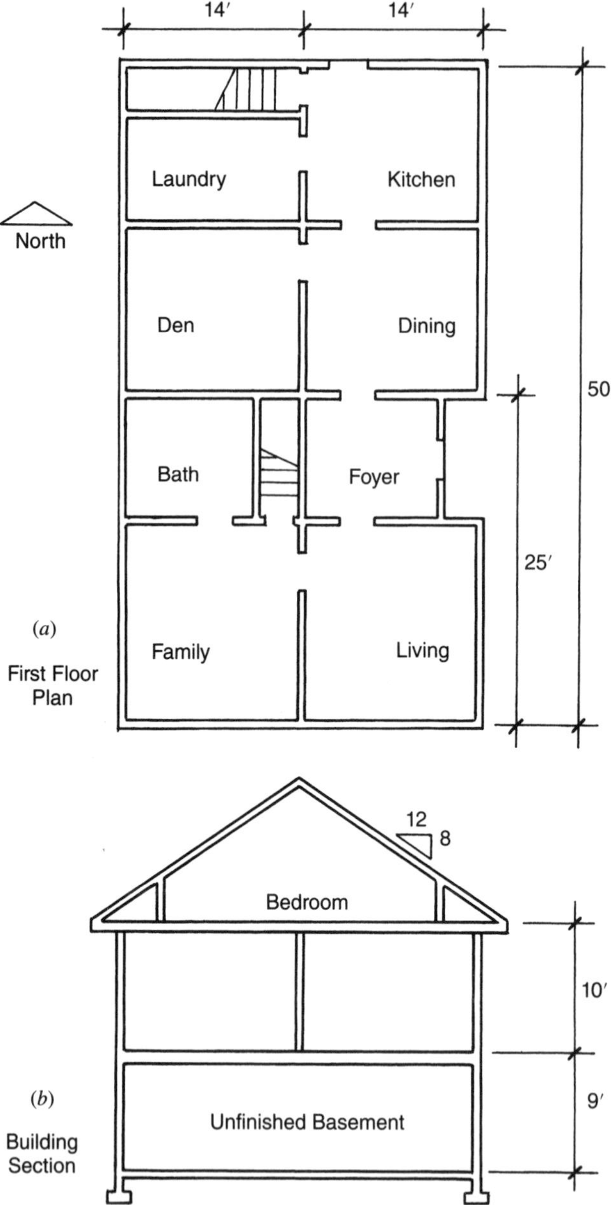

Figure 4.12 Reference for Example 1.

from the basement to the first floor. The beam under the bearing wall in the basement will be divided into three beams of length 16 ft 8 in. Table 4.2 tells us that the beam will be a nominal 10 in. in depth. We will make the columns holding the beam steel pipe columns.

The floor and roof joists have spans of 14 ft. Therefore, the floor joist will be made of 2 × 10's and the roof joists of 2 × 8's using Table 4.1. The roof joist should probably be increased to 2 × 10's to accommodate additional insulation.

Problem 4.5.A A wood structure is proposed for a speculative small office building. The floor plan for the building is illustrated in Figure 4.13*a* with partial elevation in Figure 4.13*b* and section in Figure 4.13*c*. The design intention is to have an 8-ft-wide corridor running east–west down the center of the building with offices built on each side to suit the tenants' needs. The corridor walls are not to be used as bearing walls to provide flexibility but should be where columns are placed. The floor will be made of reinforced concrete. Determine a preliminary design for the roof joists and beams, exterior walls, and columns with the roof joists running north to south.

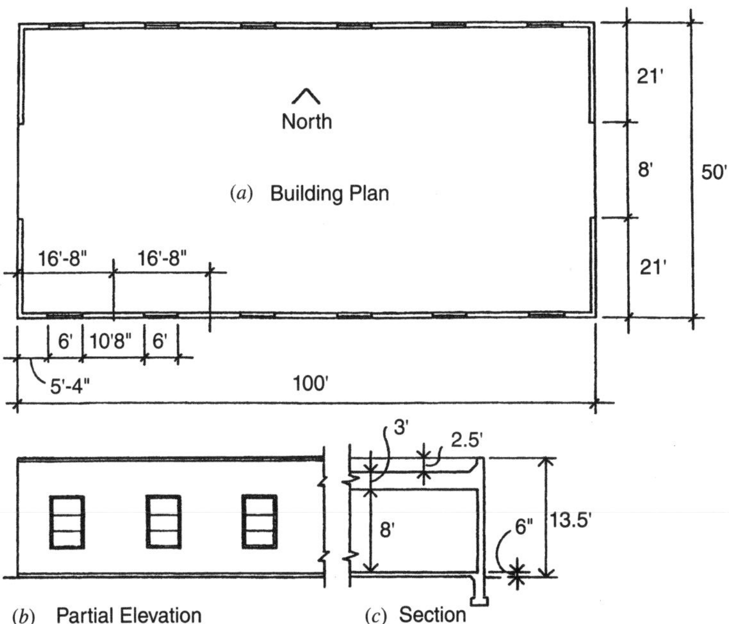

Figure 4.13 Reference for Problem 4.5.

Problem 4.5.B Using the same building as in Problem 4.5.A, determine a preliminary design for the roof joists and beams with the roof joists running east to west.

4.6 STEEL CONSTRUCTION

Steel construction is typically paired with either moment frame or braced frame systems. From a gravitational perspective, these systems work similarly (Figure 4.14). The floor or roof is held up by joists, which are held up by the primary beams. The columns hold the primary and secondary beams and transfer the loads to the foundation. The secondary beams carry a small amount of the floor or roof load in both systems, but their primary purpose in the moment frame system is to help transfer lateral loads to the foundation.

The approximations given in this chapter are not for the selection of sizes of the steel members but for determining critical dimensions for the members, that is, depth for beams and joists and width and depth for columns. Steel is available in various strengths, which may affect these dimensions. The approximations given in this section are for steel with a yield stress of 50 ksi. The sizing of the actual steel members should follow the procedures outlined in Part III.

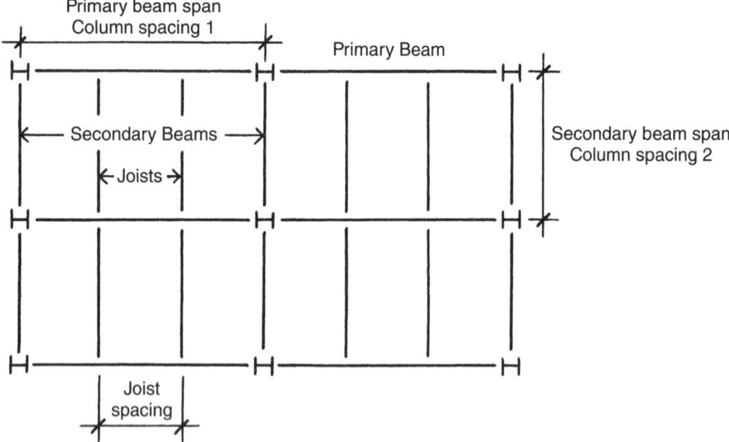

Figure 4.14 Plan layout for steel structure with braced frame or moment frame.

Column Spacing

Columns in steel construction can be spaced at large distances from each other, which is why steel is a common material for long-span systems. The greater the distance between columns, the larger the columns and the deeper the beams. Beam depth is often the controlling factor in the column spacing in multistory buildings where column spacing is limited to about 45 ft. Likewise, columns should not be spaced closer than about 18 ft. Columns in steel construction are rarely spaced equally in both directions. The column spacing in the direction of the primary beams is larger than that in the direction of the secondary beams, thus making the primary beams deeper than the secondary beams and joists. This extra space under the secondary beams and joists may be used for the placement of mechanical, electrical, and plumbing systems.

Column Sizing

I-shaped steel members (called wide-flange shapes) used for columns have a generally square cross section, meaning that the depth of the beam is approximately the same as the width of the flanges (Figure 4.15). Columns used in moment frame and braced frame systems generally have a cross-sectional dimension of 8, 10, 12, or 14 in. Columns with dimensions less than 8 in. are available and are often used in conjunction with wood construction systems.

The dimension of a column is based on how closely the columns are spaced, the number of stories in the building, the length of the columns between floors, and the magnitude of the loads being carried by the column. Table 4.3 outlines the determinates that help to select

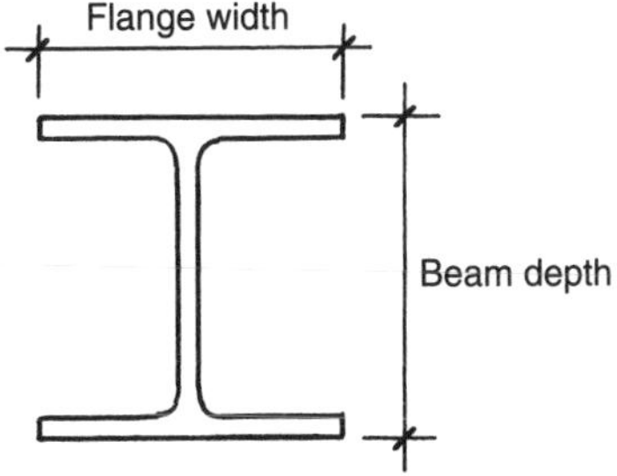

Figure 4.15 Cross section of typical wide-flange steel column.

TABLE 4.3 Determinates for Column Dimensions[a]

Modest-Size Structure	Moderate-Size Structure	Large-Size Structure
Column spacing ≤ 20 ft	20 ft < Column spacing ≤ 36 ft	36 ft < Column spacing
Building stories ≤ 2	2 < Building stories ≤ 10	10 < Building stories
Column lengths ≤ 20 ft	20 ft < Column lengths ≤ 30 ft	30 ft < Column lengths
Live load ≤ 40 psf	40 psf < Live load ≤ 100 psf	100 psf < Live load

[a]For steel or concrete structures of the categories of size as shown.

column dimensions: 8-in. columns are used for modest-sized projects and 14-in. columns are used for large projects. Moderate-sized projects fall between modest and large projects, with column spacing of about 30 ft, or have attributes that are contained in both the modest and the large project lists, that is, column spacing of 40 ft and column lengths of 15 ft. Moderate-sized columns are usually either 10 or 12 in. in dimension. If the column design has more in common with modest-sized projects than large projects, use the 10-in. dimension; if the opposite is true, use the 12-in. dimension. The majority of column applications are for moderate projects, which means the majority of steel columns are 10 or 12 in. in dimension.

Beam and Joist Layout and Depths

Beam layout from column to column was discussed in the previous section. The joists in a steel floor system are required when the distance between two primary beams is greater than 15 ft. In roof systems where larger floor deflection is acceptable, joists are required when the primary beams are spaced more than 20 ft apart from one another. Open web steel joists (light trusses) can be used in both floor and roof systems when deflection or bounciness is not considered to be a problem. When joists are required, they should be spaced at 15 ft or less apart when using steel wide-flange members and no more than 5 ft apart when using open-web steel joists.

The minimum depth of a beam or joist member is based on the length of the member and its allowable deflection, not on the strength of the member (Chapter 9). Table 4.4 outlines the minimum depths a beam needs based on these criteria. If minimizing beam depth is not an issue on the given project, planning for a beam one or two sizes greater than minimum will result in a more economical beam. Open-web steel joists may be more economical in some cases

TABLE 4.4 Minimum Depths for Steel Wide-Flange Beams and Joists[a]

Beam or Joist Span (ft)	Minimum Depth Required (in.)
Span ≤ 18	12
18 < Span ≤ 21	14
21 < Span ≤ 24	16
24 < Span ≤ 27	18
27 < Span ≤ 31	21
31 < Span ≤ 36	24
36 < Span ≤ 40	27
40 < Span ≤ 45	30

[a]Approximate minimum depths for the span ranges indicated.

but will require more depth than wide-flange members. Minimum open-web joist depths are approximately 1.5–2 times that of wide-flange beams.

Hollow Steel Sections

Hollow steel sections (HSSs), also known as tube steel, have gained popularity in structures where the steel is exposed. These sections are available in both round and rectangular (including square) cross sections. The yield strength of these sections is usually close to but slightly less than the 50 ksi assumed for the wide-flange members discussed in this section. When working on a preliminary structural plan using HSSs, the sizing approximations used for wide-flange members can still be used.

Example 2. Determine the preliminary structural plan for a 10-story (plus a roof) steel residential building which has a rectangular floor plan with a distance between outside columns of 60 ft × 120 ft (Figure 4.16). The floor-to-floor height for the building will be 14 ft. The lateral system will be concentric K braces in the center bays of the exterior frames.

Solution: The floors and roof of this building will be the diaphragms, which are used to transfer the lateral loads to the vertical cross bracing, which in turn will transfer the loads to the foundation. First, we need to check the dimensions of the diaphragms in order to determine if interior cross-bracing will be required. The floor plates are 60 ft ×

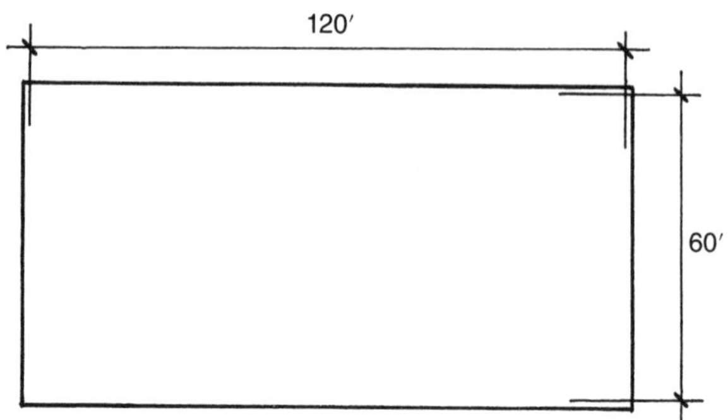

Figure 4.16 Floor plan for Example 2.

120 ft, which makes them have a length-to-width ratio of 1 : 2, which is less than the 1 : 3 maximum required for an optimal diaphragm. This means that no vertical interior braces will be needed.

The column–beam layout needs to be determined before sizing the members. The goal is usually to determine the minimum number of columns required that meet the parameters of the design and the suggested spacing outlined earlier. The 120-ft side of the building can be divided into three equal bays of 40 ft each. The 60-ft side of the building could be divided into two equal bays of 30 ft each. However, since the cross-bracing is to be in the center bay, an odd number of bays is required; therefore, it will be divided into three bays of 20 ft each (Figure 4.17). Joists will be required to carry the floor and roof

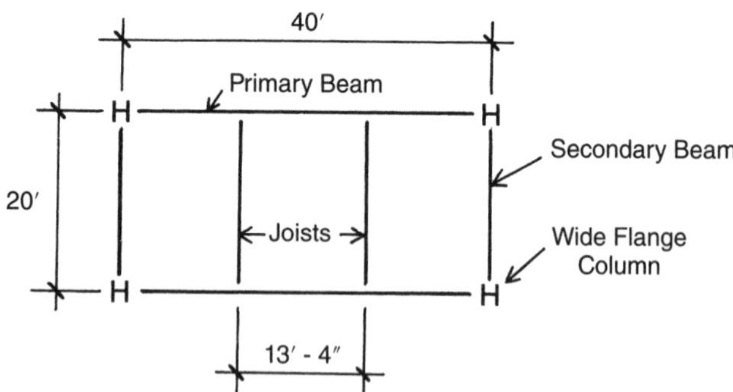

Figure 4.17 Partial structural framing plan for Example 2.

loads to the primary beams and will be spaced every 13 ft 4 in. (40 ft divided by 3) apart from each other.

This project is neither a modest nor a large project as outlined in Table 4.3. The column spacing of 40 ft makes it a large project, the 10 stories makes it a moderate project, and the column length of 14 ft makes it a small project. Finally, the live load for apartments and hotels given in Table 1 is that private rooms and corridors to them have a live load of 40 psf, which would be a modest project, and public rooms and corridors to them have a live load of 80 psf, which would be a moderate project. The column dimensions for this project will be either 10 or 12 in. If all of the public rooms are at the ground level of the building, 10-in. columns may be most prudent; otherwise one would go with 12-in. columns.

The primary beams are 40 ft long, which would require a minimum of 27 in. in depth. If the design will allow for the extra space, 30-in.-deep beams should be considered for economic reasons. The secondary beams and joists are 20 ft long; therefore, 14-in.-deep beams should suffice.

The finished structural elevations for this design are illustrated in Figure 4.18.

Problem 4.6.A Determine a preliminary structural layout and member sizing for a four-story (plus a roof) steel office building. The floor plan is rectangular with distances between the outside columns of 250 ft on one side and 350 ft on the other (similar to Figure 4.16). The lateral resistance for the building should be provided by concentric K braces (two pairs per elevation) in the corner bays on each elevation.

Problem 4.6.B Determine a preliminary structural layout and member sizing for a steel wide-flange moment frame version of the building in Example 3 in Section 4.7 (Figure 4.21*a*).

4.7 CONCRETE CONSTRUCTION

Concrete construction can be achieved with moment frames, shear walls and diaphragms, or frames with infill shear walls. Concrete can be precast in a controlled environment and then shipped to the site for erection or it can formed on-site. This section deals largely with on-site reinforced concrete, though some of the approximations are appropriate for precast concrete.

Concrete structural systems compete directly with steel in building structures. The decision whether to go with steel or concrete is

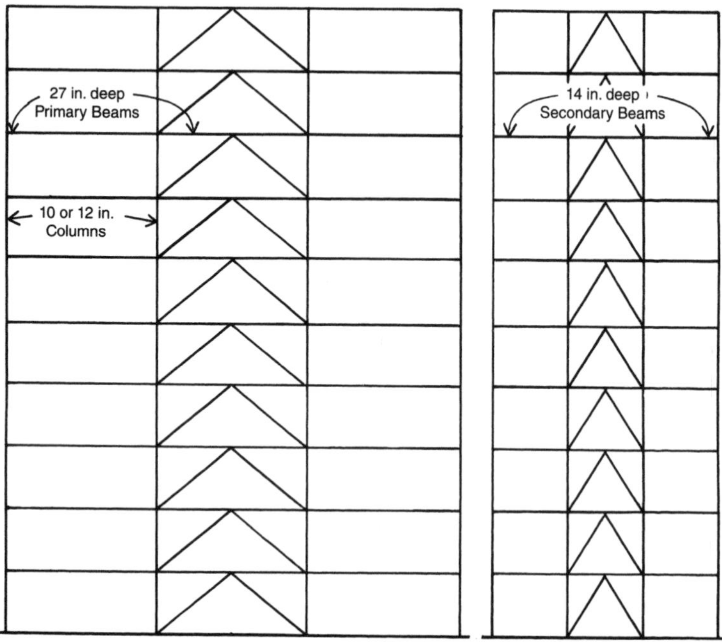

Figure 4.18 Elevations of the exterior framing for Example 2.

largely an economic decision made early in the design process based on local economic conditions and preliminary structural layouts similar to those presented in this chapter. Reinforced concrete dimensions are largely based on minimum clearances between steel reinforcing bars and between outer steel bars and formwork. The approximations given here attempt to determine not the size or placement of reinforcing steel but only the overall dimensions to aid in the preliminary design process. Selection of reinforcing steel and the final dimensions of the concrete are discussed in Part IV.

Column Layout

The frame systems for concrete are similar to those outlined in steel with floor slabs, joists, primary and secondary beams, and columns (Figure 4.19). Concrete frame systems can have more variation than steel, such as a frame (one set of columns and beams) being replaced by a bearing/shear wall, joists being replaced with a thicker slab, and moment frames being replaced with smaller frames with infill shear

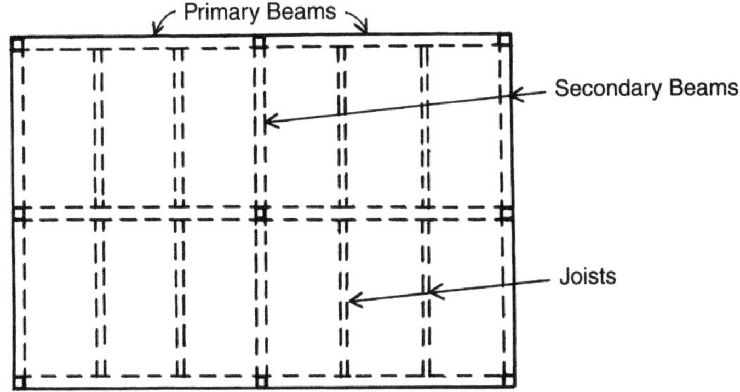

Figure 4.19 Example of slab and beam framing system.

walls to handle lateral loads. These variations make concrete systems somewhat harder to lay out since many of these variations are made based upon local economic and building practices, which one will not know until one has practiced in an area for a length of time.

Column spacing for concrete systems is similar to that of steel. The further the columns are spaced, the deeper the beams. Column spacing in most concrete frame systems is between 18 and 45 ft.

Column and Wall Dimensions

Concrete columns are usually round, square, or rectangular in cross section, with square and round being the most common. The dimensions discussed in this section will be the side of square columns or the diameter of round ones. Practically, reinforced concrete columns have a minimum dimension of 10 in. and a maximum dimension of 24 in. Table 4.3 applies to concrete columns as it did for steel columns. Modest column sizes are 10 or 12 in., moderate sizes are 14, 16, or 18 in., and large sizes are 20, 22, or 24 in. Concrete moment frames will use the larger sizes in each category, while columns that are not directly handling lateral forces will use the smaller sizes. Bearing/shear wall thicknesses will be the same dimensions as a column of similar project size and typically will use the minimum dimension.

Beam, Joist, and Floor Slab Dimensions

Beams, joists, and floor slabs are integrated systems with reinforcement that goes between them and are generally placed in a single

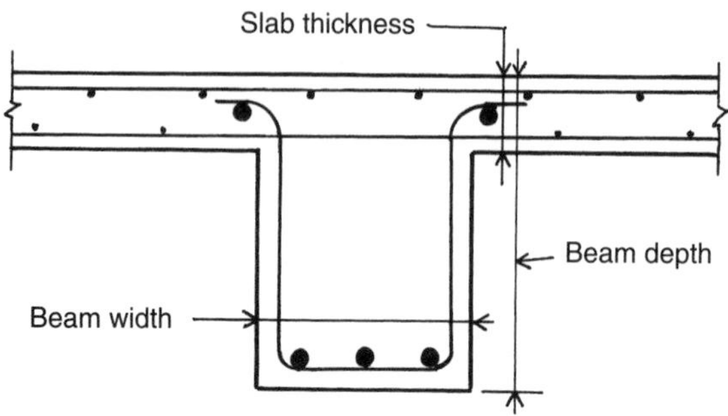

Figure 4.20 Cross section of a slab and beam system.

pour or lift of concrete. In steel, the slab generally sits on top of the beam, and there is a clear differentiation where one starts and the other stops. In concrete construction, there is no clear differentiation and generally the thickness of each system is measured from the top of the slab (Figure 4.20).

Beam width is typically the same as the width of the columns that hold them up. The width can be 2–4 in. greater than the width of the column in longer beam spans, which requires more steel reinforcing bars. Joist widths are usually equal to 2 in. less than the secondary beam widths. The depth of a concrete beam is usually 1.5–2.5 times its width. The more load and the longer the beam span, the more it will be likely the depth will be deeper. In general, joist depths will approximately be equal to 1.5 times their width and beam depth will be 2 times their width.

In concrete construction, it is sometimes cheaper to build a structure with joists supporting a thinner slab, and sometimes it is cheaper to omit the joists and use a thicker slab supported directly by the beams. This will vary with local building practices and the economics of the cost of material versus labor. If one is using joists, then they should be spaced less than 15 ft apart from each other with a slab thickness of 8 in., which is the minimum slab thickness for fire-rated construction. If joists are omitted, then the slab thickness can be approximated based on distance between primary beams (Table 4.5).

Steel Reinforcement

If the purpose of approximating the dimensions of a concrete structure is to make sure of enough clearances or to begin sizing members,

TABLE 4.5 Approximate Thicknesses for Concrete Slabs[a]

Spacing of Supports (ft)	Slab Thickness (in.)
10 < Spacing ≤ 15	6
15 < Spacing ≤ 20	8
20 < Spacing ≤ 25	10
25 < Spacing ≤ 30	12
30 < Spacing ≤ 35	14
35 < Spacing ≤ 40	16

[a]Thicknesses based on spacing of supports as indicated.

then there is no reason to approximate the amount of steel reinforcement that will be necessary. If the purpose is to obtain a preliminary cost estimate for the structure, then approximating the quantity of reinforcement steel may be necessary. The amount of steel is often given as a percentage of the concrete used in a given structural member. For most columns, the percentage of steel is between 1 and 3% of the concrete with an average of 2%. For beams and joists the percentage of steel reinforcement is between 2 and 5.5% with the average being 4%. Finally, the average amount of steel reinforcement in structural slabs is approximately 0.25%.

Example 3. Determine a preliminary concrete moment frame system for the office building in Figure 4.21*a*. The building has four stories plus a roof. The floor-to-floor height between the first and second stories is 24 ft with 15 ft for the upper floors.

Solution: The L-shaped floor plan will create a change in lateral stiffness, which will not affect the preliminary planning but will need to be considered later in the design process. The diaphragm for the system will be investigated as two rectangular diaphragms to determine whether either is problematic. The first will be the 100 × 210-ft section, which has a ratio of 1 : 2.1. The second will be 80 × 140 ft, which has a ratio of 1 : 1.75. Both of these ratios are less than the optimal 1 : 3; therefore, the diaphragm should work.

The column and beam layout will not produce equal structural bays due to the irregularities of the floor plan. The goal to produce a floor plan with the fewest number of columns is appropriate as long as the floor-to-floor heights are adequate to handle deeper beams, which seems to be the case in this example. The 100 × 130-ft segment can

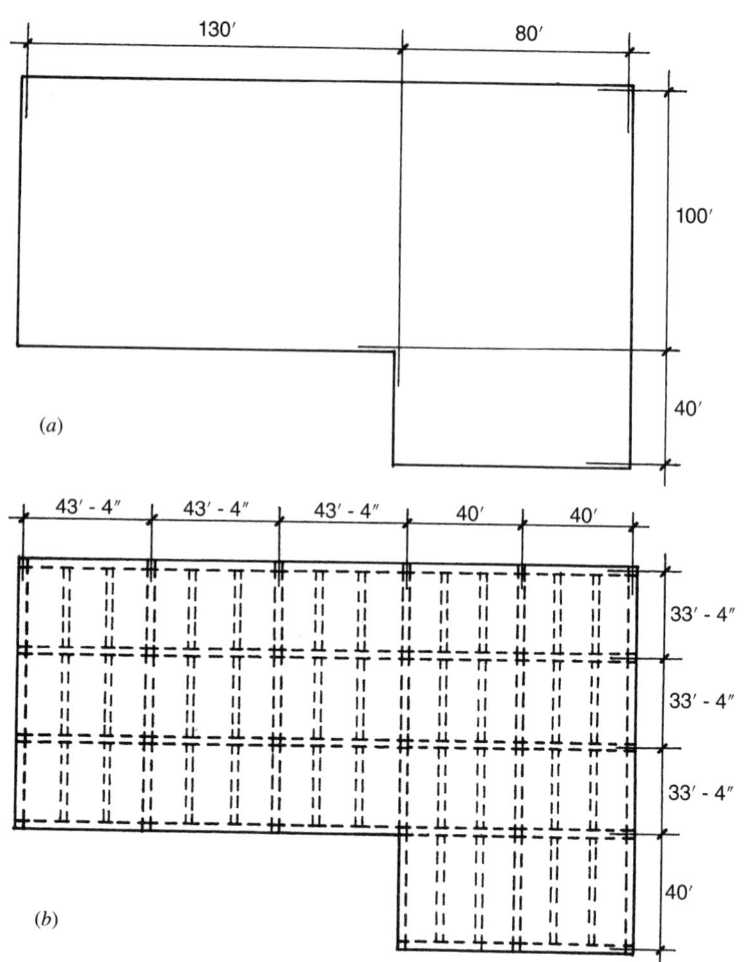

Figure 4.21 Example 3: (*a*) basic floor plan and (*b*) framing layout for the slab and beam system.

be divided into three bays of 33 ft 4 in. by three bays of 43 ft 4 in. (Figure 4.21*b*). The 100 × 80 ft will be divided into three bays by two bays of 33 ft 4 in. × 40 ft. The 40 × 80-ft section will be divided into one bay by two bays of 40 × 40 ft each. Finally, the floor will be supported on three concrete joists per primary beam.

The columns will be preliminarily sized based on the parameters listed in Table 4.3. Most of the column spacing is above 36 ft, which associates them with large projects. Four stories are associated with moderate-sized projects. The 15- and 24-ft column heights are associated with a moderate-sized project. Finally, according to Table 1, office building live loads are 50 psf for offices, 100 psf for first-floor lobbies and corridors, and 80 psf for corridors above the first floor, which are all associated with moderate-size projects. Since this is a moment frame system, the preliminary column size should be 18 in.

Since the project is using beams and joists, the floor slab thickness will be 8 in., the beams will be 18 × 36 in., and the joists will by 16 × 24 in.

Table 4.6 is used to determine the total amount of concrete and reinforcing steel used in this preliminary design and assumes that the first floor is not slab on grade and the roof member sizes are approximately the same as the floor members. Using these numbers we can work on a preliminary pricing based on 38,400 ft^3 of concrete and 270 tons of reinforcing steel. One can also determine the total structural dead load of 5760 kips or 50 psf of floor or roof, assuming reinforced concrete weighs 150 lb/ft^3. There is a fair amount of overlap in the numbers at the intersection of various members such as columns and beams, which should be investigated during further development of the project.

Problem 4.7.A Rework Example 3 assuming the use of a reinforced concrete frame with concrete masonry units (CMU) infill shear walls on the exterior of

TABLE 4.6 Estimates of Concrete and Steel Quantities for Example 3

Element	Quantity of Concrete	Average % of Steel	Quantity of Steel
Columns	$2.25\ ft^2 \times 108\ ft \times 27$ columns $= 6560\ ft^3$	2%	$131\ ft^3$ (32 tons)
Beams	$4.5\ ft^2 \times 920\ ft \times 5$ floors $= 20{,}700\ ft^3$	4%	$828\ ft^3$ (202 tons)
Joists	$2.37\ ft^2 \times 1160\ ft \times 5$ floors $= 3090\ ft^3$	4%	$124\ ft^3$ (30 tons)
Slabs	$0.67\ ft^3 \times 24{,}200\ ft^2 \times 5$ floors $= 8070\ ft^3$	0.25%	$20.2\ ft^3$ (5 tons)
Totals	$38{,}400\ ft^3$ (2765 tons)		$1100\ ft^3$ (267 tons)

the building. How much concrete and steel was saved? How much additional structural dead load was added by the CMU, assuming that the CMU is an 8-in. reinforced block that covers 60% of the building's exterior walls between the columns and the beams.

Problem 4.7.B Rework Example 2 in Section 4.6 as concrete moment frame construction.

II

WOOD CONSTRUCTION

Wood has long been the structural material of choice in the United States whenever conditions permit its use. For small buildings—where fire codes permit—it is extensively used. As with other building products, elements of wood used for building structures are produced in a highly developed industrialized production system and the quality of the materials and products is considerably controlled. In addition to the many building codes, there are several organizations in the United States that provide standards for the design of wood products. The work in this part is based primarily on one of these standards, the *National Design Specification for Wood Construction*, published by the American Forest and Paper Association (Ref. 3), herein referred to as the *NDS*.

5

WOOD SPANNING ELEMENTS

This chapter deals with applications of wood products for the development of spanning structures for building roofs and floors. Spanning systems used for roofs and floors commonly employ a variety of wood products. The solid wood material, cut directly from logs (here called *solid sawn*), is used as standard-sized structural lumber, such as the all-purpose 2 by 4 (here also called 2 × 4). Solid pieces can be mechanically connected and assembled to form various structures and can be glued together to form glued-laminated products.

A widely used product is the plywood panel, formed by gluing together three or more very thin plies of wood and used extensively for wall sheathing or for roof and floor decking. This reconstitution of the basic wood material essentially retains the grain character both structurally and for surface appearance.

A more extensive reconstitution involves the use of the basic wood fiber reduced to small pieces and adhered with a binding matrix to form fiber products such as paper, cardboard, and particleboard. Because of the major commercial use of paper products, the fiber industry is strongly established and application of wood fiber products for building construction advances steadily. Wood fiber paneling is replacing plywood and wood boards in many applications.

5.1 STRUCTURAL LUMBER

Structural lumber consists of solid-sawn, standard-sized elements produced for various construction applications. Individual pieces of lumber are marked for identification as to wood species (type of tree of origin), grade (quality), size, usage classification, and grading authority. On the basis of this identity, various structural properties are established for engineering design.

Aside from the natural properties of the species (tree), the most important factors that influence structural grading are density (unit weight), basic grain pattern (as sawn from the log), natural defects (knots, checks, splits, pitch pockets, etc.), and moisture content. Because the relative effects of natural defects vary with the size of sawn pieces and the usage application, structural lumber is classified with respect to its size and use. Incorporating these and other considerations, the four major classifications are given below. (Note that *nominal dimensions*, as explained in Appendix A, are used here.)

1. *Dimension Lumber*. Sections with thickness of 2–4 in. and width of 2 in. or more (includes most studs, rafters, joists, and planks).
2. *Beams and Stringers*. Rectangular sections 5 in. or more in thickness with width 2 in. or more greater than thickness, graded for strength in bending when loaded on the narrow face.
3. *Posts and Timbers*. Square or nearly square sections, 5×5 or larger, with width not more than 2 in. greater than thickness, graded primarily for use as compression elements with bending strength not especially important.
4. *Decking*. Lumber from 2 to 4 in. thick, tongued and grooved or splined on the narrow face, and graded for flat application (mostly as plank deck).

Design data and standards for structural lumber are provided in the publications of the wood industry, such as the *NDS*.

A broad grouping of trees identifies them as *softwoods* or *hardwoods*. Softwoods such as pine, fir, and redwood mostly come from trees that are coniferous or cone bearing, whereas hardwoods come mostly from trees that have broad leaves, as exemplified by oaks and maples. Two species of trees used extensively for structural lumber in the United States are Douglas fir and Southern pine, both of which are classified among softwoods.

Dimensions

As discussed in Appendix A, structural lumber is described in terms of a nominal size, which is slightly larger than the true dimensions of pieces. However, properties for structural computations, as given in Table A.8, are based on the true dimensions, which are also listed in the table.

For the sake of brevity, we have omitted metric units from the text and the tabular data.

5.2 REFERENCE DESIGN VALUES FOR ALLOWABLE STRESS DESIGN

There are many factors to be considered in determining the unit stresses to be used for the design of wood structures. Extensive testing has produced values known as *reference design values*. To obtain values for design work, the base values are modified for considerations such as loss of strength from defects, the size and position of knots, the size of standard dimension members, the degree of density of the wood, and the condition of seasoning or specific value of moisture content of the wood at the time of use. For specific uses of the structure, modifications may be made for considerations such as load duration, direction of stress with respect to the wood grain, and the type of structural element. Separate groups of design values are established for decks, closely spaced rafters and joists, large-dimension beams, and posts (columns).

Tables 5.1*a* and 5.1*b* give reference design values to be used for ordinary allowable stress design. The tables are adapted from the *NDS* (Ref. 3). Table 5.1*a* gives data for Douglas fir–larch and Table 5.1*b* gives it for Hem Fir. To obtain values from the table the following is determined:

1. *Species*. The *NDS* publication lists values for several different species, only two of which are included in Table 5.1.

TABLE 5.1a Reference Design Values for Visually Graded Lumber of Douglas Fir–Larch[a] (Values in psi)

Species and Commercial Grade	Member Size and Use Classification	Bending, F_b	Tension Parallel to Grain, F_t	Shear Parallel to Grain, F_v	Compression Perpendicular to Grain, $F_{c\perp}$	Compression Parallel to Grain, F_c	Modulus of Elasticity	
							E	E_{min}
Dimension Lumber 2–4 in. thick								
Select structural	2 in. and wider	1500	1000	180	625	1700	1,900,000	690,000
No. 1 and better		1200	800	180	625	1550	1,800,000	660,000
No. 1		1000	675	180	625	1500	1,700,000	620,000
No. 2		900	575	180	625	1350	1,600,000	580,000
No. 3		525	325	180	625	775	1,400,000	510,000
Stud		700	450	180	625	850	1,400,000	510,000
Timbers								
Dense select structural	Beams and stringers	1900	1100	170	730	1300	1,700,000	620,000
Select structural		1600	950	170	625	1100	1,600,000	580,000
Dense No. 1		1550	775	170	730	1100	1,700,000	620,000
No. 1		1350	675	170	625	925	1,600,000	580,000
No. 2		875	425	170	625	600	1,300,000	470,000
Dense select structural	Posts and timbers	1750	1150	170	730	1350	1,700,000	620,000
Select structural		1500	1000	170	625	1150	1,600,000	580,000
Dense No. 1		1400	950	170	730	1200	1.700,000	620,000
No. 1		1200	825	170	625	1000	1,600,000	580,000
No. 2		750	475	170	625	700	1,300,000	470,000
Decking								
Select Dex		1750	—	—	625	—	1,800,000	660,000
Commercial Dex		1450	—	—	625	—	1,700,000	620,000

Source: Data adapted from *National Design Specification*® (NDS®) *for Wood Construction*, 2005 edition (Ref. 3), with permission of the publisher, American Forest and Paper Association. The tables in the reference document have data for many other species and also have extensive footnotes.

[a]Values listed are for reference in determining design use values subject to various modifications. See Table 5.2 for size adjustment factors for dimension lumber (2–4 in. thick). For design purposes, values are subject to various modifications.

TABLE 5.1b Reference Design Values for Visually Graded Lumber of Hem Fir[a] (Values in psi)

Species and Commercial Grade	Member Size and Use Classification	Bending, F_b	Tension Parallel to Grain, F_t	Shear Parallel to Grain, F_v	Compression Perpendicular to Grain, $F_{c\perp}$	Compression Parallel to Grain, F_c	Modulus of Elasticity	
							E	E_{min}
Dimension Lumber 2–4 in. thick								
Select structural	2 in. and wider	1400	925	150	405	1500	1,600,000	580,000
No. 1 and better		1100	725	150	405	1350	1,500,000	550,000
No. 1		975	625	150	405	1350	1,500,000	550,000
No. 2		850	525	150	405	1300	1,300,000	470,000
No. 3		500	400	150	405	725	1,200,000	440,000
Stud		675	400	150	405	800	1,200,000	440,000
Timbers								
Select structural	Beams and stringers	1300	750	140	405	925	1,300,000	470,000
No. 1		1050	525	140	405	750	1,300,000	470,000
No. 2		675	350	140	405	500	1,100,000	400,000
Select structural	Posts and timbers	1200	800	140	405	975	1,300,000	470,000
No. 1		975	650	140	405	850	1,300,000	470,000
No. 2		575	375	140	405	575	1,100,000	400,000
Decking								
Select Dex		1400	—	—	405	—	1,500,000	550,000
Commercial Dex		1150	—	—	405	—	1,400,000	510,000

Source: Data adapted from *National Design Specification®* (NDS®) *for Wood Construction*, 2005 edition (Ref. 3), with permission of the publisher, American Forest and Paper Association. The tables in the reference document have data for many other species and also have extensive footnotes.

[a]Values listed are for reference in determining design use values subject to various modifications. See Table 5.2 for size adjustment factors for dimension lumber (2–4 in. thick). For design purposes, values are subject to various modifications.

2. *Moisture Condition at Time of Use*. The moisture condition corresponding to the table values is given with the species designation in the table. Adjustments for other conditions are described in the table footnotes or in various specifications in the *NDS*.
3. *Grade*. This is indicated in the first column of the table and is based on visual grading standards.
4. *Size and Use*. The second column of the table identifies size ranges or usages of the lumber.
5. *Structural Function*. Individual columns in the table yield values for various stress conditions. The last two columns yield the material modulus of elasticity.

In the reference document there are extensive footnotes for this table. Data from Tables 5.1*a* and *b* are used in various example computations in this book and some issues treated in the document footnotes are explained. In many situations there are modifications (or adjustments, as they are called in the *NDS*) to the design values, as will be explained later. Referred to in a footnote to Table 5.1, Table 5.2 yields adjustment factors for dimension lumber and decking based on the dimensions of the piece.

Bearing Stress

There are various situations in which a wood member may develop a contact bearing stress, essentially, a surface compression stress. Some examples are the following:

1. At the base of a wood column supported in direct bearing. This is a case of bearing stress that is in a direction parallel to the grain.
2. At the end of a beam that is supported by bearing on a support. This is a case of bearing stress that is perpendicular to the grain.
3. Within a bolted connection at the contact surface between the bolt and the wood at the edge of the bolt hole.
4. In a timber truss where a compression force is developed by direct bearing between the two members. This is frequently a situation involving bearing stress that is at some angle to the grain other than parallel or perpendicular. Common in the past, this form of joint is seldom used today.

TABLE 5.2 Size Adjustment Factors (*CF*) for Dimension Lumber, Decking, and Timber

Dimension Lumber					
		Thickness (breadth), F_b			
Grades	Width (depth)	2 and 3 in.	4 in.	F_t	F_c
Select structural, No. 1 and better, No. 1, No. 2, No. 3	2, 3, 4 in.	1.5	1.5	1.5	1.15
	5 in.	1.4	1.4	1.4	1.1
	6 in.	1.3	1.3	1.3	1.1
	8 in.	1.2	1.3	1.2	1.05
	10 in.	1.1	1.2	1.1	1.0
	12 in.	1.0	1.1	1.0	1.0
	14 in. and wider	0.9	1.0	0.9	0.9
Stud	2, 3, and 4 in.	1.1	1.1	1.1	1.05
	5 and 6 in.	1.0	1.0	1.0	1.0
	8 in. and wider	Use No. 3 grade values and size factors			

Decking	
2 in.	3 in.
1.10	1.04

Beams and Stringers—Loads Applied to Wide Face			
Grade	F_b	E and E_{min}	Other Properties
Select structural	0.86	1.0	1.0
No. 1	0.74	0.9	1.0
No. 2	1.0	1.0	1.0
Timbers, $d > 12$ in. CF = (12/*d*)1/9 for F_b only			

Source: Data adapted from *National Design Specification*® (NDS®) *for Wood Construction*, 2005 edition (Ref. 3), with permission of the publisher, American Forest and Paper Association.

For connections, the bearing condition is usually incorporated into the general assessment of the unit value of connecting devices. The situation of stress at an angle to the grain requires the determination of a compromise value somewhere between the allowable values for the two limiting stress conditions for stresses parallel and perpendicular to the wood grain.

Adjustment of Design Values

The values given in Tables 5.1*a* and 5.1*b* are basic references for establishing the allowable values to be used for design. The table

TABLE 5.3 Applicability of Adjustment Factors for Sawn Lumber, ASD

	F_b	F_t	F_v	$F_{c\perp}$	F_c	E	E_{min}
ASD Only							
Load duration	C_D	C_D	C_D	—	C_D	—	—
ASD and LRFD							
Wet service	C_M	C_M	C_M	C_M	C_M	C_M	C_M
Temperature	C_t	C_t	C_t	C_t	C_t	C_t	C_t
Beam stability	C_L	—	—	—	—	—	—
Size	C_F	C_F	—	—	C_F	—	—
Flat use	C_{fu}	—	—	—	—	—	—
Incising	C_i	C_i	C_i	C_i	C_i	C_i	C_i
Repetitive member	C_r	—	—	—	—	—	—
Column stability	—	—	—	—	C_P	—	—
Buckling stiffness	—	—	—	—	—	—	C_T
Bearing area	—	—	—	C_b	—	—	—
LRFD Only							
Format conversion	K_F	K_F	K_F	K_F	K_F	—	K_F
Resistance	φ_b	φ_t	φ_v	φ_c	φ_c	—	φ_s
Time effect	λ	λ	λ	λ	λ	—	—

values are based on some defined norms, and in many cases the design values will be adjusted for actual use in structural computations. In some cases the form of the modification is a simple increase or decrease achieved by a percentage factor. Table 5.3 lists the various types of adjustment and indicates their applicability to various reference design values. The types of adjustment factors are described in the following discussions.

Load Duration Factor, C_D. The values in Tables 5.1*a* and 5.1*b* are based on so-called *normal* duration loading, which is actually somewhat meaningless. Increases are permitted for very short duration loading, such as wind and earthquakes. A decrease is required when the critical design loading is a long time in duration (such as a major dead load). Table 5.4 gives a summary of the *NDS* requirements for modifications for load duration.

Wet Service Factor, C_M. The *NDS* document on which Tables 5.1*a* and 5.1*b* are based defines a specific assumed moisture content on which the table values are based. Increases may be allowed for wood that is specially cured to a lower moisture content.

TABLE 5.4 Adjustment Factors for Design Values for Structural Lumber Due to Load Duration, C_D[a]

Load Duration	Multiply Design Values by:	Typical Design Loads
Permanent	0.9	Dead load
Ten years	1.0	Occupancy live load
Two months	1.15	Snow load
Seven days	1.25	Construction load
Ten minutes	1.6	Wind or earthquake load
Impact[b]	2.0	Impact load

Source: Data adapted from *National Design Specification*® (NDS®) *for Wood Construction, 2005 edition* (Ref. 3), with permission of the publisher, American Forest and Paper Association.

[a]These factors shall not apply to reference modulus of elasticity E, to reference modulus of elasticity for beam and column stability E_{min}, or to compression perpendicular to the grain reference design values $F_{c\zeta}$ based on a deformation limit.

[b]Load duration factors greater than 1.6 shall not apply to structural member pressure treated with water-borne preservatives or fire-retardant chemicals. The impact load duration factor shall not apply to connections.

If exposed to weather or other high-moisture conditions, a reduction may be required.

Temperature Factor, C_t. Where prolonged exposure to temperatures over 150°F exists, design values must be reduced. The adjustment factor varies for different reference values and includes consideration for moisture condition and exposure of the wood.

Beam Stability Factor, C_L. Design flexural stress must be adjusted for conditions of potential buckling. The general situation of buckling and the remedies for its prevention are discussed in Section 3.11.

Size Factor, C_F. For dimension lumber, adjustments for size are made for design stresses of bending, shear, and tension as described in Table 5.2. For beams 5 in. or thicker, with depth exceeding 12 in., adjustment of bending stress is made as described in Section 5.4. Other adjustments may be required for columns and for beams loaded for bending on their wide face.

Flat Use Factor, C_{fu}. When sawn lumber 2–4 in. thick is loaded on the wide face (as a plank), adjustments are required as described in the NDS references for Table 5.1.

Incising Factor, C_i. Incising refers to small indentation-form cuts made on the surface of lumber that is treated by impregnation of chemicals for the enhancement of resistance to fire or rot. A reduction of all reference values is required for this condition.

Repetitive Member Factor, C_r. When wood beams of dimension lumber (mostly joists and rafters) are closely spaced and share a load, they may be eligible for an increase of 15% in reference design values; this condition is described as *repetitive member use*. To qualify, the members must be not less than three in number, must support a continuous deck, must be not over 24 in. on center, and must be joined by construction that makes them share deflections (usually bridging or blocking). This increase is also permitted for built-up beams formed by direct attachment of multiple-dimension lumber elements.

The following example illustrates the application of the beam design procedure for the case of a roof rafter.

Example 1. Rafters of Douglas fir–larch, No. 2 grade, are to be used at 16 in. spacing for a span of 20 ft. Solid wood blocking is provided for nailing of the plywood deck panels. Live load without snow is 20 psf and the total dead load, including the rafters, is 15 psf. Find the minimum size for the rafters, based only on bending stress.

Solution: At this spacing the rafters qualify for the increased bending stress described as repetitive member use. For the No. 2 grade rafters, the reference value from Table 5.1*a* for F_b is 900 psi. The loading condition as described (live load without snow) qualifies the situation with regard to load duration for an adjustment factor of $C_D = 1.25$ (see Table 5.4). This live load is usually considered to provide for temporary conditions during roof construction or maintenance and is of short duration. The allowable bending stress for design is thus modified as

$$F'_b = C_r C_D F_b = (1.15)(1.25)F_b = (1.15)(1.25)(900) = 1294 \text{ psi}$$

For the rafters at 16-in. spacing, the maximum bending moment is

$$M = \frac{wL^2}{8} = \frac{\left(\frac{16}{12}\right)(20 + 15)(20)^2}{8} = 2333 \text{ ft-lb}$$

and the required section modulus is

$$S = \frac{M}{F'_b} = \frac{2333 \times 12}{1294} = 21.64 \text{ in.}^3$$

From Table A.8, the smallest section with this property is a 2×12, with an S of 31.64 in.^{3} Note that the allowable stress is not changed by Table 5.2 as the table factor is 1.0. Lateral bracing for rafters is discussed in Section 5.4.

Column Stability Factor, C_P. This adjustment is performed in the typical processes of investigation and design of wood columns, which is discussed in Chapter 6. Most often, an adjustment consisting of a reduction of permissible compression stress parallel to the grain is required for relatively slender columns.

Buckling Stiffness Factor, C_T. This adjustment is made only for the modified modulus of elasticity, E_{min}, in certain situations involving wood members subjected to combined compression and bending. Its principal application is in the design of the top chords of wood trusses.

Bearing Area Factor, C_b. This factor is provided for the special case of bearing perpendicular to the grain when the length of bearing is very small. This applies primarily to situations where bearing is transferred from a wood member to a steel plate or washer. The *NDS* provides a formula for determination of an adjusted design stress for these situations.

Adjustments for the LRFD Method. The group of adjustments given at the bottom of Table 5.3 is designated as being for use with the LRFD method only. Use of these adjustments is described in Section 5.3.

Modulus of Elasticity

As discussed in Section 3.6, the modulus of elasticity is a measure of the relative stiffness of a material. For wood, two reference values are used for the modulus of elasticity. The basic reference value is designated E and is the value used for ordinary deformations—primarily the deflection of beams. The other value is designated E_{min} and it is used for stability computations involving the buckling of beams and columns. Values for the stability modulus of elasticity are given in the last column in Table 5.1*a* and 5.1*b* Applications for determination of buckling effects in columns are presented in Chapter 6.

5.3 DESIGN CONTROLS FOR LOAD AND RESISTANCE FACTOR DESIGN

For the allowable stress design (ASD) method, the concentration—as its name implies—is on maximum developed stress levels as produced by the service loads. The maximum permissible stresses are specified and adjusted for various circumstances. These limiting stress levels are then used to produce a limiting force action (shear, bending, bearing, etc.), and the result is compared to the maximum action produced by the loading.

For the LRFD method, the same relationships are used, except that the loads are quantified as factored loads (ultimate loads) and the limiting force action is a defined failure limit. The failure limit is defined as the *adjusted resistance* and is designated with the superscript prime, such as M' for adjusted moment resistance. For design purposes, the relationship between the resistance and the load effect is defined as

$$\lambda \varphi_b M' \geq M_u$$

where λ = time effect factor, see Table 5.7
φ_b = resistance factor, 0.85 for bending, see Table 5.6
M' = adjusted resisting moment
M_u = maximum moment, due to the factored loading

One approach—as defined by the *NDS* (Ref. 3)—is to define the resisting moment (or shear, bearing, etc.) as one produced in the same manner as that derived by the ASD method, except that an adjusted level of stress is used. The adjustment factors (K_F) are those given in Table 5.5, referring to the defined stress values listed in Table 5.1. It is customary to use stress units of pounds with the ASD method but to use units of kips with the LRFD method, thus the inclusion of the 1000 factor in the adjustment.

Resistance factors φ are as given in Table 5.6. They vary, depending on the type of behavior being considered.

TABLE 5.5 Adjustment Factors for LRFD Using ASD Reference Design Values

Property for	Conversion Factor[a] (ASD to LRFD) KF
Bending, F_b	$2.16/1000\varphi_b$
Tension, F_t	$2.16/1000\varphi_t$
Shear, F_v	$2.16/1000\varphi_v$
Compression parallel to grain, F_c	$2.16/1000\varphi_c$
Compression perpendicular to grain, $F_{c\perp}$	$1.875/1000\varphi_c$
Connections	$2.16/1000\varphi_z$
Modulus of elasticity for stability, $E_{\min}$	$1.5/1000\varphi_s$

Source: Adapted from data in the *National Design Specification*® (NDS®) *for Wood Construction* (Ref. 3), with permission of the publisher, American Forest and Paper Association.
[a]Produces unit values in kips when ASD values are in pounds.

TABLE 5.6 Resistance Factors for Wood Structures, LRFD

Symbol	Property	Value
φ_b	Flexure (bending)	0.85
φ_c	Compression, bearing	0.90
φ_t	Tension	0.80
φ_v	Shear	0.75
φ_s	Stability, $E_{\min}$	0.85
φ_z	Connections	0.65

Source: Adapted from data in the *National Design Specification*® (NDS®) *for Wood Construction* (Ref. 3), with permission of the publisher, American Forest and Paper Association.

The time effect factor λ is given in Table 5.7. It varies for the different load combinations used to find the required ultimate resistance. Care must be taken when using the adjustment factors to assure that the ASD reference design value is not modified by the factor for load duration (C_D). For the LRFD method this adjustment is made by use of the λ factor.

Use of this process is discussed in Section 5.8. Uses of the LRFD procedures for other applications are discussed in later chapters of this book.

TABLE 5.7 Load Combinations and Time Effect Factors, LRFD

Load Combination	Time Effect Factor, λ
1.4(Dead load)	0.6
1.2(Dead load) + 1.6(Live load):	
When live load is from storage	0.7
When live load is from occupancy	0.8
When live load is from impact	1.25
1.2(Dead load) + 1.6(Wind load) + Live load + 0.5(Roof load)	1.0
1.2(Dead load) + 1.6(Roof load) + 0.8(Wind load)	0.8
1.2(Dead load) + Earthquake load + Live load	1.0

Source: Adapted from data in the *National Design Specification®* (NDS®) *for Wood Construction* (Ref. 3), with permission of the publisher, American Forest and Paper Association.
Note: The reference document contains several additional combinations.

5.4 DESIGN FOR BENDING

The design of a wood beam for strength in bending is accomplished by use of the flexure formula (Section 3.6). The form of this equation used in design is

$$S = \frac{M}{F_b}$$

where M = maximum bending moment
F_b = allowable bending stress
S = required beam section modulus

Beams must be considered for shear, deflection, end bearing, and lateral buckling, as well as for bending stress. However, a common procedure is to first find the beam size required for bending and then to investigate for other conditions. Such a procedure is as follows:

1. Determine the maximum bending moment.
2. Select the wood species and grade of lumber to be used.
3. From Table 5.1 determine the basic allowable bending stress.
4. Consider appropriate modifications for the design stress value to be used.
5. Using the allowable bending stress in the flexure formula, find the required section modulus.
6. Select a beam size from Table A.8.
7. Investigate for applicable concerns, other than bending.

Example 2. A simple beam has a span of 16 ft and supports a total uniformly distributed load, including its own weight, of 6500 lb. Using Hem Fir, structural grade, determine the size of the beam with the least cross-sectional area on the basis of limiting bending stress.

Solution: The maximum bending moment for this condition is

$$M = \frac{WL}{8} = \frac{6500 \times 16}{8} = 13{,}000 \text{ ft-lb}$$

The next step is to use the flexure formula with the allowable stress to determine the required section modulus. A problem with this is that there are two different size/use groups in Table 5.1*b*, yielding two different values for the allowable bending stress. Assuming single-member use, the part listed under "Dimension Lumber" yields a stress of 1400 psi for the chosen grade, while the part under "Timbers, Beams and Stringers" yields a stress of 1300 psi. Using the latter category, the required value for the section modulus is

$$S = \frac{M}{F_b} = \frac{13{,}000 \times 12}{1300} = 120 \text{ in.}^3$$

while the value for F_b = 1400 psi may be determined by proportion as

$$S = \frac{1300}{1400} \times 120 = 111 \text{ in.}^3$$

From Table A.8, the smallest members in these two size categories are 4 × 16 (S = 135.661 in.^{3}, A = 53.375 in.^{2}) and 6 × 12 (S = 121.229 in.^{3}, A = 63.25 in.^{2}). For the 4 × 16 the allowable stress is not changed by Table 5.2 as the factor from the table is 1.0. Thus, the 4 × 16 is the choice for the least cross-sectional area.

Size Factors for Beams

Beams greater than 12 in. in depth with thickness of 5 in. or more have reduced values for the maximum allowable bending stress. This reduction is achieved with a reduction factor determined as

$$C_F = \left(\frac{12}{d}\right)^{1/9}$$

Values for this factor for standard lumber sizes are given in Table 5.8. For the preceding example, neither section qualifies for size reduction modification.

TABLE 5.8 Size Factors for Solid-Sawn Beams, C_F

Actual Beam Depth (in.)	C_F
13.5	0.987
15.5	0.972
17.5	0.959
19.5	0.947
21.5	0.937
23.5	0.928

Lateral Bracing

Design specifications provide for the adjustment of bending capacity or allowable bending stress when a member is vulnerable to a compression buckling failure. To reduce this effect, thin beams (mostly joists and rafters) are often provided with bracing that is adequate to prevent both lateral (sideways) buckling and torsional (rollover) buckling. The *NDS* requirements for bracing are given in Table 5.9. If bracing is not provided, a reduced bending capacity must be determined from rules given in the specifications.

TABLE 5.9 Lateral Support Requirements for Rectangular Sawn-Wood Beams

Ratio of Depth to Breadth $(d/b)^a$	Required Conditions to Avoid Reduction of Bending Stress
$d/b \leq 2$	No lateral support required.
$2 < d/b \leq 4$	Ends held in position to prevent rotation or lateral displacement.
$4 < d/b \leq 5$	Compression edge held in position for entire span and ends held in position to prevent rotation or lateral displacement.
$5 < d/b \leq 6$	Compression edge held in position for entire span, ends held in position to prevent rotation or lateral displacement, and bridging or blocking at intervals not exceeding 8 xce
$6 < d/b \leq 7$	Both edges held in position for entire span and ends held in position to prevent rotation or lateral displacement.

Source: Adapted from data in *National Design Specification*® (NDS®) *for Wood Construction* (Ref. 3), with permission of the publisher, American Forest and Paper Association.

[a]Ratio of nominal dimensions for standard sections.

Common forms of bracing consist of bridging and blocking. Bridging consists of crisscrossed wood or metal members in rows. Blocking consists of solid, short pieces of lumber the same size as the framing; these are fit tightly between the members in rows.

Problem 5.4.A No. 1 grade of Douglas fir–larch is to be used for a series of floor beams 6 ft on center, spanning 14 ft. If the total uniformly distributed load on each beam, including the beam weight, is 3200 lb, select the section with the least cross-sectional area based on bending stress.

Problem 5.4.B A simple beam of Hem Fir, select structural grade, has a span of 18 ft with two concentrated loads of 4 kips each placed at the third points of the span. Neglecting its own weight, determine the size of the beam with the least cross-sectional area based on bending stress.

Problem 5.4.C Rafters are to be used on 24-in. centers for a roof span of 16 ft. Live load is 20 psf (without snow) and the dead load is 15 psf, including the weight of the rafters. Find the rafter size required for Douglas fir–larch of (a) No. 1 grade and (b) No. 2 grade, based on bending stress.

5.5 BEAM SHEAR

As discussed in Section 3.7, the maximum beam shear stress for the rectangular sections ordinarily used for wood beams is expressed as

$$f_v = \frac{1.5V}{A}$$

where f_v = maximum unit horizontal shear stress, in psi
V = total vertical shear force at the section, in lb
A = cross-sectional area of the beam, in in.^{2}

Wood is relatively weak in shear resistance, with the typical failure producing a horizontal splitting of the beam ends. This is most frequently only a problem with heavily loaded beams of short span, for which bending moment may be low but the shear force is high. Because the failure is one of horizontal splitting, it is common to describe this stress as horizontal shear in wood design, which is how the allowable shear stress is labeled in Tables 5.1*a* and 5.1*b*.

Example 3. A 6 × 10 beam of Douglas fir–larch, No. 2 grade, has a total horizontally distributed load of 6000 lb. Investigate for shear stress.

Solution: For this loading condition the maximum shear at the beam end is one-half of the total load, or 3000 lb. Using the true dimensions of the section from Table 5.1*b*, the maximum stress is

$$f_v = \frac{1.5V}{A} = \frac{1.5 \times 3000}{52.3} = 86.1 \text{ psi}$$

Referring to Table 5.1*a*, under the classification "Beams and Stringers," the allowable stress is 170 psi. The beam is therefore adequate for this loading condition.

For uniformly loaded beams that are supported by end bearing, the code permits a reduction in the design shear force to that which occurs at a distance from the support equal to the depth of the beam.

Problem 5.5.A A 10 × 10 beam of Douglas fir–larch select structural grade supports a single concentrated load of 10 kips at the center of the span. Investigate the beam for shear.

Problem 5.5.B A 10 × 14 beam of Hem Fir select structural grade is loaded symmetrically with three concentrated loads of 4300 lb, each placed at the quarter points of the span. Is the beam safe for shear?

Problem 5.5.C A 10 × 12 beam of Douglas fir–larch No. 2 dense grade is 8 ft long and has a concentrated load of 8 kips located 3 ft from one end. Investigate the beam for shear.

Problem 5.5.D What should be the nominal cross-sectional dimensions for the beam of least weight that supports a total uniformly distributed load of 12 kips on a simple span and consists of Hem Fir No. 1 grade? Consider only the limiting shear stress.

5.6 BEARING

Bearing occurs at beam ends when a beam sits on a support or when a concentrated load is placed on top of a beam within the span. The stress developed at the bearing contact area is compression that is perpendicular to the grain, for which an allowable value ($F_{c\perp}$) is given in Tables 5.1*a* and 5.1*b*.

Although the design values given in the tables may be safely used, when the bearing length is quite short, the maximum permitted level of stress may produce some indentation in the edge of the wood member. If the appearance of such a condition is objectionable, a reduced stress is recommended. Excessive deformation may also produce some significant vertical movement, which may be a problem for the construction.

Example 4. An 8 × 14 beam of Hem Fir, No. 1 grade, has an end bearing length of 6 in. If the end reaction is 7400 lb, is the beam safe for bearing?

Solution: The developed bearing stress is equal to the end reaction divided by the product of the beam width and the length of bearing. Thus

$$f_c = \frac{\text{bearing force}}{\text{contact area}} = \frac{7400}{7.5 \times 6} = 164 \text{ psi}$$

This is compared to the allowable stress of 405 psi from Table 5.1*b*, which shows the beam to be quite safe.

Example 5. A 2 × 10 rafter cantilevers over and is supported by the 2 × 4 top plate of a stud wall. The load from the rafter is 800 lb. If both the rafter and the plate are Douglas fir–larch No. 2 grade, is the situation adequate for bearing?

Solution: The bearing stress is determined as

$$f = \frac{800}{1.5 \times 3.5} = 152 \text{ psi}$$

This is considerably less than the allowable stress of 625 psi from Table 5.1*a*, so the bearing is safe.

Example 6. A two-span 3 × 12 beam of Hem Fir, No. 1 grade, bears on a 3 × 14 beam at its center support. If the reaction force is 4200 lb, is this safe for bearing?

Solution: Assuming the bearing to be at right angles, the stress is

$$f = \frac{4200}{2.5 \times 2.5} = 672 \text{ psi}$$

This is in excess of the allowable stress of 405 psi; therefore it is not safe for bearing.

Problem 5.6.A A 6 × 12 beam of Douglas fir–larch, No. 1 grade, has 3 in. of end bearing to develop a reaction force of 5000 lb. Is the situation adequate for bearing?

Problem 5.6.B A 3 × 16 rafter cantilevers over a 3 × 16 support beam. If both members are of Hem Fir, No. 1 grade, is the situation adequate for bearing? The rafter load on the support beam is 3000 lb.

5.7 DEFLECTION

Deflections in wood structures tend to be most critical for rafters and joists, where span-to-depth ratios are often pushed to the limit. However, long-term high levels of bending stress can also produce sag, which may be visually objectionable or cause problems with the construction. In general, it is wise to be conservative with deflections of wood structures. Push the limits and you will surely get sagging floors and roofs and possibly very bouncy floors. This may in some cases make a strong argument for use of glued-laminated beams or even steel beams.

For the common uniformly loaded beam, the deflection takes the form of the equation

$$\Delta = \frac{5WL^3}{384EI}$$

Substitutions of relations between W, M, and flexural stress in this equation can result in the form

$$\Delta = \frac{5L^2 f_b}{24Ed}$$

Using average values of 1500 psi for f_b and 1500 ksi for E, the expression reduces to

$$\Delta = \frac{0.03L^2}{d}$$

where Δ = deflection, in in.
L = span, in ft
d = beam depth, in in.

Figure 5.1 is a plot of this expression with curves for nominal dimensions of depth for standard lumber. For reference the lines on the graph corresponding to ratios of deflection of L/180, L/240, and L/360 are shown. These are commonly used design limitations for total load and live-load deflections, respectively. Also shown for reference is the limiting span-to-depth ratio of 25 to 1, which is commonly considered to be a practical span limit for general

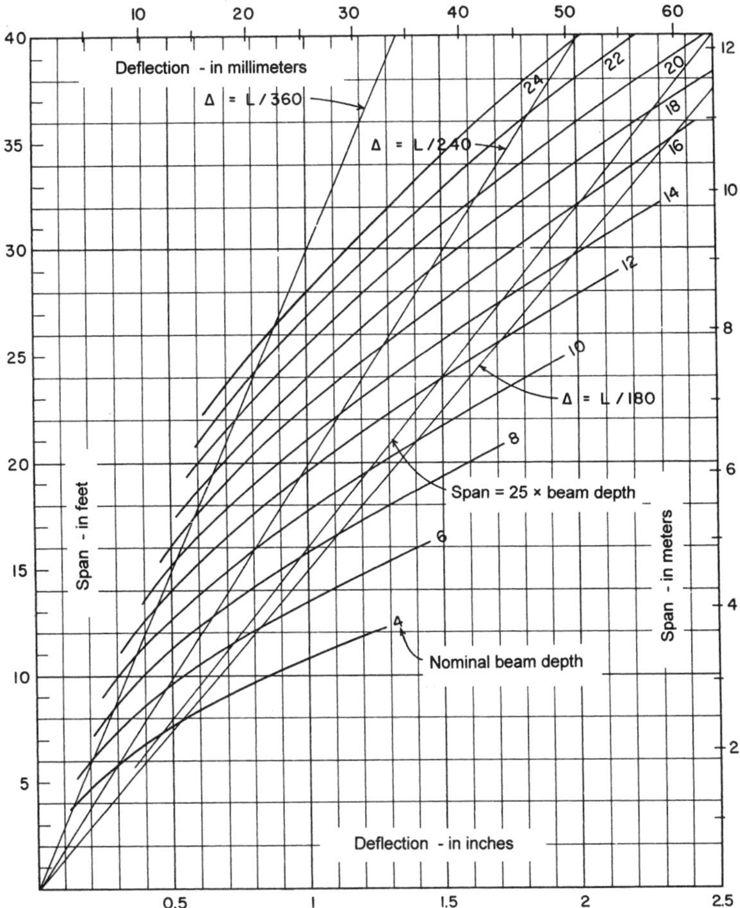

Figure 5.1 Deflection of wood beams. Assumed conditions: maximum bending stress of 1500 psi and modulus of elasticity of 1,500,000 psi.

purposes. For beams with other values for bending stress and modulus of elasticity, true deflections can be obtained as

$$\text{True } \Delta = \frac{\text{true } f_b}{1500} \times \frac{1{,}500{,}000}{\text{true } E} \times \Delta \text{ from graph}$$

The following examples illustrate problems involving deflection. Douglas fir–larch is used for these examples and for the problems that follow them.

Example 7. A Douglas fir–larch select structural 8 × 12 wood beam with $E = 1{,}600{,}000$ psi is used to carry a total uniformly distributed load of 10 kips on a simple span of 16 ft. Find the maximum deflection of the beam.

Solution: From Table A.8 find the value of $I = 950$ in.^{4} for the 8 × 12 section. Then, using the deflection formula for this loading

$$\Delta = \frac{5WL^3}{384EI} = \frac{5 \times 10{,}000 \times (16 \times 12)^3}{384 \times 1{,}600{,}000 \times 950} = 0.61 \text{ in.}$$

Or, using the graph in Figure 5.1,

$$M = \frac{WL}{8} = \frac{10{,}000 \times 16}{8} = 20{,}000 \text{ ft-lb}$$

$$f_b = \frac{M}{S} = \frac{20{,}000 \times 12}{165} = 1455 \text{ psi}$$

From Figure 5.1, Δ = approximately 0.66 in. Then

$$\text{True } \Delta = \frac{1455}{1500} \times \frac{1{,}500{,}000}{1{,}600{,}000} \times 0.66 = 0.60 \text{ in.}$$

which shows reasonable agreement with the computed value.

Example 8. A beam consisting of a Hem Fir No. 1 6 × 10 section with $E = 1{,}300{,}000$ psi spans 18 ft and carries two concentrated loads. One load is 1800 lb and is placed at 3 ft from one end of the beam, and the other load is 1200 lb, placed at 6 ft from the opposite end of the beam. Find the maximum deflection due only to the concentrated loads.

Solution: For an approximate computation, use the equivalent uniform load method, consisting of finding the hypothetical total uniform load that will produce a moment equal to the actual maximum moment in the beam. Then the deflection for uniformly distributed load may be used with this hypothetical (equivalent uniform) load. Thus:

$$\text{If} \quad M = \frac{WL}{8} \quad \text{then} \quad W = \frac{8M}{L}$$

For this loading the maximum bending moment is 6600 ft-lb (the reader should verify this by the usual procedures), and the equivalent uniform load is thus

$$W = \frac{8M}{L} = \frac{8 \times 6600}{18} = 2930 \text{ lb}$$

and the approximate deflection is

$$\Delta = \frac{5WL^3}{384EI} = \frac{5 \times 2930 \times (18 \times 12)^3}{384 \times 1{,}300{,}000 \times 393} = 0.75 \text{ in.}$$

As in the previous example, the deflection could also be found by using Figure 5.1, with adjustments made for the true maximum bending stress and the true modulus of elasticity.

Note: For the following problems, neglect the beam weight and consider deflection to be limited to $^1/_{240}$ of the beam span.

Problem 5.7.A A Douglas fir–larch 6 × 14 beam of No. 1 grade is 16 ft long and supports a total uniformly distributed load of 6000 lb. Investigate the deflection.

Problem 5.7.B An 8 × 12 beam of Hem Fir No. 1 grade is 12 ft in length and has a concentrated load of 5 kips at the center of the span. Investigate the deflection.

Problem 5.7.C Two concentrated loads of 3500 lb each are located at the third points of a 15-ft beam. The 10 × 14 beam is of Douglas fir–larch select structural grade. Investigate the deflection.

Problem 5.7.D An 8 × 14 beam of Hem Fir select structural grade has a span of 16 ft and a total uniformly distributed load of 8 kips. Investigate the deflection.

Problem 5.7.E Find the least weight section that can be used for a simple span of 18 ft with a total uniformly distributed load of 10 kips based on deflection. The wood is Douglas fir–larch No. 1 grade.

5.8 BEHAVIOR CONSIDERATIONS FOR LRFD

Investigation of beams in the LRFD method uses many of the relationships derived for the ASD method. One difference, of course, is the use of ultimate load for design rather than service load. The other main difference has to do with the basic form of expression for the resistance of the structural member to force effects (bending, shear, bearing). The ASD method uses a limiting resistance expressed in terms of a safe limiting stress condition. Thus, the limiting resistance is directly related to service loads. In the LRFD method, resistance is derived in total force effect form, based on some stress analysis, but expressed in pounds, foot-pounds, and the like.

One method for handling the LRFD analysis is to use the basic relationships of the ASD method with adjusted values for stress and some additional modification factors. Thus, the process is basically an altered version of the ASD method, but the answers are expressed in LRFD terms.

Because the numbers tend to get larger in LRFD analysis, it is customary to use force in kips and stress in kips per square inch, rather than the pounds and pounds per square inch used in the ASD method. This is somewhat arbitrary, but references are developed in this form so the reader should become accustomed to the practice.

The following discussions treat the topics in the earlier sections of this chapter, illustrating the applications in the LRFD method.

Shear in Beams

As described in Section 5.5, the total usable shear in a beam, at a given stress level, may be expressed as

$$V = \frac{2f_v A}{3}$$

where f_v is the defined value for shear stress and A is the area of the beam cross section for a rectangular beam. For a limiting condition in the ASD method, the stress used is that obtained from Tables 5.1*a* and 5.1*b*, designated F_v. This value for stress is actually only the so-called reference value and is frequently modified for size of the member, moisture content, and so on. The modified stress is designated F_v', and thus the true usable shear force is expressed as

$$V = \frac{2F_v' A}{3}$$

For the LRFD method, this expression is modified in two ways. First, the usable stress is adjusted, as explained in Section 5.3, by using the adjustment factor from Table 5.5. For shear this factor is $2.16/1000\varphi_v$, for which φ_v is 0.75 (see Table 5.6). The fully adjusted and modified shear resistance is then expressed as

$$\lambda\phi_v V' = \lambda\phi_v \left[\frac{2}{3}\left(\frac{2.16}{1000\phi_v}\right)(F_v')(A)\right]$$

In this formula the term λ is the *time effect factor* described in Section 5.3. This factor depends on the load combination for which the shear is being determined, and its values are given in Table 5.7 for seven common load combinations.

The formula for shear is typically used in one of two ways. The first way involves a beam of determined size for which the usable ultimate shear resistance is to be established, usually for comparison to the actual shear in a beam. For this problem the formula is used in the form above, with λ chosen for the appropriate load combination. In the design of beams the beam size is often first established on the basis of flexure, and consideration for shear involves a check on the beam's adequacy.

The second way the formula is used is in a design process if it is desired to find the appropriate size beam based on shear. For this problem the formula is transformed into an expression of the required area. This process may be used when a high shear force makes it possible that shear, and not flexure, may be critical for the beam.

The following two examples illustrate these two problems. The first example is the same situation as that described for Example 3.

Example 9. A simple beam with a span of 14 ft supports a uniformly distributed dead load of 300 lb/ft and a uniformly distributed live load of 500 lb/ft. A 10 × 14 beam of Douglas fir–larch of select structural grade is used. Is the beam safe with respect to shear stress?

Solution: The first step is to determine the required ultimate shear force. For the combined dead load and live load the ultimate combination is

$$w_u = 1.2(300) + 1.6(500) = 1160 \text{ lb/ft} \quad \text{or} \quad 1.16 \text{ kips/ft}$$

and the total load on the beam is 1.16(14) = 16.24 kips.

The maximum shear V_u is one-half the load, or 8.12 kips. The test for adequacy involves the satisfaction of the basic LRFD formula

$$\lambda \phi_v V' \geq V_u$$

which is to say the factored shear resistance must be equal to or greater than the ultimate shear. To find the factored shear, we first determine the adjusted shear stress. Using the formula given above, we find the reference shear stress from Table 5.1*a* to be 170 psi, or

0.170 ksi. For λ, with the combined dead and live loading, Table 5.7 yields a value of 0.8. The factored shear is thus found as

$$\lambda \phi_v V' = \lambda \phi_v \left[\frac{2}{3} \left(\frac{2.16}{1000 \phi_v} \right) (F_v{}')(A) \right]$$

$$= (0.8)(0.75) \left[\frac{2}{3} \left(\frac{2.16}{1000\,(0.75)} \right) (170)(128) \right]$$

$$= 25.1 \text{ kips}$$

As this is greater than the ultimate shear of 8.12 kips, the section is adequate (or safe).

As with the ASD method, a reduced value of shear may be used with uniformly loaded beams. This permits the exclusion of the shear within a distance from the support equal to the beam depth. This will result in a design value for ultimate shear force of less than 8.12 kips. Since the section has been demonstrated to be far from critical, this investigation is not indicated in this situation.

The following example illustrates the second type of shear problem, that in which it is desired to determine the size of beam required for a given shear performance.

Example 10. A 20-ft-span beam carries a dead load of 160 lb/ft and a live load of 240 lb/ft and is to consist of Hem Fir, select structural grade. Find the minimum size for the beam, based on shear resistance.

Solution: The first step involves the determination of the required ultimate shear force at the end of the beam. The loads are thus adjusted to factored ones as follows:

$$w_u = 1.2(160) + 1.6(240) = 576 \text{ lb/ft} \quad \text{or} \quad 576/1000 = 0.576 \text{ kip/ft}$$

The maximum shear for the simple beam is determined as

$$V_u = \frac{wL}{2} = \frac{0.576(20)}{2} = 5.76 \text{ kips}$$

Assuming a beam depth of 12 in., the reduced shear at beam depth distance from the support becomes

$$V_u = 5.76 - 0.576 = 5.184 \text{ kips}$$

Using the adjustments for load time and resistance reduction, the required shear resistance becomes

$$V' = \frac{V_u}{\lambda\phi} = \frac{5.184}{(0.8)(0.75)} = 8.64 \text{ kips}$$

Using this value for the required shear, we can transform the shear equation into one for finding A, as follows:

$$V' = \frac{2}{3}\left[\frac{2.16}{1000\phi_v}(F'_v)(A)\right]$$

from which

$$A = \frac{3}{2}\left[\frac{V'}{\frac{2.16}{1000\phi_v}(F_v')}\right] = \frac{3}{2}\left[\frac{8.64}{\frac{2.16}{1000(0.75)}(170)}\right] = 26.5 \text{ in.}^2$$

From Table A.8, sections of 4×10 or 6×8 may be used to obtain this area. The $4 \times$ member is in the class of "Dimension Lumber" in Table 5.1, and thus a slightly larger shear stress (150 psi) is permitted—still not a critical issue. However, as in many situations, considerations for bending and deflection require larger (mostly deeper) sections. Thus, initial design for shear is only justified when an exceptionally high load is carried on a relatively short span. In this example, if the load is doubled and the span is cut in half, the critical shear is approximately the same but bending and deflection are much less critical.

Problem 5.8.A A simple beam with a span of 18 ft supports a uniformly distributed dead load of 240 lb/ft and a uniformly distributed live load of 480 lb/ft. An 8×16 beam of Douglas fir–larch of No. 1 grade is used. Is the beam safe with regard to shear force?

Problem 5.8.B Same as Problem 5.8.A, except span is 24 ft, dead load is 360 lb/ft, live load is 560 lb/ft, and the section is a Hem Fir No. 1 10×20.

Problem 5.8.C A 10×18-in. beam of Douglas–fir larch, dense No. 1 grade, is used for a 16-ft span. The beam supports a uniformly distributed dead load of 160 lb/ft and three concentrated live loads of 4 kips each at the quarter points of the span (4 ft on center). Is the beam safe with respect to shear?

Problem 5.8.D Same as Problem 5.8.C, except the dead load is 240 lb/ft, the live load is 6 kips, the span is 20 ft, and the section is a Hem Fir No. 1 12 × 20.

Bending in Beams

For bending, the procedures are essentially the same as described for shear. The relationship between beam resistance and load-generated moment is expressed as

$$\lambda \phi_b M' \geq M_u$$

where M' is expressed as the product of a limiting bending stress and the beam's section modulus: $M' = F'_b S$, where F'_b is the reference value stress for the ASD method, F_b, multiplied by any applicable adjustment factors. With these substitutions, the moment equation becomes

$$\lambda \phi_b \left(\frac{2.16}{1000\phi_b} \right) (F'_b)(S) \geq M_u$$

As with shear, there are two common types of problems. The first is an investigation, or a *design check*, which consists of determining the moment resistance capacity of a given member and comparing it to a required, load-generated, ultimate moment. The second problem is a basic design situation, where the member required for a given ultimate moment is to be determined. The following examples demonstrate these problems.

Example 11. A simple beam with a span of 14 ft supports a uniformly distributed dead load of 300 lb/ft and a uniformly distributed live load of 500 lb/ft. A 10 × 14 beam of Hem Fir of select structural grade is used. Is the beam safe with respect to bending?

Solution: First, the required ultimate moment is determined. Thus,

$$w_u = 1.2(300) + 1.6(500) = 1160 \text{ lb/ft} \quad \text{or} \quad 1.16 \text{ kips/ft}$$

$$M_u = \frac{wL^2}{8} = \frac{1.16(14)^2}{8} = 28.42 \text{ kip-ft}$$

For the resistance of the beam, the value of S from Table A-8 is 289 in.^{3} From Table 5.1b the design value of F_b is 1300 psi. This must be modified by any appropriate factors, but it is assumed for this example that no modification is required. For the situation of dead load plus live load, λ from Table 5.7 is 0.8. The resistance factor for bending is 0.85 from Table 5.6. The moment capacity is thus

$$\lambda\phi_b M' = \lambda\phi_b \left[\left(\frac{2.16}{1000\phi_b}\right)(F_b{}')(S)\right]$$

$$= (0.8)(0.85)\left[\left(\frac{2.16}{1000\,(0.85)}\right)(1300)(289)\right]$$

$$= 648 \text{ kip-in.} \quad \text{or} \quad 54.0 \text{ kip-ft}$$

Since this is considerably greater than the required moment, the beam is more than adequate.

Example 12. A 20-ft-span beam carries a dead load of 160 lb/ft and a live load of 240 lb/ft and is to consist of Douglas fir–larch, select structural grade. Find the minimum size for the beam, based on bending resistance.

Solution: For the ultimate moment:

$$w_u = 1.2(160) + 1.6(240) = 576 \text{ lb/ft} \quad \text{or} \quad 576/1000 = 0.576 \text{ kip/ft}$$

and the ultimate moment is

$$M_u = \frac{w_u L^2}{8} = \frac{0.576(20)^2}{8} = 28.8 \text{ kip-ft}$$

From Table 5.1 $F_b = 1600$ psi, and

$$\lambda\phi_b M' = \lambda\phi_b \left(\frac{2.16}{1000\phi_b}F_b'\right)(S)$$

$$= (0.8)(0.85)\left[\frac{2.16}{1000\,(0.85)}(1600)\right](S)$$

$$= 2.7648S \text{ (in kip-in.)}$$

Equating this to M_u,

$$M_u = 28.8 \text{ kip-ft} = 28.8(12) = 345.6 \text{ kip-in.} = 2.7648S$$

$$S = \frac{345.6}{2.7648} = 125 \text{ in.}^3$$

From Table A.8, possible choices are 6×14, 8×12, and 10×10.

Problem 5.8.E A simple beam with a span of 18 ft supports a uniformly distributed dead load of 240 lb/ft and a uniformly distributed live load of 480 lb/ft. An 8 × 16 beam of Douglas fir–larch of No. 1 grade is used. Is the beam safe with regard to bending?

Problem 5.8.F Same as Problem 5.8.E, except span is 24 ft, dead load is 360 lb/ft, live load is 560 lb/ft, and the section is a Hem Fir select structural 10 × 20.

Problem 5.8.G A 10 × 18-in. beam of Douglas fir–larch, dense No. 1 grade, is used for a 16-ft span. The beam supports a uniformly distributed dead load of 160 lb/ft and three concentrated live loads of 4 kips each at the quarter points of the span (4 ft on center). Is the beam safe with regard to bending?

Problem 5.8.H Same as Problem 5.8.G, except the dead load is 240 lb/ft, the live load is 6 kips, the span is 20 ft, and the section is a Hem Fir No. 1 12 × 20.

Problem 5.8.I A simple beam with a span of 22 ft supports a uniformly distributed dead load of 200 lb/ft and a uniformly distributed live load of 600 lb/ft. Douglas fir–larch of No. 1 grade is used. Design the beam for bending ignoring the weight of the beam.

Problem 5.8.J Same as Problem 5.8.I, except the span is 26 ft, dead load is 300 lb/ft, live load is 400 lb/ft, and the section is Hem Fir select structural.

Deflection

In both the ASD and LRFD methods deflections are considered to occur under the service, unfactored, loads. Computations are thus the same for both methods.

Bearing

Bearing is treated in the same manner as shear and bending. The bearing load is a factored load, adjusted stress is used, and the process is similar in form to that used for shear and bending.

Design Process

The general beam design process is discussed in Section 5.4. The broadest context for beam design is illustrated in the building design examples in Chapters 18 and 19.

5.9 JOISTS AND RAFTERS

Floor joists and roof rafters are closely spaced beams that support floor or roof decks. They are common elements of the structural system described as the *light wood frame*. These may consist of sawn lumber, light trusses, laminated pieces, or composite elements achieved with combinations of sawn lumber, laminated pieces, plywood, or particleboard. The discussion in this section deals only with sawn lumber, typically in the class called dimension lumber having nominal thickness of 2–4 in. Although the strength of the structural deck is a factor, spacing of joists and rafters is typically related to the dimensions of the panels used for decking. The most used panel size is 48 × 96 in., from which are derived spacings of 12, 16, 19.2, 24, and 32 in.

Floor Joists

A common form of floor construction is shown in Figure 5.2. The structural deck shown in the figure is plywood, which produces a top surface not generally usable as a finished surface. Thus, some finish must be used, such as the hardwood flooring shown here. More common now for most interiors is carpet or thin tile, both of which require some smoother surface than the structural plywood panels, resulting in the use of *underlayment* typically consisting of wood fiber panels.

A drywall panel finish (paper-faced gypsum plaster board) is shown here for the ceiling directly attached to the underside of the joists. Since the floor surface above is usually required to be

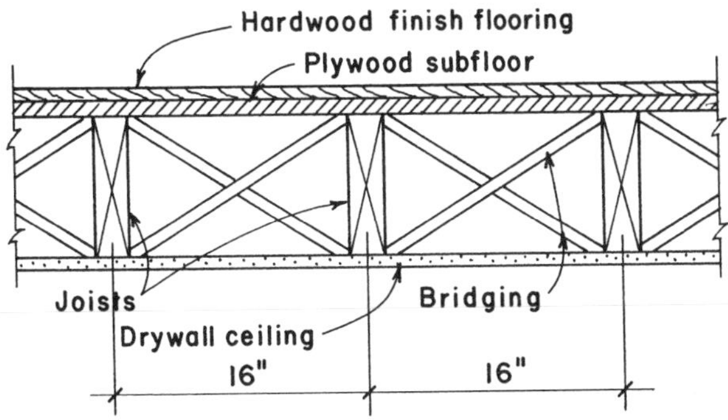

Figure 5.2 Typical wood joist floor construction.

horizontally flat, the same surface can thus be developed for the ceiling. With rafter construction for roofs that must normally be sloped, or when more space is required in the floor/ceiling construction, a separate *suspended* structure may be provided for the ceiling.

Lateral bracing for joists is usually provided by the attached deck. If additional bracing is required (see Table 5.9), it may consist of bridging, as shown in Figure 5.2, or of solid blocking consisting of short pieces of the joist elements aligned in rows between the joists. If blocking is used, it will normally be located so as to provide for edge nailing of the deck panels. This nailing of all the edges of panels is especially critical when the joist and deck construction is required to serve as a horizontal diaphragm for wind or seismic forces. (See discussion of diaphragms in Chapter 18.)

Solid blocking is also used under any supported walls perpendicular to the joists or under walls parallel to the joists but not directly above a joist. Any loading on the joist construction other than the usual assumed dispersed load on the deck should be considered for reinforcement of the regular joist system. A simple way to give extra local strength to the system is to double up joists. Doubling of joists is a common practice at the edges of large openings in the floor.

With the continuous, multiple-span effect of decking and possible inclusion of bridging or blocking, there is typically a potential for load sharing by adjacent joists. This is the basis for classification as a *repetitive member*, permitting an increase of 15% in the allowable bending stress.

Floor joists may be designed as beams by the procedure illustrated in Section 5.4. However, the most frequent use of joists in light wood framing systems is in situations that are well defined in a short range of conditions. Spans are usually quite short, both dead and live loads are predictable, and a relatively few wood species and grades are most commonly used. This allows for the development of tabulated lists of joist sizes from which appropriate choices can be made. Table 5.10 is an abbreviated sample of such a table.

Example 13. Using Table 5.10, select joists to carry a live load of 40 psf and a dead load of 10 psf on a span of 15 ft 6 in. Wood is Douglas fir–larch, No. 2 grade.

Solution: From Table 5.10, possible choices are 2 × 10 at 12 in., 2 × 12 at 16 in., or 2 × 12 at 19.2 in.

TABLE 5.10 Maximum Spans for Floor Joists (ft-in.)[a]

	Joist Size			
Spacing (in.)	2 × 6	2 × 8	2 × 10	2 × 12
Live load = 40 psf, Dead load = 10 psf, Maximum live-load deflection = *L*/360				
12	10-9	14-2	17-9	20-7
16	9-9	12-7	15-5	17-10
19.2	9-1	11-6	14-1	16-3
24	8-1	10-3	12-7	14-7
Live load = 40 psf, Dead load = 20 psf, Maximum live-load deflection = *L*/360				
12	10-6	13-3	16-3	18-10
16	9-1	11-6	14-1	16-3
19.2	8-3	10-6	12-10	14-10
24	7-5	9-50	11-6	13-4

Source: Compiled from data in the *International Building Code* (Ref. 4), with permission of the publisher, International Code Council.

[a]Joists are Douglas fir–larch, No. 2 grade. Assumed maximum available length of single piece is 26 ft.

Note that the values in Table 5.10 are based on a maximum deflection of $^1/_{360}$ of the span under live load. In using the table for the example, it is also assumed that there is no modification of the reference stress values. If true conditions are significantly different from those assumed for Table 5.10, the full design procedure for a beam is required.

Rafters

Rafters are used for roof decks in a manner similar to floor joists. While floor joists are typically installed dead flat, rafters are commonly sloped to achieve roof drainage. For structural design it is common to consider the rafter span to be the horizontal projection, as indicated in Figure 5.3.

As with floor joists, rafter design is frequently accomplished with the use of safe load tables. Table 5.11 is representative of such tables and has been developed from data in the *International Building Code* (IBC) (Ref. 4). Organization of the table is similar to that for Table 5.10. The following example illustrates the use of the data in Table 5.11.

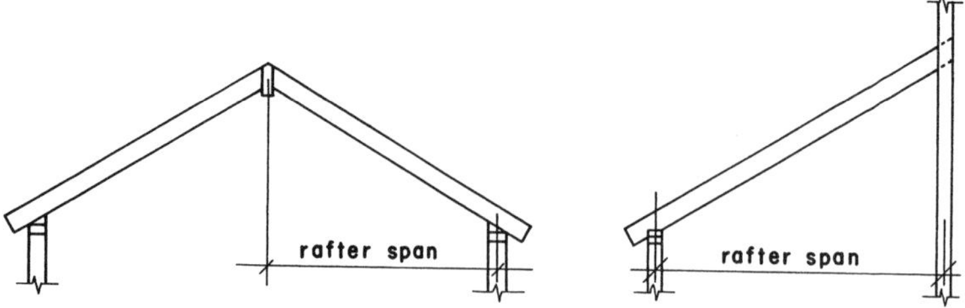

Figure 5.3 Span of sloping rafters.

TABLE 5.11 Maximum Spans for Rafters (ft-in.)[a]

	Rafter Size				
Spacing (in.)	2 × 4	2 × 6	2 × 8	2 × 10	2 × 12
Live load = 20 psf, Dead load = 10 psf, Maximum live-load deflection = *L*/240					
12	9-10	15-6	20-5	25-8	26-0
16	8-11	14-1	18-2	22-3	25-9
19.2	8-5	13-1	16-7	20-3	23-6
24	7-10	11-9	14-10	18-2	21-0
Live load = 20 psf, Dead load = 20 psf, Maximum live-load deflection = *L*/240					
12	9-10	14-4	18-2	22-3	25-9
16	8-6	12-5	15-9	19-3	22-4
19.2	7-9	11-4	14-4	17-7	20-4
24	6-11	10-2	12-10	15-8	18-3

Source: Compiled from data in the *International Building Code* (Ref. 4), with permission of the publisher, International Code Council.

[a]Rafters are Douglas fir–larch, No. 2 grade. Ceiling is not attached to rafters. Assumed maximum available length of single piece is 26 ft.

Example 14. Rafters are to be used for a roof span of 16 ft. Live load is 20 psf; total dead load is 10 psf; live-load deflection is limited to 1/240 of the span. Find the rafter size required for Douglas fir–larch of No. 2 grade.

Solution: From Table 5.11, possible choices are for 2 × 8 at 16 in., 2 × 8 at 19.2 in., or 2 × 10 at 24 in.

Problems 5.9.A–D Using Douglas fir–larch, No. 2 grade, pick the joist size required from Table 5.10 for the stated conditions. Live load is 40 psf, dead load is 10 psf, and deflection is limited to *L*/360 under live load only.

	Joist Spacing (in.)	Joist Span (ft)
A	16	14
B	12	14
C	16	16
D	12	20

Problems 5.9.E–H Using Douglas fir–larch, No. 2 grade, pick the rafter size required from Table 5.11 for the stated conditions. Live load is 20 psf, dead load is 20 psf, and deflection is limited to *L*/240 under live load only.

	Rafter Spacing (in.)	Rafter Span (ft)
E	16	12
F	24	12
G	16	18
H	24	18

5.10 DECKING FOR ROOFS AND FLOORS

Materials used to produce roof and floor surfaces include the following:

1. Boards of nominal 1-in.-thick solid-sawn wood, typically with tongue-and-groove edges
2. Solid-sawn wood elements thicker than 1-in. nominal dimension (usually called planks or planking) with tongue-and-groove or other edge development to prevent vertical slipping between adjacent units
3. Plywood of appropriate thickness for the span and the construction
4. Other panel materials, including those of compressed wood fibers or particles most notably oriented strand board (OSB).

Plank deck is especially popular for roof decks that are exposed to view from below. A variety of forms of products used for this construction are shown in Figure 5.4. Widely used is a nominal 2-in.-thick unit, which may be of solid-sawn form (Figure 5.4*a*) but is

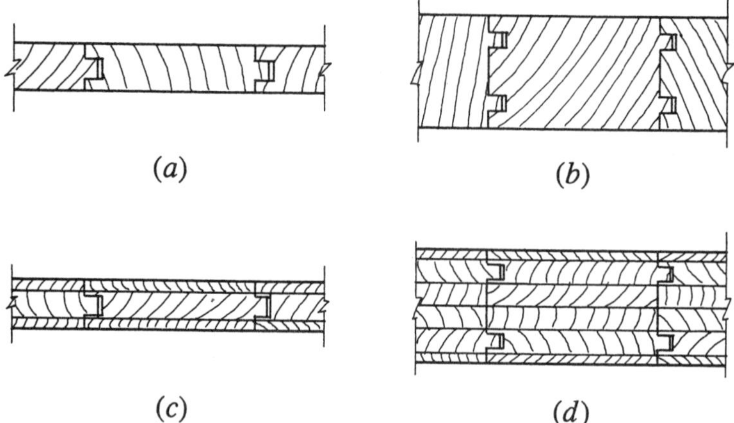

Figure 5.4 Units for board and plank decks.

now more likely to be of glue-laminated form (Figure 5.4*c*). Thicker units can be obtained for considerable spans between supporting members, but the thinner plank units are most popular.

Plank decks and other special decks are fabricated products produced by individual manufacturers. Information about their properties should be obtained from suppliers or the manufacturers. Plywood decks are widely used where their structural properties are critical. Plywood is an immensely variable material, although a few selected types are commonly used for structural purposes.

5.11 PLYWOOD

Plywood is the term used to designate structural wood panels made by gluing together multiple layers of thin wood veneer (called *plies*) with alternate layers having their grain direction at right angles. The outside layers are called the *faces* and the others *inner plies*. Inner plies with the grain perpendicular to the faces are called *crossbands*. There is usually an odd number of plies so that the faces have the grain in the same direction. For structural applications in building construction, the common range of panel thickness is from $^5/_{16}$ to $1\,^1/_8$ in.

The alternating grain direction of the plies gives the panels considerable resistance to splitting, and as the number of plies increases, the panels become approximately equal in strength in both directions.

Thin panels may have only three plies, but for most structural applications plies will number from 5 to 9.

Types and Grades of Plywood

Many different kinds of panels are produced. For structural applications, the principal distinctions other than panel thickness are the following:

1. *Exposure Classification*. Panels described as being *exterior* are for use where high-moisture conditions are enduring, such as outdoor uses and bathrooms, laundry rooms, and other high-moisture interior spaces. A classification of *exterior 1* is for panels where the end usage is for interior conditions but the panels may be exposed to the weather during construction.
2. *Structural Rating*. Code-approved rated sheathing is identified as to class for the purpose of establishing reference design values. Identification is established by marking panels with an indelible stamp. Information in the stamp designates several properties of the panel, including basic structural capabilities.

Design Usage Data for Plywood

Data for structural design of plywood may be obtained from industry publications or from individual plywood manufacturers. Data are also provided in most building codes. Tables 5.12 and 5.13 are reproductions of tables in the *International Building Code* (Ref. 4). These provide data for the loading and span capabilities of rated plywood panels. Table 5.12 treats panels with the panel face grain perpendicular to the supports, and Table 5.13 treats panels with the face grain parallel to the supports. Footnotes to these tables present various qualifications, including some of the loading and deflection criteria.

Plywood Diaphragms

Plywood deck-and-wall sheathing is frequently utilized to develop diaphragm actions for resistance to lateral loads from wind or earthquakes. Considerations for design of both horizontal deck diaphragms and vertical wall diaphragms (shear walls) are discussed in Chapter 18. Where both gravity loading and lateral loading must be considered, choices for the construction must relate to both problems.

TABLE 5.12 Data for Plywood Roof and Floor Deck, Face Grain Perpendicular to Supports

Panel Span Rating	Panel Thickness (inches)	Maximum Span (inches) × 25.4 for mm		Load[5] (pounds per square foot) × 0.0479 for kN/m²		Maximum Span (inches)
Roof/Floor Span	× 25.4 for mm	With Edge Support[6]	Without Edge Support	Total Load	Live Load	× 25.4 for mm
12/0	$5/16$	12	12	40	30	0
16/0	$5/16$, $3/8$	16	16	40	30	0
20/0	$5/16$, $3/8$	20	20	40	30	0
24/0	$3/8$, $7/16$, $1/2$	24	20[7]	40	30	0
24/16	$7/16$, $1/2$	24	24	50	40	16
32/16	$15/32$, $1/2$, $5/8$	32	28	40	30	16[8]
40/20	$19/32$, $5/8$, $3/4$, $7/8$	40	32	40	30	20[8,9]
48/24	$23/32$, $3/4$, $7/8$	48	36	45	35	24
54/32	$7/8$, 1	54	40	45	35	32
60/48	$7/8$, 1, $1\,1/8$	60	48	45	35	48
SINGLE-FLOOR GRADES		ROOF[3]				FLOOR[4]
Panel Span Rating (inches)	Panel Thickness (inches)	Maximum Span (inches) × 25.4 for mm		Load[5] (pounds per square foot) × 0.0479 for kN/m²		Maximum Span (inches)
× 25.4 for mm		With Edge Support[6]	Without Edge Support	Total Load	Live Load	× 25.4 for mm
16 oc	$1/2$, $19/32$, $5/8$	24	24	50	40	16[8]
20 oc	$19/32$, $5/8$, $3/4$	32	32	40	30	20[8,9]
24 oc	$23/32$, $3/4$	48	36	35	25	24
32 oc	$7/8$, 1	48	40	50	40	32
48 oc	$1\,3/32$, $1\,1/8$	60	48	50	50	48

[1]Applies to panels 24 inches (610 mm) or wider.
[2]Floor and roof sheathing conforming with this table shall be deemed to meet the design criteria of Section 2312.
[3]Uniform load deflection limitations $1/180$ of span under live load plus dead load, $1/240$ under live load only.
[4]Panel edges shall have approved tongue-and-groove joints or shall be supported with blocking unless $1/4$-inch (6.4 mm) minimum thickness underlayment or $1\,1/2$ inches (38 mm) of approved cellular or lightweight concrete is placed over the subfloor, or finish floor is $3/4$-inch (19 mm) wood strip. Allowable uniform load based on deflection of $1/360$ of span is 100 pounds per square foot (psf) (4.79 kN/m²) except the span rating of 48 inches on center is based on a total load of 65 psf (3.11 kN/m).
[5]Allowable load at maximum span.
[6]Tongue-and-groove edges, panel edge clips [one midway between each support, except two equally spaced between supports 48 inches (1219 mm) on center], lumber blocking, or other. Only lumber blocking shall satisfy blocked diaphragms requirements.
[7]For $1/2$-inch (12.7 mm) panel, maximum span shall be 24 inches (610 mm).
[8]May be 24 inches (610 mm) on center where $3/4$-inch (19 mm) wood strip flooring is installed at right angles to joist.
[9]May be 24 inches (610 mm) on center for floors where $1\,1/2$ inches (38 mm) of cellular or lightweight concrete is applied over the panels.

Source: Reproduced from *International Building Code* (Ref. 4), with permission of the publisher, International Code Council, Inc.

TABLE 5.13 Data for Plywood Roof Deck, Face Grain Parallel to Supports

ALLOWABLE LOAD (PSF) FOR WOOD STRUCTURAL PANEL ROOF SHEATHING CONTINUOUS OVER TWO OR MORE SPANS AND STRENGTH AXIS PARALLEL TO SUPPORTS (Plywood Structural Panels Are Five-Ply, Five-Layer Unless Otherwise Noted)[a,b]

PANEL GRADE	THICKNESS (inch)	MAXIMUM SPAN (inches)	LOAD AT MAXIMUM SPAN (psf)	
			Live	Total
Structural I sheathing	$^7/_{16}$	24	20	30
	$^{15}/_{32}$	24	35[c]	45[c]
	$^1/_2$	24	40[c]	50[c]
	$^{19}/_{32}$, $^5/_8$	24	70	80
	$^{23}/_{32}$, $^3/_4$	24	90	100
Sheathing, other grades covered in DOC PS 1 or DOC PS 2	$^7/_{16}$	16	40	50
	$^{15}/_{32}$	24	20	25
	$^1/_2$	24	25	30
	$^{19}/_{32}$	24	40[c]	50[c]
	$^5/_8$	24	45[c]	55[c]
	$^{23}/_{32}$, $^3/_4$	24	60[c]	65[c]

For SI: 1 inch = 25.4 mm, 1 pound per square foot = 0.0479 kN/m^2.

a. Roof sheathing conforming with this table shall be deemed to meet the design criteria of Section 2304.7.

b. Uniform load deflection limitations $^1/_{180}$ of span under live load plus dead load, $^1/_{240}$ under live load only. Edges shall be blocked with lumber or other approved type of edge supports.

c. For composite and four-ply plywood structural panel, load shall be reduced by 15 pounds per square foot.

Source: Reproduced from *International Building Code* (Ref. 4), with permission of the publisher, International Code Council, Inc.

Usage Considerations for Structural Panels

The following are some of the principal usage considerations for ordinary applications of structural panels:

1. *Choice of Thickness and Grade*. This is largely a matter of common usage and building code acceptability. For economy the thinnest, lowest grade panels will always be used unless various concerns require otherwise. In addition to structural spanning capabilities, concerns may include moisture resistance, appearance of face plies, and fastener holding capability.
2. *Modular Supports*. With the usual common panel size of 4 ft × 8 ft, logical spacing for studs, rafters, and joists become even-number divisions of the 48- or 96-in. dimensions: 12, 16, 24, 32, or 48. However, spacing of framing must often relate to what is attached on the other side of a wall or as a directly attached ceiling.
3. *Panel Edge Supports*. Panel edges not falling on a support may need some provision for nailing, especially for roof and floor decks. Solid blocking is the common answer, although thick deck panels may have tongue-and-groove edges.
4. *Attachment to Supports*. For reference design values for shear loads in diaphragms, attachment is usually considered to be achieved with common wire nails. Required nail size and spacing relate to panel thickness and code minimums as well as to shear capacities in diaphragms. Attachment is now mostly achieved with mechanically driven fasteners rather than old-fashioned pounding with a hand-held hammer. These means of attachment and the actual fasteners used are usually rated for capacity in terms of equivalency to ordinary nailing.

5.12 GLUED-LAMINATED PRODUCTS

In addition to plywood panels, there are a number of other products used for wood construction that are fabricated by gluing together pieces of wood into solid form. Girders, framed bents, and arch ribs of large size are produced by assembling standard 2-in. nominal lumber (2 × 6, etc.). The resulting thickness of such elements is essentially the width of the standard lumber used, with a small dimensional loss due to finishing. The depth is a multiple of the lumber thickness of 1.5 in.

Availability of large glued-laminated products should be investigated on a regional basis, as shipping to job sites is a major cost factor. Information about these products can be obtained from local suppliers or from the product manufacturers in the region. As with other widely used products, there are industry standards and usually some building code data for design.

5.13 WOOD FIBER PRODUCTS

Various products are produced with wood that is reduced to fiber form from the logs of trees. Major considerations are those for the size and shape of the wood fiber elements and their arrangement in the finished products. For paper, cardboard, and some fine hardboard products, the wood is reduced to very fine particles and generally randomly placed in the mass of the products. This results in little orientation of the material, other than that produced by the manufacturing process of the particular products.

For structural products, somewhat larger wood particle elements are used and some degree of orientation is obtained. Two types of products with this character are the following:

Wafer Board or Flake Board. These are panel products produced with wood chips in wafer form. The wafers are laid randomly on top of each other, producing a panel with a two-way fiber-oriented nature that simulates the character of plywood panels. Applications include wall sheathing and some structural decks.

Strip or Strand Elements. These are produced from long strands that are shredded from the logs. These are bundled with the strands all in the same direction to produce elements that have something approaching the character of the linear orientation in solid-sawn wood. Applications include studs, rafters, joists, and small beams.

For decking or wall sheathing, these products are generally used in thicknesses greater than that of plywood for the same spans. Other construction issues must be considered, such as nail holding for materials attached to the deck. Consideration must also be made for the type and magnitude of loads, the type of finished flooring for floor decks, and need for diaphragm action for lateral loads. Code approval is an important issue and must be determined on a local basis.

Information about these products should be obtained from the manufacturers or suppliers of particular proprietary products. Some data are now included in general references, such as the IBC, but particular competitive products are marketed by individual companies.

This is definitely a growth area, as plywood becomes increasingly expensive and logs for producing plywood are harder to find. Resources for fiber products include small trees, smaller sections from large trees, and even some recycled wood. A general trend to use of composite materials certainly indicates the likelihood of more types of products for future applications.

5.14 ASSEMBLED WOOD STRUCTURAL PRODUCTS

Various types of structural components can be produced with assembled combinations of plywood, fiber panels, laminated products, and solid-sawn lumber. Figure 5.5 shows some commonly used elements that can serve as structural components for buildings.

The unit shown in Figure 5.5*a* consists of two panels of plywood attached to a frame of solid-sawn lumber elements. This is generally described as a *sandwich panel*; however, when used for structural purposes it is called a *stressed-skin panel*. For spanning actions, the plywood panels serve as bending stress-resisting flanges and the lumber elements as beam webs for shear development.

Another common type of product takes the form of the box beam (Figure 5.5*b*) or the built-up I beam (Figure 5.5*c*). In this case the roles defined for the sandwich panel are reversed, with the solid-sawn elements serving as flanges and the panel material as the web. These elements are highly variable, using both plywood and fiber products for the panels and solid-sawn lumber or glued-laminated products for the flange elements. It is also possible to produce various profiles, with a flat chord opposed to a sloped or curved one on the opposite side. Use of these elements allows for production of relatively large components from small trees, resulting in a saving of large solid-sawn lumber and old-growth forests.

The box beam shown in Figure 5.5*b* can be assembled with attachments of ordinary nails or screws. The I beam uses glued joints to attach the web and flanges. Box beams may be custom assembled at the building site, but the I beams are produced in highly controlled factory conditions.

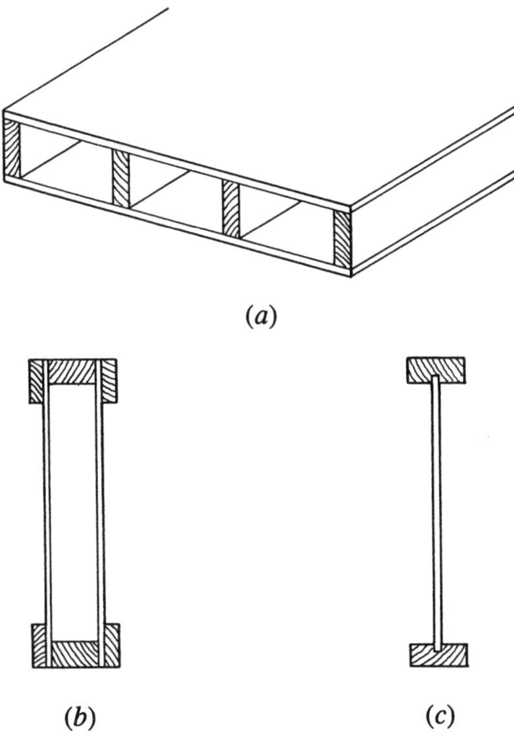

Figure 5.5 Composite, built-up components with elements of solid-sawn lumber and panels of plywood or wood fiber.

I-beam products have become highly popular for use in the range of spans just beyond the feasibility for solid-sawn lumber joists and rafters, that is, over about 15 ft for joists and about 20 ft for rafters.

Two types of light wood trusses are widely used. The W truss, shown in Figure 5.6*a*, is widely used for short-span gable-form roofs. Achieved with a single layer of 2 × lumber members, and with simple gusset-plated joints (Figure 5.6*b*), this has been the form of the roof structure for small wood-framed buildings for many years. Gussets may consist of pieces of plywood, attached with nails, but are now mostly factory assembled with metal connector plates.

For flat spanning structures—both roofs and floors—the truss shown in Figure 5.6*c* is used, mostly for spans just beyond the spanning length feasible for solid-sawn wood rafters or joists.

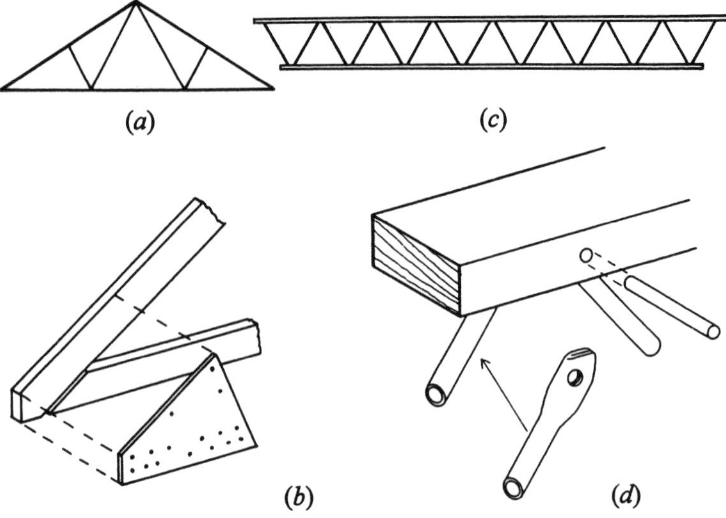

Figure 5.6 Light wood trusses.

One possible assembly is shown in Figure 5.6*d*, using steel tubes with flattened ends, connected to the chords with pins driven through drilled holes. Chords may be simple solid-sawn lumber elements but are also made of proprietary laminated elements that permit virtually unlimited length for single-piece members.

6

WOOD COLUMNS

A column is a compression member, the length of which is several times greater than its least lateral dimension. The term *column* is generally applied to relatively heavy vertical members, and the term *strut* is given to shorter compression members not necessarily in a vertical position. The type of wood column used most frequently is the *simple solid column*, which consists of a single sawn piece of wood that is square or oblong in cross section. Solid columns of circular cross section are also considered simple solid columns and typically consist of trimmed, but not sawn, tree trunks called *poles*. A *spaced column* is an assembly of two or more sawn pieces with their longitudinal axes parallel and separated at their ends and at middle points of their length by blocking. Two other types are *built-up columns* consisting of multiple sawn pieces bound by mechanical fasteners and *glued-laminated columns*. The *studs* in light wood framing are columns.

6.1 SLENDERNESS RATIO FOR COLUMNS

In wood construction the slenderness ratio of a free-standing simple solid column with a rectangular cross section is the ratio of its unbraced (laterally unsupported) length to the dimension of its least side, expressed as L/d. (See Figure 6.1*a*.) When members are braced so that the laterally unsupported length with respect to one face is less than that with respect to the other, L is the distance between the supports that prevent lateral movement in the direction along which the least dimension is measured. This is illustrated in Figure 6.1*b*. If the section is not square or round, it may be necessary to investigate two L/d conditions for such a column to determine which is the limiting one. The slenderness ratio for simple solid columns is limited to $L/d \leq 50$.

6.2 COMPRESSION CAPACITY OF SIMPLE SOLID COLUMNS, ASD METHOD

Figure 2.17 illustrates the typical form of the relationship between axial compression capacity and slenderness ratio for a linear

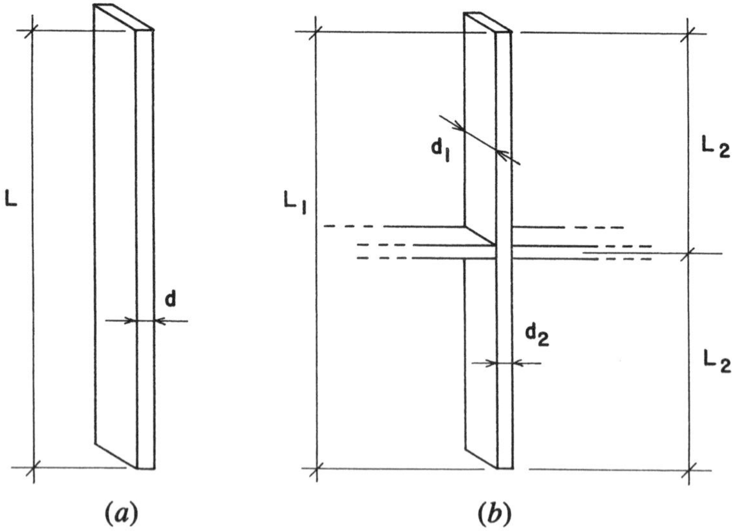

Figure 6.1 Determination of unbraced height for a column as related to the critical column thickness dimension.

compression member (column). The limiting conditions are those of the very short member and the very long member. The short member—such as a block of wood—fails in crushing, which is limited by the mass of material and the stress limit for compression. The very long member—such as a yardstick—fails in elastic buckling, which is determined by the stiffness of the member in bending resistance and the material stiffness property (modulus of elasticity). Between these two extremes, which is where most wood columns fall, the behavior is determined by a transition between the two distinctly different responses.

Over the years, several methods have been employed to deal with this situation in the design of wood columns, or of any column for that matter. Earlier editions of the *NDS* (Ref. 3) used three separate formulas to cover the entire range of slenderness, reflecting the three distinct forms of response, as described in Section 2.8. In recent editions, however, a single formula is employed, effectively covering the whole range of the graph. The formula and its various factors are complex, and its use involves considerable computation; nevertheless, the basic process is essentially simplified through the use of a single defined relationship.

In design practice, use is commonly made of either tabulated data or computer-aided processes. The *NDS* formulas are basically analytical and inverting them to produce design formulas for direct use is not practical. Direct use of the formulas for design involves a trial-and-error process, with many runs through the complex formula before a good fit is found. Once the basic relationships are understood, design aids are very useful.

Column Load Capacity

The following discussion presents materials from the *NDS* (Ref. 3) for design of axially loaded columns. The basic formula for determination of the capacity of a wood column, based on the working stress method, is

$$P = (F_c^*)(C_p)(A)$$

where A = area of column cross section
F_c^* = design value for compression, modified
C_p = column stability factor
P = allowable column axial compression load

The column stability factor is determined as follows:

$$C_p = \frac{1 + (F_{cE}/F_c^*)}{2c} - \sqrt{\left[\frac{1 + (F_{cE}/F_c^*)}{2c}\right]^2 - \frac{F_{cE}/F_c^*}{c}}$$

where F_{cE} = Euler buckling stress, as defined below
c = 0.8 for sawn lumber, 0.85 for poles, 0.9 for glued-laminated timbers

For the buckling stress,

$$F_{cE} = \frac{0.822\ E'_{\min}}{(L_e/d)^2}$$

where $E'_{\min}$ = modulus of elasticity for stability
L_e = effective unbraced length (height) of the column
d = critical column thickness for buckling

The values to be used for the effective column length and the corresponding column width should be considered as discussed for the conditions displayed in Figure 6.1. For a basic reference, the buckling phenomenon typically uses a member that is pinned at both ends and prevented from lateral movement only at the ends, for which no modification for support conditions is made; this is a common condition for wood columns. The *NDS* presents methods for modified buckling lengths that are essentially similar to those used for steel design.

For solid-sawn columns, the formula for C_p is simply a function of the value of F_{cE}/F_c^* with the value of c being a constant of 0.8. It is, therefore, possible to plot a graph of the value for C_p as a function of the value of F_{cE}/F_c^*, as is done in three parts in Figure 6.2. Accuracy of values obtained from Figure 6.2 is low but is usually acceptable for column design work. Of course, greater accuracy can always be obtained with the use of the formula.

The following examples illustrate the use of the *NDS* formulas for columns.

Example 1. A wood column consists of a 6 × 6 of Hem Fir, No. 1 grade. Using the ASD method, find the safe axial compression load for unbraced lengths of (1) 2 ft, (2) 8 ft, and (3) 16 ft.

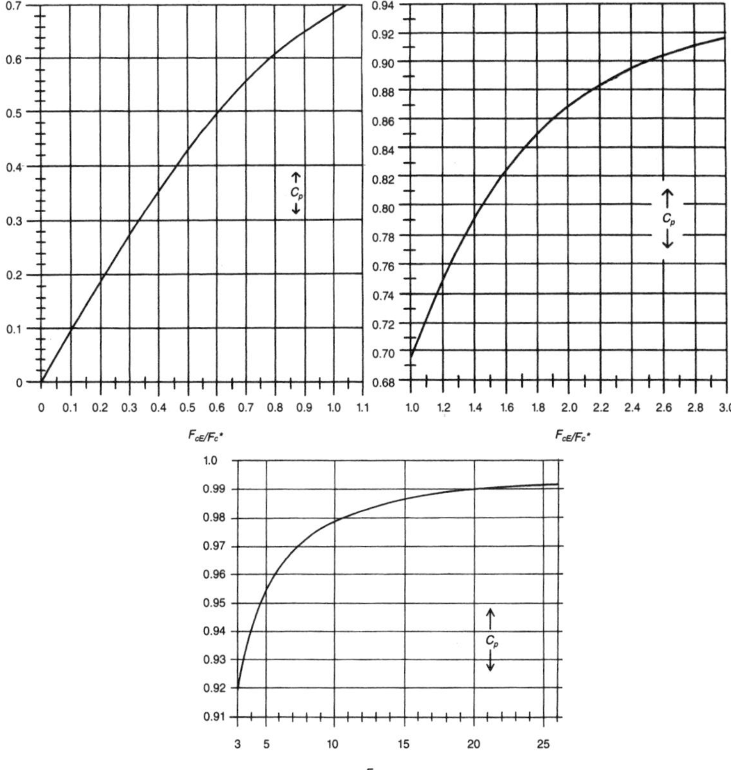

Figure 6.2 Column stability factor C_p as a function of F_{cE}/F_c^*.

Solution: From Table 5.1b find values of $F_c = 850$ psi and $E_{\min} =$ 470,000 psi. With no basis for adjustment given, the F_c value is used directly as the F_c^* value in the column formulas.

For (1): $L/d = 2(12)/5.5 = 4.36$. Then

$$F_{cE} = \frac{0.822\ E_{\min}}{(L_c/d)^2} = \frac{0.822(470{,}000)}{(4.36)^2} = 20{,}300 \text{ psi}$$

$$\frac{F_{cE}}{F_c^*} = \frac{20{,}300}{850} = 23.9$$

$$C_p = \frac{1 + 23.9}{1.6} - \sqrt{\left(\frac{1 + 23.9}{1.6}\right)^2 - \frac{23.9}{0.8}} = 0.985$$

and the allowable compression load is

$$P = (F_c^*)(C_p)(A) = (850)(0.985)(5.5)^2 = 25{,}300 \text{ lb}$$

As mentioned previously, Figure 6.2 may be used to simplify the computation.

For (2): $L/d = 8(12)/5.5 = 17.45$, for which $F_{cE} = 1270$ psi, $F_{cE}/F_c^* = 1.49$, from Figure 6.2, $C_p = 0.81$, and thus

$$P = (850)(0.81)(5.5)^2 = 20{,}800 \text{ lb}$$

For (3): $L/d = 16(12)/5.5 = 34.9$, for which $F_{cE} = 317$ psi, $F_{cE}/F_c^* = 0.373$, from Figure 6.2, $C_p = 0.30$, and thus

$$P = (850)(0.3)(5.5)^2 = 7710 \text{ lb}$$

Example 2. Wood 2 × 4 elements are to be used as vertical compression members to form a wall (ordinary stud construction). If the wood is Douglas fir–larch, stud grade, and the wall is 8.5 ft high, what is the column load capacity of a single stud?

Solution: It is assumed that the wall has a covering attached to the studs or blocking between the studs to brace them on their weak (1.5-in.-dimension) axis. Otherwise, the practical limit for the height of the wall based on a maximum slenderness ratio or 50 is 50 × 1.5 = 75 in (6 ft 3 in). Therefore, using the larger dimension,

$$\frac{L}{d} = \frac{8.5 \times 12}{3.5} = 29.14$$

From Table 5.1a $F_c = 850$ psi, $E_{\min} = 510{,}000$ psi. From Table 5.2, the value for F_c is adjusted by a size factor to 1.05(850) = 892.5 psi. Then

$$F_{cE} = \frac{0.822(510{,}000)}{(29.14)^2} = 494 \text{ psi}$$

$$\frac{F_{cE}}{F_c^*} = \frac{494}{892.5} = 0.554$$

From Figure 6.2, $C_p = 0.47$, and the column capacity is

$$P = (F_c^*)(C_p)(A) = (892.5)(0.47)(1.5 \times 3.5) = 2202 \text{ lb}$$

Problems 6.2.A–D Using the ASD method, find the allowable axial compression load for the following wood columns. Use Douglas fir–larch, No. 2 grade.

	Nominal Size, Species, and Grade	Unbraced Length (ft)
A	4 × 4 Douglas fir–larch No. 2	8
B	6 × 6 Hem Fir No. 1	10
C	8 × 8 Douglas fir–larch No. 2	18
D	10 × 10 Hem Fir select structural	14

Design of Wood Columns

The design of columns is complicated in the column formulas. The allowable stress for the column is dependent upon the actual column dimensions, which are not known at the beginning of the design process. This does not allow for simply inverting the column formulas to derive required properties for the column. A trial-and-error process is therefore indicated. For this reason, designers typically use various design aids: graphs, tables, or computer-aided processes.

Because of the large number of wood species, resulting in many different values for allowable stress and modulus of elasticity, precisely tabulated capacities become impractical. Nevertheless, aids using average values are available and simple to use for design. Figure 6.3 is a graph on which the axial compression load capacity of some square column sections of a single species and grade are plotted. Table 6.1 yields the capacity for a range of columns. Note that the smaller size column sections fall into the classification in Tables 5.1a and 5.1b for "Dimension Lumber," rather than for "Timbers." This makes for one more complication in the column design process.

Problems 6.2.E–H Select square column sections of Douglas fir–larch, No. 1 grade, for the following data:

	Required Axial Load (kips)	Unbraced Length (ft)
E	20	8
F	50	12
G	50	20
H	100	16

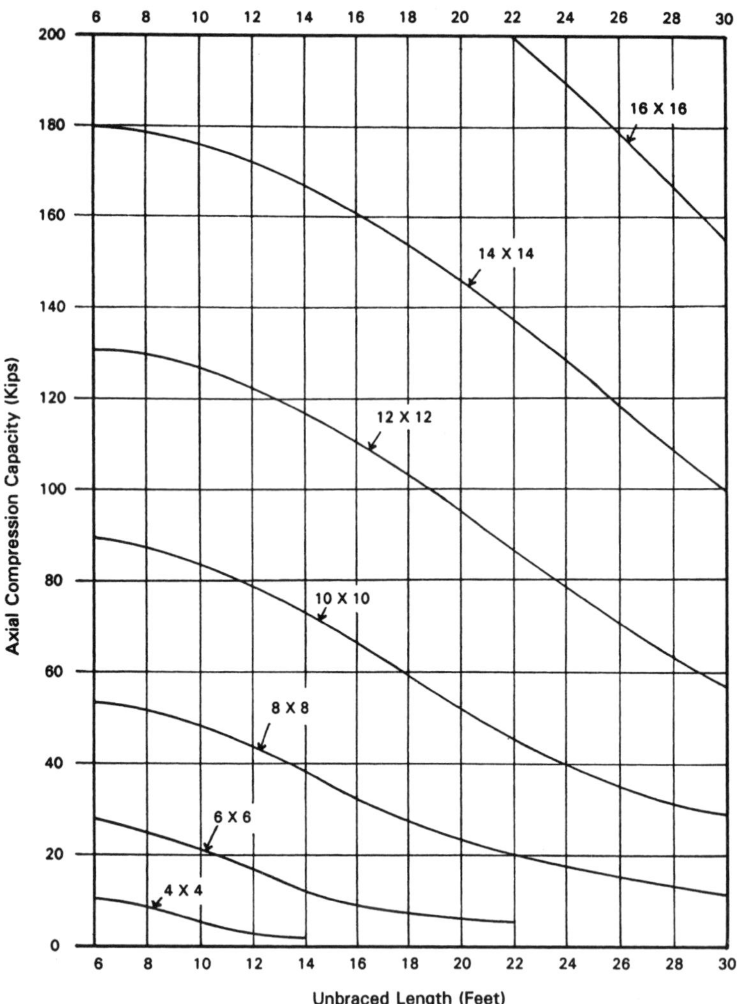

Figure 6.3 Axial compression load capacity for wood members of square cross section. Derived from NDS requirements for Douglas fir–larch, No. 1 grade.

TABLE 6.1 Safe Service Loads for Douglas Fir–Larch No. 1 Wood Columns, ASD Method[a]

Column Section		Unbraced Length (ft)										
Nominal Size	Area (in.^{2})	6	8	10	12	14	16	18	20	22	24	26
4 × 4	12.25	11.1	7.28	4.94	3.50	2.63						
4 × 6	19.25	17.4	11.4	7.76	5.51	4.14						
4 × 8	25.375	22.9	15.1	10.2	7.26	6.46						
6 × 6	30.25	27.6	24.8	20.9	16.9	13.4	10.7	8.71	7.17	6.53		
6 × 8	41.25	37.6	33.9	28.5	23.1	18.3	14.6	11.9	9.78	8.91		
6 × 10	52.25	47.6	43.0	36.1	29.2	23.1	18.5	15.0	13.4	11.3		
8 × 8	56.25	54.0	51.5	48.1	43.5	38.0	32.3	27.4	23.1	19.7	16.9	14.6
8 × 10	71.25	68.4	65.3	61.0	55.1	48.1	41.0	34.7	29.3	24.9	21.4	18.4
8 × 12	86.25	82.8	79.0	73.8	66.7	58.2	49.6	42.0	35.4	30.2	26.0	22.3
10 × 10	90.25	88.4	85.9	83.0	79.0	73.6	67.0	60.0	52.9	46.4	40.4	35.5
10 × 12	109.25	107	104	100	95.6	89.1	81.2	72.6	64.0	56.1	48.9	42.9
10 × 14	128.25	126	122	118	112	105	95.3	85.3	75.1	65.9	57.5	50.4
12 × 12	132.25	130	128	125	122	117	111	104	95.6	86.9	78.3	70.2
14 × 14	182.25	180	178	176	172	168	163	156	148	139	129	119
16 × 16	240.25	238	236	234	230	226	222	216	208	200	190	179

[a]Load capacity in kips for solid-sawn sections of No. 1 grade Douglas fir–larch with no adjustment for moisture or load duration conditions.

6.3 COLUMN LOAD CAPACITY, LRFD METHOD

For the LRFD process, the steps are essentially the same as for the ASD method described in the preceding sections. The principal differences consist of adjustments of values as achieved by various factors. The adjustments are as follows:

For loads: Load factors from Section 1.9 for various combinations.

Time effect factor, λ (lambda): See Table 5.7.

Resistance factor: Table 5.6, $\varphi_c = 0.90$ for compression, $\varphi_s = 0.85$ for stability (E_{min}).

Reference values: For stress, $2.16/1000\varphi_c$; for stability, $1.5/1000\varphi_s$.

Computations for investigation:

$$E'_{min} = \phi_s \frac{1.5}{1000\phi_s} E_{min}$$

$$F_{cE} = \frac{0.822\ E'_{min}}{(L/d)^2}$$

$$F_c^* = \lambda\phi_c \left(\frac{2.16}{1000\phi_c}\right) F_C$$

$$C_p = \frac{1 + F_{cE}/F_c *}{1.6} - \sqrt{\left(\frac{1 + F_{cE}/F_c *}{1.6}\right)^2 - \frac{F_{cE}/F_c *}{0.8}}$$

For factored usable compression capacity:

$$P' = \lambda\phi_c \frac{2.16}{1000\phi_c} C_p F_c A$$

The following example illustrates this process. It uses the same data as in Example 1, part 2, Example 2, which treats the ASD method.

Example 3. A wood column consists of a 6 × 6 of Hem Fir, No. 1 grade. Using the LRFD method, find the factored usable compression capacity (factored resistance) for an unbraced length of 8 ft.

Solution: From Table 5.1b find values of $F_c = 975$ psi and $E_{min} =$ 470,000 psi. With no other information about conditions for modification, these values are subject only to the necessary adjustments for the LRFD method. Assume that the load is a typical combination of dead and live load, which yields a value for λ of 0.8 (Table 5.7).

$$E'_{\min} = \phi_s \frac{1.5}{1000\phi_s} E_{\min} = 0.85 \frac{1.5}{1000(0.85)}(470{,}000) = 705 \text{ ksi}$$

$$\frac{L}{d} = \frac{8 \times 12}{5.5} = 17.45$$

$$F_{cE} = \frac{0.822\, E'_{\min}}{(L/d)^2} = \frac{0.822(705)}{(17.45)^2} = 1.90 \text{ ksi}$$

$$F_c^* = \lambda\phi_c \left(\frac{2.16}{1000\phi_c}\right) F_C = 0.8(0.9)\left(\frac{2.16}{1000(0.9)}\right) 975 = 1.68 \text{ ksi}$$

$$\frac{F_{cE}}{F_c^*} = \frac{1.9}{1.68} = 1.13$$

Using Figure 6.2, $C_p = 0.72$, and the capacity is

$$\begin{aligned}\lambda\phi_c P' &= \lambda\phi_c \left(\frac{2.16}{1000\phi_c} C_p F_c A\right)\\ &= 0.8(0.9)\left(\frac{2.16}{1000(0.9)}(0.72)(975)(5.5)^2\right)\\ &= 36.7 \text{ kips}\end{aligned}$$

As with the ASD method, the column design process is quite laborious, unless some design aid or a computer-assisted procedure is used. These aids are indeed available, although they are not described in this book.

Problems 6.3.A–D Using the LRFD method, find the factored usable compression capacity (factored resistance) for the following wood columns:

	Nominal Size, Species, and Grade (in.)	Unbraced Length (ft)
A	4 × 4 Douglas fir–larch No. 2	8
B	6 × 6 Hem Fir select structural	10
C	8 × 8 Douglas fir–larch No. 2	18
D	10 × 10 Hem Fir No. 1	14

6.4 STUD WALL CONSTRUCTION

Studs are the vertical elements used for wall framing in light wood construction. Studs serve utilitarian purposes of providing for attachment of wall surfacing but also serve as columns when the wall provides support for roof or floor systems. The most common stud is a 2 × 4 spaced at intervals of 12, 16, or 24 in., the spacing derived from the common 4 ft × 8 ft panels of wall coverings.

Studs of nominal 2 in. thickness must be braced on the weak axis when used for story-high walls; a simple requirement deriving from the limiting ratio of *L*/*d* of 50 for columns. If the wall is surfaced on both sides, the studs are usually considered to be adequately braced by the surfacing. If the wall is not surfaced or is surfaced on only one side, horizontal blocking between studs must be provided, as shown in Figure 6.4. The number of rows of blocking and the spacing of the blocking will depend on the wall height and the need for column action by the studs.

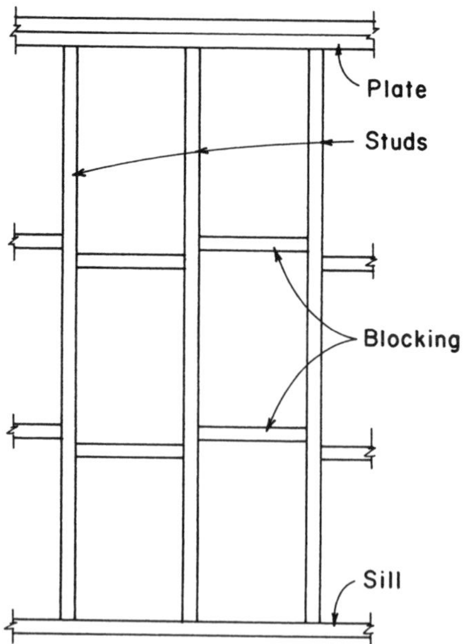

Figure 6.4 Stud wall construction with blocking.

Studs may also serve other functions, as in the case of an exterior wall subjected to wind forces. For this situation the studs must be designed for the combined actions of bending plus compression, as discussed in Section 6.5.

In colder climates it is now common to use studs with widths greater than the nominal 4 in. in order to create a larger void space within the wall to accommodate insulation. This often results in studs with redundant strength for the ordinary tasks of one- and two-story buildings. Wider studs may also be required structurally for very tall walls.

If vertical loads are high or bending is great, it may be necessary to strengthen a stud wall. This can be done in a number of ways, such as:

1. Decreasing stud spacing from the usual 16 in. to 12 in.
2. Increasing the stud width from 4 in. nominal to 6 in. nominal or greater
3. Using doubled studs or large timber sections as posts at locations of concentrated loads

It is also sometimes necessary to use thicker studs or to restrict stud spacing for walls that function as shear walls.

In general, studs are columns and must comply to the various requirements for design of solid-sawn sections. Any appropriate grade of wood may be used, although special stud grades are commonly used for ordinary 2 × 4 or 2 × 6 studs.

Table 6.2, which is adapted from a table in the *International Building Code* (Ref. 4), provides data for the selection of studs for both bearing and nonbearing walls. The code stipulates that these data may be used in lieu of any engineering design for the studs, which means that other possibilities may be considered if computations can support a case for them.

Stud wall construction is often used as part of a general light construction system described as *light wood frame construction*. The joist and rafter construction discussed in Chapter 5 and the stud wall construction discussed here are the primary structural elements of this system. In most applications the system is almost entirely comprised of 2-in.-nominal-dimension lumber. Timber elements are sometimes used for free-standing columns and for heavily loaded or long-span beams.

TABLE 6.2 Requirements for Stud Wall Construction

	Bearing Walls				Nonbearing Walls	
Stud Size (in.)	Laterally Unsupported Stud Height[a] (ft)	Supporting Roof and Ceiling Only	Supporting One Floor, Roof, and Ceiling	Supporting Two Floors, Roof, and Ceiling	Laterally Unsupported Stud Height[a] (ft)	Spacing (in.)
		Spacing (in.)				
2 × 4	10	24	16	—	14	24
2 × 6	10	24	24	16	20	24

Source: Compiled from data in the *International Building Code* (Ref. 4), with permission of the publisher, International Code Council.

[a]Listed heights are distances between points of lateral support placed perpendicular to the plane of the wall. Increases in unsupported height are permitted where justified by analysis.

6.5 COLUMNS WITH BENDING

There are a number of situations in which structural members are subjected to combined effects of axial compression and bending. Studs in exterior walls represent the situation shown in Figure 6.5*a*, with a loading consisting of vertical gravity plus horizontal wind loads. Due to use of common construction details, columns carrying the ends of beams may sometimes be loaded eccentrically, as shown in Figure 6.5*b*.

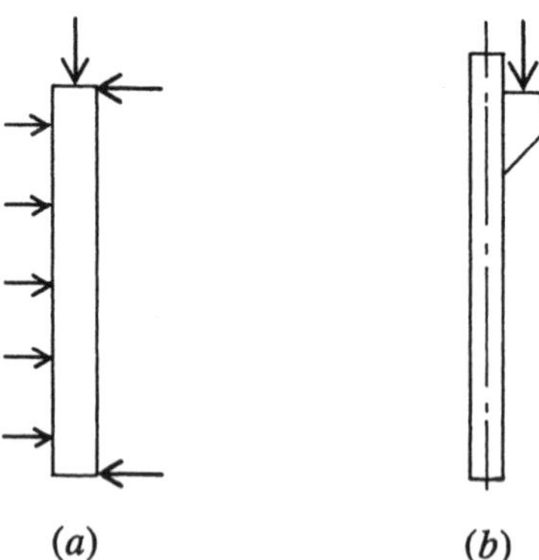

Figure 6.5 Common cases involving combined axial compression and bending in columns: (*a*) exterior stud or truss chord and (*b*) column with bracketed support for spanning member.

Stresses developed by these two actions are both of the direct type (tension and compression) and can be combined for consideration of a net stress condition. However, the basic actions of a column and a bending member are essentially different in character, and it is therefore customary to consider this combined activity by what is called *interaction*. The classic form of interaction is represented by the graph in Figure 3.37*a*.

The interaction relationship is expressed in the formula

$$\frac{P_n}{P_o} + \frac{M_n}{M_o} \leq 1$$

The plot of this equation is the straight line connecting P_o and M_o as shown in Figure 3.37*a*.

An interaction graph similar to that in Figure 3.37*a* can also be produced using stresses rather than loads and moments since the stresses are directly proportional to the loads and moments. This is the procedure generally used in wood and steel design, with the graph taking a form expressed as

$$\frac{f_a}{F_a} + \frac{f_b}{F_b} \leq 1$$

where f_a = computed stress due to load
F_a = allowable column action stress
f_b = computed stress due to bending
F_b = allowable bending stress

Investigation of Columns with Bending, ASD Method

Present design of wood columns uses the straight-line interaction relationship and then adds considerations for buckling due to bending, *P*–delta effects, and so on. For solid-sawn columns the *NDS* provides the following formula for investigation:

$$\left(\frac{f_c}{F_c'}\right)^2 + \frac{f_b}{F_b\left(1 - \frac{f_c}{F_{cE}}\right)} \leq 1$$

where f_c = computed compressive stress
F_c' = adjusted reference design value for compressive stress
f_b = computed bending stress

F_b = reference design stress for bending
F_{cE} = value determined for buckling, described in Section 6.2

The following examples demonstrate some applications for the procedure.

Example 4. An exterior wall stud of Douglas fir–larch, stud grade, is loaded as shown in Figure 6.6*a*. Investigate the stud for the combined loading. (*Note:* This is the wall stud from the building example in Chapter 18.)

Solution: From Table 5.1a, $F_b = 700$ psi, $F_c = 850$ psi, and $E_{min} =$ 510,000 psi. Note that the allowable stresses are not changed by Table 5.2, as the table factors are 1.0. With inclusion of the wind loading, the stress values (but not E) may be increased by a factor of 1.6 (see Table 5.3).

Assume that wall surfacing braces the 2 × 6 studs adequately on their weak axis ($d = 1.5$ in.), so the critical value for d is 5.5 in. Thus,

$$\frac{L}{d} = \frac{11 \times 12}{5.5} = 24$$

$$F_{cE} = \frac{0.822E_{min}}{\left(\frac{L}{d}\right)^2} = \frac{0.822 \times 510{,}000}{(24)^2} = 728 \text{ psi}$$

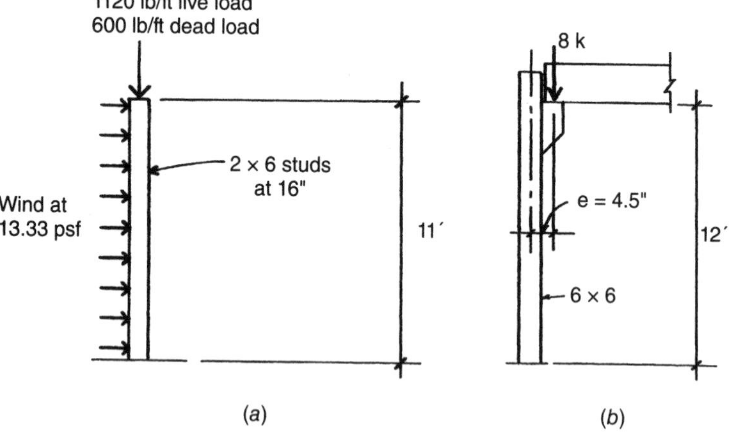

Figure 6.6 Reference for Examples 4 and 5.

The first investigation involves the gravity load without the wind, for which the stress increase factor of 1.6 is omitted. Thus,

$$F_c^* = 850 \text{ psi}$$
$$\frac{F_{cE}}{F_c^*} = \frac{728}{850} = 0.856$$

From Figure 6.2, $C_p = 0.63$, and the stud compression capacity is

$$P = (F_c^*)(C_p)(A) = (850)(0.63)(8.25) = 4418 \text{ lb}$$

This is compared to the given load for the 16-in. stud spacing, which is

$$P = \left(\frac{16}{12}\right)(1720) = 2290 \text{ lb}$$

which demonstrates that the gravity-only load is not a critical concern.

Proceeding with consideration for the combined loading, we determine that

$$F_c^* = 1.6F_c = 1.6(850) = 1360 \text{ psi}$$
$$\frac{F_{cE}}{F_c^*} = \frac{728}{1360} = 0.535$$

From Figure 6.2, $C_p = 0.45$.

For the load combination with wind, the adjusted vertical load is

$$P = \frac{16}{12}\text{ [Dead load + 0.75(Live load)]}$$
$$= \frac{16}{12}(600 + 840) = 1920 \text{ lb}$$
$$F_c' = C_p F_c^* = 0.45 \times 1360 = 612 \text{ psi}$$
$$f_c = \frac{P}{A} = \frac{1920}{8.25} = 233 \text{ psi}$$

For the wind load use $w = 0.75(13.33) = 10$ psf. Then

$$M = \frac{16}{12}\frac{wL^2}{8} = \frac{16}{12}\frac{10(11)^2}{8} = 202 \text{ lb-ft}$$

$$f_b = \frac{M}{S} = \frac{202 \times 12}{7.563} = 320 \text{ psi}$$

$$\frac{f_c}{F_{cE}} = \frac{233}{728} = 0.320$$

Then, using the code formula for the interaction,

$$\left(\frac{f_c}{F'_c}\right)^2 + \frac{f_b}{F_b\left(1 - \frac{f_c}{F_{cE}}\right)} \leq 1$$

$$\left(\frac{233}{612}\right)^2 + \frac{320}{1.6 \times 700(1 - 0.320)} = 0.145 + 0.420 = 0.565$$

As the result is less than 1, the stud is adequate.

Example 5. The column shown in Figure 6.6*b* is of Douglas fir–larch, dense No. 1 grade. Investigate the column for combined column action and bending.

Solution: From Table 5.1a, $F_b = 1400400$, $F_c = 1200$ psi, $E_{min} =$ 620,000 psi. From Table A.8, $A = 30.25$ in.^{2}, and $S = 27.7$ in.^{3}. Then

$$\frac{L}{d} = \frac{12 \times 12}{5.5} = 26.18$$

$$F_{cE} = \frac{0.822 \times 620{,}000}{(26.18)^2} = 744 \text{ psi}$$

$$\frac{F_{cE}}{F_c} = \frac{744}{1200} = 0.62$$

From Figure 6.2, $C_p = 0.51$.

$$f_c = \frac{8000}{30.25} = 264 \text{ psi}$$

$$F'_c = C_p F_c = (0.51)(1200) = 612 \text{ psi}$$

$$\frac{f_c}{F_{cE}} = \frac{264}{744} = 0.355$$

$$f_b = \frac{M}{S} = \frac{8000 \times 4.5}{27.7} = 1300 \text{ psi}$$

and for the column interaction

$$\left(\frac{f_c}{F'_c}\right)^2 + \frac{f_b}{F_b\left(1 - \frac{f_c}{F_{cE}}\right)} \leq 1$$

$$\left(\frac{264}{612}\right)^2 + \frac{1300}{1400(1 - 0.355)} = 0.186 + 1.440 = 1.626$$

As this exceeds 1.0, the column is inadequate. Since bending is the main problem, a second try might be for a 6 × 8 or a 6 × 10, or for an 8 × 8 if a square section is required.

Problem 6.5.A Nine-feet-high 2 × 4 studs of Douglas fir–larch, No. 1 grade, are used in an exterior wall. Wind load is 17 psf on the wall surface; studs are 24 in. on center; the gravity load on the wall is 400 lb/ft of wall length. Investigate the studs for combined action of compression plus bending using the ASD method.

Problem 6.5.B Ten-feet-high 2 × 4 studs of Hem Fir, No. 1 grade, are used in an exterior wall. Wind load is 25 psf on the wall surface; studs are 16 in. on center; the gravity load on the wall is 500 lb/ft of wall length. Investigate the studs for combined action of compression plus bending using the ASD method.

Problem 6.5.C A 10 × 10 column of Douglas fir–larch, No. 1 grade, is 9 ft high and carries a compression load of 20 kips that is 7.5 in. eccentric from the column axis. Investigate the column for combined compression and bending using the ASD method.

Problem 6.5.D A 12 × 12 column of Hem Fir, No. 1 grade, is 12 ft high and carries a compression load of 24 kips that is 9.5 in. eccentric from the column axis. Investigate the column for combined compression plus bending using the ASD method.

Investigation of Columns with Bending, LRFD Method

The process for investigation of columns with bending in the LRFD method uses essentially the same steps as in the ASD method. The usual adjustments are made with load factors, resistance factors, and conversions of reference values. The following example illustrates the process using the same data as in the ASD work for Example 4.

Example 6. The column shown in Figure 6.6*b* is of Douglas fir–larch, dense No. 1 grade. Investigate the column for combined column action and bending using the LRFD method. The applied compression load is one-half live load and one-half dead load.

Solution: From Table 5.1a, $F_b = 1400 = 140 F_c = 1200 = 120 E_{\min} =$ 620,000 psi. From Table A.8, $A = 30.25$ in.^{2}, $S = 27.7$ in.^{3}

$$P_u = 1.2(\text{DL}) + 1.6(\text{LL}) = 1.2(4000) + 1.6(4000)$$

$$= 11{,}200 \text{ lb} \quad \text{or} \quad 11.2 \text{ kips}$$

$$M_u = 11.2 \times 4.5 = 50.4 \text{ kip} - \text{in.}$$

$$\frac{L}{d} = \frac{12 \times 12}{5.5} = 26.18$$

$$E'_{min} = \frac{1.5}{\phi_s}\phi_s E_{min} = \frac{1.5}{0.85}0.85(620{,}000) = 930{,}000 \text{ psi}$$

$$F_{cE} = \frac{0.822 E'_{min}}{(L/d)^2} = \frac{0.822(930{,}000)}{(26.18)^2} = 1115 \text{ psi} \quad \text{or} \quad 1.115 \text{ ksi}$$

$$F_c^* = \lambda\phi_c\left(\frac{2.16}{1000\phi_c})F_c\right)$$

$$= 0.8(0.90)\left[\frac{2.16}{1000\,(0.90)}\right]1200 = 2.074 \text{ ksi}$$

$$\frac{F_{cE}}{F_c^*} = \frac{1.115}{2.074} = 0.5376$$

From Figure 6.2, $C_p = 0.46$.

$$F'_c = \lambda\phi_c\left(\frac{2.16}{1000\phi_c}\right)C_p F_c$$

$$= 0.8(0.90)\left[\frac{2.16}{1000\,(0.90)}\right](0.46)(1200) = 0.954 \text{ ksi}$$

For a consideration of the value to be used for bending stress, an investigation should ordinarily be done of the effects of lateral and torsional buckling, as in the case of a beam. This is usually not critical unless the depth-to-width ratio of the section is greater than 3. In this case the square section has a ratio of 1.0 and the issue is not a concern. What remains to be done to establish the limit for bending stress is simply to make the appropriate adjustments. Thus,

$$F'_b = \lambda\phi_b \frac{2.16}{1000\phi_b} F_b = 0.8(0.85)\frac{2.16}{1000(0.85)}1400 = 2.42 \text{ ksi}$$

$$f_c = \frac{P}{A} = \frac{11.2}{(5.5)^2} = 0.370 \text{ ksi}$$

$$f_b = \frac{M}{S} = \frac{50.4}{27.7} = 1.819 \text{ ksi}$$

And, for the interaction analysis,

$$\left(\frac{f_c}{F'_c}\right)^2 + \left[\frac{f_b}{F'_b\left(1 - \frac{f_c}{F_{cE}}\right)}\right] = \left(\frac{0.370}{0.954}\right)^2 + \left[\frac{1.819}{2.42\left(1 - \frac{0.370}{1.115}\right)}\right]$$

$$= 0.150 + 1.126 = 1.28$$

As this exceeds 1.0, the column is not adequate. A second try might use a 6 × 8, which has a significantly larger section modulus to reduce the bending stress.

Problem 6.5.E A 10 × 10 column of Douglas fir–larch, No. 1 grade, is 9 ft high and carries a compression load of 10 kips dead load plus 10 kips live load that is 7.5 in. eccentric from the column axis. Investigate the column for combined compression and bending using the LRFD method.

Problem 6.5.F A 12 × 12 column of Hem Fir, No. 1 grade, is 12 ft high and carries a compression load of 12 kips dead load plus 12 kips live load that is 9.5 in. eccentric from the column axis. Investigate the column for combined compression plus bending using the LRFD method.

7

CONNECTIONS FOR WOOD STRUCTURES

Structures of wood typically consist of large numbers of separate pieces that must be joined together. For assemblage of building construction, fastening is most often achieved by using some steel device, common ones being nails, screws, bolts, and specially formed steel fasteners.

7.1 BOLTED JOINTS

When steel bolts are used to connect wood members, there are several design concerns. Some of the principal concerns are the following:

1. *Net Cross Section in Member.* Holes made for the placing of bolts reduce the wood member cross section. For this investigation, the hole diameter is assumed to be $^1/_{16}$ in. larger than that of the bolt. Common situations are shown in Figure 7.1.

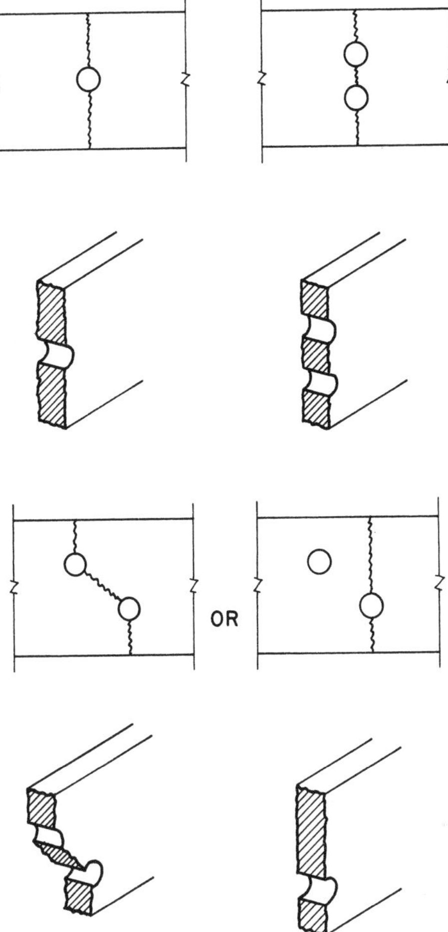

Figure 7.1 Effect of bolt holes on reduction of cross section for tension members.

When bolts in multiple rows are staggered, it may be necessary to make two investigations, as shown in the illustration.

2. *Bearing of the Bolt on the Wood.* This compressive stress limit varies with the angle of the wood grain to the load direction.

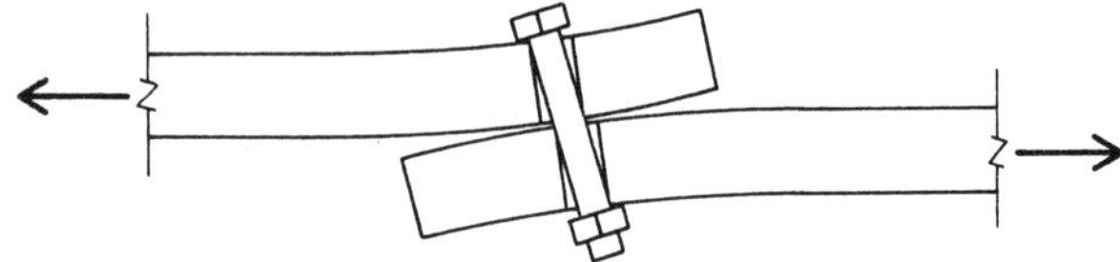

Figure 7.2 Twisting in the two-member bolted joint.

3. *Bending of the Bolt.* Long thin bolts in thick wood members will bend considerably, causing a concentration of bearing at the edge of the hole.
4. *Number of Members Bolted at a Single Joint.* The worst case, as shown in Figure 7.2, is that of the two-member joint. In this case the lack of symmetry in the joint produces considerable twisting. This situation is referred to as *single shear* since the bolt is subjected to shear on a single cross section of the bolt. With more members in the joint, twisting may be eliminated and the bolt is sheared at multiple cross sections.
5. *Ripping Out the Bolt When Too Close to an Edge.* This problem, together with that of the minimum spacing of bolts, is dealt with by using criteria given in the *NDS.*

7.2 NAILED JOINTS

Nails are used in great variety in building construction. For structural fastening, the nail most commonly used is called—appropriately—the *common wire nail.* As shown in Figure 7.3, the critical concerns for such nails are the following:

1. *Nail Size.* Critical dimensions are the diameter and length (see Figure 7.3*a*). Sizes are specified in pennyweight units, designated as 4d, 6d, and so on, and referred to as four penny, six penny, and so on.
2. *Load Direction.* Pullout loading in the direction of the nail shaft is called *withdrawal*; shear loading perpendicular to the nail shaft is called *lateral load.*
3. *Penetration.* Nailing is typically done through one element and into another, and the load capacity is essentially limited by the amount of the length of embedment of the nail in the second

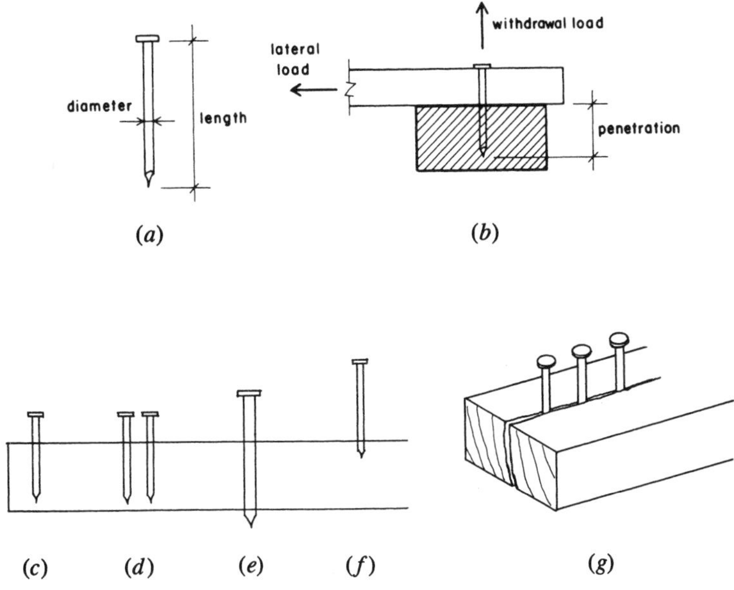

Figure 7.3 Use of common wire nails: (*a*) critical dimensions, (*b*) loading considerations; (*c–g*) poor nailing practices: (*c*) too close to edge, (*d*) nails too close together, (*e*) nails too large for wood piece, (*f*) too little penetration of nail into holding piece of wood, and (*g*) too many closely spaced nails in a single row parallel to the wood grain and/or nails too close to end of piece.

member (see Figure 7.3*b*). The length of this embedment is called the penetration.

4. *Species and Grade of Wood.* The heavier the wood (indicating generally harder, tougher material), the greater is the load resistance capability.

Design of good nailed joints requires a little engineering and a lot of good carpentry. Some obvious situations to avoid are those shown in Figures 7.3*c–g*.

Withdrawal load capacities of nails are given in units of force per inch of nail penetration length. This unit load is multiplied by the actual penetration length to obtain the total force capacity of the nail. For structural connections, withdrawal resistance is relied on only when the nails are perpendicular to the wood grain direction.

Lateral load capacities for common wire nails are given in Table 7.1 for joints with both plywood and lumber side pieces. The *NDS* contains very extensive tables for many wood types as well as metal side pieces. The following example illustrates the design of a nailed joint using the ASD method with data from Table 7.1.

TABLE 7.1 Reference Lateral Load Values for Common Wire Nails (lb/in.)

Side Member Thickness, t_s (in.)	Nail Length, L (in.)	Nail Diameter, D (in.)	Nail Pennyweight	Load per Nail, Z (lb)
Part 1 — With Wood Structural Panel Side Members[a] ($G = 0.42$)				
	2	0.113	6d	48
3/8	2½	0.131	8d	63
	3	0.148	10d	76
	2	0.113	6d	50
	2½	0.131	8d	65
	3	0.148	10d	78
	3½	0.162	16d	92
	2	0.113	6d	58
	2½	0.131	8d	73
	3	0.148	10d	86
	3½	0.162	16d	100
Part 2 — With Sawn-Lumber Side Members[b] ($G = 0.50$)				
	2½	0.131	8d	90
	3	0.148	10d	105
	3½	0.162	16d	121
	4	0.192	20d	138
	3	0.148	10d	118
	3½	0.162	16d	141
	4	0.192	20d	170
	4½	0.207	30d	186
	5	0.225	40d	205
	5½	0.244	50d	211

Source: Adapted from the *National Design Specification*® (NDS®) *for Wood Construction*, 2015 edition (Ref. 3), with permission of the publisher, American Forest and Paper Association.

[a]Values for single-shear joints with wood structural panel side members with $G = 0.42$ and nails anchored in sawn lumber of Douglas fir–larch with $G = 0.50$.

[b]Values for single-shear joints with both members of sawn lumber of Douglas fir–larch with $G = 0.50$.

Example 1. A structural joint is formed as shown in Figure 7.4, with the wood members connected by 16d common wire nails. Wood is Douglas fir–larch. What is the maximum value for the compression force in the two side members?

Solution: From Table 7.1, we read a value of 141 lb per nail (side member thickness of 1.5 in., 16d nails). As shown in the illustration, there are five nails on each side, or a total of 10 nails in the joint. The total joint load capacity is thus

$$C = (10)(141) = 1410 \text{ lb}$$

No adjustment is made for direction of load to the grain. However, the basic form of nailing assumed here is so-called side grain nailing in which the nail is inserted at 90° to the grain direction and the load is perpendicular (lateral) to the nails.

Minimum adequate penetration of the nails into the supporting member is a necessity, but use of the combinations given in Table 7.1 assures adequate penetration if the nails are fully buried in the members.

Problem 7.2.A A joint similar to that in Figure 7.4 is formed with outer members of 1-in. nominal thickness (3/4-in. actual thickness) and 10d common wire nails. Find the compression force that can be transferred to the two side members.

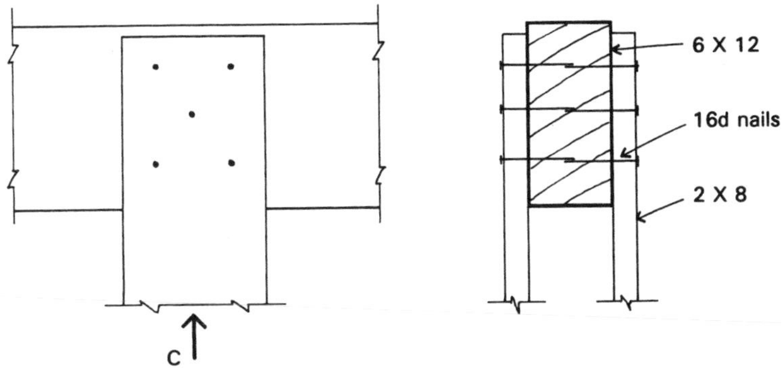

Figure 7.4 Reference for Example 1.

Problem 7.2.B Same as Problem 7.2.A, except outer members are 2 × 10, middle member is 4 × 10, and nails are 20d.

7.3 PLYWOOD GUSSETS

Cut pieces of plywood are sometimes used as connecting devices, although the availability of manufactured metal devices is widespread. Light trusses consisting of a single plane of wood members of 2-in. nominal thickness are sometimes assembled with gussets of plywood. Although such connections may have considerable load resistance, it is best to be conservative in using them for computed structural forces, especially with regard to tension stress in the plywood. The following example treats a joint for a light truss using lumber members for the truss and connecting panels of structural grade plywood.

Example 2. The truss heel joint shown in Figure 7.5 is made with 2-in.-nominal-thickness lumber and gusset plates of ½-in.-thick plywood. Nails are 6d common wire with the nail layout shown occurring on both sides of the joint. Find the tension load capacity for the bottom chord member (load 3 in the figure).

Solution: From Table 7.1 the capacity of one nail is 50 lb. With 12 nails on each side of the joint, the total capacity of the joint is thus

$$T = (24)(50) = 1200 \text{ lb}$$

Problem 7.3.A A truss heel joint similar to that in Figure 7.5 is made with gusset plates of ½-in. plywood and 8d nails. Find the tension force limit for the bottom chord.

Problem 7.3.B A truss heel joint similar to that in Figure 7.5 is made with ¾-in. plywood and 10d nails. Find the tension force limit for the bottom chord.

7.4 INVESTIGATION OF CONNECTIONS, LRFD METHOD

Use of the LRFD method for connections involves the same basic procedures as for the ASD method. Reference values are adjusted by the format conversion factor, $K_F = 2.16/1000\varphi_Z$, for which $\varphi_Z = 0.65$. Other adjustment factors for loads and resistance are used appropriate to the load combinations, moisture conditions, and the like.

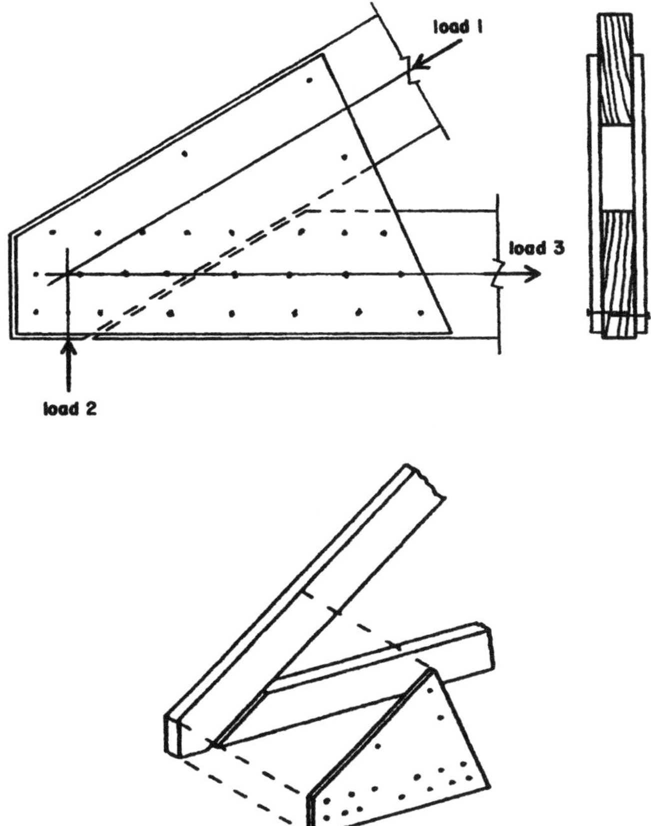

Figure 7.5 Reference for Example 2.

7.5 FORMED STEEL FRAMING ELEMENTS

Formed metal framing devices have been used for many centuries for the assembly of structures of heavy timber. In ancient times elements were formed of bronze or cast iron or wrought iron. Later they were formed of forged or bent and welded steel elements. Some of the devices commonly used today are essentially the same in function and detail to those used long ago.

For large timber members, connecting elements are now mostly formed of steel plate that is bent and welded to produce the desired shape. (See Figure 7.6.) The ordinary tasks of attaching beams to

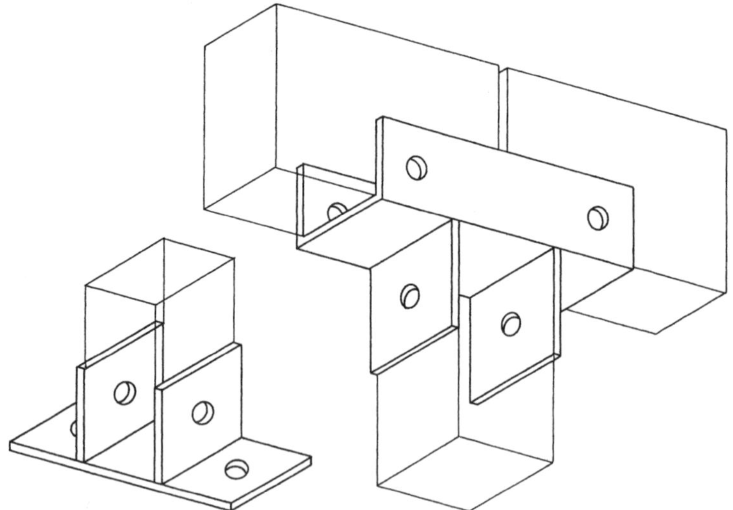

Figure 7.6 Simple connecting devices formed from bent and welded steel plates.

columns and columns to foundations continue to be required, and the simple means of achieving the tasks evolve from practical concerns.

For resistance to gravity loads, connections such as those shown in Figure 7.6 sometimes have no direct structural functions. In theory, it is possible to simply rest a beam on top of a column, as is done in some rustic construction. However, for resistance to lateral loads from wind or earthquakes, the tying and anchoring functions of these connecting devices are often quite essential. They also serve a practical function of simply holding the parts together during the construction process.

A development of more recent times is the extension of the use of metal devices for the assembly of light wood frame construction. Devices of thin sheet metal, such as those shown in Figure 7.7, are now commonly used for stud and joist construction employing predominantly wood members of 2 in. nominal dimension thickness. As with the devices used for heavy timber construction, these lighter connectors often serve useful functions of tying and anchoring the structural members. Load transfers between basic elements of a building's lateral bracing system are often achieved with these elements.

Commonly used connection devices of both the light sheet steel type and the heavier steel plate type are readily available from building

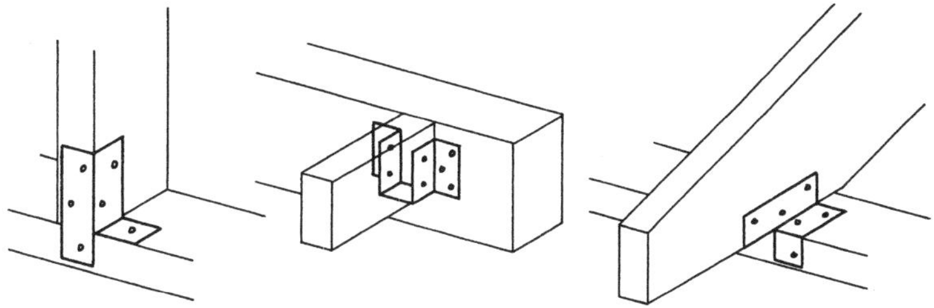

Figure 7.7 Common connection devices used for light wood frame construction, formed from bent sheet steel.

material suppliers. Many of these devices are approved by building codes for rated structural capacity functions.

Concrete and Masonry Anchors

Wood members supported by concrete or masonry structures must usually be anchored through some intermediate device. The most common attachment is with steel bolts cast into the concrete or masonry. However, there is also a wide variety of devices that may be directly cast into the supports or attached with drilled-in, dynamically anchored, or other elements.

Two common situations are shown in Figure 7.8. The sill member for a wood stud in Figure 7.8*a* is typically attached directly with steel anchor bolts that are cast into the supports. These bolts serve to hold

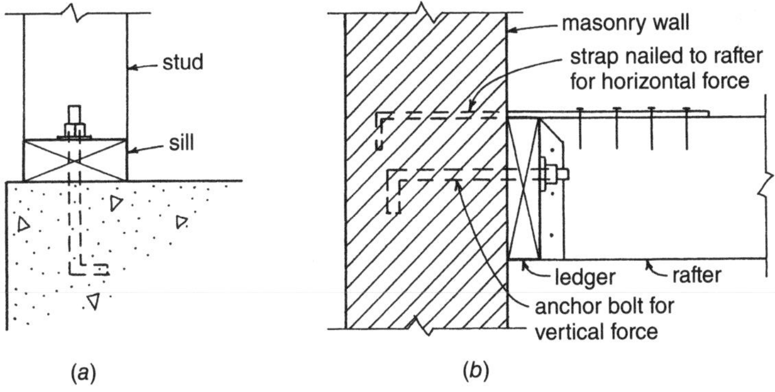

Figure 7.8 Devices for anchoring wood structures to concrete and masonry supports.

the wall securely in position during the construction process. However, they may also serve to anchor the wall against lateral or uplift forces.

Figure 7.8*b* shows a common situation in which a wood-framed roof or floor is attached to a masonry wall through a member bolted to the face of the wall, called a *ledger*. Vertical load transfer develops a shear effect on the bolt. For lateral force a problem is the pullout or tension effect on the bolt, although another problem may be the cross-grain bending in the ledger. In zones of high seismic risk, it is usually required to have a separate *horizontal anchor*, such as the strap shown in Figure 7.8*b*.

STEEL CONSTRUCTION

Steel is used in a wide variety of forms for many tasks in building construction. Wood, concrete, and masonry structures require many steel objects. This part of the book, however, deals with steel as a material for the production of components and systems for steel structures. This usage includes some common ones and an endless range of special possibilities. The concentration here is on ordinary uses.

The industry organization that provides much of the material used in analyzing and designing structures made of rolled steel in the United States is the American Institute of Steel Construction (AISC). Steel open-web joists and joist girders are represented by the Steel Joist Institute (SJI) and steel decking by the Steel Deck Institute (SDI). Reference manuals for design in steel include the *Steel Construction Manual* (Ref. 5), *Standard Specifications, Load Tables, and Weight Tables for Steel Joists and Joist Girders* (Ref. 6), and *Steel Deck Institute Design Manual for Composite Decks, Form Decks, and Roof Decks* (Ref. 9).

8

STEEL STRUCTURAL PRODUCTS

For the assemblage of building structures, components of steel consist mostly of standard forms of industrially produced products. The most common and widely used of these products are in forms that have been developed and produced for a long time. Modifications of production and assemblage methods are made continuously, but the basic forms of most steel structures are pretty much as they have been for many years.

8.1 DESIGN METHODS FOR STEEL STRUCTURES

Presently, two fundamentally different methods for structural investigation and design are in use. The first of these, traditionally used for many years by designers and researchers, is referred to as the *working stress method* or the *allowable stress method*. At present, this method is called the *allowable stress design* method, designated ASD. The second method, first adopted in 1986, is called the *ultimate strength method* or simply the *strength method.* At present, this method is called the *load and resistance factor design* method, designated LRFD.

In general, the techniques and operational procedures of the ASD method are simpler to use. They are based largely on direct use of classical analytical formulas for stress and strain and a direct use of the actual working loads (called *service loads*) assumed for the structure. In fact, it is often useful to explain the analytical procedures of the LRFD method by comparing them to those used for the ASD method.

The ASD method for steel design is similar to the method used for wood design demonstrated in Part II. Allowable stresses for steel design are mostly based on the yield point of the steel adjusted with a safety factor. The basis for design is the limiting of the conditions in the structure under service loads to an acceptable level of stress and strain.

With the LRFD method, the basis for design is a visualization of the mode of failure of the structure under the given loads. The failure loading is considered as the ultimate resistance of the structure, designated as the *nominal resistance* of the structure. For design purposes, the true resistance is reduced by some percentage through the use of a *resistance factor* designated φ. The design process consists of comparing this modified resistance to a loading that is also modified by use of factors to increase the load above the level of the service loads. The design strength must be equal to or greater than the resistance required by the design loading. The basic process is expressed mathematically as

$$\varphi R_n \geq \sum \delta_i Q_i$$

where φ = resistance factor ($\varphi < 1$)
R_n = nominal resistance of the member
δ_i = load factor ($\delta > 1$)
Q_i = different loading effects

Recently steel design has another method of design, which in simplistic terms is a combination of ASD and LRFD. This method is known as allowable strength design. This is a strength design method using service loads, not factored loads. The basic process is mathematically expressed as

$$\frac{R_n}{\Omega} \geq \sum Q_i$$

where R_n = nominal resistance of the member
Ω = safety factor ($\Omega > 1$ and $\Omega = 1.5/\varphi$)
Q_i = different loading effects

The ASD is still in most building codes as an acceptable method of design; however, any reference to ASD in the steel design manuals refers to the allowable strength design, not stress design.

The chief source for information for design of steel structures—the American Institute of Steel Construction (AISC)—publishes design references supporting both the ASD and LRFD methods. These references are used extensively throughout this book, with a major source being the AISC *Steel Construction Manual* (Ref. 5).

8.2 MATERIALS FOR STEEL PRODUCTS

The strength, hardness, corrosion resistance, and various other physical properties of steel can be varied through a considerable range by changes in the material production process. Literally hundreds of different steels are produced, although only a few standard products are used for the elements of building structures. Working and forming processes, such as rolling, drawing, forging, and machining, may also alter some properties. However, some properties, such as density (unit weight), stiffness (modulus of elasticity), thermal expansion, and fire resistance, tend to remain constant for all steels.

For various applications, other properties may be significant. Hardness affects the ease with which cutting, drilling, planing, and other working can be done. For welded connections, the weldability of the base material must be considered. Resistance to rusting is normally low but can be enhanced by various materials added to the steel, producing special steels, such as stainless steel and so-called rusting steel, which rusts at a very slow rate.

These various properties of steel must be considered when working with the material and when designing for its use. However, in this book, we are most concerned with the unique structural nature of steel.

Structural Properties of Steel

Basic structural properties, such as strength, stiffness, ductility, and brittleness, can be interpreted from laboratory load tests on specimens of the material. Figure 8.1 displays characteristic forms of curves that are obtained by plotting stress (load resistance) and strain (deformation) values from such tests. An important property

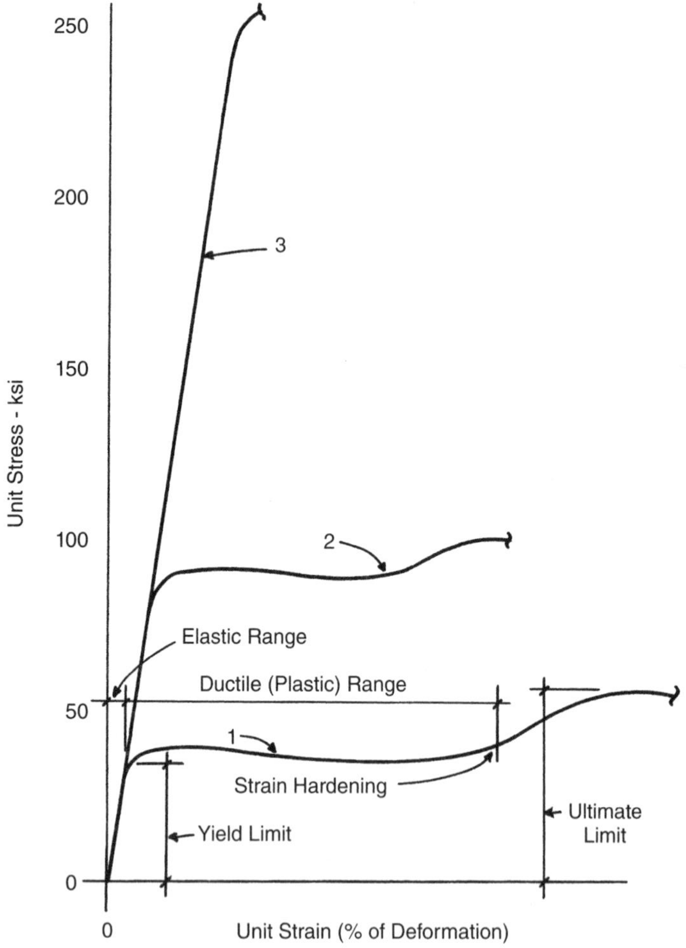

Figure 8.1 Stress–strain response of steel: (1) ordinary structural steel, (2) high-strength steel for rolled shapes, and (3) superstrength steel, mostly in wire form.

of most structural steels is the plastic deformation (ductility or yield) phenomenon. This is demonstrated by curve 1 in Figure 8.1. For steels with this character, there are two different stress values of significance: the yield limit and the ultimate failure limit.

Generally, the higher the yield limit, the less the degree of ductility. The extent of ductility is measured as the ratio of the plastic deformation between first yield and strain hardening (see Figure 8.1, curve 1) to the elastic deformation at the point of yield. Curve 1 in Figure 8.1 is representative of ordinary structural steel [American Society for Testing and Materials (ASTM) A36], and curve 2 indicates the typical effect when the yield strength is raised a significant amount. Eventually, the significance of the yield phenomenon becomes virtually negligible when the yield strength approaches as much as three times the yield of ordinary steel (36 ksi for ASTM A36 steel).

Some of the highest strength steels are produced only in thin-sheet or drawn-wire forms. Bridge strand is made from wire with strength as high as 300 ksi. At this level, yield is almost nonexistent, and the wires approach the brittle nature of glass rods (see curve 3 in Figure 8.1).

For economical use of the expensive material, steel structures are generally composed of elements with relatively thin parts. This results in many situations in which the limiting ultimate strength of elements in bending, compression, and shear is determined by buckling, rather than by the stress limits of the material. Because buckling is a function of stiffness (modulus of elasticity) of the material, and because this property remains the same for all steels, there is limited opportunity to make effective use of higher-strength steels in many situations. The grades of steel commonly used are to some extent ones that have the optimal effective strength for most tasks.

Because many structural elements are produced as some manufacturer's product line, choices of basic materials are often mostly out of the hands of individual building designers. The proper steel for a given task—on the basis of many properties—is determined as part of the product design, although a range of grades may be available for some products.

Steel that meets the requirements of ASTM Specification 36 was the grade of structural steel commonly used to produce rolled steel elements for building construction in the past. It had an ultimate tensile strength of 58–80 ksi and a minimum yield point of 36 ksi. This material was largely produce using virgin iron ore in a smelter furnace. This steel is referred to simply as A36 steel.

High-strength steel is now the steel most commonly used to produce rolled steel elements for buildings. Steel designated as high strength meets ASTM Specification A992. It is available mainly because steel produced by electric arc furnaces is made largely of recycled steel, and one of the outcomes of using recycled steel is that the steel's yield strength increases. Electric arc furnaces produce the steel for all domestically made wide-flange beam shapes. Thus, the use of structural steel with 50 ksi yield strength has replaced A36 steel in the North American construction industry. All work in this text will be done using high-strength steel.

Other Uses of Steel

Steel used for other purposes than the production of rolled products generally conforms to standards developed for the specific product. This is generally true for steel connectors, wire, cast and forged elements, and very-high-strength steels produced in sheet, bar, and rod forms for fabricated products. The properties and design stresses for some of these product applications are discussed in other places in this book. Standards used typically conform to those established by industrywide organizations, such as the Steel Joist Institute (SJI) and the Steel Deck Institute (SDI). In some cases, larger fabricated products make use of ordinary rolled products, produced from A36 steel or other grades of steel from which hot-rolled products can be obtained.

8.3 TYPES OF STEEL STRUCTURAL PRODUCTS

As a material, steel itself is formless, used basically for production as a molten material or a heat-softened lump. The structural products produced derive their basic forms from the general potentialities and limitations of the industrial processes of forming and fabricating. A major process used for structural products is that of *hot rolling*, which is used to produce the familiar cross-sectional forms (called *shapes*) of I, H, L, T, U, C, and Z, as well as flat plates and round or rectangular bars. Other processes include drawing (used for wire), extrusion, casting, and forging.

Raw stock can be assembled by various means into objects of multiple parts, such as a manufactured truss, a prefabricated wall panel, or a whole building framework. Learning to design with steel begins with acquiring some familiarity with the standard industrial processes and products and with the means of reforming them and attaching them to other elements in structural assemblages.

Rolled Structural Shapes

The products of the steel rolling mills used as beams, columns, and other structural members are designated as *sections* or *shapes*, relating to the form of their cross sections. Their usage relates to the industry-developed standard cross sections that have been developed in response to common uses. American standard I beams (Figure 8.2*a*) were the first beam sections rolled in the United States and are currently produced in sizes of 3–24 in. in depth. The W shapes (Figure 8.2*b*, originally called wide-flange shapes) are a modification of the I cross section and are characterized by parallel flange surfaces (of constant thickness) as contrasted with the tapered form of the I-beam flanges. W shapes are available in depths from 4 to 44 or more inches. In addition to the standard I and W shapes, the structural steel shapes most frequently used in building construction

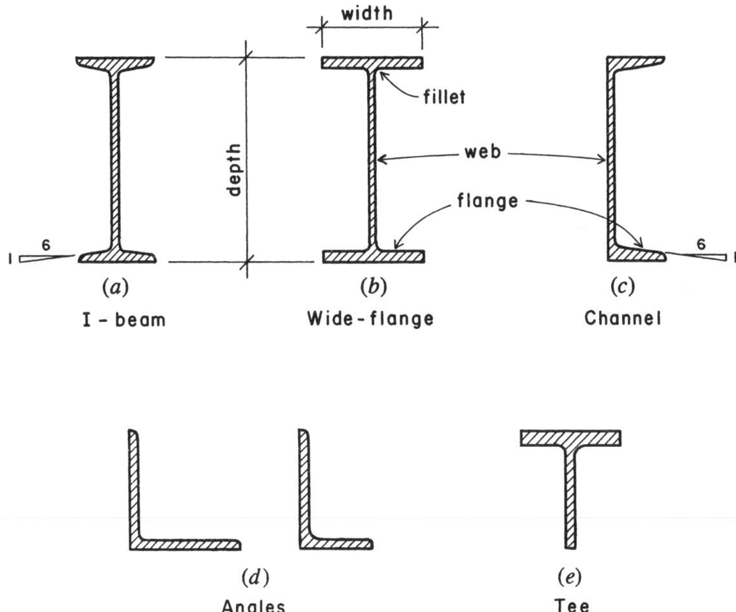

Figure 8.2 Shapes of typical hot-rolled products.

TABLE 8.1 Standard Designations for Structural Steel Elements

Elements	Designation
American standard I beams, S shapes	S 12 × 35
Wide flanges, W shapes	W 12 × 27
Miscellaneous shapes, M shapes	M 8 × 18.5
American standard channels, C shapes	C 10 × 20
Miscellaneous channels, MC shapes	MC 12 × 40
Bearing piles, HP shapes	HP 14 × 117
Angles, L shapes	L 5 × 3 × 1/2
Structural tees, WT, ST, MT	WT 9 × 38
Plates	PL 1 1/2 × 10 × 16
Structural tubing	HSS 10 × 6 × 1/2
Pipe, standard weight	Pipe 4 Std.
Pipe, extra strong	Pipe 4 X-strong
Pipe, double extra strong	Pipe 4 XX-strong

are channels, angles, tees, plates, and bars. The tables in Appendix A list the dimensions, weights, and various properties of some of these shapes. Complete tables of structural shapes are given in the AISC Steel Manual (Ref. 5). Table 8.1 lists the standard designations used for rolled shapes and for formed rectangular tubing and round steel pipe.

W Shapes

In general, W shapes have greater flange widths and relatively thinner webs than standard I beams. As noted earlier, the inner faces of the flanges are parallel to the outer faces. These sections are identified by the alphabetic symbol W, followed by the nominal depth in inches and the weight in pounds per linear foot. Thus, the designation W 12 × 26 indicates a W shape of nominal 12-in. depth, weighing 26 lb per linear foot (plf).

The actual depths of W shapes vary within the nominal depth groupings. From Table A.3, we know that a W 12 × 26 has an actual depth of 12.22 in., whereas the depth of a W 12 × 30 is 12.34 in. This is a result of the rolling process during manufacture in which the cross-sectional areas of W shapes are increased by spreading the rollers both horizontally and vertically. Additional area is thereby added to the cross section by increasing flange and web thicknesses as well as flange width (Figure 8.2*b*). The higher percentage of material in the flanges makes the W shapes more efficient for bending resistance than standard I beams. A wide variety of weights is available within each nominal depth group.

Many W shapes are rolled with flange widths approximately equal to their depth. The resulting H-shape configurations are more suitable for columns than the I-shape profiles.

Cold-Formed Steel Products

Sheet steel can be bent, punched, or rolled into a variety of forms. Structural elements so formed are called *cold-formed* or *light-gage* steel products. Steel decks and very light weight framing elements are produced in this manner. These products are described in Chapter 12.

Fabricated Structural Components

A number of structural products are produced with both hot-rolled and cold-formed elements. Open-web steel joists consist of prefabricated, light steel trusses. For short spans and light loads, a common design is that shown in Figure 8.3*a* in which the web consists of a single, continuous bent steel rod and the chords of steel rods or cold-formed elements. For larger spans or heavier loads, the forms more closely resemble those of ordinary light steel trusses with members of single angles, double angles, and structural tees. Open-web joists are discussed in Section 9.9.

Another type of fabricated joist is shown in Figure 8.3*b*. This member is formed from standard rolled shapes by cutting the web in a zigzag fashion as shown in Figure 8.3*c*. The resulting product, called a *castellated beam*, has a greatly reduced weight-to-depth ratio when compared with the lightest rolled shapes.

Other fabricated products range from those used to produce whole building structural systems to individual elements for construction of frames for windows, doors, curtain wall systems, and partitions. Many components and systems are produced as proprietary items by a single manufacturer, although some are developed under controls of industrywide standards, such as those published by the Steel Joist Institute and the Steel Deck Institute.

Development of Structural Systems

Structural systems that comprise entire roof, floor, or wall constructions—or even entire building frameworks—are typically assembled from many individual elements. These elements may be

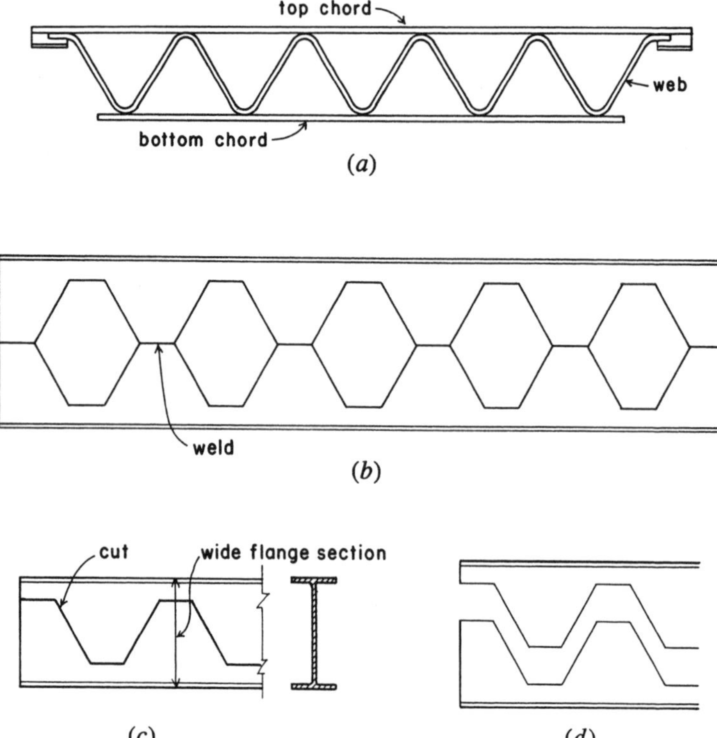

Figure 8.3 Fabricated products formed from steel elements.

of some variety, as in the case of floors using rolled shapes for beams and a formed sheet steel deck. Selection of individual elements may be made from structural investigations but is also often largely a matter of practical development of the form of construction with commonly used systems.

It is common for a building to incorporate more than a single material for its entire structural system. Various combinations, including a wood deck on steel beams or masonry bearing walls for support of steel framing, are possible. This part of the book deals primarily with structures of steel, but some of these mixed material situations are very common and are discussed in other parts of this book.

Connection Methods

Connection of structural steel members that consist of rolled steel shapes is typically achieved by direct welding or by steel rivets or bolts. Riveting for building structures, although common in the past, has become obsolete in favor of bolting with high-strength bolts. However, the forms of structural members, and even of connections, that were developed for riveted construction are still used with little change for bolted construction. The design of bolted connections is discussed in Chapter 11. In general, welding is preferred for shop (factory) fabrication and bolting for field (construction site) connections.

Thin elements of cold-formed sheet steel may be attached by welding, bolting, or using sheet metal screws. Thin deck and wall paneling elements are sometimes attached to one another by simple interlocking at their abutting edges; the interlocked parts can be folded or crimped to give further security to the connection.

A frequent structural design problem is that of the connections of columns and beams in heavy frames for multistory buildings. Design of these connections is beyond the scope of this book, although lighter framing connections of various forms are discussed in Chapter 11.

9

STEEL BEAMS AND FRAMING ELEMENTS

There are many steel elements that can be used for the basic functions of spanning, including rolled shapes, cold-formed shapes, and fabricated beams and trusses. This chapter deals with fundamental considerations for these elements, with an emphasis on rolled shapes. For simplicity, it is assumed that all the rolled shapes used for the work in this chapter is steel with $F_y = 50$ ksi.

9.1 FACTORS IN BEAM DESIGN

Various rolled shapes may serve beam functions, although the most widely used is the *wide-flange shape*, that is, the member with an I-shaped cross section that bears the standard designation of W shape. Except for those members of the W series that approach a square in cross section (flange width approximately equal to nominal depth), the proportions of the shapes in this series are developed for optimal use in flexure about their major axis (designated as x–x). Design for beam use may involve any combination of the following considerations:

Flexural Stress. Flexural stresses generated by bending moments are the primary stress concern in beams. There are many failure modes for beams that define the approach to designing with them, but the general equation for the design of bending members, using the LRFD method, is

$$\phi_b M_n \geq M_u$$

where φ_b = 0.9 for rolled sections
M_n = nominal moment capacity of the member
M_u = maximum moment due to factored loading

Buckling. In general, beams that are not adequately braced may be subject to various forms of buckling. Especially critical are beams with very thin webs or narrow flanges or with cross sections very weak in the lateral (sideways) direction, that is, with low resistance in regard to the minor axis, or y–y axis. Buckling controls the failure mechanism in inadequately braced members and greatly reduces bending capacity. The most effective solution is to provide adequate bracing to eliminate this mode of failure. See discussion in Section 3.11.

Deflection. Although steel is one of the strongest materials used for ordinary construction, steel structures tend to be quite flexible; thus, vertical deflection of beams must be carefully investigated. A significant value to monitor is the span-to-depth ratio of beams; if this is kept within certain limits, deflection is much less likely to be critical.

Connections and Supports. Framed structures contain many joints between separate pieces, and details of the connections and supports must be developed for proper construction as well as for the transfer of necessary structural forces through the joints. End connections for beams may also provide some bracing to reduce buckling failures.

System Design Concerns. Individual beams are often parts of a system in which they play an interactive role. Besides their basic beam functions, there are often design considerations that derive from the overall system actions and interactions. Discussions in this chapter focus mostly on individual beam actions, but discussions in other chapters treat the general usage and overall incorporation of beams in structural systems and, indeed, in the whole building construction system.

There are several hundred different W shapes for which properties are listed in the AISC *Steel Manual* (Ref. 5). A sampling of these is presented in Table A.3 in Appendix A. In addition, there are several other shapes that frequently serve beam functions in special circumstances. Selection of the optimal shape for a given situation involves many considerations; an overriding concern is often the choice of the most economical shape for the task. In general, the least costly shape is usually the one that weighs the least—other things being equal—because steel is priced by unit weight. In most design cases, therefore, the *least weight* selection is typically considered the most economical.

Just as a beam may be asked to develop other actions, such as tension, compression, or torsion, other structural elements (such as columns) may be asked to develop beam functions. Walls may span for bending against wind pressure, columns may receive bending moments as well as compression loads, and truss chords may span as beams as well as function for basic truss actions. The basic beam functions described in this chapter may thus be part of the design work for various structural elements besides the singular-purpose beam.

9.2 INELASTIC VERSUS ELASTIC BEHAVIOR

As discussed in Chapters 1 and 8, there are two competing methods of design for steel: stress design (ASD) and strength design (LRFD). Strength design in steel dominates current practice and can either be allowable strength design or load and resistance factor design. LRFD has been selected for this book because it has become the predominant standard for the construction industry. The basic difference between these two methods is rooted in the distinction between elastic or inelastic theory of member behavior. Stress design is rooted in elastic behavior and strength design is rooted in inelastic behavior. The purpose of this section is to compare these two theories so the reader can better understand inelastic theory and, therefore, the LRFD method of design.

The maximum resisting moment by elastic theory is predicted to occur when the stress at the extreme fiber of a cross section reaches the elastic yield value, F_y, and it may be expressed as the product of the yield stress and the section modulus of the member cross section. Thus,

$$M_y = F_y \times S$$

Beyond this condition the resisting moment can no longer be expressed by elastic theory equations because an inelastic, or *plastic*, stress condition will start to develop on the beam cross section.

Figure 9.1 represents an idealized form of a load test response for a specimen of ductile steel. The graph shows that up to the yield point the deformations are proportional to the applied stress and that beyond the yield point there is a deformation without an increase in stress. For steel, this additional deformation, called the *plastic range*, is approximately 15 times that produced just before yield occurs. This relative magnitude of the plastic range of deformation is the basis for qualification of the material as significantly ductile.

Note that beyond the plastic range the material once again stiffens, called the *strain-hardening* effect, which indicates a loss of ductility and the onset of a second range of increased stress resistance in which

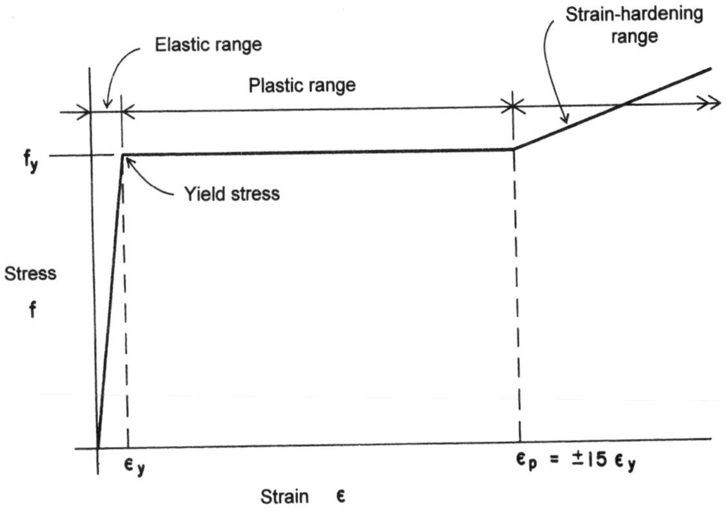

Figure 9.1 Idealized form of the stress–strain response of ductile steel.

increased deformation is produced only with additional increase in stress. The end of this range establishes the *ultimate stress* limit for the material.

The following example illustrates the application of the elastic theory and will be used for comparison with an analysis of plastic behavior.

Example 1. A simple beam has a span of 16 ft and supports a single concentrated load of 18 kips at its center. If the beam is a W 12 × 30, compute the maximum flexural stress.

Solution: See Figure 9.2. For the maximum value of the bending moment,

$$M = \frac{PL}{4} = \frac{18 \times 16}{4} = 72 \text{ kip-ft}$$

In Table A.3 find the value of S for the shape as 38.6 in.^{3} Thus, the maximum stress is

$$f = \frac{M}{S} = \frac{72 \times 12}{38.6} = 22.4 \text{ ksi}$$

and it occurs as shown in Figure 9.2*d*.

Note that this stress condition occurs only at the beam section at midspan. Figure 9.2*e* shows the form of the deformations that

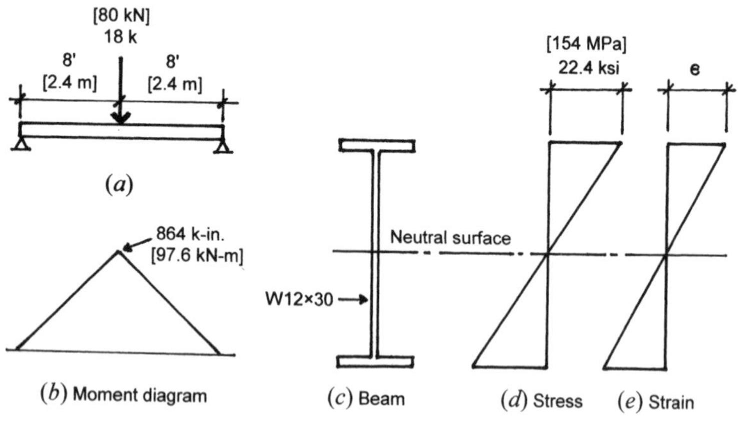

Figure 9.2 Elastic behavior of the beam.

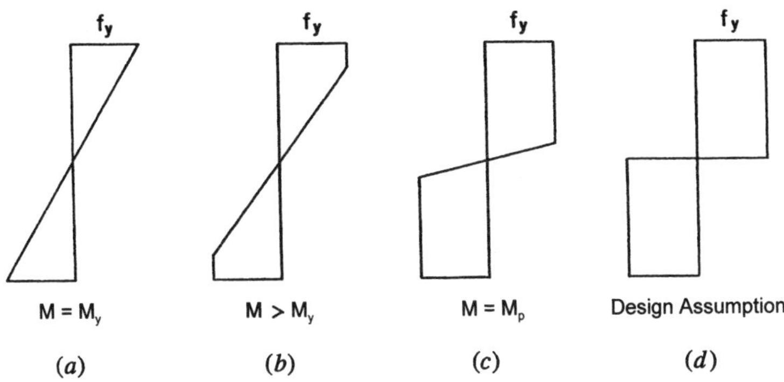

Figure 9.3 Progression of development of bending stress, from the elastic to the plastic range.

accompany the stress condition. This stress level is well below the elastic stress limit (yield point) of 50 ksi.

The limiting moment for elastic stress is that which occurs when the maximum flexural stress reaches the yield limit, as stated before in the expression for M_y. This condition is illustrated by the stress diagram in Figure 9.3*a*.

If the loading and the resulting bending moment that causes the yield limit flexural stress is increased, a stress condition such as illustrated in Figure 9.3*b* begins to develop as the ductile material deforms plastically. This spread of the yield stress level over the beam cross section indicates the development of a resisting moment in excess of M_y. With a high level of ductility, a limit for this situation takes a form as shown in Figure 9.3*c*, and the limiting resisting moment is described as the *plastic moment*, designated M_p. Although a small portion of the beam cross section near the beam's neutral axis remains in an elastic stress condition, its effect on the development of the resisting moment is quite negligible. Thus, it is assumed that the full plastic limit is developed by the condition shown in Figure 9.3*d*.

Attempts to increase the bending moment beyond the value of M_p will result in large rotational deformation, with the beam acting as though it were hinged (pinned) at this location. For practical purposes, therefore, the resisting moment capacity of the ductile beam is considered to be exhausted with the attaining of the plastic moment; additional loading will merely cause a free rotation at the location of the plastic moment. This location is thus described as a *plastic hinge*

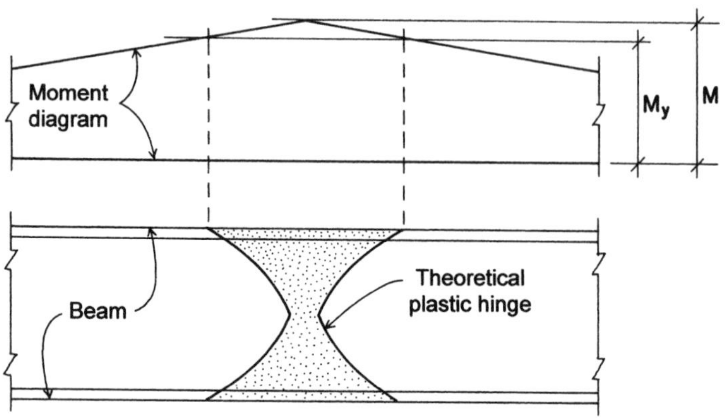

Figure 9.4 Development of the plastic hinge.

(see Figure 9.4), and its effect on beams and frames will be discussed further.

In a manner similar to that for elastic stress conditions, the value of the resisting plastic moment is expressed as

$$M_p = F_y \times Z$$

The term Z is called the *plastic section modulus*, and its value is determined as follows:

Refer to Figure 9.5, which shows a W shape subjected to a level of flexural stress corresponding to the fully plastic section (Figures 9.3*d* and 9.4), and note the following:

where A_u = upper area of the cross section, above the neutral axis
y_u = distance of the centroid of A_u from the neutral axis
A_l = lower area of the cross section, below the neutral axis
y_l = distance of the centroid of A_l from the neutral axis

For equilibrium of the internal forces on the cross section (the resulting forces C and T developed by the flexural stresses), the condition can be expressed as

$$\sum F_h = 0$$

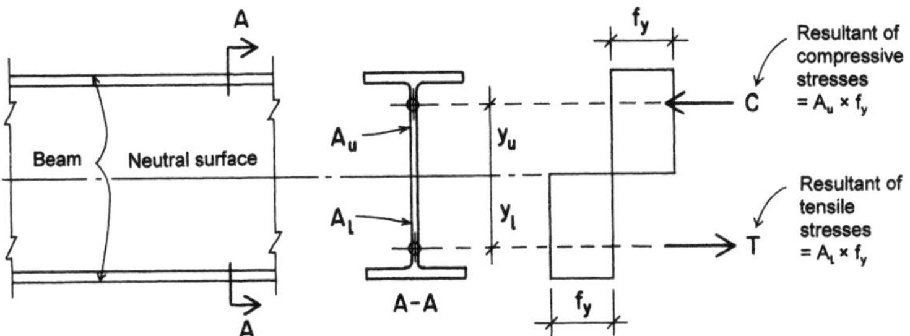

Figure 9.5 Development of the plastic bending moment.

or

$$[A_u \times (+f_y)] + [A_l \times (-f_y)] = 0$$

and, thus,

$$A_u = A_l$$

This shows that the plastic stress neutral axis divides the cross section into equal areas, which is apparent for symmetrical sections, but it applies to unsymmetrical sections as well. The resisting moment equals the sum of the moments of the stresses; thus, the value for M_p may be expressed as

$$M_p = C \times y_u + T \times y_1$$

or

$$M_p = (A_u \times f_y \times y_u) + (A_l \times f_y \times y_l)$$

or

$$M_p = f_y[(A_u \times y_u) + (A_l \times y_l)]$$

or

$$M_p = f_y \times Z$$

and the quantity $(A_u \times y_u) + (A_l \times y_l)$ is the property of the cross section defined as the plastic section modulus, designated Z.

Using the value for Z just derived, its value for any cross section can be computed. However, values for Z are tabulated in the AISC *Steel Manual* (Ref. 5) for all rolled sections used as beams. See Table A.3 in Appendix A.

Comparison of the values for S_x and Z_x for the same W shape will show that the values for Z are larger. This presents an opportunity to compare the fully plastic resisting moment to the yield stress limiting moment by elastic stress—that is, the advantage of using plastic analysis to express a beam's true limiting capacity.

Example 2. A simple beam consisting of a W 21 × 57 is subjected to bending about its major axis. Find the limiting moments based on (1) elastic stress conditions and a limiting yield strength of 50 ksi and (2) full development of the plastic moment.

Solution: For (1) the limiting moment is expressed as

$$M_y = F_y \times S_x$$

From Table A.3, $S_x = 111$ in.^{3}, so the limiting moment is

$$M_y = 50 \times 111 = 5550 \text{ kip-in.}$$

or

$$\frac{5550}{12} = 463 \text{ kip-ft}$$

For (2) the limiting plastic moment, using the value of $Z_x = 129$ in.^{3} from Table A.3, is

$$M_y = 50 \times 129 = 6450 \text{ kip-in.}$$

or

$$\frac{6450}{12} = 537 \text{ kip-ft}$$

The increase in moment resistance represented by the plastic moment indicates an increase of 537 – 463 = 74 kip-ft, or a percentage gain of (74/463)(100) = 16.0%.

Problem 9.2.A A simple-span, uniformly loaded beam consists of a W 18 × 50 with $F_y = 50$ ksi. Find the percentage of gain in the limiting bending moment if a plastic condition is assumed, instead of a condition limited by elastic stress.

Problem 9.2.B A simple-span, uniformly loaded beam consists of a W 16 × 45 with $F_y = 50$ ksi. Find the percentage of gain in the limiting bending moment if a plastic condition is assumed, instead of a condition limited by elastic stress.

9.3 NOMINAL MOMENT CAPACITY OF STEEL BEAMS

A critical step in design for bending using LRFD is the determination of the bending capacity (M_n) of a steel section. The capacity of a steel section is based on its cross-sectional properties, its yield stress, and the conditions of bracing of the member from out-of-plane buckling. Each of these parameters affect how the beam will ultimately fail and thus how much capacity it will have for bending.

Ideally, the failure mode that every beam will be controlled by is the inelastic failure described in Section 9.2. If a member section is capable of failing in a plastic hinge, it is considered a "compact" cross section. A compact shape is one that meets the following criteria:

$$\frac{b_f}{2t_f} \le \frac{65}{\sqrt{F_y}} \quad \text{and} \quad \frac{h_c}{t_w} \le \frac{640}{\sqrt{F_y}}$$

where b_f = flange width, in.
t_f = flange thickness, in.
F_y = minimum yield stress, ksi
h_c = height of the web, in.
t_w = web thickness, in.

For steel with a yield stress of 50 ksi, it translates to

$$\frac{b_f}{2t_f} \le 9.19 \quad \text{and} \quad \frac{h_c}{t_w} \le 90.5$$

Table 9.1 contains data for several properties of selected W shapes. In the AISC *Steel Manual* (Ref. 5) and in Table 9.1, the ratios needed to determine compactness are computed for each structural shape, and in many of the AISC tables, the noncompact sections are clearly labeled. It should be noted that most shapes used for beams are compact by design.

Compact beams that are adequately laterally supported will fail with a plastic hinge, and, therefore, the moment capacity is the yield moment or *plastic moment* (M_p) for the section.

Example 3. Determine the moment capacity of a W 24 × 76 steel beam that has adequate lateral support.

Solution: First, check to make sure that the section is compact. Compact section criteria are taken from Table 9.1. For the ratio of flange dimensions, the table yields a value of 6.61, which is less than the

TABLE 9.1 Load Factor Resistance Design Selection for Shapes Used as Beams

		F_y = 50 ksi								
Designation	Z_x (in.^{3})	L_p (ft)	L_r (ft)	M_p (kip-ft)	M_r (kip-ft)	r_y (in.)	$b_f/2t_f$	h/t_w	X_1 (ksi)	$X_2 \times 10^6$ [(1/ksi)^2]
W 33 × 141	**514**	**8.59**	**23.1**	**2,142**	**1,493**	**2.43**	**6.01**	**49.6**	**1,800**	**17,800**
W 30 × 148	500	8.06	22.8	2,083	1,453	2.28	4.44	41.6	2,310	6,270
W 24 × 162	468	10.8	32.4	1,950	1,380	3.05	5.31	30.6	2,870	2,260
W 24 × 146	418	10.6	30.6	1,742	1,237	3.01	5.92	33.2	2,590	3,420
W 33 × 118	**415**	**8.20**	**21.7**	**1,729**	**1,197**	**2.32**	**7.76**	**54.5**	**1,510**	**37,700**
W 30 × 124	408	7.88	21.5	1,700	1,183	2.23	5.65	46.2	1,930	13,500
W 21 × 147	373	10.4	32.8	1,554	1,097	2.95	5.44	26.1	3,140	1,590
W 24 × 131	370	10.5	29.1	1,542	1,097	2.97	6.70	35.6	2,330	5,290
W 18 × 158	356	9.69	38.0	1,483	1,033	2.74	3.92	19.8	4,410	403
W 30 × 108	**346**	**7.60**	**20.3**	**1,442**	**997**	**2.15**	**6.89**	**49.6**	**1,680**	**24,200**
W 27 × 114	343	7.71	21.3	1,429	997	2.18	5.41	42.5	2,100	9,220
W 24 × 117	327	10.4	27.9	1,363	970	2.94	7.53	39.2	2,090	8,190
W 21 × 122	307	10.3	29.8	1,279	910	2.92	6.45	31.3	2,630	3,160
W 18 × 130	290	9.55	32.8	1,208	853	2.7	4.65	23.9	3,680	810
W 30 × 90	**283**	**7.39**	**19.4**	**1,179**	**817**	**2.09**	**8.52**	**57.5**	**1,410**	**49,600**
W 24 × 103	280	7.04	20.0	1,167	817	1.99	4.59	39.2	2,390	5,310
W 27 × 94	278	7.50	19.9	1,158	810	2.12	6.70	49.5	1,740	19,900
W 14 × 145	260	14.1	54.7	1,083	773	3.98	7.11	16.8	4,400	348
W 24 × 94	254	7.00	19.4	1,058	740	1.98	5.18	41.9	2,180	7,800
W 21 × 101	253	10.2	27.6	1,054	757	2.89	7.68	37.5	2,200	6,400
W 12 × 152	243	11.3	62.1	1,013	697	3.19	4.46	11.2	6,510	79
W 18 × 106	230	9.40	28.7	958	680	2.66	5.96	27.2	2,990	1,880
W 14 × 120	212	13.2	46.2	883	633	3.74	7.80	19.3	3,830	601
W 24 × 76	**200**	**6.79**	**18.0**	**833**	**587**	**1.92**	**6.61**	**49.0**	**1,760**	**18,600**
W 16 × 100	200	8.84	29.6	833	590	2.5	5.29	23.2	3,530	947
W 21 × 83	196	6.47	18.5	817	570	1.83	5.00	36.4	2,400	5,250
W 18 × 86	186	9.30	26.1	775	553	2.63	7.20	33.4	2,460	4,060
W 12 × 120	186	11.1	50.0	775	543	3.13	5.57	13.7	5,240	184
W 21 × 68	**160**	**6.36**	**17.3**	**667**	**467**	**1.8**	**6.04**	**43.6**	**2,000**	**10,900**
W 24 × 62	**154**	**4.84**	**13.3**	**642**	**440**	**1.37**	**5.97**	**49.7**	**1,730**	**23,800**
W 16 × 77	152	8.70	25.5	633	453	2.46	6.77	29.9	2,770	2,460
W 12 × 96	147	10.9	41.4	613	437	3.09	6.76	17.7	4,250	407
W 10 × 112	147	9.48	56.5	613	420	2.68	4.17	10.4	7,080	57
W 18 × 71	146	6.01	17.8	608	423	1.7	4.71	32.4	2,690	3,290
W 14 × 82	139	8.77	29.5	579	410	2.48	5.92	22.4	3,590	849
W 24 × 55	**135**	**4.74**	**12.9**	**563**	**383**	**1.34**	**6.94**	**54.1**	**1,570**	**36,500**
W 21 × 57	129	4.77	13.1	538	370	1.35	5.04	46.3	1,960	13,100
W 18 × 60	123	5.94	16.6	513	360	1.68	5.44	38.7	2,290	6,080
W 12 × 79	119	10.8	35.7	496	357	3.05	8.22	20.7	3,530	839
W 14 × 68	115	8.70	26.4	479	343	2.46	6.97	27.5	3,020	1,660

TABLE 9.1 (*Continued*)

		F_y = 50 ksi								
Designation	Z_x (in.^{3})	L_p (ft)	L_r (ft)	M_p (kip-ft)	M_r (kip-ft)	r_y (in.)	$b_f/2t_f$	h/t_w	X_1 (ksi)	$X_2 \times 10^6$ [(1/ksi)^2]
W 10 × 88	113	9.30	45.1	471	328	2.63	5.18	13.0	5,680	132
W 21 × 50	**110**	**4.60**	**12.5**	**458**	**315**	**1.3**	**6.10**	**49.4**	**1,730**	**22,600**
W 16 × 57	105	5.66	16.6	438	307	1.6	4.98	33.0	2,650	3,400
W 18 × 50	**101**	**5.83**	**15.6**	**421**	**296**	**1.65**	**6.57**	**45.2**	**1,920**	**12,400**
W 21 × 44	**95.4**	**4.45**	**12.0**	**398**	**272**	**1.26**	**7.22**	**53.6**	**1,550**	**36,600**
W 18 × 46	90.7	4.56	12.6	378	263	1.29	5.01	44.6	2,060	10,100
W 14 × 53	87.1	6.79	20.1	363	259	1.92	6.11	30.9	2,830	2,250
W 10 × 68	85.3	9.16	36.0	355	252	2.59	6.58	16.7	4,460	334
W 16 × 45	82.3	5.55	15.2	343	242	1.57	6.23	41.1	2,120	8,280
W 18 × 40	**78.4**	**4.49**	**12.1**	**327**	**228**	**1.27**	**5.73**	**50.9**	**1,810**	**17,200**
W 12 × 53	77.9	8.77	25.6	325	235	2.48	8.69	28.1	2,820	2,100
W 14 × 43	69.6	6.68	18.3	290	209	1.89	7.54	37.4	2,330	4,880
W 10 × 54	66.6	9.05	30.2	278	200	2.56	8.15	21.2	3,580	778
W 12 × 45	64.2	6.89	20.3	268	192	1.95	7.00	29.6	2,820	2,210
W 16 × 36	**64.0**	**5.37**	**14.0**	**267**	**188**	**1.52**	**8.12**	**48.1**	**1,700**	**20,400**
W 10 × 45	54.9	7.11	24.1	229	164	2.01	6.47	22.5	3,650	758
W 14 × 34	**54.6**	**5.41**	**14.4**	**228**	**162**	**1.53**	**7.41**	**43.1**	**1,970**	**10,600**
W 12 × 35	51.2	5.44	15.2	213	152	1.54	6.31	36.2	2,430	4,330
W 16 × 26	**44.2**	**3.96**	**10.4**	**184**	**128**	**1.12**	**7.97**	**56.8**	**1,480**	**40,300**
W 14 × 26	**40.2**	**3.82**	**10.2**	**168**	**118**	**1.08**	**5.98**	**48.1**	**1,880**	**14,100**
W 10 × 33	38.8	6.86	19.8	162	117	1.94	9.15	27.1	2,720	2,480
W 12 × 26	**37.2**	**5.34**	**13.8**	**155**	**111**	**1.51**	**8.54**	**47.2**	**1,820**	**13,900**
W 10 × 26	**31.3**	**4.81**	**13.6**	**130**	**93**	**1.36**	**6.56**	**34.0**	**2,510**	**3,760**
W 12 × 22	**29.3**	**3.00**	**8.41**	**122**	**85**	**0.848**	**4.74**	**41.8**	**2,170**	**8,460**
W 10 × 19	**21.6**	**3.09**	**8.89**	**90**	**63**	**0.874**	**5.09**	**35.4**	**2,440**	**5,030**

Source: Compiled from data in the *Steel Construction Manual* (Ref. 5) with permission of the publisher, American Institute of Steel Construction.
Note: Designations in boldface typerefer to shapes with sections that have an especially efficient bending moment resistance, indicated by the fact that there are other shapes of greater weight but the same or smaller section modulus.

limit of 9.19. For the ratio of the web dimensions, the table yields a value of 49, which is less than the limit of 90.5. The shape is, therefore, compact and the moment capacity will be equal to the plastic moment:

$$M_n = M_y = F_y \times Z_x = 50 \times 200 = 10{,}000 \text{ kip-in.}$$

$$= \frac{10{,}000}{12} = 833 \text{ kip-ft}$$

To ensure that a compact section fails plastically, the maximum spacing between lateral supports of the beam (L_b) must be less than a limiting laterally unbraced length for fully plastic flexural strength (L_p), which is defined as

$$L_p = \frac{300 \times r_y}{\sqrt{F_y}}$$

where r_y = radius of gyration about the y axis, in.
F_y = minimum yield stress, ksi

If the actual unbraced length (L_b) is greater than the limiting value of Lp, the beam will fail in buckling at a moment less than the plastic moment. The plastic moment will attempt to develop but cannot be attained. However, the stress condition at failure will be in the plastic range, so the form of buckling is described as *inelastic lateral-torsional buckling*. This form of buckling will occur for unbraced lengths up to a second limiting length called L_r. Unbraced lengths greater than L_r will result in a different form of buckling with the beam in the elastic stress range; this limiting unbraced length causes the form of buckling to change to one described as *elastic buckling*. The limiting length L_r is defined as

$$L_r = \left(\frac{r_y \times X_1}{F_y - F_r}\right) \times \sqrt{1 + \sqrt{1 + X_2 \times (F_y - F_r)^2}}$$

where r_y = radius of gyration about the y axis, in.
X_1 = beam buckling factor
F_y = minimum yield stress, ksi
F_r = compressive residual stress (10 ksi for rolled shapes)
X_2 = beam buckling factor

Example 4. Determine the limiting lateral bracing lengths L_p and L_r for a W 24 × 76 steel beam.

Solution: From Table 9.1, for the shape: r_y = 1.92 in., X_1 = 1760 ksi, $X_2 = 0.0186\ (1/\text{ksi})^2$. *Note:* Values given in Table 9.1 for X_2 are 1,000,000 times the actual value; therefore, move the decimal point six places to the left for computations (0.0816, not 18,600):

$$L_p = \frac{300 \times r_y}{\sqrt{F_y}} = \frac{300 \times 1.92}{\sqrt{50}} = 81.5 \text{ in.} = 6.79 \text{ ft}$$

and

$$L_r = \left(\frac{r_y \times X_1}{F_y - F_r}\right) \times \sqrt{1 + \sqrt{1 + X_2 \times (F_y - F_r)^2}}$$

$$= \left(\frac{1.92 \times 1760}{50 - 10}\right) \times \sqrt{1 + \sqrt{1 + 0.0186 \times (50 - 10)^2}}$$

$$= 216 \text{ in.} = 18.0 \text{ ft}$$

These may be verified by the values listed for the shape in Table 9.1.

Knowing the relationship between the unbraced length (L_b) and the limiting lateral unbraced lengths (L_p and L_r), it is possible to determine the nominal moment capacity (M_n) for any beam. Figure 9.6 shows the form of the relation between M_n and L_b using the example of a W 18 × 50 with yield stress of 50 ksi. The AISC *Steel Manual* (Ref. 5) contains a series of such graphs for shapes commonly used as beams.

For the three cases of lateral unsupported length (L_b), relating to the three parts of the graph in Figure 9.6, the nominal resisting moment (M_n) is determined as follows:

Case 1. If $L_b \le L_p$, then

$$M_n = M_p = F_y \times Z_x$$

Case 2. If $L_p < L_b \le L_r$, then

$$M_n = M_p - (M_p - M_r) \times \left(\frac{L_b - L_p}{L_r - L_p}\right)$$

where $M_r = (F_y - F_r) \times S_x$
F_r = compressive residual stress, 10 ksi
S_x = section modulus for the x axis

Case 3. If $L_b > L_r$, then

$$M_n = \left(\frac{S_x \times X_1 \times \sqrt{2}}{L_b/r_y}\right) \times \sqrt{1 + \frac{(X_1)^2 \times X_2}{2 \times (L_b/r_y)^2}}$$

where X_1, X_2 = beam buckling factors from Table 9.1
r_y = radius of gyration for the y axis

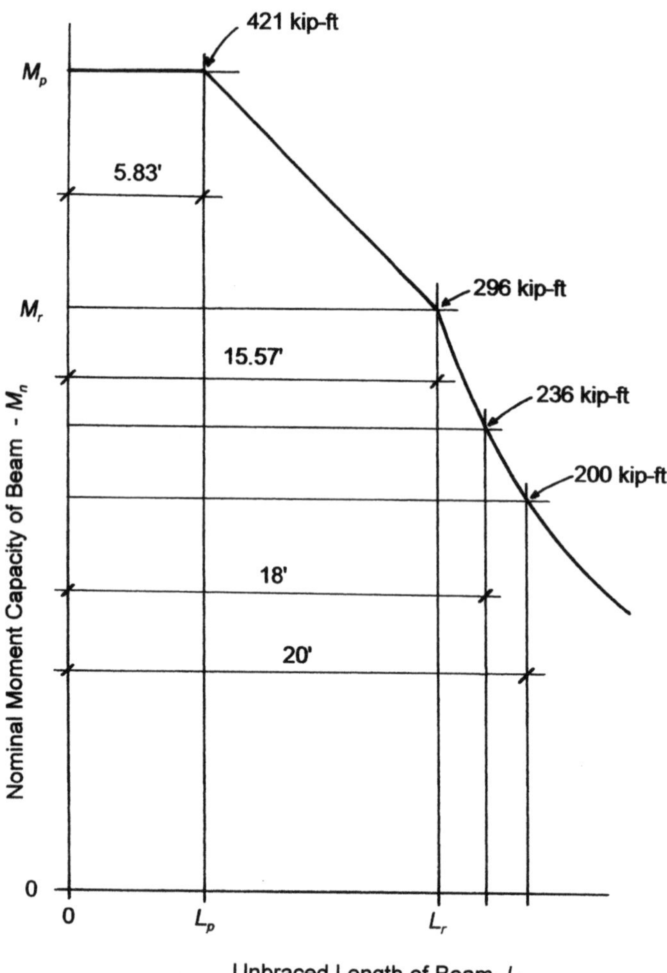

Figure 9.6 Relation between nominal capacity M_n and unbraced length L_b for a W 18 × 50 steel beam with F_y = 50 ksi.

Example 5. Determine the nominal resisting moment capacity of a W 24 × 76 steel beam that is laterally supported every 10 ft. (This is the same shape that is used in Example 4.)

Solution: From Table 9.1, $L_p = 6.79$ ft, $L_r = 18.0$ ft, $M_p = 833$ kip-ft, $M_r = 587$ kip-ft.

Note that this is case 2, where $L_p < L_b < L_r$. Then

$$M_n = M_p - (M_p - M_r) \times \left(\frac{L_b - L_p}{L_r - L_p}\right)$$

$$= 833 - (833 - 587) \times \left(\frac{10 - 6.79}{18 - 6.79}\right)$$

$$= 763 \text{ kip-ft}$$

Example 6. Determine the nominal resisting moment capacity of a W 24 × 76 steel beam that is laterally unsupported on a 25-ft span.

Solution: Note that this is case 3 since the unsupported length exceeds L_r.

From Table 9.1, $S_x = 176$ in.^{3}, $r_y = 1.92$ in., $X_1 = 1760$ ksi, $X_2 = 18{,}600 \times 10^{-6}$ (1/ksi). *Note:* As discussed for Example 4, the value of X_2 is actually 0.0186. Then

$$M_n = \left(\frac{S_x \times X_1 \times \sqrt{2}}{L_b/r_y}\right) \times \sqrt{1 + \frac{(X_1)^2 \times X_2}{2 \times (L_b/r_y)^2}}$$

$$= \left(\frac{176 \times 1760 \times \sqrt{2}}{(25 \times 12)/1.92}\right) \times \sqrt{1 + \frac{(1760)^2 \times 0.0186}{2 \times ((25 \times 12)/1.92)^2}}$$

$$= \left(\frac{438{,}000}{156}\right) \times \sqrt{1 + \frac{57{,}600}{48{,}800}}$$

$$= 2800 \times 1.48 = 4150 \text{ kip-in.} = 345 \text{ kip-ft}$$

Problem 9.3.A Determine the nominal resisting moment capacity (M_n) for a W 27 × 94 for the following unbraced lengths: (1) 5 ft, (2) 15 ft, and (3) 30 ft.

Problem 9.3.B Determine the nominal resisting moment capacity (M_n) for a W 16 × 36 for the following unbraced lengths: (1) 5 ft, (2) 10 ft, and (3) 20 ft.

9.4 DESIGN FOR BENDING

Design for bending usually involves the determination of the ultimate bending moment (M_u) that the beam must resist and the use of formulas derived in Section 9.3 for definition of bending resistance of the member (M_n). Stated in equation form, the relationship is

$$\phi_b M_n \geq M_u$$

where $\varphi_b = 0.9$.

Design for Plastic Failure Mode

Using the formula for resisting moment defined for case 1 in the preceding section, a relationship can be stated that leads to the determination of the required plastic modulus of elasticity for a beam. Thus,

$$Z \geq \frac{M_u}{F_y \phi_b}$$

Assuming no additional requirements on the beam than cost, the beam chosen will be the lightest beam that has a section modulus (Z_x) greater than what is required. The following example illustrates the basic procedure.

***Example** 7.* Design a simply supported floor beam to carry a superimposed dead load of 0.5 kip/ft and a live load of 1.5 kips/ft over a span of 24 ft. (A superimposed load is any load other than the weight of the beam itself.) The beam is continuously supported against lateral buckling ($L_b = 0$ ft).

Solution: The load must first be factored to produce the maximum required moment (M_u). Thus,

$$w_u = 1.4(\text{dead load}) = 1.4(0.5) = 0.7 \text{ kip/ft}$$

or

$$\begin{aligned} w_u &= 1.2(\text{dead load}) + 1.6(\text{live load}) = 1.2(0.5) + 1.6(1.5) \\ &= 3.0 \text{ kips/ft} \end{aligned}$$

The second factored load will produce the greatest bending moment; therefore it will be used in this problem. The bending moment due to the maximum factored load is

$$M_u = \frac{wL^2}{8} = \frac{3(24)^2}{8} = 216 \text{ kip-ft}$$

The required bending resistance of the member is

$$M_n = \frac{M_u}{\phi_b} = \frac{216}{0.9} = 240 \text{ kip-ft}$$

The required plastic section modulus for this moment is

$$Z_x \geq \frac{M_n}{F_y} = \frac{240 \times 12}{50} = 57.6 \text{ in.}^3$$

Table 9.1 lists a number of shapes in descending order of the value of their plastic section modulus. Possible choices with a section modulus of at least 57.6 in.^{3} are

$$\text{W } 16 \times 36 \quad Z_x = 64 \text{ in.}^3$$

$$\text{W } 12 \times 45 \quad Z_x = 64.2 \text{ in.}^3$$

$$\text{W } 10 \times 54 \quad Z_x = 66.6 \text{ in.}^3$$

$$\text{W } 14 \times 43 \quad Z_x = 69.6 \text{ in.}^3$$

Although there are often other considerations for a complete design, for flexure alone the lightest choice is W 16 × 36. A check should be made to assure that the added weight of the beam does not push the required value for the plastic section modulus above that of the selected shape. The additional factored dead load is

$$w_u = 1.2(36) = 43.2 \text{ lb/ft} = 0.0432 \text{ kip/ft}$$

and the percentage increase in the required section modulus is

$$\frac{0.0432}{3.0}(100) = 1.44\% \quad \Rightarrow \quad Z_x \geq 58.4 \text{ in.}^3$$

This is a negligible increase for this shape, as its section modulus is considerably larger than required. However, in some cases it may make for different options if other shapes are considered.

Use of Plastic Section Modulus Tables

Selection of rolled shapes on the basis of required plastic section modulus may be achieved by the use of tables in the AISC *Steel Manual*

(Ref. 5) in which beam shapes are listed in descending order of their section modulus values. Table 9.1 presents a small sample of the reference table data. Note that certain shapes have their designations in boldface type. These are shapes with sections that have an especially efficient bending moment resistance, indicated by the fact that there are other shapes of greater weight but the same or smaller section modulus. Thus, for a savings of material cost, these *least weight* shapes offer an advantage. Consideration of other beam design factors, however, may sometimes make this a less important concern.

Data are also supplied in Table 9.1 for the consideration of lateral support for beams. Values are given for the two critical limiting lengths L_p and L_r. If the required section modulus was obtained using a minimum yield stress of 50 ksi, the lateral unsupported length needs to be equal to or less than L_p as listed under the "50 ksi" column.

A second method of using Table 9.1 for beams omits the calculation for a required plastic section modulus and refers directly to the listed values for the plastic bending moment for the shapes, given as M_p in the tables. If $M_p \geq M_u/\varphi_b$, then it is an appropriate shape if $L_b \leq L_p$.

Example 8. Rework the problem in Example 7 by using Table 9.1 directly.

Solution: As before, the required resisting bending moment is found to be 240 kip-ft. Noting that some additional bending capacity will be required because of the beam's own weight, scan the tables for shapes with M_p slightly greater than 240 kip-ft. Thus:

Shape	M_p (kip-ft)
W 16 × 36	267
W 12 × 45	268
W 10 × 54	278
W 14 × 43	290

As before, the least weight choice is the W 16 × 36.

In Examples 7 and 8 the beam was assumed to be continuously laterally supported; that is, $L_b = 0$. If the lateral unsupported length is a specific value, full development of the plastic moment will be possible only if $L_b \leq L_p$. Choices can still be made from Table 9.1, as the following examples illustrate for the 24-ft-span beam.

Lateral Supports at:	Maximum Unsupported Length	Least Weight Shape	Listed L_p for the Shape
Quarter points	6 ft	W 12 × 45	6.89 ft
Third points	8 ft	W 12 × 53	8.77 ft
Midpoint	12 ft	W 14 × 102	13.2 ft
Ends of beam only	24 ft	No shape qualifies	

It should be noted that not all available W shapes listed in Table A.3 are included in Table 9.1. Specifically excluded are the shapes that are approximately square (depth equal to flange width) and are ordinarily used for columns rather than for beams.

The following problems involve design for bending under plastic failure mode only. Use A992 steel and assume that least weight members are desired for each case.

Problem 9.4.A Design for flexure a simple beam 14 ft in length and having a total uniformly distributed dead load of 13.2 kips and a total uniformly distributed live load of 26.4 kips.

Problem 9.4.B Design for flexure a beam having a span of 16 ft with a concentrated live load of 40 kips at the center of the span.

Problem 9.4.C A beam 15 ft in length has three concentrated live loads of 6, 7.5, and 9 kips at 4, 10, and 12 ft, respectively, from the left-hand support. Design the beam for flexure.

Problem 9.4.D A beam 30 ft long has concentrated live loads of 9 kips each at the third points and also a total uniformly distributed dead load of 20 kips and a total uniformly distributed live load of 10 kips. Design the beam for flexure.

Problem 9.4.E Design for flexure a beam 12 ft in length, having a uniformly distributed dead load of 1 kip/ft, a uniformly distributed live load of 1 kip/ft, and a concentrated dead load of 8.4 kips a distance of 5 ft from one support.

Problem 9.4.F A beam 19 ft in length has concentrated live loads of 6 kips and 9 kips at 5 ft and 13 ft, respectively, from the left-hand support. In addition, there is a uniformly distributed dead load of 1.2 kips/ft beginning 5 ft from the left support and continuing to the right support. Design the beam for flexure.

Problem 9.4.G A steel beam 16 ft long has a uniformly distributed dead load of 100 lb/ft extending over the entire span and a uniformly distributed live load of 100 lb/ft extending 10 ft from the left support. In addition, there is a concentrated live load of 8 kips at 10 ft from the left support. Design the beam for flexure.

Problem 9.4.H Design for flexure a simple beam 21 ft in length, having two concentrated loads of 20 kips each, one 7 ft from the left end and the other 7 ft from the right end. The concentrated loads are each made up of equal parts of dead load and live load.

Problem 9.4.I A cantilever beam 12 ft long has a uniformly distributed dead load of 600 lb/ft and a uniformly distributed live load of 1000 lb/ft. Design the beam for flexure.

Problem 9.4.J A cantilever beam 6 ft long has a concentrated live load of 12.3 kips at its unsupported end. Design the beam for flexure.

9.5 DESIGN OF BEAMS FOR BUCKLING FAILURE

Although it is preferable to design beams to fail under the plastic hinge mode discussed in Section 9.4, it is not always possible to do so. This is commonly caused by excessive unbraced lengths for lateral support. The simplest solution is to decrease the maximum unbraced length to make it less than the plastic failure limit (L_p). If this is not possible, the solution is to accept that the beam failure mode is buckling and use the appropriate equations to determine the nominal moment capacity of the beam as described in Section 9.3. The following example demonstrates the process.

Example 9. A 14-ft-long simply supported beam has a uniformly distributed live load of 3 kips/ft and a uniformly distributed dead load of 2 kips/ft. It is laterally supported only at its ends. Determine the least weight A992 steel W shape that will work.

Solution: First, determine the appropriate load combination and the maximum factored moment. Thus,

$$w_u = 1.4(\text{dead load}) = 1.4(2) = 2.8 \text{ kips/ft}$$

or

$$w_u = 1.2(\text{dead load}) + 1.6(\text{live load}) = 1.2(2) + 1.6(3) = 7.2 \text{ kips/ft}$$

The bending moment caused by the maximum factored load is

$$M_u = \frac{wL^2}{8} = \frac{7.2 \times (14)^2}{8} = 176 \text{ kip-ft}$$

The required bending resistance of the member is

$$M_n = \frac{M_u}{\phi_b} = \frac{176}{0.9} = 196 \text{ kip-ft}$$

The required plastic modulus for this moment is

$$Z_x \geq \frac{M_n}{F_y} = \frac{196 \times 12}{50} = 47 \text{ in.}^3$$

From Table 9.1, the least weight shape not taking into account unbraced length is a W 14 × 34, but this shape has a plastic limit on unbraced length (L_p) of only 5.41 ft. However, its plastic section modulus (Z_x) is larger than required, so it still has reserve moment capacity beyond the plastic length limit. If this shape does not work, a search of Table 9.1 can be done to see if there are any shapes that have $Z_x > 47$ in.^{3} and $L_p \geq 14$ ft. There is one listed shape that matches this criteria, a W 14 × 145. As this shape is quite heavy, the search should be continued for lighter shapes. Try for a shape that has $Z_x > 47$ in.^{3} and $L_p < 14 \text{ ft} < L_r$ and whose moment capacity $M_n > 196$ kip-ft.

For a first try, check the W 14 × 34 because it would be the least weight choice if it works (from Table 9.1):

$$L_p = 5.41 \text{ ft} \qquad L_r = 14.4 \text{ ft}$$

$$M_p = 228 \text{ kip-ft} \qquad M_r = 162 \text{ kip-ft}$$

and

$$\begin{aligned} M_n &= M_p - \left[(M_p - M_r) \times \left(\frac{L_b - L_p}{L_r - L_p}\right)\right] \\ &= 228 - \left[(228 - 162) \times \left(\frac{14 - 5.41}{14.4 - 5.41}\right)\right] \\ &= 163 \text{ kip-ft} < 196 \text{ kip-ft}, \end{aligned}$$

therefore a W 14 × 34 is not acceptable.

Next, try the W 16 × 36 (the next lightest shape) and check to see if it works:

$$L_p = 5.37 \text{ ft} \qquad L_r = 14.0 \text{ ft}$$
$$M_p = 267 \text{ kip-ft} \qquad M_r = 188 \text{ kip-ft}$$

and

$$\text{Since } L_b = L_r$$
$$M_n = M_r = 188 \text{ kip-ft} < 196 \text{ kip-ft},$$

therefore a W 16 × 36 is not acceptable.

Try a W 21 × 35:

$$L_p = 5.44 \text{ ft} \qquad L_r = 15.2 \text{ ft}$$
$$M_p = 213 \text{ kip-ft} \qquad M_r = 152 \text{ kip-ft}$$

and

$$M_n = M_p - \left[(M_p - M_r) \times \left(\frac{L_b - L_p}{L_r - L_p}\right)\right]$$
$$= 213 - \left[(213 - 152) \times 45.2\left(\frac{14 - 7.88}{14.4 - 7.88}\right)\right]$$
$$= 160 \text{ kip-ft} < 196 \text{ kip-ft},$$

therefore a W 21 × 35 is not acceptable.

Try a W 18 × 40:

$$L_p = 4.49 \text{ ft} \qquad L_r = 12.1 \text{ ft}$$
$$M_p = 327 \text{ kip-ft} \qquad M_r = 228 \text{ kip-ft}$$

and

$$M_n = M_p - \left[(M_p - M_r) \times \left(\frac{L_b - L_p}{L_r - L_p}\right)\right]$$
$$= 327 - \left[(327 - 228) \times \left(\frac{14 - 4.49}{12.1 - 4.49}\right)\right]$$
$$= 203 \text{ kip-ft} > 196 \text{ kip-ft}$$

As this is larger than the moment required, the 18 × 40 appears to work. A check should be made of the increased moment due to the beam weight. This investigation will show that the factored moment including the 40 lb/ft of beam weight is 197 kip-ft and therefore the W 18 × 40 is an acceptable choice for this beam.

The following problems involve the use of Table 9.1 to choose the least weight beams when lateral bracing is a concern. Use A992 steel for all beams.

Problem 9.5.A A W shape is to be used for a uniformly loaded simple beam carrying a total dead load of 27 kips and a total live load of 50 kips on a 45-ft span. Select the lightest weight shape for unbraced lengths of (1) 10 ft, (2) 15 ft, and (3) 22.5 ft.

Problem 9.5.B A W shape is to be used for a uniformly loaded simple beam carrying a total dead load of 30 kips and a total live load of 40 kips on a 24-ft span. Select the lightest weight shape for unbraced lengths of (1) 6 ft, (2) 8 ft, and (3) 12 ft.

Problem 9.5.C A W shape is to be used for a uniformly loaded simple beam carrying a total dead load of 22 kips and a total live load of 50 kips on a 30-ft span. Select the lightest weight shape for unbraced lengths of (1) 6 ft, (2) 10 ft, and (3) 15 ft.

Problem 9.5.D A W shape is to be used for a uniformly loaded simple beam carrying a total dead load of 26 kips and a total live load of 26 kips on a 36-ft span. Select the lightest weight shape for unbraced lengths of (1) 9 ft, (2) 12 ft, and (3) 18 ft.

9.6 SHEAR IN STEEL BEAMS

Investigation and design for shear forces in beams with the LRFD method is similar to that for bending moment in that the maximum factored shear force must be equal to or less than the factored shear capacity of the beam chosen. This is expressed as

$$\phi_v V_n \geq V_u$$

where φ_v = 0.90

V_n = nominal shear capacity of the beam

V_u = required (factored) shear load on the beam

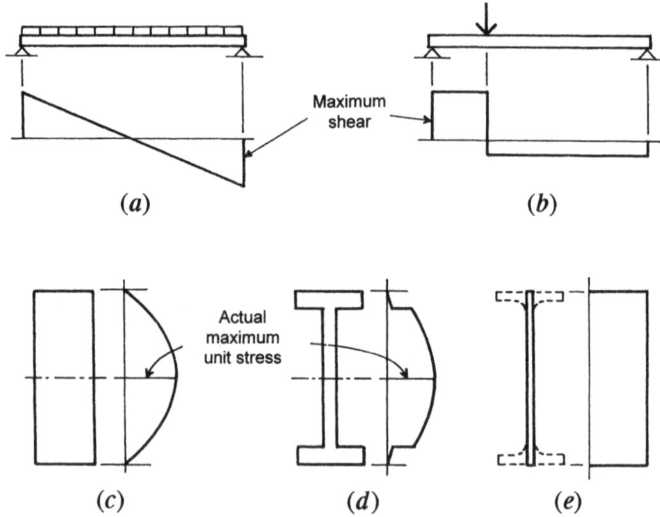

Figure 9.7 Development of shear in beams.

Shear in beams consists of the vertical slicing effect produced by the opposition of the vertical loads on the beams (downward) and the reaction forces at the beam supports (upward). The internal shear force mechanism is visualized in terms of the shear force diagram for the beam. With a uniformly loaded simple beam, this diagram takes the form of that shown in Figure 9.7*a*.

As the shear diagram for the uniformly distributed load shows, the internal shear force has a maximum value at the beam supports and decreases steadily to zero at midspan. With a beam having the same cross section throughout its length, the critical location for shear is thus at the supports and of less concern at other locations.

Figure 9.7*b* shows the form of shear force variation due to a concentrated load, a condition commonly occurring when a larger beam supports the ends of other beams. In this case, major internal shear force is generated for some length along the beam.

Internal shear force develops shear stress in the beam (see Section 3.7). The form of distribution of these stresses depends on the shape of the cross section of the beam. For a simple rectangular cross section, such as that of a wood beam, the distribution is as shown in Figure 9.7*c*, taking the form of a parabola with a maximum value at the centroidal axis of the beam and decreasing to zero at the top and bottom edges.

For the I-shaped cross section of the typical W-shape steel beam, the beam shear stress distribution takes the form shown in Figure 9.7*d* (referred to as the "derby hat" form). Again, the maximum stress occurs at the neutral axis, but the drop-off toward the edges is slower and then drops dramatically at the flanges. A traditional shear stress investigation for the W shape, therefore, is based on ignoring the flanges and assuming the shear-resisting portion of the beam to be an equivalent vertical plate (Figure 9.7*e*) with a width equal to the beam web thickness and a height equal to the full beam depth. For the stress design method, an allowable value is established for a unit shear stress on the basis of this analogy, and the stress is defined as

$$f_v = \frac{V}{t_w d_b} = \frac{V}{A_w}$$

where f_v = average shear stress, based on the distribution in Figure 9.7*e*

V = shear force at the cross section

t_w = beam web thickness

d_b = overall beam depth

A_w = area of the beam web, $t_w \times d_b$

Uniformly loaded steel beams are seldom critical with regard to shear stress on the basis just described. The most common case for beam support is that shown in Figure 9.8*a*, where a connecting device affects the transfer of the end shear force to the beam support, usually using a pair of steel angles that grasp the beam web and are turned outward to fit against another beam's web or the side of a column. If the connecting device is welded to the supported beam's web, as shown in Figure 9.8*a*, it actually reinforces the web at this location; thus, the critical section for shear stress becomes that portion of the beam web just beyond the connector. At this location, the shear force is as shown in Figure 9.8*b*, and it is assumed to operate on the effective section of the beam as discussed previously.

Some situations, however, can result in critical conditions for the transfer of the vertical force at the beam end. When the supporting beam is also a W shape and the tops of the two beams are at the same level (common in framing systems), it becomes necessary to cut back the top flange and a portion of the beam web of the supported beam to permit the end of the web to get as close as possible to the web of the supporting beam (Figure 9.8*c*). This results in some loss

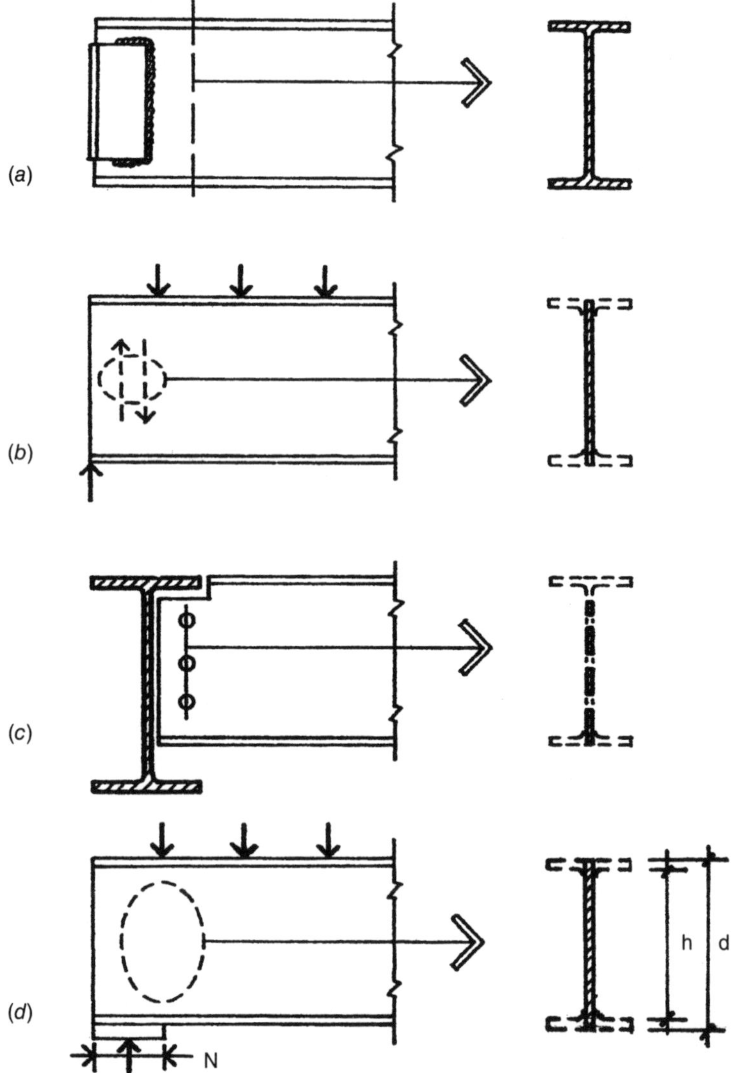

Figure 9.8 Considerations for end support in rolled beams: (*a*) development of shear by the full beam section with a framed beam connection, (*b*) design assumption for resistance to shear by the web only, (*c*) development of shear on a reduced section, and (*d*) development of vertical compression in the beam web at a bearing support.

of the shear-resisting area of the web as assumed in Figure 9.7*e* and an increase in the shear stress.

Another possible reduction of the shear-resisting area may occur when bolts, rather than welds, are used to fasten connecting angles to the supported beam's web, as shown in Figure 9.8*c*. The full reduction of the shear-resisting area will thus include the losses due to the bolt holes and the notched top of the beam. The form of failure in this situation may be one of block tearing of the beam web, as described in Section 11.1.

Shear stress as such, however, may not be the nature of the critical problem at a beam support. Figure 9.8*d* illustrates the condition of a vertical compression effect at a beam support consisting of a bearing of the beam end on top of the support, usually in this case the top of a wall or a wall ledge. The potential problem here is a squeezing of the beam end and a columnlike action in the thin beam web. As with a column, the range of possibilities for this form of failure relate to the relative slenderness of the web. For the strength design methods, three cases are defined by the slenderness ratio of the web (h/t_w) and control the shear capacity (V_n) of a beam. The dimension h, as shown in Figure 9.8*d*, is the true unbraced height of the web, equal to the beam depth minus twice the flange thickness. The three distinct cases are as follows:

1. A very stiff (thick) web that may actually reach something close to the full yield stress limit of the steel where

$$\frac{h}{t_w} \leq 418/\sqrt{F_y}$$

$$V_n = (0.6F_y)A_w$$

2. A somewhat slender web that responds with some combined yield stress and buckling effect (called an *inelastic buckling* response) where

$$\frac{418}{\sqrt{F_y}} \leq \frac{h}{t_w} \leq \frac{523}{\sqrt{F_y}}$$

$$V_n = (0.6F_y)A_w\left(\frac{418/\sqrt{F_y}}{h/t_w}\right)$$

3. A very slender web that fails essentially in *elastic buckling* in the classic Euler formula manner, basically, a deflection failure rather than a stress failure where

$$\frac{h}{t_w} > \frac{523}{\sqrt{F_y}}$$

$$V_n = 132{,}000\left(\frac{A_w}{(h/t_w)^2}\right)$$

Example 10. A simple beam of A992 steel is 6 ft long and has a concentrated live load 36 kips applied 1 ft from one end. It is found that a W 10 × 19 is adequate for the bending moment. Investigate the beam to determine if the shear capacity is adequate for the required shear.

Solution: The load must first be factored:

$$P_u = 1.2(\text{dead load}) + 1.6(\text{live load}) = 1.2(0) + 1.6(36) = 57.6 \text{ kips}$$

The two beam reactions for this loading are 48 kips and 9.6 kips. The maximum shear in the beam (V_u) is equal to the larger reaction force.

From Table A.3, for the given shape, $d = 10.24$ in., $t_w = 0.250$ in., $t_f = 0.395$ in. Then

$$h = d - 2(t_f) = 10.24 - 2(0.395) = 9.45 \text{ in.}$$

$$\frac{h}{t_w} = \frac{9.45}{0.25} = 37.8$$

$$\frac{418}{\sqrt{F_y}} = \frac{418}{\sqrt{50}} = 59.1 > \frac{h}{t_w}$$

Therefore, the shear capacity will be determined using the equation associated with the full yield stress limit of the steel:

$$A_w = d \times t_w = 10.24 \times 0.250 = 2.56 \text{ in.}^2$$

$$V_n = (0.6F_y)A_w = (0.6(50))2.56 = 76.8 \text{ kips}$$

$$\phi_v V_n = 0.9(76.8) = 69.1 \text{ kips} > 48 \text{ kips}$$

Because the factored capacity of the beam is greater than the factored shear force, the shape is acceptable.

The net effect of investigations of all the situations so far described, relating to end shear in beams, may be to influence a choice of a beam shape with a web that is sufficient. However, other criteria for selection (flexure, deflection, framing details, and so on) may indicate an ideal choice that has a vulnerable web. In the latter case, it is sometimes decided to *reinforce* the web, the usual means being to insert vertical plates on either side of the web and to fasten them to the web as well as to the beam flanges. These plates then both brace the slender web and absorb some of the vertical compression stress in the beam.

For practical purposes, the considerations for beam end shear and end support limitations of unreduced webs can be handled by data supplied in tables in the AISC *Steel Manual* (Ref. 5).

Problems 9.6.A–C Compute the shear capacity ($\varphi_v V_n$) for the following beams of A992 steel: (A) W 24 × 94, (B) W 12 × 45, and (C) W 10 × 33.

9.7 DEFLECTION OF BEAMS

Deformations of structures must often be controlled for various reasons. These reasons may relate to the proper functioning of the structure, but more often they relate to effects on the supported construction or to the overall purpose of the structure.

To steel's advantage is the relative stiffness of the material itself. With a modulus of elasticity of 29,000 ksi, it is 8–10 times as stiff as average structural concrete and 15–20 times as stiff as structural lumber. However, it is often the overall deformation of whole structural assemblages that must be controlled; in this regard, steel structures are often quite deformable and flexible. Because of its high material cost, steel is usually formed into elements with thin parts (e.g., beam webs and flanges), and because of its high strength, it is frequently formed into slender elements (e.g., beams and columns).

For a beam in a horizontal position, the critical deformation is usually the maximum sag, called the beam's *deflection*. For most beams in service, this deflection will be too small to be detected by eye. However, any load on the beam, such as that in Figure 9.9, will cause some amount of deflection, beginning with the beam's own weight. In the case of a simply supported, symmetrical, single-span beam, the maximum deflection will occur at midspan, and it usually is the only deformation value of concern for design. However, as the beam deflects, its ends rotate unless restrained, and this twisting deformation may also be of concern in some situations.

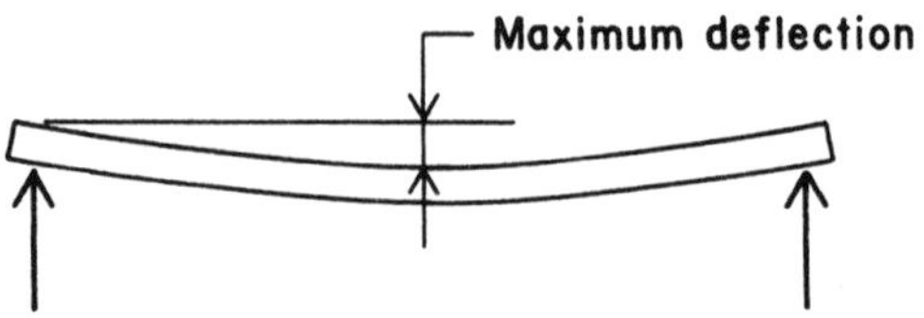

Figure 9.9 Deflection of a simple beam under symmetrical loading.

If deflection is determined to be excessive, the usual remedy is to select a deeper beam. Actually, the critical geometric property of the beam cross section is the *moment of inertia* (I) about its major axis (I_x) for a W shape), which is affected significantly by increase in depth of the beam. Formulas for deflection of beams take a typical form that involves variables as follows:

$$\Delta = C\frac{WL^3}{EI}$$

where Δ = deflection, measured vertically, usually in. or mm
C = constant related to load and support conditions
W = total load on the beam
L = span of the beam
E = modulus of elasticity of the beam material
I = moment of inertia for the beam about the bending axis

Note that the magnitude of the deflection is directly proportional to the magnitude of the total load; that is, if the load is doubled, the deflection will double. However, the deflection is proportional to the third power of the span; double the span and get 2^3 or eight times as much deflection. For resistance to deflection, increases in either the material's stiffness or the beam's geometric form, I, will cause direct proportional reduction of the deflection. Because E is constant for all steel, design modification of deflections must deal only with the beam's shape.

Excessive deflection may cause problems in buildings. Excessive sag may disrupt the intended drainage patterns for a generally flat roof surface. For floors, a common problem is the development of some perceivable bounciness. The form of the beam and its supports may also be a consideration. For the simple-span beam in Figure 9.9, the usual concern is simply for the maximum sag at midspan. For a

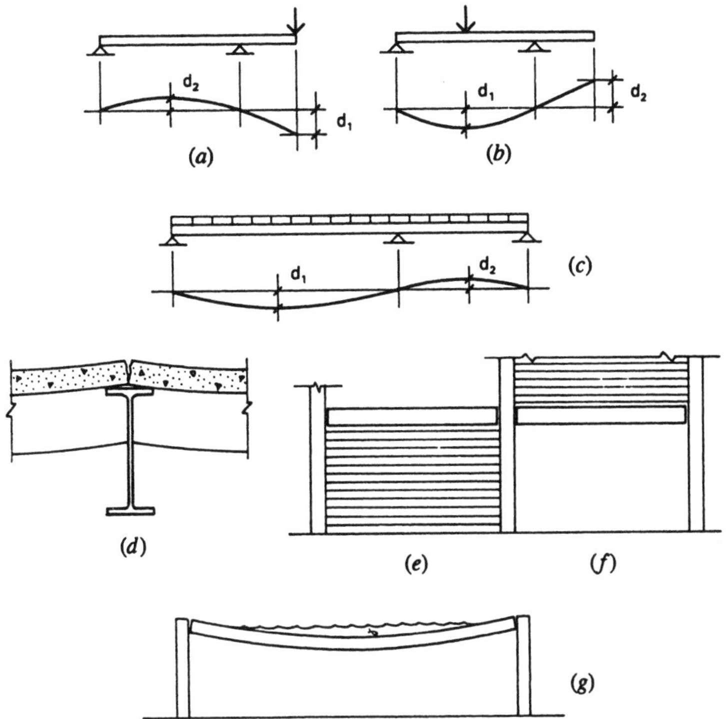

Figure 9.10 Considerations for deflection of beams.

beam with a projected (cantilevered) end, however, a problem may be created at the unsupported cantilevered end; depending on the extent of the cantilever, this may involve downward deflection (as shown in Figure 9.10*a*) or upward deflection (as shown in Figure 9.10*b*).

With continuous (multiple-span) beams, a potential problem derives from the fact that a load in any span causes some deflection in all spans. This is most critical when loads vary in different spans or the lengths of spans differ significantly (see Figure 9.10*c*).

Most deflection problems in buildings stem from the effects of the structural deformations on adjacent or supported elements of the building construction. When beams are supported by other beams, excessive deflection of the supported beams can cause rotation at the supported ends, resulting in cracking or separation of the floor deck that is continuous over the supporting beams, as shown in Figure 9.10*d*. For such a system, there is also an accumulative deflection caused by the independent deflections of the deck, beams, and support beams, which can cause problems for maintaining a flat floor surface or a desired roof profile for drainage.

An especially difficult problem related to deflections is the effect of beam deflections on nonstructural elements of the construction. Figure 9.10*e* shows the case of a beam occurring directly over a solid wall. If the wall is made to fit tightly beneath the beam, any deflection of the beam will cause it to bear on top of the wall—not an acceptable situation if the wall is especially fragile (a metal and glass curtain wall, for example). A different sort of problem occurs when relatively rigid walls (e.g., masonry or plastered) are supported by spanning beams, as shown in Figure 9.10*f*. In this case, the wall is relatively intolerant of *any* deformation, so anything significant in the form of sag of the beam is really critical.

For long-span structures (an ambiguous class, usually meaning 100 ft or more span), a special problem is that of the relatively flat roof surface. In spite of provisions for code-mandated minimum drainage, heavy rain will run off slowly from the surface and linger to cause some deflection. The sag of the spanning structure can swiftly cause formation of a pond (see Figure 9.10*g*). The pond itself then produces more deflection, resulting in a deeper pond, a progression that can accelerate into a failure condition. Building codes and design specifications (including the AISCs) provide requirements relating to this phenomenon, referred to as "ponding."

Standard Equations for Deflection

Determining deflection of a beam is usually done using a series of standard equations. These equations are listed for various loading conditions in the AISC *Steel Manual* (Ref. 5), samples of which are given in Figure 3.18. The purpose of these equations is to determine actual deflection dimensions due to service loads, so the computations are the same for both stress and strength design methods and the loads are not factored for LRFD. This is the major advantaged of the allowable strength design method (ASD for steel) where the loads are never factored. Care must be taken to assure the use of consistent units in these equations; beam section properties and modulus of elasticity are usually in inch, while spans are often in feet and uniformly distributed loads are usually in pounds per foot of beam length. Conversions to proper units should be made before entering data in the equations, or conversion must be done within the equations. The most used equations for simply supported beams are as follows:

Loading Condition	Maximum Deflection
Uniform load over entire span	$\Delta = \dfrac{5wl^4}{384EI}$ or $\Delta = \dfrac{5Wl^3}{384EI}$, where $W = wl$
Concentrated load at midspan	$\Delta = \dfrac{Pl^3}{48EI}$

Example 11. A simple beam has a span of 20 ft with a uniformly distributed load of 1.95 kips/ft. The beam consists of a W 14 × 34 shape. Find the maximum deflection.

Solution: To use the deflection equation, first convert the data to all kip and inch units. Thus,

$$\text{Total load} = 1.95 \times 20 = 39 \text{ kips}$$

$$\text{Span} = 20 \times 12 = 240 \text{ in.}$$

Then

$$\Delta = \frac{5Wl^3}{384EI} = \frac{5(39)(240)^3}{384(39{,}000)(340)} = 0.712 \text{ in.}$$

Allowable Deflections

What is permissible for beam deflection is mostly a judgment to be made by experienced designers. While code specifications must be recognized, it is difficult to provide useful guidance for specific limitations to avoid the various problems described in Figure 9.10. Each situation must be considered individually, requiring cooperation of the structural designer and the rest of the building design team.

For spanning beams in ordinary situations, some rules of thumb have been derived over many years of experience. These usually consist of establishing some maximum degree of curvature described in the form of a ratio of the deflection to the beam span, expressed as a fraction of the span. These are sometimes, although not always, specified in design codes or legally enacted building codes. Some typical limitations recognized by designers are the following:

To avoid visible sag under total load on short to medium spans, $\Delta < L/150$

For total load deflection of a roof structure, $\Delta < L/180$

For deflection under live load only for a roof structure, $\Delta < L/240$

For total load deflection of a floor structure, $\Delta < L/240$

For deflection under live load only for a floor structure, $\Delta < L/360$

Deflection of Uniformly Loaded Simple Beams

The most frequently used beam in flat roof and floor systems is the single, simply supported beam (no end restraint) with a uniformly distributed loading, as shown in Figure 3.18, Case 2. For this case, the following values may be obtained for the beam behavior:

Maximum bending moment:

$$M = \frac{wL^2}{8}$$

Maximum elastic bending stress on the beam cross section:

$$f = \frac{Mc}{I} \quad \text{or} \quad \frac{M}{S}$$

Maximum midspan deflection:

$$\Delta = \frac{5wL^4}{384EI} \quad \text{or} \quad \frac{5WL^3}{384EI}$$

Using these relationships, together with a known modulus of elasticity ($E = 29{,}000$ ksi for steel), a convenient abbreviated formula can be derived. Noting that the dimension c in the bending stress formula is one-half the beam depth ($d/2$) for symmetrical shapes and substituting the expression for M, we can say

$$f = \frac{Mc}{I} = \left(\frac{wL^2}{8}\right)\left(\frac{d/2}{I}\right) = \frac{wL^2d}{16I}$$

Then

$$\Delta = \frac{5wL^4}{384EI} = \left(\frac{wL^2d}{16I}\right)\left(\frac{5L^2}{24Ed}\right) = (f)\left(\frac{5L^2}{24Ed}\right) = \frac{5fL^2}{24Ed}$$

This is a basic formula for any beam symmetrical about its bending axis. For a shorter version the value for E for steel (29,000 ksi) may be used. Also, because deflections are only appropriate within the elastic range, it is reasonable to set the limit for f as that of the allowable bending stress for ASD, $F_b = 0.667 \times F_y$. Also, for convenience, spans are usually measured in feet, not inches, so a factor of 12 is added. For $F_y = 50$ ksi:

$$\Delta = \frac{5fL^2}{24Ed} = \left(\frac{5}{24}\right)\left(\frac{0.667 \times 50}{29{,}000}\right)\left(\frac{(12 \times L)^2}{d}\right) = \frac{0.0345 \times L^2}{d}$$

The derived deflection formula involving only span and beam depth can be used to plot a graph that displays the deflection of a beam of a constant depth for a variety of spans. Figure 9.11 consists of a series of such graphs for beams from 6 to 36 in. in depth and a yield stress of 50 ksi. Use of this graph presents yet another means for determining beam deflections. An answer within about 5% should be considered reasonable from the graphs.

The real value of the graph in Figure 9.12, however, is in assisting the design process. Once the span is known, it may be initially determined from the graphs what beam depth is required for a given deflection. The limiting deflection may be given as an actual dimension or, more commonly, as a limiting percentage of the span ($^1/_{240}$, $^1/_{360}$, etc.), as previously discussed. To aid in the latter situation, lines are drawn on the figures representing the usual percentages of $^1/_{180}$, $^1/_{240}$, and $^1/_{360}$. Thus, if a beam is to be used for a span of 36 ft and the total load deflection limit is $^1/_{240}$, it may be observed in Figure 9.11 that the lines for a span of 36 ft and a ratio $^1/_{240}$ intersect almost precisely on the graph for an 18-in.-deep beam. This means that an 18-in.-deep beam will deflect almost precisely $^1/_{240}$th of the span if it is stressed in bending to 24 ksi. Thus, any beam chosen with greater depth will be conservative for deflection, and any beam with less depth will function only if it has lower stress.

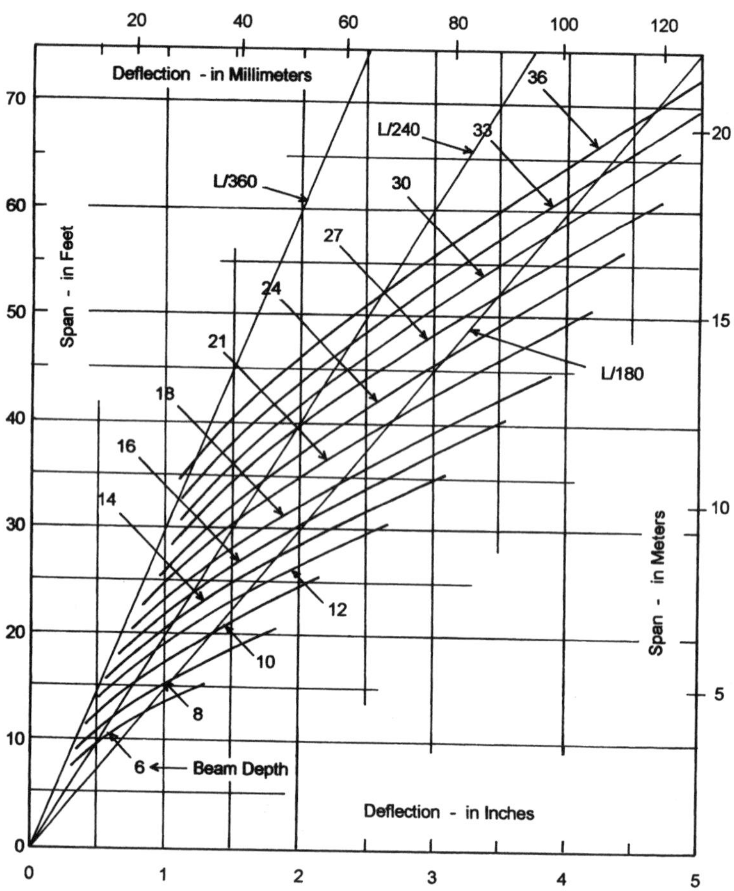

Figure 9.11 Deflection of steel beams with yield stress of 50 ksi under a constant bending stress of 33 ksi.

It must be noted that the curves in Figure 9.11 are based on an exact value of 33 ksi bending stress. When the true service load moment and bending stress is determined, the real value for deflection can be determined by direct proportion from that found from the graphs.

The minimum depth of a beam required by deflection criteria can be approximated from the span–depth equations. They are derived by placing the allowable deflection criteria into the equation derived earlier. It is important to remember here that the beam length (L) is in feet, whereas the beam depth (d) is in inches. The equations are as follows:

Yield Stress	$\Delta = L/180$ (roof)	$\Delta = L/240$ (floor)
$F_y = 50$ ksi	$d_{min} = 0.517L$ or $L/2$	$d_{min} = 0.690L$ or $2L/3$

Problems 9.7.A–D Find the maximum deflection in inches for the following simple beams of A992 steel with uniformly distributed load. Find the values using (1) the equation for deflection of a uniformly distributed load and (2) the curves in Figure 9.11.

A. W 10 × 33, span = 18 ft, total service load = 1.67 kips/ft
B. W 16 × 36, span = 20 ft, total service load = 2.5 kips/ft
C. W 18 × 46, span = 24 ft, total service load = 2.29 kips/ft
D. W 21 × 57, span = 27 ft, total service load = 2.5 kips/ft

9.8 SAFE LOAD TABLES

The simple beam with uniformly distributed load occurs so frequently that it is useful to have a rapid design method for quick selection of shapes based on knowing only the beam load and span. The AISC *Steel Manual* (Ref. 5) provides a series of such tables with data for W, M, S, and C shapes most often used for beams.

Table 9.2 presents data for selected W shapes for $F_y = 50$ ksi. Table values are for the total factored load capacity in kips, assuming lateral bracing at points not further apart than the limiting dimension of L_p (see discussion in Section 9.5). The table values are determined as the maximum factored plastic moment capacity of the shapes; thus,

$$\phi_b M_p = 0.9 Z_x F_y$$

For very short spans, loads are often limited by beam end shear or end support conditions rather than by bending or deflection limits. For this reason, table values are not shown for spans less than 12 times the beam depth.

For long spans, loads are often limited by deflection rather than by bending. Thus, table values are not shown for spans exceeding a limit of 24 times the beam depth.

TABLE 9.2 Factored Load–Span Values for 50-ksi Beams[a]

	F_y = 50 ksi				Span (ft)									
Designation	L_p (ft)	L_r (ft)	M_p (kip-ft)	M_r (kip-ft)	12	14	16	18	20	22	24	26	28	30
W 10 × 17	2.99	8.37	78	54	46.8	40.1	35.1	31.2	28.1					
W 12 × 16	2.73	7.44	84	57	50.3	43.1	37.7	33.5	30.2	27.4	25.1			
W 10 × 19	3.09	8.89	90	63	54.0	46.3	40.5	36.0	32.4					
W 12 × 22	3.00	8.41	122	85	73.3	62.8	54.9	48.8	44.0	40.0	36.6			
W 10 × 26	4.81	13.6	130	93	78.3	67.1	58.7	52.2	47.0					
W 12 × 26	5.34	13.8	155	111	93.0	79.7	69.8	62.0	55.8	50.7	46.5			
W 10 × 33	6.86	19.8	162	117	97.0	83.1	72.8	64.7	58.2					
W 14 × 26	3.82	10.2	168	118		86.1	75.4	67.0	60.3	54.8	50.3	46.4	43.1	
W 16 × 26	3.96	10.4	184	128			82.9	73.7	66.3	60.3	55.3	51.0	47.4	44.2
W 12 × 35	5.44	15.2	213	152	128	110	96.0	85.3	76.8	69.8	64.0			
W 14 × 34	5.41	14.4	228	162		117	102	91.0	81.9	74.5	68.3	63.0	58.5	
W 10 × 45	7.11	24.1	229	164	137	118	103	91.5	82.4					
W 16 × 36	5.37	14.0	267	188			120	107	96.0	87.3	80.0	73.8	68.6	64.0
W 12 × 45	6.89	20.3	268	192	161	138	120	107	96.3	87.5	80.3			
W 10 × 54	9.05	30.2	278	200	167	143	125	111	100					
W 14 × 43	6.68	18.3	290	209		149	131	116	104	94.9	87.0	80.3	74.6	
W 12 × 53	8.77	25.6	325	235	195	167	146	130	117	106	97.4			
W 18 × 40	4.49	12.1	327	228				131	118	107	98.0	90.5	84.0	78.4
W 16 × 45	5.55	15.2	343	242			154	137	123	112	103	95.0	88.2	82.3
W 10 × 68	9.16	36.0	355	252	213	183	160	142	128					

(continued)

TABLE 9.2 (Continued)

	$F_y = 50$ ksi				Span (ft)									
Designation	L_p (ft)	L_r (ft)	M_p (kip-ft)	M_r (kip-ft)	12	14	16	18	20	22	24	26	28	30
W 14 × 53	6.79	20.1	363	259		187	163	145	131	119	109	101	93.3	
W 18 × 46	4.56	12.6	378	263				151	136	124	113	105	97.2	90.7
W 18 × 50	5.83	15.6	421	296				168	152	138	126	117	108	101
W 16 × 57	5.66	16.6	438	307			197	175	158	143	131	121	113	105
W 10 × 88	9.30	45.1	471	328	283	242	212	188	170					
W 14 × 68	8.70	26.4	479	343		246	216	192	173	157	144	133	123	
W 12 × 79	10.8	35.7	496	357	298	255	223	198	179	162	149			
W 18 × 60	5.94	16.6	513	360				205	185	168	154	142	132	123
W 14 × 82	8.77	29.5	579	410		298	261	232	209	190	174	160	149	
W 18 × 71	6.01	17.8	608	423				243	219	199	183	168	156	146
W 10 × 112	9.48	56.5	613	420	368	315	276	245	221					
W 12 × 96	10.9	41.4	613	437	368	315	276	245	221	200	184			
W 16 × 77	8.70	25.5	633	453			285	253	228	207	190	175	163	152
W 12 × 120	11.1	50.0	775	543	465	399	349	310	279	254	233			
W 16 × 100	8.84	29.6	833	590			375	333	300	273	250	231	214	200
W 14 × 120	13.2	46.2	883	633		454	398	353	318	289	265	245	227	
W 12 × 152	11.3	62.1	1,013	697	608	521	456	405	365	331	304			
W 14 × 145	14.1	54.7	1,083	773		557	488	433	390	355	325	300	279	
W 21 × 44	4.45	12.0	398	272	119	110	102	95.4	89.4	84.2	79.5	75.3	71.6	68.1
W 21 × 50	4.60	12.5	458	315	138	127	118	110	103	97.1	91.7	86.8	82.5	78.6
W 21 × 57	4.77	13.1	538	370	161	149	138	129	121	114	108	102	96.8	92.1
W 24 × 55	4.74	12.9	563	383	169	156	145	135	127	119	113	107	101	96.4

W 24 × 62	4.84	13.3	642	440	193	178	165	154	144	136	128	122	116	110
W 21 × 68	6.36	17.3	667	467	200	185	171	160	150	141	133	126	120	114
W 18 × 86	9.30	26.1	775	553	233	215	199	186	174	164	155			
W 21 × 83	6.47	18.5	817	570	245	226	210	196	184	173	163	155	147	140
W 24 × 76	6.79	18.0	833	587	250	231	214	200	188	176	167	158	150	143
W 18 × 106	9.40	28.7	958	680	288	265	246	230	216	203	192			
W 21 × 101	10.2	27.6	1,054	757	316	292	271	253	237	223	211	200	190	181
W 24 × 94	7.00	19.4	1,058	740	318	293	272	254	238	224	212	201	191	181
W 27 × 94	7.50	19.9	1,158	810			298	278	261	245	232	219	209	199
W 24 × 103	7.04	20.0	1,167	817	350	323	300	280	263	247	233	221	210	200
W 30 × 90	7.39	19.4	1,179	817				283	265	250	236	223	212	202
W 24 × 104	10.3	26.8	1,204	860	361	333	310	289	271	255	241	228	217	206
W 18 × 130	9.55	32.8	1,208	853	363	335	311	290	272	256	242			
W 21 × 122	10.3	29.8	1,279	910	384	354	329	307	288	271	256	242	230	219
W 24 × 117	10.4	27.9	1,363	970	409	377	350	327	307	289	273	258	245	234
W 27 × 114	7.71	21.3	1,429	997			368	343	322	303	286	271	257	245
W 30 × 108	7.60	20.3	1,442	997				346	324	305	288	273	260	247
W 18 × 158	9.69	38.0	1,483	1,033	445	411	381	356	334	314	297			
W 24 × 131	10.5	29.1	1,542	1,097	463	427	396	370	347	326	308	292	278	264
W 21 × 147	10.4	32.8	1,554	1,097	466	430	400	373	350	329	311	294	280	266
W 30 × 124	7.88	21.5	1,700	1,183				408	383	360	340	322	306	291
W 33 × 118	8.20	21.7	1,729	1,197						366	346	328	311	296
W 24 × 146	10.6	30.6	1,742	1,237	523	482	448	418	392	369	348	330	314	299
W 24 × 162	10.8	32.4	1,950	1,380	585	540	501	468	439	413	390	369	351	334
W 30 × 148	8.06	22.8	2,083	1,453				500	469	441	417	395	375	357
W 33 × 141	8.59	23.1	2,142	1,493						454	428	406	386	367

Source: Compiled from data in the *Steel Construction Manual* (Ref. 5), with permission of the publisher, American Institute of Steel Construction.

[a]Table yields total safe uniformly distributed load in kips for indicated spans.

TABLE 9.3 Possible Choices for the Beam (Loads in kips)

Shape	Load from Table 9.2	Beam Weight	Net Safe Superimposed Load	Is Beam Acceptable?
W 16 × 36	80	1.0	79.0	Yes
W 12 × 45	80.3	1.3	79.0	Yes
W 14 × 43	87	1.2	85.8	Yes
W 18 × 40	98	1.2	96.8	Yes

The self-weight of the beam is included in the load given in these tables. Once a beam is selected, its weight must be subtracted from the table value to obtain the net allowable superimposed load.

The following example illustrates the use of Table 9.2 for a common design situation.

Example 12. Design a simply supported A992 steel beam to carry a uniformly distributed live load of 1.33 kips/ft and a superimposed uniformly distributed dead load of 0.66 kip/ft on a span of 24 ft. Find (1) the lightest shape and (2) the shallowest (least depth) shape.

Solution: First, the load must be factored:

$$w_u = 1.4(\text{dead load}) = 1.4(0.66) = 0.924\ \text{kip/ft}$$

or

$$w_u = 1.2(\text{dead load}) + 1.6(\text{live load}) = 1.2(0.66) + 1.6(1.33)$$
$$= 2.92\ \text{kip/ft}$$

and the total superimposed load is

$$W_u = w_u L = 2.92(24) = 70.1\ \text{kips}$$

Possible choices from Table 9.2 are displayed in Table 9.3. Also shown are the values for the factored beam self-weight and the net usable load. The lightest acceptable shape is the W 16 × 36 and the shallowest choice is the W 12 × 45. The 12-in.-deep beam is 25% heavier than the lightest choice, so its use is questionable, unless a really constricting dimensional problem exists.

Problems 9.8.A–H For each of the following conditions, find (1) the lightest permitted shape and (2) the shallowest permitted shape of A992 steel.

	Span	Live Load	Superimposed Dead Load
A	16 ft	3 kips/ft	3 kips/ft
B	20 ft	1 kip/ft	0.5 kip/ft
C	36 ft	1 kip/ft	0.5 kip/ft
D	40 ft	1.25 kips/ft	1.25 kips/ft
E	18 ft	0.33 kip/ft	0.625 kip/ft
F	32 ft	1.167 kips/ft	3.5 kips/ft
G	42 ft	1 kip/ft	0.238 kip/ft
H	28 ft	0.5 kip/ft	0.5 kip/ft

Equivalent Load Techniques

The safe loads in Tables 9.2 are uniformly distributed loads on simple beams. Actually, the table values are determined on the basis of bending moments and limiting bending stress so that it is possible to use the tables for other loading conditions for some purposes. Because framing systems usually contain some beams with other than simple uniformly distributed loadings, this is sometimes a useful process for design.

Consider the following situation: a beam with a load consisting of two equal concentrated loads placed at the beam third points—in other words, Case 2 in Figure 3.18. For this condition, the figure yields a maximum moment value expressed as $PL/3$. By equating this to the moment value for a uniformly distributed load, a relationship between the two loading conditions can be derived. Thus,

$$\frac{WL}{8} = \frac{PL}{3} \quad \text{or} \quad W = 2.67P$$

which shows that if the value of one of the concentrated loads in Case 3 of Figure 3.18 is multiplied by 2.67, the result would be an *equivalent uniform load* or *equivalent tabular load* (called EUL or ETL) that would produce the same maximum bending moment as the true loading condition.

Although the expression "equivalent uniform load" is the general name for this converted loading, when derived to facilitate the use of tabular materials, it is also referred to as the "equivalent tabular loading." Figure 3.18 yields the ETL factors for several common loading conditions.

It is important to remember that the EUL or ETL is based only on consideration of flexure, so that investigation for shear, bearing, and deflection must still use the true loading condition.

This method may also be used for any loading condition, not just the simple, symmetrical conditions shown in Figure 3.18. The process consists of first finding the true maximum bending moment due to the actual loading; then this is equated to the expression for the maximum moment for a theoretical uniformly distributed load, and the EUL is determined. Thus,

$$M = \frac{WL}{8} \quad \text{or} \quad W = \frac{8M}{L}$$

The expression $W = 8M/L$ is the general expression for an equivalent uniform load for any loading condition.

9.9 STEEL TRUSSES

When iron and then steel emerged as major industrial materials in the eighteenth and nineteenth centuries, one of the earliest applications to spanning structures was in the development of trusses and trussed forms of arches and bents. One reason for this was the early limit on the size of members that could be produced. In order to create a reasonably large structure, therefore, it was necessary to link together a large number of small parts.

Besides the truss elements themselves, a major technical problem to be solved for such assemblages is the achieving of the many joints. Thus, the creation of steel structural assemblages involves design of many joints, which must be both economical and practical for formation. Various connecting devices or methods have been employed, a major one used for building structures in earlier times being the use of hot-driven rivets. The process for this consists of matching up of holes in members to be connected, placing a heat-softened steel pin in the hole, and then beating the heck out of the protruding ends of the pin to form a rivet.

Basic forms developed for early connections are still widely used. Today, however, joints are mostly achieved by welding or with highly tightened bolts in place of rivets. Welding is mostly employed for connections made in the fabricating shop (called *shop connections*), while bolting is preferred for connections made at the erection site (called *field connections*). Riveting and bolting are often achieved with

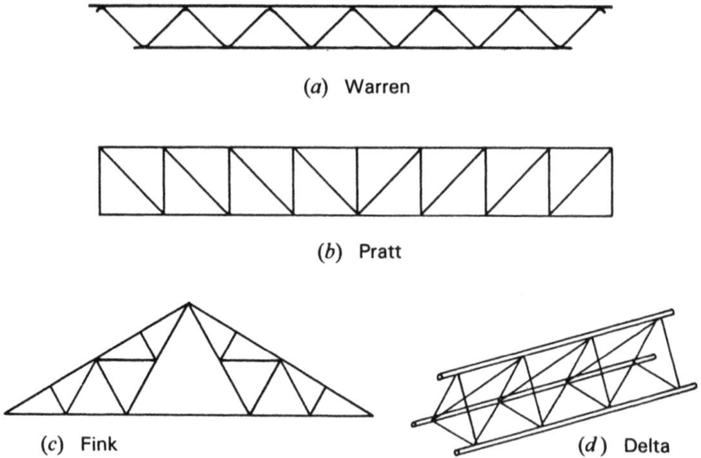

Figure 9.12 Common forms of light steel trusses.

an intermediate connecting device (gusset plate, connecting angles, etc.), while welding is frequently achieved directly between attached members.

Truss forms relate to the particular structural application (bridge, gable-form roof, arch, flat-span floor, etc.), to the magnitude of the span, and to the materials and methods of construction.

Some typical forms for individual trusses used in steel construction are shown in Figure 9.12. The forms in widest use are the parallel-chorded types, shown here in Figures 9.12*a* and *b*, these often being produced as manufactured, proprietary products by individual companies. (See discussion in Section 9.10.) Variations of the gable-form truss (Figure 9.12*c*) can be produced for a wide range of spans.

As planar elements, trusses are quite unstable in a lateral direction (perpendicular to the plane of the truss). Structural systems employing trusses require considerable attention for development of bracing. The delta truss (Figure 9.12*d*) is a unique self-stabilizing form that is frequently used for towers and columns but may also be used for a spanning truss that does not require additional bracing.

9.10 MANUFACTURED TRUSSES FOR FLAT SPANS

Factory-fabricated, parallel-chord trusses are produced in a wide range of sizes by a number of manufacturers. Most producers comply with the regulations of industrywide organizations; for light steel trusses the principal such organization is the Steel Joist Institute, called the SJI. Publications of the SJI are a chief source of general information (see Ref. 6), although the products of individual manufacturers vary, so that much valuable design information is available directly from the suppliers of a specific product. Final design and development of construction details for a particular project must be done in cooperation with the supplier of the products.

Light steel parallel-chord trusses, called *open-web joists*, have been in use for many years. Early versions used all-steel bars for the chords and the continuously bent web members (see Figure 9.13), so that they were also referred to as *bar joists*. Although other elements are now used for the chords, the bent-steel rod is still used for the diagonal web members for some of the smaller size joists. The range of size of this basic element has now been stretched considerably, resulting in members as long as 150 ft and depths of 7 ft and more. At the larger size range the members are usually more common forms for steel trusses—double angles, structural tees, and so on. Still, a considerable usage is made of the smaller sizes for both floor joists and roof rafters.

Table 9.4 is adapted from a standard table in a publication of the SJI (Ref. 6). This table lists a number of sizes available in the K series, which is the lightest group of joists. Joists are identified by a three-unit designation. The first number indicates the overall nominal depth of the joist, the letter indicates the series, and the second number indicates the class of size of the members—the higher the number, the heavier and stronger the joist.

Table 9.4 can be used to select the proper joist for a determined load and span situation. Figure 9.14 shows the basis for determination of the span for a joist. There are two entries in the table for each span; the first number represents the total factored load capacity of

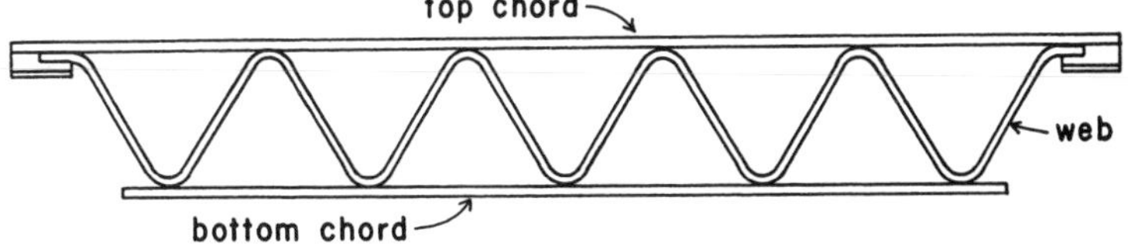

Figure 9.13 Form of a short-span open-web joist.

TABLE 9.4 Safe Factored Loads for K-Series Open-Web Joists[a]

Joist Designation:	12K1	12K3	12K5	14K1	14K3	14K6	16K2	16K4	16K6	18K3	18K5	18K7	20K3	20K5	20K7
Weight (lb/ft):	5.0	5.7	7.1	5.2	6.0	7.7	5.5	7.0	8.1	6.6	7.7	9.0	6.7	8.2	9.3
Span (ft)															
20	357	448	607	421	528	729	545	732	816	687	816	816	767	816	816
	(142)	(177)	(230)	(197)	(246)	(347)	(297)	(386)	(426)	(423)	(490)	(490)	(517)	(550)	(550)
22	295	369	500	347	435	641	449	602	739	567	769	816	632	816	816
	(106)	(132)	(172)	(147)	(184)	(259)	(222)	(289)	(351)	(316)	(414)	(438)	(393)	(490)	(490)
24	246	308	418	291	363	537	377	504	620	475	644	781	530	720	816
	(81)	(101)	(132)	(113)	(141)	(199)	(170)	(221)	(269)	(242)	(318)	(382)	(302)	(396)	(448)
26				246	310	457	320	429	527	403	547	665	451	611	742
				(88)	(110)	(156)	(133)	(173)	(211)	(190)	(249)	(299)	(236)	(310)	(373)
28				212	267	393	276	369	454	347	472	571	387	527	638
				(70)	(88)	(124)	(106)	(138)	(168)	(151)	(199)	(239)	(189)	(248)	(298)
30							239	320	395	301	409	497	337	457	555
							(86)	(112)	(137)	(123)	(161)	(194)	(153)	(201)	(242)
32							210	282	346	264	359	436	295	402	487
							(71)	(92)	(112)	(101)	(132)	(159)	(126)	(165)	(199)
36										209	283	344	233	316	384
										(70)	(92)	(111)	(88)	(115)	(139)
40													188	255	310
													(64)	(84)	(101)

28	516	634	816	565	693	816	692	816	816	813	816	816			
	(270)	(328)	(413)	(323)	(393)	(456)	(427)	(501)	(541)	(543)	(543)				
30	448	550	738	491	602	807	601	730	816	708	816	816	816	816	816
	(219)	(266)	(349)	(262)	(319)	(419)	(346)	(417)	(459)	(439)	(500)	(500)	(543)	(543)	(543)
32	393	484	647	430	530	709	528	641	770	620	764	815	743	815	815
	(180)	(219)	(287)	(215)	(262)	(344)	(285)	(343)	(407)	(361)	(438)	(463)	(461)	(500)	(500)
36	310	381	510	340	417	559	415	504	607	490	602	723	586	705	723
	(126)	(153)	(201)	(150)	(183)	(241)	(199)	(240)	(284)	(252)	(306)	(366)	(323)	(383)	(392)
40	250	307	412	274	337	451	337	408	491	395	487	629	473	570	650
	(91)	(111)	(146)	(109)	(133)	(175)	(145)	(174)	(207)	(183)	(222)	(284)	(234)	(278)	(315)
44	206	253	340	227	277	372	277	337	405	326	402	519	390	470	591
	(68)	(83)	(109)	(82)	(100)	(131)	(108)	(131)	(155)	(137)	(167)	(212)	(176)	(208)	(258)
48				190	233	313	233	282	340	273	337	436	328	395	542
				(63)	(77)	(101)	(83)	(100)	(119)	(105)	(128)	(163)	(135)	(160)	(216)
52							197	240	289	233	286	371	279	335	498
							(65)	(79)	(102)	(83)	(100)	(128)	(106)	(126)	(184)
56										200	246	319	240	289	446
										(66)	(80)	(102)	(84)	(100)	(153)
60													209	250	389
													(69)	(81)	(124)

Source: Data adapted from more extensive tables in the *Guide for Specifying Steel Joists with LRFD, 2000* (Ref. 7), with permission of the publisher, Steel Joist Institute. The Steel Joist Institute publishes both specifications and load tables; each of these contains standards that are to be used in conjunction with one another.

[a]Loads in pounds per foot of joist span. First entry represents the total factored joist capacity; entry in parentheses is the load that produces a deflection of 1/360 of the span. See Figure 9.15 for definition of span.

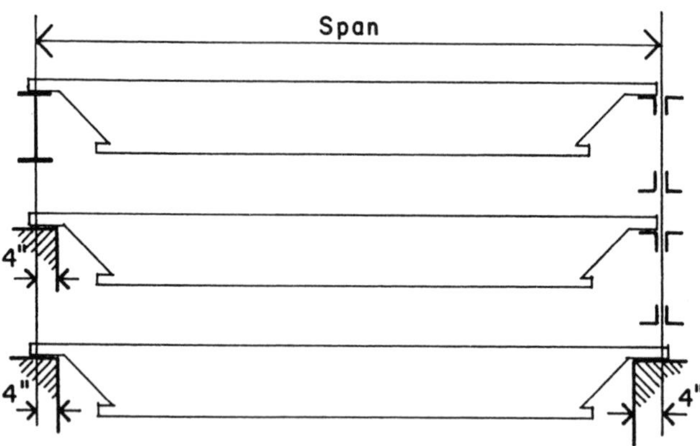

Figure 9.14 Definition of span for open-web steel joists, as given in Ref. 7. Reprinted with permission of the Steel Joist Institute

TABLE 9.5 Possible Choices for the Roof Joist

Load Condition	Required Capacity (lb/ft)	Capacity of the Indicated Joists (lb/ft) 22K9	24K6	26K5
Factored total capacity		412	337	337
Joist weight from Table 9.4		11.3	9.7	9.8
Factored joist weight		14	12	12
Net usable capacity	300	398	325	325
Load for deflection of $^{1}/_{360}$	120	146	133	145

the joist in pounds per feet of the joist length (lb/ft); the number in parentheses is the load that will produce a deflection of $^{1}/_{360}$ of the span. The following examples illustrate the use of the table data for some common design situations. For the purpose of illustration the examples use data from Table 9.5. However, more joist sizes are available and their capacities are given in the reference for Table 9.4.

Example 13. Open-web steel joists are to be used to support a roof with a unit live load of 20 psf and a unit dead load of 15 psf (not including the weight of the joists) on a span of 40 ft. Joists are spaced at 6 ft center to center. Select the lightest joist if deflection under live load is limited to $^{1}/_{360}$ of the span.

Solution: The first step is to determine the unit load per foot on the joists. Thus:

Live load: 6(20) = 120 lb/ft
Dead load: 6(15) = 90 lb/ft (not including joist weight)
Total factored load: 1.2(90) + 1.6(120) = 108 + 192 = 300 lb/ft

This yields the two numbers (total factored load and live load only) that can be used to scan the data for the given span in Table 9.4. Note that the joist weight—so far excluded in the computation—is included in the total load entries in the table. Once a joist is selected, therefore, the actual joist weight (given in the table) must be deducted from the table entry for comparison with the computed values. We thus note from the table the possible choices listed in Table 9.5. Although the joist weights are all very close, the 24K6 is the lightest choice.

Example 14. Open-web steel joists are to be used for a floor with a unit live load of 75 psf and a unit dead load of 40 psf (not including the joist weight) on a span of 30 ft. Joists are 2 ft on center, and deflection is limited to $^1/_{240}$ of the span under total load and $^1/_{360}$ of the span under live load only. Determine the lightest possible joist and the lightest joist of least depth possible.

Solution: As in the previous example, the unit loads are first determined. Thus:

Live load: 2(75) = 150 lb/ft (for limiting deflection of L/360)
Dead load: 2(40) = 80 lb/ft (not including joist weight)
Total service load: 150 + 80 = 230 lb/ft
Total factored load: 1.2(80) + 1.6(150) = 96 + 240 = 336 lb/ft

To satisfy the deflection criteria for total load, the limiting value for deflection in parentheses in the table should be not less than (240/360)(230) = 153 lb/ft. Since this is slightly larger than the live load, it becomes the value to look for in the Table 9.4. Possible choices obtained from Table 9.4 are listed in Table 9.6, from which it may be observed:

The lightest joist is the 18K5.
The shallowest depth joist is the 18K5.

TABLE 9.6 Possible Choices for the Floor Joist

Load Condition	Required Capacity (lb/ft)	Capacity of the Indicated Joists (lb/ft)		
		18K5	20K5	22K4
Factored total capacity		409	457	448
Joist weight from Table 9.4		7.7	8.2	8.0
Factored joist weight		10	10	10
Net usable capacity	336	399	447	438
Load for deflection	153	161	201	219

In some situations it may be desirable to select a deeper joist, even though its load capacity may be somewhat redundant. Total sag, rather than an abstract curvature limit, may be of more significance for a flat roof structure. For example, for the 40-ft span in Example 13, a sag of $^1/_{360}$ of the span = $(^1/_{360})(40 \times 12) = 1.33$ in. The actual effect of this dimension on roof drainage or in relation to interior partition walls must be considered. For floors, a major concern is for bounciness, and this very light structure is highly vulnerable in this regard. Designers therefore sometimes deliberately choose the deepest feasible joist for floor structures in order to get all the help possible to reduce deflection as a means of stiffening the structure in general against bouncing effects.

As mentioned previously, joists are available in other series for heavier loads and longer spans. The SJI, as well as individual suppliers, also have considerably more information regarding installation details, suggested specifications, bracing, and safety during erection for these products.

Stability is a major concern for these elements since they have very little lateral or torsional resistance. Other construction elements, such as decks and ceiling framing, may help, but the whole bracing situation must be carefully studied. Lateral bracing in the form of X braces or horizontal ties is generally required for all steel joist construction, and the reference source for Table 9.4 (Ref. 7) has considerable information on this topic.

One means of assisting stability has to do with the typical end support detail, as shown in Figure 9.14. The common method of support consists of hanging the trusses by the ends of their top chords, which is a general means of avoiding the rollover type of rotational buckling at the supports that is illustrated in Figure 3.48*c*. For construction detailing, however, this adds a dimension to the overall depth

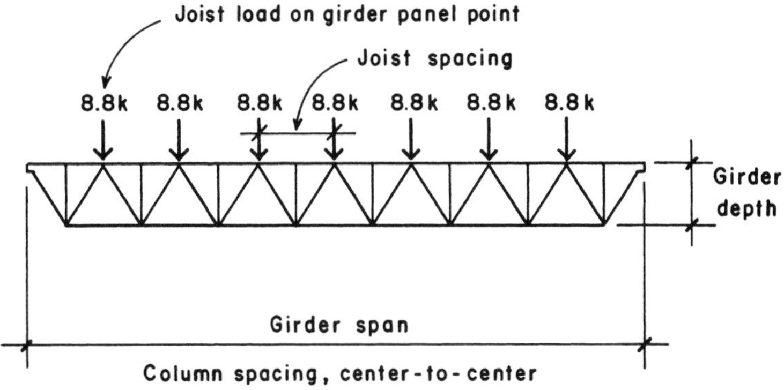

Standard Designation:

48G	8N	8.8 K
Depth in inches	Number of joist spaces	Load on each panel point in kips

Specify: 48G8N8.8K

Figure 9.15 Considerations for layout of the joist girder.

of the construction, in comparison to an all-beam system with the joist/beams and supporting girders all having their tops level. This added dimension (the depth of the end of the joist) is typically 2.5 in. for small joists and 4 in. for larger joists.

For development of a complete truss system, a special type of prefabricated truss available is that described as a *joist girder*. This truss is specifically designed to carry the regularly spaced, concentrated loads consisting of the end support reactions of joists. A common form of joist girder is shown in Figure 9.15. Also shown in the figure is the form of standard designation for a joist girder, which includes indications of the nominal girder depth, the number of spaces between joists (called the girder *panel unit*), and the end reaction force from the joists—which is the unit concentrated load on the girder.

Predesigned joist girders (i.e., girders actually designed for fabrication by the joist suppliers) may be selected from catalogs in a manner similar to that for open-web joists. The procedure is usually as follows:

1. The designer determines the joist spacing, joist load, and girder span. (The joist spacing should be a full number division of the girder span.)
2. This information is used to specify the girder by the standard designation.
3. The girder may be chosen from a catalog or simply specified for the supplier.

Illustrations of use of joists and complete truss systems are given in the building design examples in Part V.

Problem 9.10.A Open-web steel joists are to be used for a roof with a live load of 25 psf and a dead load of 20 psf (not including the joist weight) on a span of 48 ft. Joists are 4 ft on center, and deflection under live load is limited to $^1/_{360}$ of the span. Select the lightest joist.

Problem 9.10.B Open-web steel joists are to be used for a roof with a live load of 30 psf and a dead load of 18 psf (not including the joist weight) on a span of 44 ft. Joists are 5 ft on center, and deflection is limited to $^1/_{360}$ of the span. Select the lightest joist.

Problem 9.10.C Open-web steel joists are to be used for a floor with a live load of 50 psf and a dead load of 45 psf (not including the joist weight) on a span of 36 ft. Joists are 2 ft on center, and deflection is limited to $^1/_{360}$ of the span under live load only and to $^1/_{240}$ of the span under total load. Select (a) the lightest possible joist and (b) the shallowest depth possible joist.

Problem 9.10.D Repeat Problem 9.10.C, except that the live load is 100 psf, the dead load is 35 psf, and the span is 26 ft.

9.11 DECKS WITH STEEL FRAMING

Figure 9.16 shows four possibilities for a floor deck used in conjunction with a framing system of rolled steel beams. When a wood deck is used (Figure 9.16*a*), it is usually supported by and nailed to a series of wood joists, which are in turn supported by the steel beams. However, in some cases the deck may be nailed to wood members that are bolted to the tops of the steel beams, as shown in the figure. For floor

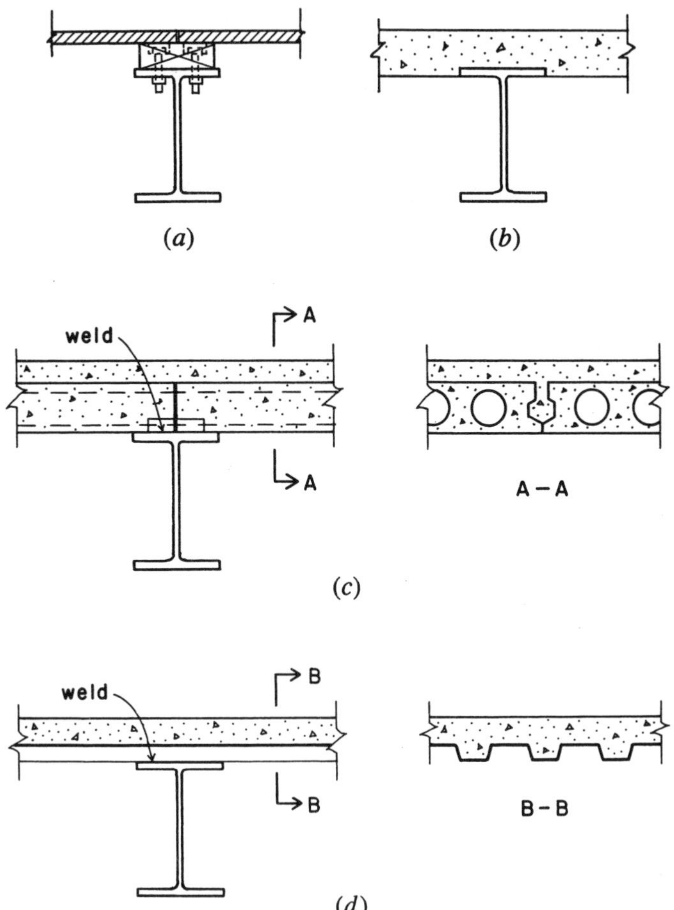

Figure 9.16 Typical forms of floor deck construction used with steel framing.

construction, it is now also common to use a concrete fill on top of the wood deck, for added stiffness, fire protection, and improved acoustic behavior.

A site-cast concrete deck (Figure 9.16*b*) is typically formed with plywood panels placed against the bottoms of the top flanges of the beams. This helps to lock the slab and beams together for lateral effects, although steel lugs are also typically welded to the tops of the beams for composite construction.

Concrete may also be used in the form of precast deck units (Figure 9.16*c*). In this case, steel elements are imbedded in the ends of the precast units and are welded to the beams. A site-poured concrete fill is typically used to provide a smooth top surface and is bonded to the precast units for added structural performance.

Formed sheet steel units may be used in one of three ways: as the primary structure, as strictly forming for the concrete deck, or as a composite element in conjunction with the concrete (Figure 9.16*d*). Attachment of this type of deck to the steel beams is usually achieved by welding the steel units to the beams before the concrete is placed.

Three possibilities for roof decks using steel elements are shown in Figure 9.17. Roof loads are typically lighter than floor loads and bounciness of the deck is usually not a major concern. (A possible exception to this is the situation where suspended elements may be hung from the deck and can create a problem with vertical movements during an earthquake.) A fourth possibility for the roof is the plywood deck shown in Figure 9.16*a*. This is, in fact, probably a wider use of this form of construction. Decks of formed sheet steel are discussed in Chapter 12, together with other structural products and systems developed from formed sheet stock.

9.12 CONCENTRATED LOAD EFFECTS ON BEAMS

An excessive bearing reaction on a beam, or an excessive concentrated load at some point in the beam span, may cause either localized yielding or *web crippling* (buckling of the thin beam web). The Steel Specification within the AISC *Steel Manual* (Ref. 5) requires that beam webs be investigated for these effects and that web stiffeners be used if the concentrated load exceeds limiting values.

The three common situations for this effect occur as shown in Figure 9.18. Figure 9.18*a* shows the beam end bearing on a support (commonly a masonry or concrete wall), with the reaction force transferred to the beam bottom flange through a steel bearing plate. Figure 9.18*b* shows a column load applied to the top of the beam at some point within the beam span. Figure 9.18*c* shows what may be the most frequent occurrence of this condition—that of a beam supported in bearing on top of a column with the beam continuous through the joint.

Figure 9.18*d* shows the development of the effective portion of the web length (along the beam span) that is assumed to resist

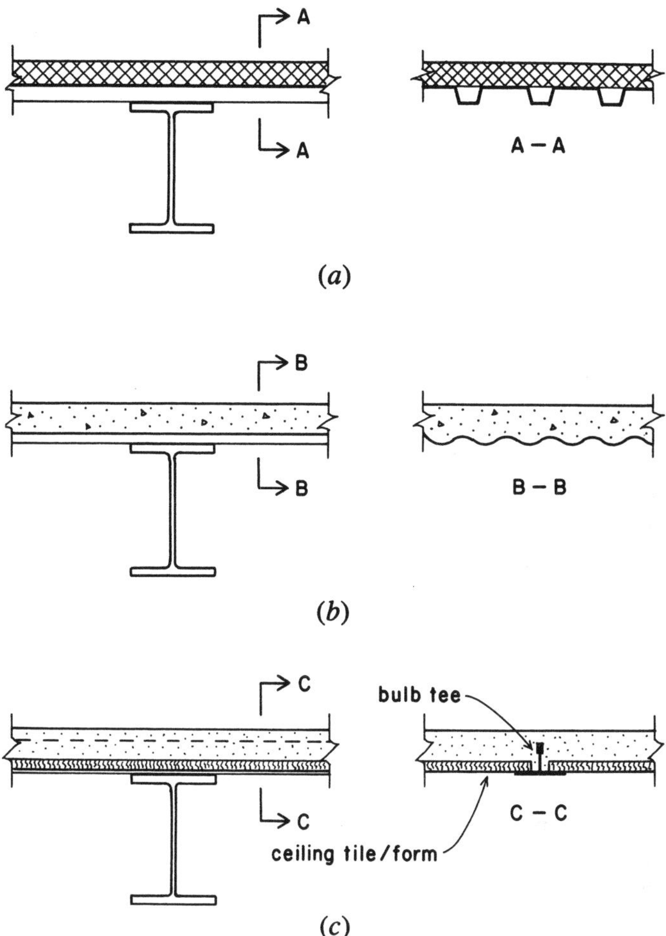

Figure 9.17 Typical forms of roof deck construction used with steel framing.

bearing forces. For yield resistance, the maximum end reaction and the maximum load within the beam span are defined as follows (see Figure 9.18*d*):

$$\text{Maximum end reaction} = (0.66F_y)(t_w)[N + 2.5(k)]$$

$$\text{Maximum interior load} = (0.66F_y)(t_w)[N + 5(k)]$$

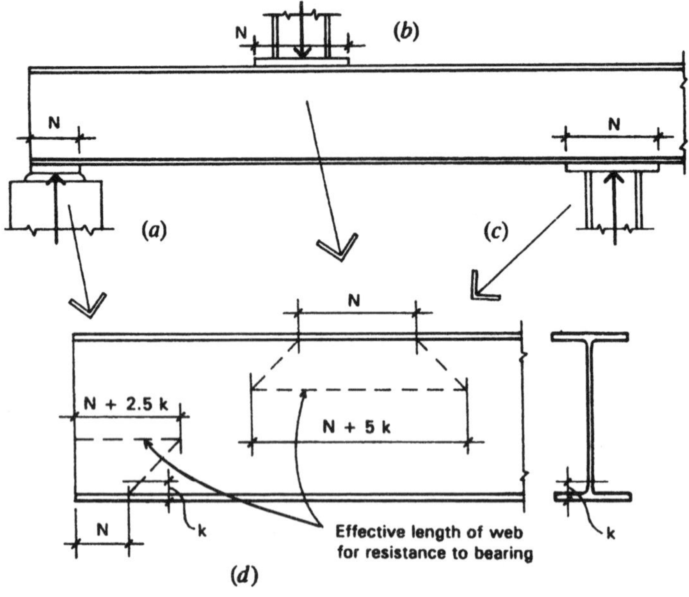

Figure 9.18 Considerations for concentrated bearing in steel beams with thin webs, as related to web crippling (compression buckling).

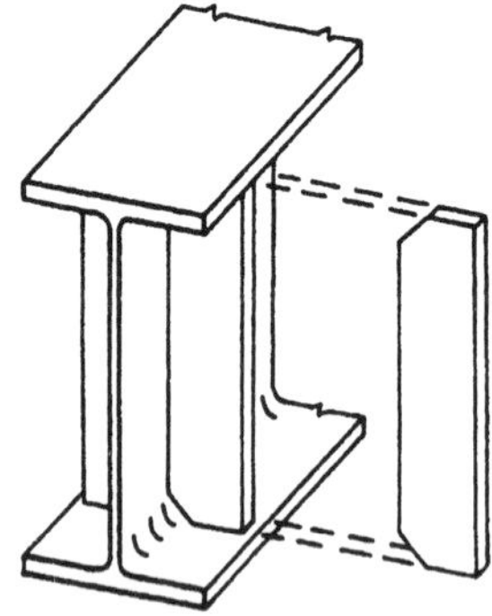

Figure 9.19 Use of stiffeners to prevent buckling of a thin beam web.

where t_w = thickness of the beam web
N = length of the bearing
k = distance from the outer face of the beam flange to the web toe of the fillet (radius) of the corner between the web and the flange

For W shapes, the dimensions t_w and k are provided in the AISC *Steel Manual* (Ref. 5) tables of properties for rolled shapes.

When these values are exceeded, it is recommended that web stiffeners be provided at the locations of the concentrated loads, as shown in Figure 9.19. These stiffeners add to bearing resistance and also brace the web for buckling in general.

The AISC also provides additional information for computation of limiting loads due to web crippling. However, the AISC *Steel Manual* (Ref. 5) tables also provide shortcuts for such computations.

10

STEEL COLUMNS AND FRAMES

Steel compression members range from small, single-piece columns and truss members to huge, built-up sections for high-rise buildings and large tower structures. The basic column function is one of simple compressive force resistance, but it is often complicated by the effects of buckling and the possible presence of bending actions. This chapter deals with various issues relating to the design of individual compression members and with the development of building structural frameworks.

10.1 COLUMN SHAPES

For modest load combinations, the most frequently used shapes are the round pipe, the rectangular tube, and the wide-flange shapes where the depth of the cross section is approximately equal to its width. (See Figure 10.1.) Accommodation with beams for framing is most easily achieved with W shapes of 10 in. or greater depth.

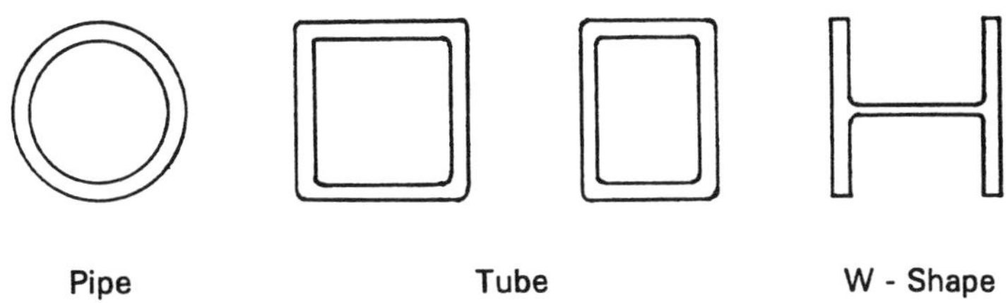

Figure 10.1 Common cross-sectional shapes for steel columns.

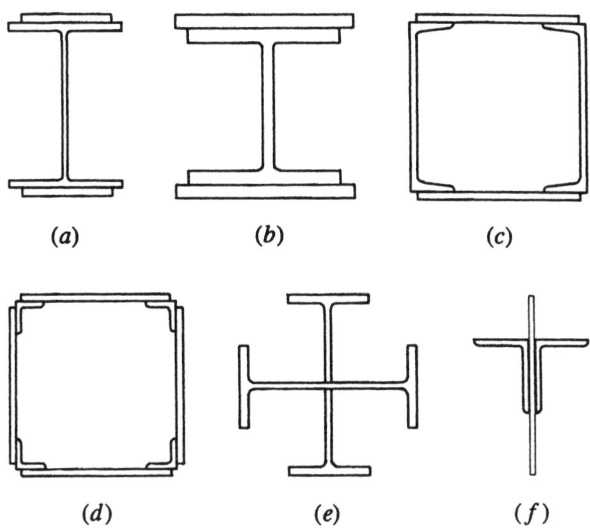

Figure 10.2 Various forms of combined, built-up shapes for steel columns.

For various reasons, it is sometimes necessary to make up a column section by assembling two or more individual steel elements. Figure 10.2 shows some such shapes that are used for special purposes. The custom assemblage of built-up sections is costly, so a single piece is typically favored if one is available.

One widely used built-up section is the double angle, as shown in Figure 10.2*f*. This occurs most often as a member of a truss or as a bracing member in a frame, the general stability against buckling being much better than that for a single-angle member. This section is not used, however, for a building column.

10.2 COLUMN SLENDERNESS AND END CONDITIONS

For steel columns, the value of the critical stress (F_c) in compression is determined from formulas in the AISC specification, contained in the AISC *Steel Manual* (Ref. 5); it includes variables of the steel yield stress and modulus of elasticity, the relative slenderness of the column, and special considerations for the restraint of the column ends.

Column slenderness (or stiffness) is determined as the ratio of the column unbraced length to the radius of gyration of the column section: L/r. Effects of end restraint are considered by use of a modifying factor (K). (See Figure 10.3.) The modified slenderness is thus expressed as KL/r.

Figure 10.4 is a graph of the critical compressive stress for a column with two grades of steel with yield stress (F_y) of 36 and 50 ksi. Values for full-number increments of KL/r for A992 steel, derived from the AISC specification formulas, are also given in Table 10.1.

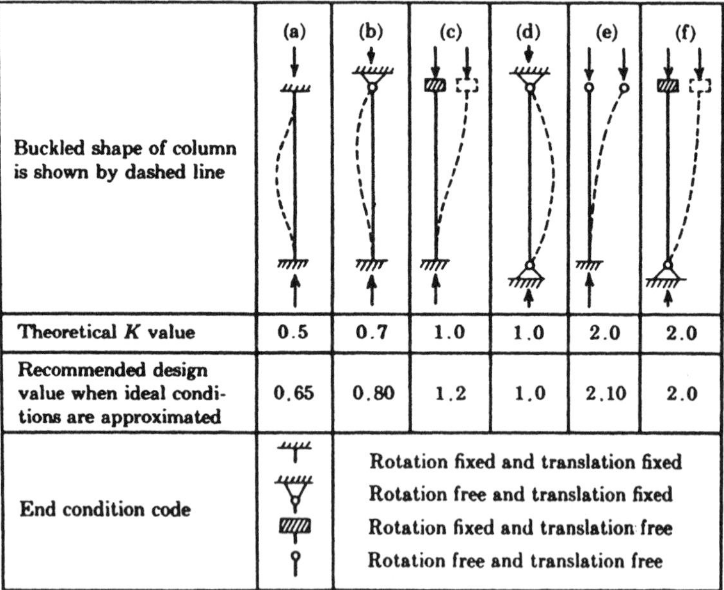

	(a)	(b)	(c)	(d)	(e)	(f)
Buckled shape of column is shown by dashed line						
Theoretical K value	0.5	0.7	1.0	1.0	2.0	2.0
Recommended design value when ideal conditions are approximated	0.65	0.80	1.2	1.0	2.10	2.0

End condition code	
	Rotation fixed and translation fixed
	Rotation free and translation fixed
	Rotation fixed and translation free
	Rotation free and translation free

Figure 10.3 Determination of effective column length (KL) for buckling. Reprinted from the *Steel Construction Manual* (Ref. 5), with permission of the publishers, the American Institute of Steel Construction

TABLE 10.1 Critical Unfactored Compressive Stress for Columns, F_c[a], where $F_y = 50$ ksi

KL/r		*KL/r*		*KL/r*	
1	50.0	41	44.2	81	30.9
2	50.0	42	43.9	82	30.6
3	50.0	43	43.7	83	30.2
4	49.9	44	43.4	84	29.8
5	49.9	45	43.1	85	29.5
6	49.9	46	42.8	86	29.1
7	49.8	47	42.5	87	28.7
8	49.8	48	42.2	88	28.4
9	49.7	49	41.9	89	28.0
10	49.6	50	41.6	90	27.7
11	49.6	51	41.3	91	27.3
12	49.5	52	41.0	92	26.9
13	49.4	53	40.7	93	26.6
14	49.3	54	40.4	94	26.2
15	49.2	55	40.1	95	25.8
16	49.1	56	39.8	96	25.5
17	49.0	57	39.4	97	25.1
18	48.8	58	39.1	98	24.8
19	48.7	59	38.8	99	24.4
20	48.6	60	38.4	100	24.1
21	48.4	61	38.1	101	23.7
22	48.3	62	37.7	102	23.4
23	48.1	63	37.4	103	23.0
24	47.9	64	37.1	104	22.7
25	47.8	65	36.7	105	22.3
26	47.6	66	36.4	106	22.0
27	47.4	67	36.0	107	21.6
28	47.2	68	35.7	108	21.3
29	47.0	69	35.3	109	21.0
30	46.8	70	34.9	110	20.6
31	46.6	71	34.6	111	20.3
32	46.4	72	34.2	112	20.0
33	46.2	73	33.9	113	19.7
34	45.9	74	33.5	114	19.3
35	45.7	75	33.1	115	19.0
36	45.5	76	32.8	116	18.7
37	45.2	77	32.4	117	18.3
38	45.0	78	32.0	118	18.0
39	44.7	79	31.7	119	17.7
40	44.5	80	31.3	120	17.4

Source: Developed from data in the *Steel Construction Manual* (Ref. 5), with permission of the publisher, American Institute of Steel Construction.

[a]Usable design stress limit in ksi for obtaining nominal strength (unfactored) of steel columns.

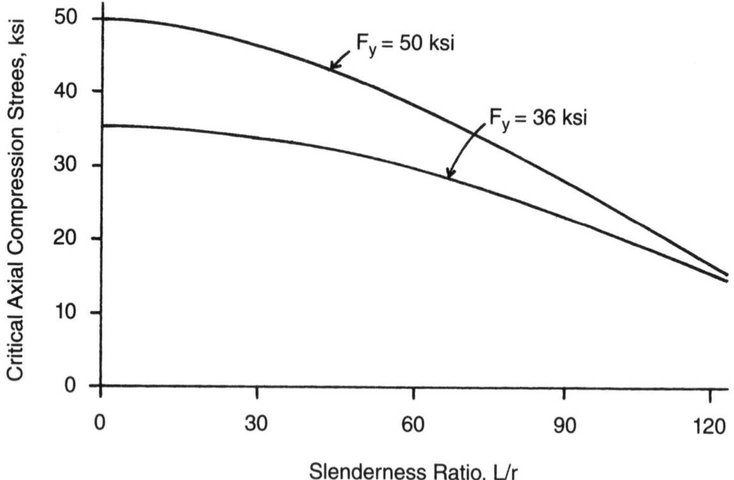

Figure 10.4 Critical unfactored compressive stress for steel columns as a function of yield stress and column slenderness.

For practical reasons, most building columns tend to have a value for relative stiffness between about 50 and 100, with only very heavily loaded columns being stiffer and with designers avoiding very slender columns. The AISC specification for steel discourages use of any compression members with a slenderness ratio greater than 200. For columns, a practical limit is 120.

10.3 SAFE AXIAL LOADS FOR STEEL COLUMNS

The design strength in axial compression for a steel column is computed by multiplying the design stress ($\varphi_c F_c$) by the cross-sectional area of the column, where $\varphi_c = 0.9$. Thus,

$$P_u = \phi_c \times P_n = \phi_c \times F_c \times A$$

where P_u = maximum factored load
φ_c = resistance factor, 0.9 for columns
P_n = nominal load resistance (unfactored) of the column
F_c = critical compressive stress for the column, based on KL/r
A = area of the column cross section

The following examples demonstrate the process for column load investigation. For a more direct design process for single-piece columns, load tables are frequently used. For built-up sections, however, it is necessary to compute the required column properties.

Example 1. A W 12 × 53 shape of A992 steel is used as a column with an unbraced length of 16 ft. Compute the maximum factored load.

Solution: Referring to Table A.3, $A = 15.6$ in.^{2}, $r_x = 5.23$ in., and $r_y = 2.48$ in. If the column is unbraced on both axes, it is limited by the lower r value for the weak axis. With no stated end conditions, case (d) in Figure 10.3 is assumed, for which $K = 1.0$; that is, no modification is made. Thus, the relative stiffness is computed as

$$\frac{KL}{r} = \frac{1 \times 16 \times 12}{2.48} = 77.4$$

It is usually considered acceptable to round the slenderness ratio off to the nearest whole number. Thus, with a KL/r value of 77, Table 10.1 yields a value for F_c of 32.4 ksi. The maximum factored load for the column is then

$$P_u = \phi_c F_c A = 0.9 \times 32.4 \times 15.6 = 455 \text{ kips}$$

Example 2. Compute the maximum factored load for the column in Example 1 if the top is pinned but prevented from lateral movement and the bottom is fully fixed.

Solution: Referring to Figure 10.3, note that for this case—(b) in the figure—the modifying factor K is 0.8. Then

$$\frac{KL}{r} = \frac{0.8 \times 16 \times 12}{2.48} = 62$$

From Table 10.1, $F_c = 37.7$ ksi. Then

$$P_u = \phi_c F_c A = 0.9 \times 37.7 \times 15.6 = 530 \text{ kips}$$

The following example illustrates the situation where a W shape is braced differently on its two axes.

Example 3. Figure 10.5*a* shows an elevation of the steel framing at the location of an exterior wall. The column is laterally restrained but rotation free at the top and bottom in both directions (on both the *x* and *y* axes). With respect to the *x* axis, the column is laterally unbraced for its full height. However, the existence of the horizontal framing in the wall plane provides lateral bracing with respect to the *y* axis of the section; thus, the buckling of the column in this direction takes the form shown in Figure 10.5*b*. If the column is a W 12 × 53 of A992 steel, L_1 is 30 ft, and L_2 is 18 ft, what is the maximum factored compression load?

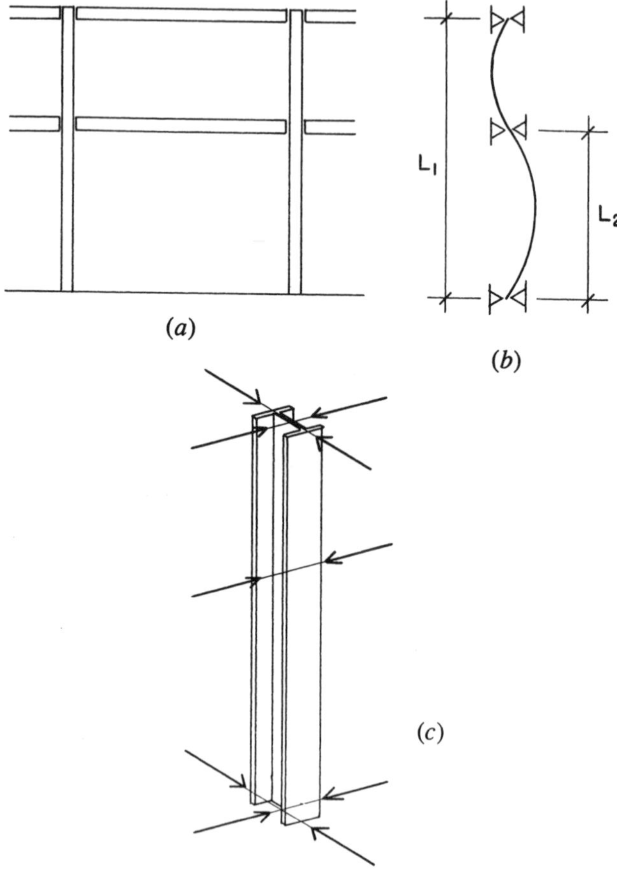

Figure 10.5 Biaxial bracing for steel columns.

Solution: Note that this is the shape used in Example 1 for which the properties are $A = 15.6$ in.^{2}, $r_x = 5.23$ in., and $r_y = 2.48$ in. For the x axis, the situation is case (d) from Figure 10.3. Thus,

$$x\ axis:\quad \frac{KL}{r} = \frac{1 \times 30 \times 12}{5.23} = 68.8 \quad \text{say } 69$$

For the y axis, the situation is also assumed to be case (d) from Figure 10.3, except that the deformation occurs in two parts (see Figure 10.5b). The lower part is used as it has the greater unbraced length. Thus,

$$y\ axis:\quad \frac{KL}{r} = \frac{1 \times 18 \times 12}{2.48} = 87.1 \quad \text{say } 87$$

Despite the bracing, the column is still critical on its weak (y) axis. From Table 10.1, the value of $F_c = 28.7$ ksi, and the maximum factored load is thus

$$P_u = \phi_c F_c A = 0.9 \times 28.7 \times 15.6 = 403 \text{ kips}$$

For the following problems, use A996 steel with $F_y = 50$ ksi.

Problem 10.3.A Determine the maximum factored axial compression load for a W 10 × 49 column with an unbraced height of 15 ft. Assume $K = 1.0$.

Problem 10.3.B Determine the maximum factored axial compression load for a W 12 × 120 column with an unbraced height of 22 ft if both ends are fixed against rotation and horizontal movement.

Problem 10.3.C Determine the maximum factored axial compression load in Problem 10.3.A if the conditions are as shown in Figure 10.5 with $L_1 =$ 15 ft and $L_2 = 8$ ft.

Problem 10.3.D Determine the maximum factored axial compression load in Problem 10.3.B if the conditions are as shown in Figure 10.5 with $L_1 =$ 40 ft and $L_2 = 22$ ft.

10.4 DESIGN OF STEEL COLUMNS

Unless a computer-supported procedure is used, design of steel columns is mostly accomplished through the use of tabulated data.

The following discussions in this section consider the latter process using materials from the AISC *Steel Manual* (Ref. 5). Design of a column without using load tables is hampered by the fact that the critical stress (F_c) is not able to be precisely determined until *after* the column shape is selected. That is, KL/r and the associated value for F_c cannot be determined until the critical r value for the column is known. This leads to a trial-and-error approach, which can become laborious in even simple circumstances. This process is unavoidable with built-up sections, so their design is unavoidably tedious.

The real value of the safe load tables for single rolled shapes is the ability to use them directly, once only the factored load, column unbraced height, and K factor are determined. The AISC *Steel Manual* (Ref. 5) provides load tables for W shapes for $F_y = 50$ ksi. The work that follows in this section provides examples of the use of tables similar to those in the AISC *Steel Manual* (Ref. 5) for a selection of columns of a variety of shapes.

In many cases, the simple, axial compression capacity is all that is involved in design selection. However, columns are also sometimes subjected to bending and shear and in some cases to torsion. In combined actions, however, the axial load capacity is usually included, so its singular determination is still a factor—thus, the safe compression load under simple axial application conditions is included in some way in just about every column design situation. The case of combined axial compression and bending is discussed in Section 10.5.

Single Rolled Shapes as Columns

The single rolled shape used most commonly for steel columns is the squarish, H-shaped element, with a nominal depth of 8 in. or more. Responding to this, steel rolling mills produce a wide range of these shapes. Most are in the W-shape series, although some are designated as M shapes.

Table 10.2 summarizes data from the series of AISC tables for shapes ranging from the W 8 × 24 to the W 14 × 211 for steel with a yield of 50 ksi. Table values are based on the r value for the y axis with $K = 1.0$. Also included in this table are values for the bending factor (m), which is used for approximate design of columns with bending, as discussed in Section 10.5.

To illustrate one use of Table 10.2, refer to Example 1 of Section 10.3. For the safe load for the W 12 × 53 with unbraced

TABLE 10.2 Safe Factored Loads for Selected 50-ksi-Yield-Stress W Shapes[a]

	Ratio	Effective Length (*KL*) in feet																		
F_y = 50 ksi	*rx*/*ry*	0	6	8	10	12	14	16	18	20	22	24	26	28	30	32	34	36	38	40
W 14 × 211	1.61	2790	2726	2679	2618	2546	2463	2371	2271	2163	2051	1934	1816	1695	1574	1455	1338	1224	1114	1007
W 14 × 176	1.60	2332	2276	2236	2184	2122	2052	1973	1888	1796	1700	1602	1500	1399	1297	1196	1098	1002	910	821
W 14 × 145	1.59	1922	1876	1841	1798	1746	1687	1621	1549	1473	1393	1310	1226	1141	1057	973	892	812	735	663
W 14 × 120	1.67	1588	1546	1514	1473	1425	1370	1310	1245	1175	1103	1029	955	881	807	735	665	598	537	484
W 14 × 82	2.44	1080	1015	968	911	844	772	697	620	544	471	402	343	295	257	227	200	179	160	145
W 14 × 68	2.44	900	845	805	756	701	640	576	512	449	388	329	281	242	211	185	164	146	131	119
W 14 × 53	3.07	702	633	584	527	465	401	338	278	226	186	157	133	115	101	88	0	0	0	0
W 12 × 336	1.85	4446	4308	4204	4073	3920	3746	3554	3349	3134	2912	2686	2462	2240	2023	1816	1615	1440	1293	1167
W 12 × 279	1.82	3686	3565	3474	3361	3227	3077	2911	2734	2550	2359	2167	1977	1789	1608	1434	1270	1133	1016	917
W 12 × 230	1.75	3046	2942	2865	2768	2652	2523	2382	2231	2074	1913	1751	1591	1434	1283	1136	1007	898	806	727
W 12 × 190	1.79	2512	2423	2356	2273	2175	2066	1945	1818	1686	1550	1415	1280	1149	1024	903	799	714	641	578
W 12 × 152	1.77	2012	1938	1883	1814	1733	1642	1544	1439	1330	1219	1109	1000	894	793	697	617	551	494	446
W 12 × 120	1.76	1588	1528	1483	1426	1361	1286	1206	1121	1033	944	856	769	684	602	529	469	418	376	339
W 12 × 96	1.76	1270	1220	1183	1136	1083	1023	957	887	816	744	672	602	535	469	413	365	326	292	264
W 12 × 79	1.75	1044	1003	971	932	887	836	781	723	664	604	544	486	430	376	330	293	262	234	212
W 12 × 53	2.11	702	660	629	592	548	502	453	403	354	307	262	222	192	167	147	130	116	104	94
W 12 × 45	2.64	590	534	493	447	396	343	290	240	196	161	136	115	100	87	76				
W 10 × 112	1.74	1480	1404	1348	1279	1199	1111	1018	921	824	728	636	548	473	412	362	321	286	256	232
W 10 × 88	1.73	1166	1103	1058	1001	936	865	790	712	634	558	485	416	359	312	274	244	217	195	176
W 10 × 68	1.71	900	850	814	770	718	662	602	541	481	421	364	311	269	234	205	182	162	146	131
W 10 × 54	1.71	712	671	642	606	564	519	471	422	374	327	282	240	208	180	159	141	125	112	102
W 10 × 45	2.15	598	545	506	462	411	359	307	257	211	174	146	125	108	93	83				
W 10 × 33	2.16	437	395	365	330	292	252	214	177	143	119	100	85	73	64	56				
W 8 × 58	1.74	770	706	661	606	545	482	417	355	295	245	205	175	151	131	115	103			
W 8 × 40	1.73	526	481	448	409	365	321	275	232	191	158	132	113	97	85	74	66			
W 8 × 31	1.72	411	374	348	317	283	248	212	178	146	121	102	87	74	65	57				
W 8 × 24	2.12	319	275	246	212	178	144	112	89	72	59	50	42							
Bending Factor (m)			2.1	2.0	1.9	1.8	1.7	1.6	1.4	1.3	1.2	1.2	1.2	1.2	1.2	1.2	1.2	1.2	1.2	1.2

Source: Developed from data in the *Steel Construction Manual* (Ref. 5), with permission of the publisher, American Institute of Steel Construction.

[a]Factored nominal strength of columns in kips.

height of 16 ft, the table yields a value of 453 kips, which agrees closely with the computed load of 455 kips found in the example.

The real value of Table 10.2, however, is for quick design selections. The following examples illustrate this use of the table.

Example 4. Using Table 10.2, select a W shape A992 steel column for an axial load of 100 kips dead load and 150 kips live load. The unbraced height is 24 ft and the end conditions are pinned at the top and bottom.

Solution: The load must be factored using the load combinations to determine the maximum factored load on the column. Thus,

$$P_u = 1.4(\text{dead load}) = 1.4(100) = 140 \text{ kips}$$

or

$$P_u = 1.2(\text{dead load}) + 1.6(\text{live load}) = 1.2(100) + 1.6(150) = 360 \text{ kips}$$

360 kips is the value for the maximum factored column resistance that must be found in Table 10.2 for the unbraced height of 24 ft. Some possible choices are:

Shape	Table Safe Load
W 10 × 68	363 kips
W 12 × 79	544 kips
W 14 × 82	402 kips

The 10-in.-deep shape is the lightest choice, but details of the framing system may indicate the desirability of other column dimensions.

Table 10.2 is set up to work when the y axis has the least bending resistance—that is, when $K_y L_y / r_y > K_x L_x / r_x$. If the x axis is critical for slenderness, another step is required to use the table values. For this purpose, use is made of the ratio of r_x / r_y. This ratio varies with each section but is usually in a range of 1.6–3.1, with an average value being 1.75. If $K_y L_x / L_y < K_x r_x / r_y$, then the slenderness ratio about the y axis still controls and the unbraced length about the y axis is used. If $K_y L_x / L_y > K_x r_x / r_y$, then the slenderness ratio about the x axis controls. When this happens, a new equivalent length KL'_y must be used to find the most appropriate steel section for a column.

The new equivalent length can be found by using the following equation:

$$KL'_y = \frac{KL_x}{r_x/r_y}$$

Example 5. Using Table 10.2, select an A992 steel column section for an axial load of 100 kips dead load and 150 kips live load if the unbraced height is 24 ft about the x axis and the unbraced height is 8 ft about the y axis. The end conditions are pinned at top and bottom for both axes.

Solution: The load is the same as in Example 4, for which the maximum factored load was determined as 360 kips. Next, determine the ratio of the unbraced lengths for the two axes:

$$\frac{K_x L_x}{K_y L_y} = \frac{24}{8} = 3$$

Looking at Table 10.2, note that only one shape has a ratio of r values greater than 3; therefore, it is safe to assume that the x axis will control. Next, determine a new effective length for the column and use it to find the most appropriate columns. Use an average value for the r ratio of 1.75; it can be verified once a choice is made:

$$KL'_y = \frac{KL_x}{r_x/r_y} = \frac{24}{1.75} = 13.7 \quad \text{say } 14$$

Options from Table 10.2 are as follows:

Shape	Design Load ($\varphi_c P_n$)	r_x/r_y	Actual New Equivalent Length
W 8 × 58	482 kips	1.74	13.8 ft
W 10 × 54	519 kips	1.71	14.0 ft
W 12 × 53	502 kips	2.11	11.4 ft
W 14 × 53	401 kips	3.07	7.81 ft

In looking at the options from our initial iteration, it looks like the lightest column will either be a W 12 × 53 or a W 14 × 53 since they both weigh 53 pounds per foot of member. A closer examination of the actual new equivalent lengths tells us that the W 14 × 53 will be controlled by its y axis since its actual new equivalent length is less than the 8 ft unbraced length in the y axis. This column still works, but its design load is 584 kips using $KL_y = 8.0$ ft from Table 10.2. For the W 12×53 column in the first iteration, its actual equivalent length is 11.4 ft. If one takes $KL'_y = 12$ ft into Table 10.2, a W 12×45 has a design load of 396 kips, a r_x/r_y ratio of 2.64, and an actual new equivalent length of 9.09 ft. This is now the lightest member with the capacity to handle the 360-kip load and therefore the most economical section for this column is a W 12×45.

Problem 10.4.A Using Table 10.2, select a column shape for an axial dead load of 60 kips and an axial live load of 88 kips if the unbraced height about both axes is 12 ft. A992 steel is to be used and K is assumed as 1.0.

Problem 10.4.B Select a column shape using the same data as in Problem 10.4.A, except the dead load is 103 kips and the live load is 155 kips. The unbraced height about the x axis is 16 ft and the unbraced height about the y axis is 12 ft.

Problem 10.4.C Select a column shape using the same data as in Problem 10.4.A, except the dead load is 142 kips and the live load is 213 kips. The unbraced height about the x axis is 20 ft and the unbraced height about the y axis is 10 ft.

Problem 10.4.D Using Table 10.2, select a column shape for an axial dead load of 400 kips and a live load of 600 kips. The unbraced height about the x axis is 16 ft and about the y axis is 4 ft. The steel is to have a yield stress of 50 ksi and K is assumed as 1.0.

Steel Pipe Columns

Round steel pipe columns most frequently occur as single-story columns, supporting either wood or steel beams. Pipe is available in three weight categories: *standard* (Std), *extra strong* (XS), and *double-extra strong* (XXS). Pipe is designated with a nominal diameter slightly less than the outside diameter. The outside diameter is the same for all three weights, with variation occurring in terms of the wall thickness and interior diameter. See Table A.7 in Appendix A for properties of standard weight pipe. Table 10.3 gives safe loads for pipe columns of steel with a yield stress of 35 ksi.

Example 6. Using Table 10.3, select a standard weight steel pipe column to carry a dead load of 15 kips and a live load of 26 kips if the unbraced height is 12 ft.

TABLE 10.3 Safe Factored Loads for Selected 35-ksi-Yield-Stress Pipe Columns[a]

$F_y = 35$ ksi			Area (in.^{2})	Effective Length (KL) in feet																		
				0	6	8	10	12	14	16	18	20	22	24	26	28	30	32	34	36	38	40
Pipe	12	XS	19.2	605	596	590	581	572	560	547	533	517	500	482	464	445	425	404	384	363	343	323
Pipe	12	Std.	14.6	460	453	449	443	435	427	417	407	394	382	368	355	340	325	310	295	280	264	249
Pipe	10	XS	16.1	507	497	489	480	468	454	439	424	406	386	367	347	327	307	286	266	246	227	208
Pipe	10	Std.	11.9	375	367	362	355	346	337	326	314	301	288	273	259	245	229	214	199	184	170	157
Pipe	8	XXS	21.3	671	648	631	609	583	555	524	490	455	420	384	348	314	281	249	220	197	176	159
Pipe	8	XS	12.8	403	391	381	368	355	339	321	303	283	263	241	221	201	181	162	144	128	115	104
Pipe	8	Std.	8.4	265	256	251	242	234	223	213	201	188	175	162	148	136	123	110	98	88	79	71
Pipe	6	XXS	15.6	491	462	439	413	382	349	316	280	246	212	180	154	132	115	102	90			
Pipe	6	XS	8.4	265	250	239	227	212	196	179	161	143	126	109	93	80	70	61	55	49		
Pipe	6	Std.	5.6	176	167	160	152	143	132	121	110	98	87	76	66	56	50	43	38	34		
Pipe	5	XXS	11.3	356	325	304	277	249	218	188	159	131	108	91	77	67						
Pipe	5	XS	6.1	193	178	167	155	141	126	110	95	80	67	56	48	41	36					
Pipe	5	Std.	4.3	136	126	119	110	101	90	79	69	59	50	41	35	31	26					
Pipe	4	XXS	8.1	255	221	198	173	145	119	93	74	59	50									
Pipe	4	XS	4.4	139	123	112	100	86	72	58	47	38	32	26								
Pipe	4	Std.	3.2	100	89	82	72	62	53	43	35	29	23	20								
Pipe	3.5	XS	3.7	115	100	88	75	62	50	39	31	24										
Pipe	3.5	Std.	2.7	85	73	65	56	47	38	30	23	19	16									
Pipe	3	XXS	5.5	173	136	112	88	66	49	37												
Pipe	3	XS	3.0	95	77	66	54	42	32	24	19											
Pipe	3	Std.	2.2	70	57	50	40	32	24	18	15											

Source: Developed from data in the *Steel Construction Manual* (Ref. 5), with permission of the publisher, American Institute of Steel Construction.

[a]Factored nominal strength of columns in kips.

Solution: The load must be factored to determine the maximum factored load on the column:

$$P_u = 1.4(\text{dead load}) = 1.4(15) = 21 \text{ kips}$$

or

$$P_u = 1.2(\text{dead load}) + 1.6(\text{live load}) = 1.2(15) + 1.6(26) = 59.6 \text{ kips}$$

For the height of 12 ft, the table yields a value of 62 kips as the design load for a 4-in. Std. pipe, which is greater than the required 59.6 kips and therefore will work.

Problems 10.4.E–H Select the minimum size standard weight pipe column for an axial dead load of 20 kips, a live load of 30 kips, and the following unbraced heights: (E) 8 ft, (F) 12 ft, (G) 18 ft, and (H) 25 ft.

Structural Tubing Columns (HSS)

Structural tubing, designated HSS for *hollow structural sections*, is used for building columns and for members of trusses. Members are available in a range of designated nominal sizes that indicate the actual outer dimensions of the rectangular tube shapes. Within these sizes, various wall thicknesses (the thickness of the steel plates used to make the shapes) are available. For building structures, sizes used range upward from the 3-in. square tube to the largest sizes fabricated (60 in. square at present). Tubing can be specified in various grades of steel.

Table 10.4 yields factored design strengths for square tubes from 3 to 12 in. The steel grade for the shapes in the table has a yield stress of 46 ksi. Use of the table is similar to that for other design strength tables.

Problem 10.4.I A structural tubing column, designated HSS $4 \times 4 \times {}^3/_8$, is used with an unbraced height of 12 ft. Find the maximum factored axial load.

Problem 10.4.J A structural tubing column, designated HSS $3 \times 3 \times > {}^5/_{16}$, is used with an unbraced height of 15 ft. Find the maximum factored axial load.

Problem 10.4.K Using Table 10.5, select the lightest tubing column to carry an axial dead load of 30 kips and a live load of 34 kips if the unbraced height is 10 ft.

TABLE 10.4 Safe Factored Loads for Selected 46-ksi-Yield-Stress Tube Steel Columns[a]

		Effective Length (KL) in feet																		
F_y = 46 ksi	Area (in.^{2})	0	6	8	10	12	14	16	18	20	22	24	26	28	30	32	34	36	38	40
HSS 12 × 12 × 5/8	25.7	1064	1047	1033	1016	996	973	948	918	887	854	820	782	745	707	668	630	591	553	515
HSS 12 × 12 × 3/8	16	663	652	645	634	623	609	593	576	557	537	517	494	472	449	426	401	378	355	331
HSS 12 × 12 × 1/4	10.8	447	440	435	429	420	412	401	390	378	364	350	336	321	306	290	274	258	244	228
HSS 10 × 10 × 1/2	17.2	713	696	683	667	648	627	602	577	550	520	489	458	428	397	366	336	307	278	251
HSS 10 × 10 × > P_u/P'_u < 0.2	11.1	460	449	442	432	420	407	392	376	358	340	321	302	282	263	242	223	204	186	169
HSS 10 × 10 × > P'_u	6.76	280	274	269	264	256	249	239	230	219	209	197	186	174	162	150	139	128	116	106
HSS 8 × 8 × 1/2	13.5	559	538	523	503	481	455	428	398	367	337	306	275	246	217	192	169	151	136	123
HSS 8 × 8 × > P'_u	8.76	363	350	341	328	314	299	282	264	245	224	205	186	167	149	131	116	104	93	84
HSS 8 × 8 × > 5/8	5.37	222	215	209	202	194	184	174	163	151	140	128	116	105	94	84	74	66	59	53
HSS 7 × 7 × 5/8	14	579	550	528	501	470	436	399	362	324	287	251	216	186	162	143	126	113	102	91
HSS 7 × 7 × 3/8	8.97	372	354	341	325	306	286	264	240	217	194	172	150	130	113	100	88	78	71	64
HSS 7 × 7 × 1/4	6.17	255	244	235	224	213	199	184	168	154	138	122	107	93	82	72	64	56	51	46
HSS 6 × 6 × 1/2	9.74	403	376	356	331	305	275	245	215	185	157	132	112	97	85	74	66	58		
HSS 6 × 6 × > 3/8	6.43	266	250	237	222	205	186	167	148	129	110	93	79	69	59	53	47	41	37	
HSS 6 × 6 × > 1/4	3.98	165	155	147	139	128	118	106	94	83	72	61	52	44	39	34	31	28	24	
HSS 5 × 5 × 3/8	6.18	256	232	214	194	172	148	126	104	85	70	59	50	43	38					
HSS 5 × 5 × 1/4	4.3	178	162	150	138	123	107	91	76	62	52	43	37	32	28	24				
HSS 5 × 5 × 1/8	2.23	92	85	79	72	65	57	50	41	35	29	24	20	18	16	14				
HSS 4 × 4 × 3/8	4.78	198	168	148	126	104	83	64	50	40	34	29								
HSS 4 × 4 × 1/4	3.37	140	120	107	92	76	61	48	38	31	25	21								
HSS 4 × 4 × 1/8	1.77	73	64	57	50	42	34	28	21	17	15	12	11							
HSS 3 × 3 × > 5/16	3.52	146	123	107	90	73	57	43	34	28	23									
HSS 3 × 3 × > 3/16	2.24	93	79	70	60	50	39	31	24	19	16	14								

Source: Developed from data in the *Steel Construction Manual* (Ref. 5), with permission of the publisher, American Institute of Steel Construction.

[a]Factored nominal strength of columns in kips.

TABLE 10.5 Safe Factored Loads for Double-Angle Compression Members of A36 Steel[a]

Size (in.)		4 × 3			3.5 × 2.5			3 × 2		2.5 × 2	
Thickness (in.)		$>^5/_{16}$	$>^3/_{16}$		$>^5/_8$	$^1/_4$		$>^3/_8$	$^1/_4$	$>^5/_{16}$	$^1/_4$
Weight (lb/ft)		16.9	14.2		12.2	9.88		10.1	8.18	8.97	7.3
Area (in.2)		4.98	4.19		3.58	2.9		2.96	2.4	2.64	2.14
rx		1.26	1.27		1.11	1.12		0.95	0.95	0.77	0.78
ry		1.3	1.29		1.09	1.08		0.9	0.88	0.94	0.93
Effective Buckling Length in Feet (*L*/*r*) with Respect to Indicated Axis: *X*–*X* Axis	0	161	136	0	116	91	0	96	78	86	69
	4	149	126	2	113	89	2	93	75	81	66
	6	136	114	4	105	83	3	89	72	76	62
	8	119	101	6	93	73	4	84	68	70	57
	10	100	85	8	78	62	5	78	63	62	51
	12	82	69	10	62	51	6	71	58	54	44
	14	64	54	12	48	39	8	56	46	38	31
	16	49	41	14	35	29	10	41	34	25	21
	18	38	33	16	26	22	12	29	24	17	14
	20	31	26	18	21	18	14	21	17	—	—
Y–*Y* Axis	0	161	136	0	116	91	0	96	78	86	69
	4	138	110	2	102	75	2	85	66	79	62
	6	125	101	4	94	70	3	81	63	75	59
	8	110	89	6	83	62	4	76	59	70	55
	10	93	76	8	69	53	5	69	54	65	51
	12	75	61	10	54	42	6	62	48	58	46
	14	58	49	12	40	32	8	46	36	45	35
	16	46	37	14	31	23	10	34	26	33	26
	18	36	30	16	24	18	12	24	19	23	18
	20	30	24	18	19	15	14	17	14	17	14

Size (in.)		8 × 6			6 × 4			5 × 3.5		5 × 3	
Thickness (in.)		$^3/_4$	$^1/_2$		$^1/_2$	$^3/_8$		$^1/_2$	$^3/_8$	$^3/_8$	$>^3/_{16}$
Weight (lb/ft)		68	46.3		32.3	24.6		27.2	20.8	19.5	16.4
Area (in.2)		20	13.6		9.5	7.22		8.01	6.1	5.73	4.81
rx		2.52	2.55		1.91	1.93		1.58	1.59	1.6	1.61
ry		2.47	2.43		1.64	1.61		1.48	1.46	1.22	1.21
Effective Buckling Length in Feet (*L*/*r*) with Respect to Indicated Axis: *X*–*X* Axis	0	648	401	0	308	213	0	259	194	182	142
	10	575	361	10	250	177	6	233	175	164	129
	12	545	344	12	229	163	8	214	161	151	120
	14	512	326	14	204	148	10	192	145	137	109
	16	478	306	16	181	132	12	167	127	120	96
	20	402	263	18	157	116	14	143	109	103	84
	24	326	218	20	134	102	16	120	91	87	72
	28	254	175	22	112	87	18	96	74	71	60
	32	195	136	26	80	62	22	65	50	48	40
	36	154	107	30	60	47	26	47	36	34	29
Y–*Y* Axis	0	648	401	0	308	213	0	259	194	182	142
	10	522	292	6	245	157	4	226	158	143	106
	12	494	280	10	205	136	6	211	148	130	97
	14	464	266	12	182	122	8	191	134	114	87
	16	430	250	14	157	107	10	167	120	95	74
	20	358	215	16	132	92	12	143	103	76	60
	24	285	177	18	108	77	14	118	86	58	48
	28	217	139	20	88	64	16	94	69	46	37
	32	167	108	22	73	53	18	75	55	36	30
	36	133	87	26	53	38	22	51	37		

Source: Developed from data in the *Manual of Steel Construction* (Ref. 5) with permission of the publisher, American Institute of Steel Construction.

[a]Factored nominal axial compression strength for members in kips.

Problem 10.4.L Using Table 10.5, select the lightest structural column to carry an axial dead load of 90 kips and a live load of 60 kips if the unbraced height is 12 ft.

Double-Angle Compression Members

Matched pairs of angles are frequently used for trusses or for braces in steel frames. The common form consists of two angles placed back to back but separated a short distance to achieve end connections by use of gusset plates or by sandwiching the angles around the web of a structural tee. Compression members that are not columns are frequently called *struts*.

The AISC *Steel Manual* (Ref. 5) contains safe load tables for double angles with an assumed average separation distance of 3/8 in. [9.5 mm]. For angles with unequal legs, two back-to-back arrangements are possible, described either as long legs back to back or as *short legs back to back*. Table 10.5 presents data for selected pairs of double angles (F_y = 36 ksi) with long legs back to back. Note that separate data are provided for the variable situation of either axis being used for the determination of the unbraced length. If conditions relating to the unbraced length are the same for both axes, then the lower value for safe load from the table must be used. Properties for selected double angles are given in Table A.6 in Appendix A.

Like other members that lack biaxial symmetry, such as the structural tee, there may be some reduction applicable due to the slenderness of the thin elements of the cross section. This reduction is incorporated in the values provided in Table 10.5.

Problem 10.4.M A double-angle compression member 8 ft long is composed of two A36 steel angles 4 × 3 × 3/8 in., with the long legs back to back. Determine the maximum factored axial compression load for the angles.

Problem 10.4.N A double-angle compression member 8 ft long is composed of two A36 steel angles 6 × 4 × 1/2 in., with the long legs back to back. Determine the maximum factored axial compression load for the angles.

Problem 10.4.O Using Table 10.5, select a double-angle compression member for an axial compression dead load of 25 kips and a live load of 25 kips if the unbraced length is 10 ft.

Problem 10.4.P Using Table 10.5, select a double-angle compression member for an axial compression dead load of 75 kips and a live load of 100 kips if the unbraced length is 16 ft.

10.5 COLUMNS WITH BENDING

Steel columns must frequently sustain bending in addition to the usual axial compression. Figures 10.6*a–c* show three of the most common situations that result in this combined effect. When loads are supported by connection at the column face, the eccentricity of the compression adds a bending effect (Figure 10.6*a*). When moment-resistive connections are used to produce a rigid frame, any load on the beams will induce a twisting (bending) effect on the columns (Figure 10.6*b*). Columns built into exterior walls (a common occurrence) may become involved in the spanning effect of the wall in resisting wind forces (Figure 10.6*c*).

Adding bending to a direct compression effect results in a combined stress, or net stress, distribution on the member cross section. The two separate effects may be analyzed separately and their stresses added to consider this effect. However, the two *actions*—compression and bending—are essentially different, so that a combination of the separate actions, not just the stresses, is more significant. Consideration of this combination is accomplished with the so-called *interaction* analysis that takes the form of

$$\frac{P_u}{\phi_c P_n} + \frac{M_{ux}}{\phi_b M_{nx}} + \frac{M_{uy}}{\phi_b M_{ny}} \leq 1$$

On a graph, the interaction formula describes a straight line, which is the classic form of the relationship in elastic theory. However,

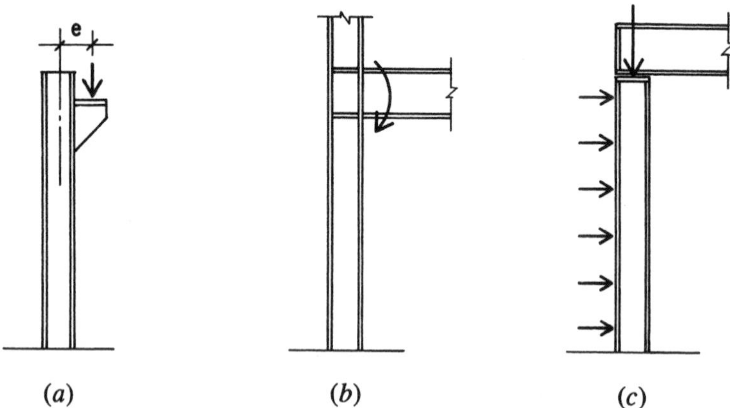

Figure 10.6 Considerations for development of bending in steel columns.

variations from the straight-line form occur because of special conditions having to do with the nature of the materials, the usual forms of columns, inaccuracies in fabrication and construction, and the different effects of buckling for columns and beams (bending members).

For steel columns, major issues include slenderness of column flanges and webs (for W shapes), ductility of the steel, and overall column slenderness that effects potential buckling in both axial compression and bending. Understandably, the AISC formulas are considerably more complex than the simple straight-line formula above. The AISC interaction formulas for compression and bending are as follows:

If $P_u/(\phi_c P_n) \geq 0.2$

$$\frac{P_u}{\phi_c P_n} + \frac{8}{9}\left(\frac{M_{ux}}{\phi_b M_{nx}} + \frac{M_{uy}}{\phi_b M_{ny}}\right) \leq 1.0$$

If $P_u/(\phi_c P_n) < 0.2$

$$\frac{P_u}{2\phi_c P_n} + \left(\frac{M_{ux}}{\phi_b M_{nx}} + \frac{M_{uy}}{\phi_b M_{ny}}\right) \leq 1.0$$

Another potential problem with combined compression and bending is that of the *P*–delta effect. This occurs when a relatively slender column is subjected to bending and the resulting deflection of the curved, bent, column produces an additional bending in conjunction with the compression force. This problem should be carefully considered with very slender columns.

For use in preliminary design work or to quickly obtain a first trial shape for use in a more extensive design investigation, a procedure developed by Uang, Wattar, and Leet (Ref. 8) may be used that involves the determination of an equivalent axial load that incorporates the bending effect. This is accomplished by use of a bending factor (m), which is listed for the W shapes in Table 10.2 at the bottom. Using this factor, the equivalent axial load is obtained as

$$P'_u = P_u + (m \times M_{ux}) + [(2 \times m) \times M_{uy}]$$

If $P_u/P'_u < 0.2$, recalculate P'_u using the following equation:

$$P'_u = \frac{P_u}{2} + \frac{9}{8} \times [m \times M_{ux} + (2 \times m) \times M_{uy}]$$

where P'_u = equivalent factored axial compression load
P_u = actual factored compression load
m = bending factor
M_{ux} = factored bending moment, x axis
M_{uy} = factored bending moment, y axis

The following examples illustrate the use of this approximation method.

Example 7. A 10-in. W shape is desired for a column in a situation such as that shown in Figure 10.7. The factored axial load from above on the column is 175 kips, and the factored beam load at the column face is 35 kips. The column has an unbraced height of 16 ft and a K factor of 1.0. Select a trial shape for the column.

Solution: From Table 10.2, the bending factor (m) for a 16-ft column is 1.6. Then

$$P_u' = P_u + m \times M_{ux} = (175 + 35) + 1.6 \times (35 \times 5/12)$$
$$= 210 + 23.3 = 233 \text{ kips}$$

Check to see if the correct equation was used:

$$\frac{P_u}{P_u'} = \frac{210}{233} = 0.90$$

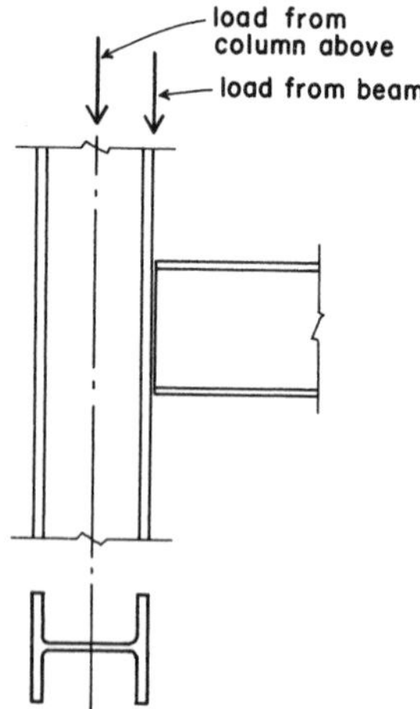

Figure 10.7 Development of an eccentric loading condition with steel framing.

As this is greater than 0.2, the correct equation was used. Using the value of 233 kips and the unbraced height of 16 ft, Table 10.2 yields a W 10×45 with a table value of 307 kips. For a more assured selection, the next step is to use the AISC equations to investigate the chosen shape. To determine which equation to use, we find

$$\frac{P_u}{\phi_c P_n} = \frac{210}{307} = 0.684$$

As this is greater than 0.2, the appropriate AISC equation is

$$\frac{P_u}{\phi_c P_n} + \frac{8}{9}\left(\frac{M_{ux}}{\phi_b M_{nx}} + \frac{M_{uy}}{\phi_b M_{ny}}\right) \leq 1.0$$

The value for $\phi_b M_{nx}$ is determined by the process described for beams in Section 9.5. With the example data and using data from Table 9.1, this value is found to be 175 kip-ft. The combined interaction equation for the example is thus

$$\frac{P_u}{\phi_c P_n} + \frac{8}{9}\left(\frac{M_{ux}}{\phi_b M_{nx}} + \frac{M_{uy}}{\phi_b M_{ny}}\right) = \frac{210}{307_n} + \frac{8}{9}\left(\frac{14.6}{175} + 0\right)$$

$$= 0.754 < 1.0$$

which indicates that the chosen shape is acceptable.

Although this process is laborious, without the use of the approximation it is even more laborious. This is one of those design situations for which a computer-aided process is highly desirable.

When bending occurs about both axes, as it does in most three-dimensional rigid frames, all parts of the combined action formula must be used.

For the following problems use A992 steel.

Problem 10.5.A It is desired to use a 12-in. W shape for a column to support a beam as shown in Figure 10.6. Select a trial size for the column for the following data: column factored axial load from above is 200 kips, factored beam reaction is 30 kips, and unbraced column height is 14 ft.

Problem 10.5.B Check the section found in Problem 10.5.A to see if it complies with the AISC interaction equations for axial compression plus bending.

Problem 10.5.C Same as Problem 10.5.A, except factored axial load is 485 kips, factored beam reaction is 100 kips, and unbraced height is 18 ft.

Problem 10.5.D Check the section found in Problem 10.5.C to see if it complies with the AISC interaction equations for axial compression plus bending.

10.6 COLUMN FRAMING AND CONNECTIONS

Connection details for columns must be developed with considerations of the column form and size; with the form, size, and orientation of other framing; and with the particular structural functions of the joints. Some common forms of simple connections for light frames are shown in Figure 10.8. The usual means for attachment are by welding, by bolting with high-strength bolts, or with anchor bolts embedded in concrete or masonry.

When beams sit directly on top of a column (Figure 10.8*a*), the usual solution is to weld a bearing plate on top of the column and bolt the bottom flange of the beam to the plate. For this, and for all connections, it is necessary to consider what parts of the connection are achieved in the fabrication shop and what is achieved as part of

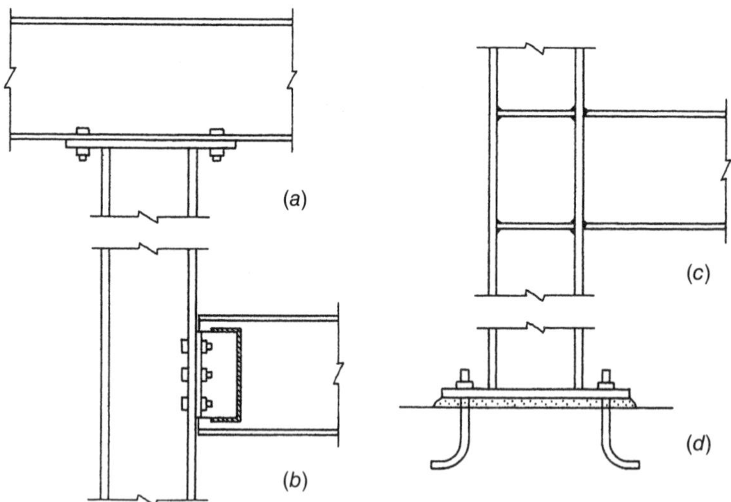

Figure 10.8 Typical fabrication details for steel columns in lightly loaded frames.

the erection of the frame at the job site (called the *field*). In this case, it is likely that the plate will be attached to the column in the shop (where welding is preferred) and the beam will be attached in the field (where bolting is preferred). In this joint the plate serves no particular structural function, because the beam could theoretically bear directly on the column. However, field assembly of the frame works better with the plate serving as an intermediate connecting device. Incidentally, the plate also helps to spread the bearing from the beam more fully on the column cross section.

In many situations, beams must frame into the side of a column. If simple transfer of vertical force is all that is basically required, a common solution is the connection shown in Figure 10.8*b*, in which a pair of steel angles is used to connect the beam web to the column face. With minor variation, this form of connection can also be used to connect a beam to the column web when framing intersects the column differently. When the latter is the case, the outspread legs of the angles must fit between the column flanges, which generally requires at least a 10-in. W-shape column—thus, the popularity of the 10-, 12-, and 14-in. W shapes for columns.

If bending moment must be transferred between a beam's end and its supporting column, a common solution is to weld the cut ends of the beam flanges directly to the column face, as shown in Figure 10.8*c*. Because the bending must be developed in both column flanges and the connected beam grabs only one flange, the filler plates shown—welded to the column web and the insides of the flanges—are often used for a more effective transfer of the bending moment from the beam. This leaves the beam web as yet not connected, so some attachment must also be made there because the beam web essentially carries the beam shear force. Although common for many years and still widely used for gravity and wind loads, this form of connection has recently received a lot of scrutiny because of its poor performance in earthquakes, and some refinements are now required for transfer of major earthquake forces.

At the column bottom, where bearing is usually on top of a concrete pier or footing, the major concern is for reduction of the bearing pressure on the much softer concrete. With upward of 20 ksi or more of compression in the column steel and possibly a little over 1 ksi resistance in the concrete, the contact bearing must be quite spread out. For this reason, as well as the simple practical one of holding the column bottom in place, the common solution is a steel bearing plate attached to the column in the shop and made to bear on a leveling filler material between the rough concrete surface and the smooth plate (see Figure 10.8*d*). This form of connection is adequate for lightly loaded columns. For transfer of very large column loads, development of uplift forces or bending moment, or other special concerns, this joint can receive a lot of special modification. Still, the simple joint shown here is the most common form.

11

BOLTED CONNECTIONS FOR STEEL STRUCTURES

Making a steel structure for a building typically involves the connecting of many parts. The technology available for achieving connections is subject to considerable variety, depending on the form and size of the connected parts, the structural forces transmitted between parts, and the nature of the connecting materials. At the scale of building structures, the primary connecting methods utilized presently are those using electric arc welding and high-strength steel bolts. Considerations for the design of bolted connections are treated in this chapter.

11.1 BOLTED CONNECTIONS

Elements of steel are often connected by mating flat parts with common holes and inserting a pin-type device to hold them together. In times past the device was a rivet; today it is usually a bolt. Many types and sizes of bolt are available, as are many connections in which they are used.

Structural Actions of Bolted Connections

Figures 11.1*a* and *b* show the plan and section of a simple connection between two steel bars that function to transfer a tension force from one bar to the other. Although this is a tension transfer connection, it is also referred to as a shear connection because of the manner in which the connecting device (the bolt) works in the connection (see Figure 11.1*c*). For structural connections, this type of joint is now achieved mostly with so-called *high-strength bolts*, which are special bolts that are tightened in a controlled manner that induces development of yield stress in the bolt shaft. For a connection using such bolts, there are many possible forms of failure that must be considered, including the following:

Bolt Shear. In the connection shown in Figures 11.1*a* and *b*, the failure of the bolt involves a slicing (shear) failure that is developed as a shear stress on the bolt cross section. The resistance factor (ϕ_v) is taken as 0.75. The design shear strength ($\phi_v R_n$) of the bolt can be expressed as a nominal shear stress (F_v) times the nominal cross-sectional area of the bolt, or

$$\phi_v R_n = \phi_v F_v A_b$$

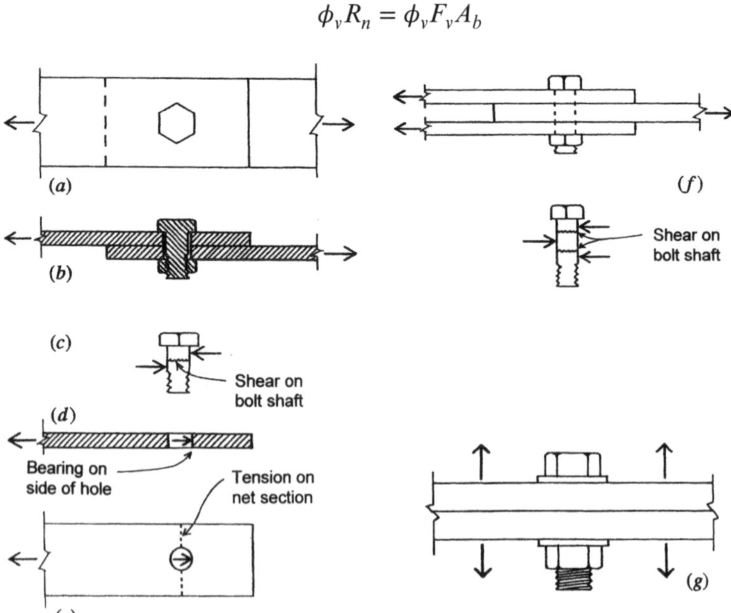

Figure 11.1 Actions of bolted connections.

With the size of the bolt and the grade of steel known, it is a simple matter to establish this limit. In some types of connections, it may be necessary to slice the same bolt more than once to separate the connected parts. This is the case in the connection shown in Figure 11.1*f*, in which it may be observed that the bolt must be sliced twice to make the joint fail. When the bolt develops shear on only one section (Figure 11.1*c*), it is said to be in *single shear*; when it develops shear on two sections (Figure 11.1*f*), it is said to be in *double shear*.

Bearing. If the bolt tension (due to tightening of the nut) is relatively low, the bolt serves primarily as a pin in the matched holes, bearing against the sides of the holes, as shown in Figure 11.1*d*. When the bolt diameter is larger or the bolt is made of very strong steel, the connected parts must be sufficiently thick if they are to develop the full capacity of the bolts. The factored design bearing strength ($\varphi_v R_n$) permitted for this situation by the AISC specification (see the *AISC Steel Manual*, Ref. 5) is

$$\phi_v R_n = \phi_v \times 1.5 \times L_c \times t \times F_u \leq \phi_v \times 3.0 \times d \times t \times F_u$$

where φ_v = 0.75
R_n = nominal bearing strength
L_c = distance between edge of hole and edge of next hole in material, in.
t = thickness of connected material, in.
F_u = ultimate tensile strength of connected material, ksi
d = diameter of bolt, in.

Tension on Net Section of Connected Parts. For the connected bars in Figure 11.1*b*, the tension stress in the bars will be a maximum at a section across the bar at the location of the hole. This reduced section is called the *net section* for tension resistance. Although this is indeed a location of critical stress, it is possible to achieve yield here without serious deformation of the connected parts. For this reason, design strength ($\varphi_t P_n$) at the net section is based on the ultimate—rather than the yield—strength of the connected parts. The value used for the design tensile strength is

$$\phi_t P_n = \phi_t \times F_u \times A_e$$

where φ_t = 0.75 for tension
F_u = ultimate tensile strength of the connected part
A_e = reduced (net) area of the part

Bolt Tension. While the shear (slip-resisting) connection shown in Figures 11.1*a* and *b* is common, some joints employ bolts for their resistance in tension, as shown in Figure 11.1*g*. For the threaded bolt, the maximum tension stress is developed at the net section through the cut threads. However, it is also possible for the bolt to have extensive elongation if yield stress develops in the bolt shaft (at an unreduced section). However stress is computed, bolt tension resistance is established on the basis of data from destructive tests.

Bending in the Connection. Whenever possible, bolted connections are designed to have a bolt layout that is symmetrical with regard to the directly applied forces. This is not always possible, so that in addition to the direct force actions, the connection may be subjected to twisting due to a bending moment or torsion induced by the loads. Figure 11.2 shows some examples of this situation.

In Figure 11.2*a* two bars are connected by bolts, but the bars are not aligned in a way to transmit tension directly between the bars. This may induce a rotational effect on the bolts, with a torsional twist equal to the product of the tension force and the eccentricity due to misalignment of the bars. Shearing forces on individual bolts will be increased by this twisting action. And, of course, the ends of the bars will also be twisted.

Figure 11.2*b* shows the single-shear joint, as shown in Figures 11.1*a* and *b*. When viewed from the top, such a joint may appear to have the bars aligned; however, the side view shows that the basic nature of the single-shear joint is such that a twisting action is inherent in the joint. This twisting increases with thicker bars. It is usually not highly critical for steel structures, where connected elements are usually relatively thin; for connecting of wood elements, however, it is not a favored form of joint.

Figure 11.2*c* shows a side view of a beam end with a typical form of connection that employs a pair of angles. As shown, the angles grasp the beam web between their legs and turn the other legs out to fit flat against a column or the web of another beam. Vertical load from the beam, vested in the shear in the beam web, is transferred

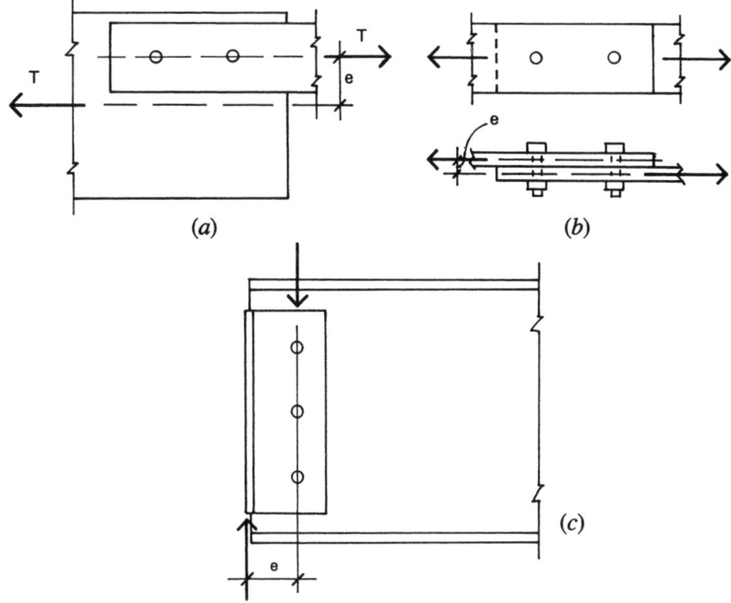

Figure 11.2 Development of bending in bolted connections.

to the angles by the connection of the angles to the beam web—with bolts as shown here. This load is then transferred from the angles at their outward-turned face, resulting in a separated set of forces due to the eccentricity shown. This action must be considered with others in design of these connections.

Slipping of Connected Parts. Highly tensioned, high-strength bolts develop a very strong clamping action on the mated flat parts being connected, analogous to the situation shown in Figure 11.3*a*. As a result there is a strong development of friction at the slip face, which is the initial form of resistance in the shear-type joint. Development of bolt shear, bearing, and even tension on the net section will not occur until this slipping is allowed. For service-level loads, therefore, this is the *usual* form of resistance, and the bolted joint with high-strength bolts is considered to be a very rigid form of joint.

Block Shear. One possible form of failure in a bolted connection is that of tearing out the edge of one of the attached members.

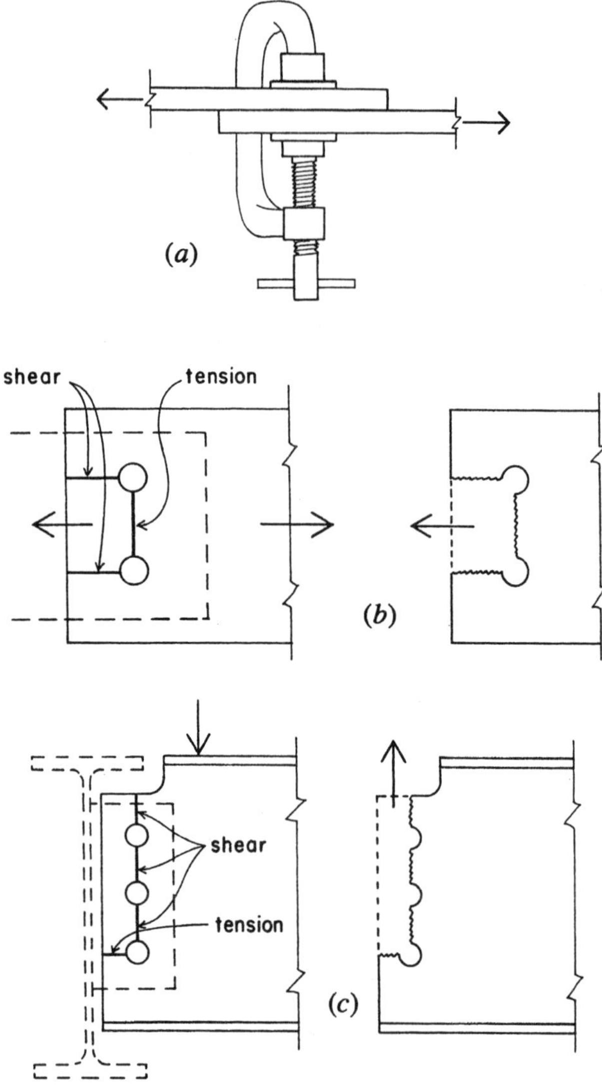

Figure 11.3 Special actions of bolted connections.

This is called a *block shear* failure. The diagrams in Figure 11.3*b* show this potentiality in a connection between two plates. The failure in this case involves a combination of shear and tension to produce the torn-out form shown. The total tearing force is computed as the sum required to cause both forms of failure. The design strength ($\varphi_t P_n$) of the net tension area is computed as described before for net cross sections. The design strength ($\varphi_v R_n$) of the shear areas is specified as $0.75 F_v A_c$, where A_c is the cross-sectional area experiencing shear stress.

With the edge distance, hole spacing, and diameter of the holes known, the net widths for tension and shear are determined and multiplied by the thickness of the part in which the tearing occurs. These areas are then multiplied by the appropriate stress to find the total tearing force that can be resisted. If this force is greater than the connection design load, the tearing problem is not critical.

Another case of potential tearing is shown in Figure 11.3*c*. This is the common situation for the end framing of a beam in which support is provided by another beam, whose top is aligned with that of the supported beam. The end portion of the top flange of the supported beam must be cut back to allow the beam web to extend to the side of the supporting beam. With the use of a bolted connection, the tearing condition shown is developed.

Types of Steel Bolts

Bolts used for the connection of structural steel members come in two basic types. Bolts designated A307 and called *unfinished bolts* have the lowest load capacity of the structural bolts. The nuts for these bolts are tightened just enough to secure a snug fit of the attached parts; because of this low resistance to slipping, plus the oversizing of the holes to achieve practical assemblage, there is some movement in the development of full resistance. These bolts are generally not used for major connections, especially when joint movement or loosening under vibration or repeated loading may be a problem. They are, however, used extensively for temporary connections during erection of frames.

Bolts designated A325, F1852, or A490 are called high-strength bolts. The nuts of these bolts are tightened to produce a considerable tension force, which results in a high degree of friction resistance between the attached parts. Different specifications for installation of these bolts result in different classifications of their strength, relating generally to the critical mode of failure.

When loaded in shear-type connections, bolt capacities are based on the development of shearing action in the connection. The shear capacity of a single bolt is further designated as *S* for single shear (Figure 11.1*c*) or *D* for double shear (Figure 11.1*f*). In high-strength bolts, the shear capacity is affected by the bolt threads. If the threads are present in the shear plane being considered, the cross-sectional area is reduced and therefore the capacity of the bolt is also reduced. The capacities of structural bolts in both tension and shear are given in Table 11.1. These bolts range in size from 1/2 to 1 1/2 in. in diameter, and capacities for these sizes are given in tables in the AISC *Steel Manual* (Ref. 5). However, the most commonly used sizes for light structural steel framing are 3/4 and 7/8 in. However, for larger connections and large frameworks, sizes of 1–1 1/4 are also used. This is the size range for which data are given in Table 11.1: 3/4–1 1/4.

Bolts are ordinarily installed with a washer under both head and nut. Some manufactured high-strength bolts have specially formed heads or nuts that in effect have self-forming washers, eliminating the

TABLE 11.1 Design Strength of Structural Bolts (kips)[a]

ASTM Designation	Loading Condition[b]	Thread Condition	Nominal Diameter of Bolts (in.)				
			3/4	7/8	1	1 1/8	1/4
A307	S		7.95	10.8	14.1	**17.9**	22.1
	D		15.9	21.6	28.3	35.8	44.2
	T		14.9	20.3	26.5	33.5	41.4
A325	S	Included	15.9	21.6	28.3	35.8	44.2
		Excluded	19.9	27.1	35.3	44.7	55.2
	D	Included	31.8	43.3	56.5	71.6	88.4
		Excluded	39.8	54.1	70.7	89.5	110
	T		29.8	40.6	53	67.1	82.8
A490	S	Included	19.9	27.1	35.3	44.7	55.2
		Excluded	24.9	33.8	44.2	55.9	69
	D	Included	39.8	54.1	70.7	89.5	110
		Excluded	49.7	67.6	88.4	112	138
	T		37.4	51	66.6	84.2	104

Source: Compiled from data in the *Steel Construction Manual* (Ref. 5), with permission of the publisher, American Institute of Steel Construction.

[a]Slip-critical connections; assuming there is no bending in the connection and that bearing on connected materials is not critical.

[b]S = single shear, D = double shear, and T = tension.

need for a separate, loose washer. When a washer is used, it is sometimes the limiting dimensional factor in detailing for bolt placement in tight locations, such as close to the fillet (inside radius) of angles or other rolled shapes.

For a given diameter of bolt, there is a minimum thickness required for the bolted parts in order to develop the full shear capacity of the bolt. This thickness is based on the bearing stress between the bolt and the side of the hole. The stress limit for this situation may be established by either the bolt steel or the steel of the bolted parts.

Steel rods are sometimes threaded for use as anchor bolts or tie rods. When they are loaded in tension, their capacities are usually limited by the stress on the reduced section at the threads. Tie rods are sometimes made with *upset ends*, which consist of larger diameter portions at the ends. When these enlarged ends are threaded, the net section at the thread is the same as the gross section in the remainder of the rods; the result is no loss of capacity for the rod.

Layout of Bolted Connections

Design of bolted connections generally involves a number of considerations in the dimensional layout of the bolt hole patterns for the attached structural members. The material in this section presents some basic factors that often must be included in the design of bolted connections. In some situations, the ease or difficulty of achieving a connection may affect the choice for the form of the connected members.

Figure 11.4*a* shows the layout of a bolt pattern with bolts placed in two parallel rows. Two basic dimensions for this layout are limited by the size (nominal diameter) of the bolt. The first is the center-to-center spacing of the bolts, usually called the *pitch*. The AISC specification limits this dimension to an absolute minimum of 2.5 times the bolt diameter. The preferred minimum, however, which is used in this book, is 3 times the diameter.

The second critical layout dimension is the *edge distance*, which is the distance from the center line of the bolt to the nearest edge of the member containing the bolt hole. There is also a specified limit for this as a function of bolt size and the nature of the edge, the latter referring to whether the edge is formed by rolling or is cut. Edge distance may also be limited by edge tearing in block shear, as previously discussed.

Table 11.2 gives the recommended limits for pitch and edge distance for the bolt sizes used in ordinary steel construction.

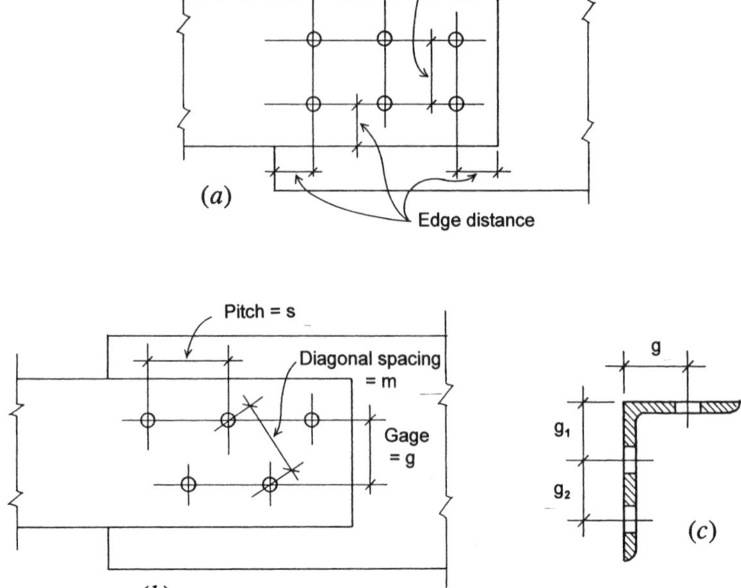

Figure 11.4 Layout considerations for bolted connections: (*a*) pitch and end distances, (*b*) bolt spacing, and (c) gage distance for angle legs.

In some cases bolts are staggered in parallel rows (Figure 11.4*b*). In this case the diagonal distance, labeled *m* in the illustration, must also be considered. For staggered bolts the spacing in the direction of the rows is usually referred to as the pitch; the spacing of the rows is called the *gage*. The usual reason for staggering the bolts is that sometimes the rows must be spaced closer (gage spacing) than the minimum spacing required for the bolts selected. However, staggering the bolt holes also helps to create a slightly less critical net section for tension stress in the steel member with the holes.

Location of bolt lines is often related to the size and type of structural members being attached. This is especially true of bolts placed in the legs of angles or in the flanges of W, M, S, C, and structural tee shapes. Figure 11.4*c* shows the placement of bolts in the legs of angles. When a single row is placed in a leg, its recommended location is at

TABLE 11.2 Pitch and Edge Distances for Bolts

Rivet or Bolt Diameter d (in.)	Minimum Edge Distance for Punched, Reamed, or Drilled Holes (in.): At Sheared Edges	Minimum Edge Distance for Punched, Reamed, or Drilled Holes (in.): At Rolled Edges of Plates, Shapes, or Bars, or Gas-Cut Edges[a]	Minimum Recommended Pitch, Center to Center (in.): $2.667d$	Minimum Recommended Pitch, Center to Center (in.): $3d$
0.625	1.125	0.875	1.67	1.875
0.750	1.25	1.0	2.0	2.25
0.875	1.5[b]	1.125	2.33	2.625
1.000	1.75[b]	1.25	2.67	3.0

Source: Adapted from data in the *Steel Construction Manual* (Ref. 5), with permission of the publisher, American Institute of Steel Construction.

[a]May be reduced 0.125 in. when the hole is at a point where stress does not exceed 25% of the maximum allowed in the connected element.

[b]May be 1.25 in. at the ends of beam connection angles.

TABLE 11.3 Usual Gage Dimensions for Angles (in.)

Gage Dimension	Width of Angle Leg: 8	7	6	5	4	3.5	3	2.5	2
g	4.5	4.0	3.5	3.0	2.5	2.0	1.75	1.375	1.125
g_1	3.0	2.5	2.25	2.0					
g_2	3.0	3.0	2.5	1.75					

Source: Adapted from data in the *Steel Construction Manual* (Ref. 5), with permission of the publisher, American Institute of Steel Construction.

the distance labeled g from the back of the angle. When two rows are used, the first row is placed at the distance g_1, and the second row is spaced a distance g_2 from the first. Table 11.3 gives the recommended values for these distances.

When placed at the recommended locations in rolled shapes, bolts will end up a certain distance from the edge of the part. Based on the recommended edge distance for rolled edges given in Table 11.2, it is thus possible to determine the maximum size of bolt that can be accommodated. For angles, the maximum fastener may be limited by the edge distance, especially when two rows are used; however, other factors may in some cases be more critical. The distance from the center of the bolts to the inside fillet of the angle may limit the use of a large washer where one is required. Another consideration may be the stress on the net section of the angle, especially if the member load is taken entirely by the attached leg.

Tension Connections

When tension members have reduced cross sections, two stress investigations must be considered. This is the case for members with holes for bolts. For the member with a hole, the design tension strength at the reduced cross section through the hole is

$$\phi_t P_n = \phi_t \times F_u \times A_e$$

where φ_t = 0.75 for tension
F_u = ultimate strength of the steel
A_e = reduced (net) cross-sectional area

The resistance at the net section must be compared with the resistance at the unreduced section of the member for which the resistance factor is 0.90. For steel bolts the design strength is specified as a value based on the type of bolt.

Angles used as tension members are usually connected by only one leg. In a conservative design, the effective net area is only that of the connected leg less the reduction caused by holes.

Rivet and bolt holes are punched larger in diameter than the nominal diameter of the fastener. The punching damages a small amount of the steel around the perimeter of the hole; consequently the diameter of the hole to be deducted in determining the net section is $^1/_8$ in. greater than the nominal diameter of the fastener.

When only one hole is involved, as with a single row of fasteners along the line of stress, the net area of the cross section of one of the plates is found by multiplying the plate thickness by its net width (width of member minus diameter of hole).

When holes are staggered in two rows along the line of stress (Figure 11.5), the net section is determined somewhat differently. The AISC specification reads:

> In the case of a chain of holes extending across a part in any diagonal or zigzag line, the net width of the part shall be obtained by deducting from the gross width the sum of the diameters of all the holes in the

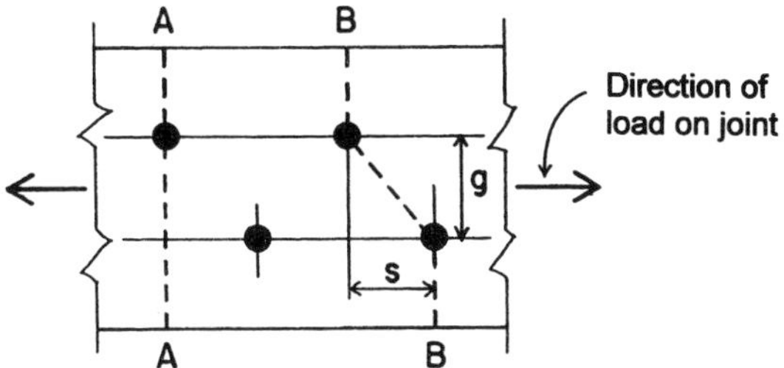

Figure 11.5 Determination of net cross-sectional area for connected members in a bolted connection.

chain and adding, for each gage space in the chain, the quantity $s^2/4\,g$, where

s = longitudinal spacing (pitch) in inches or any two successive holes.
g = transverse spacing (gage) in inches for the same two holes.

The critical net section of the part is obtained from that chain which gives the least net width.

11.2 DESIGN OF A BOLTED CONNECTION

The issues raised in the preceding sections are illustrated in the following design example.

Example 1. The connection shown in Figure 11.6 consists of a pair of narrow plates that transfer a tension force that is produced by a dead load of 50 kips and a live load of 90 kips. This load is transferred to a single middle plate. All plates are of A992 steel with $F_y = 50$ ksi and $F_u = 65$ ksi and are attached with 3/4-in. A325 bolts placed in two rows with threads included in the planes of shear. Using data from Table 11.1, determine the number of bolts required, the width and thickness of the narrow plates, the thickness of the wide plate, and the layout for the connection.

Solution: The process begins with the determination of the ultimate load:

$$P_u = 1.2(\text{dead load}) + 1.6(\text{live load}) = 1.2(50) + 1.6(90)$$
$$= 204 \text{ kips}$$

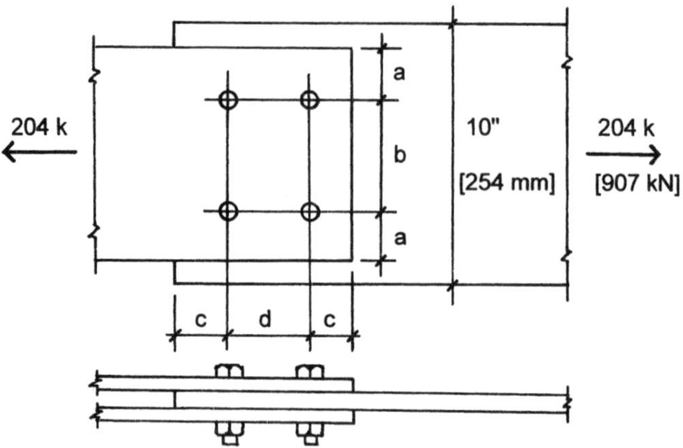

Figure 11.6 Reference figure for Example 1.

From Table 11.1, the capacity of a single bolt in double shear is found as 31.8 kips. The required number of bolts for the connection is thus

$$n = \frac{204}{31.8} = 6.41 \quad \text{or} \quad 7$$

Although placement of seven bolts in the connection is possible, most designers would choose to have a symmetrical arrangement with eight bolts, four to a row. The average bolt load is thus

$$V_u = \frac{204}{8} = 25.5 \text{ kips}$$

From Table 11.2, for the 3/4-in. bolts, minimum edge distance for a cut edge is 1.25 in. and minimum recommended spacing is 2.25 in. The minimum required width for the plates is thus (see Figure 11.6)

$$w = b + 2(a) = 2.25 + 2(1.25) = 4.75 \text{ in.}$$

If space is tightly constrained, this actual width could be specified for the narrow plates. For this example a width of 6 in. is used. Checking for the requirement of stress on the gross area of the plate cross section

$$A_g = \frac{P_u}{\phi_t F_y} = \frac{204}{0.9(50)} = 4.53 \text{ in.}^2$$

and, with the 6-in. width, the required thickness is

$$t = \frac{4.53}{2 \times 6} = .378 \text{ in.}$$

This permits the use of a minimum thickness of $\frac{7}{16}$ in. (0.438 in.). The next step is to check the stress on the net section. For the computations it is recommended to use a bolt hole size at least $\frac{1}{8}$ in. larger than the bolt diameter. This allows for the true oversize (usually $\frac{1}{16}$-in.) and some loss due to the roughness of the hole edges. Thus the hole is assumed to be $\frac{7}{8}$ in. (0.875) in diameter, and the net width and area are

$$w = 6 - 2(0.875) = 4.25 \text{ in.}$$

$$A_e = w \times t = 4.25 \times \tfrac{7}{16} = 1.86 \text{ in.}^2$$

and the design strength at the net section is

$$\phi_t P_n = 0.75 \times F_u \times A_e = 0.75 \times 65 \times (2 \times 1.86) = 181 \text{ kips}$$

Since $\phi_t P_n < P_u$, the section is not adequately sized. An increase of thickness to $\frac{1}{2}$ in. needs to be investigated:

$$A_e = w \times t = 4.25 \times \tfrac{1}{2} = 2.13 \text{ in.}^2$$

and the design strength at the net section is

$$\phi_t P_n = 0.75 \times F_u \times A_e = 0.75 \times 65 \times (2 \times 2.13) = 207 \text{ kips}$$

As this is greater than the factored load (P_u), the $\frac{1}{2}$-in.-thick plates are adequate for tension stress.

The bolt capacities in Table 11.1 are based on a slip-critical condition, which assumes a design failure limit to be that of the friction resistance (slip resistance) of the bolts. However, the back-up failure mode is the one in which the plates slip to permit development of the pin action of the bolts against the sides of the holes; this then involves the shear capacity of the bolts and the bearing resistance of the plates. Bolt shear capacities are higher than the slip failures, so the only concern for this is the bearing on the plates.

Bearing design strength is computed for a single bolt as

$$\phi_v R_n = 2(\phi_v \times 1.5 \times L_c \times t \times F_u) = 2\left(0.75 \times 1.5 \times 1.5 \times \tfrac{1}{2} \times 65\right)$$
$$= 110 \text{ kips}$$

which is clearly not a critical concern since the factored load on each bolt is 25.5 kips.

For the middle plate the procedure is essentially the same, except that the width is given and there is a single plate. As before, the stress on the unreduced cross section requires an area of 4.53 in.^{2}, so the required thickness of the 10-in.-wide plate is

$$t = \frac{4.53}{10} = 0.453 \text{ in.}$$

which indicates the use of a 1/2 in. thickness.

For the middle plate the width and cross-sectional area at the net section are

$$w = 10 - (2 \times 0.875) = 8.25 \text{ in.}$$
$$A_e = w_t \times t = 8.25 \times \tfrac{1}{2} = 4.13 \text{ in.}^2$$

and the design strength at the net section is

$$\phi_t P_n = 0.75 \times F_u \times A_e = 0.75 \times 65 \times 4.13 = 201 \text{ kips}$$

Since $\phi_t P_n < P_u$, the section is not adequately sized. An increase of thickness to 5/8 in. needs to be investigated:

$$A_e = w_t \times t = 8.25 \times \tfrac{5}{8} = 5.16 \text{ in.}^2$$

and the design strength at the net section is

$$\phi_t P_n = 0.75 \times F_u \times A_e = 0.75 \times 65 \times 5.16 = 251 \text{ kips}$$

which is greater than the factored load of 204 kips.

The computed bearing design strength on the sides of the holes in the middle plate is

$$\phi_v R_n = \phi_v \times 1.5 \times L_c \times t \times F_u = 0.75 \times 1.5 \times 1.5 \times \tfrac{5}{8} \times 65 = 68.3 \text{ kips}$$

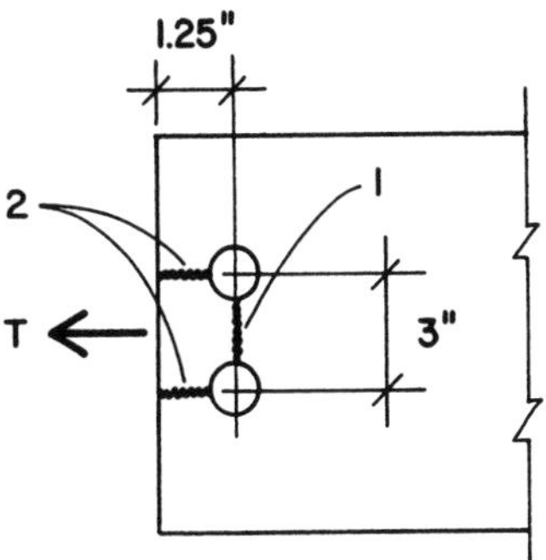

Figure 11.7 Tearing in example problem.

which is greater than the factored load of 25.5 kips, as determined previously.

A final problem that must be considered is the possibility for tearing out of the two bolts at the end of a plate in a block shear failure (Figure 11.3*a*). Because the combined thicknesses of the outer plates is greater than that of the middle plate, the critical case for this connection is that of the middle plate. Figure 11.7 shows the condition for tearing, which involves a combination of tension on the section labeled 1 and shear on the two sections labeled 2. For the tension section

$$\text{Net } w = 3 - 0.875 = 2.13 \text{ in.}$$

$$A_e = w_t \times t = 2.13 \times \tfrac{5}{8} = 1.33 \text{ in.}^2$$

and the design strength for tension is

$$\phi_t P_n = \phi_t \times F_u \times A_e = 0.75 \times 65 \times 1.33 = 64.8 \text{ kips}$$

For the two shear sections

$$\text{Net } w = 2 \times \left(1.25 - \frac{0.875}{2}\right) = 1.63 \text{ in.}$$

$$A_e = w_t \times t = 1.63 \times \tfrac{5}{8} = 1.02 \text{ in.}^2$$

and the design strength for shear is

$$\phi_v R_n = \phi_t \times F_u \times A_e = 0.75 \times 65 \times 1.02 = 49.5 \text{ kips}$$

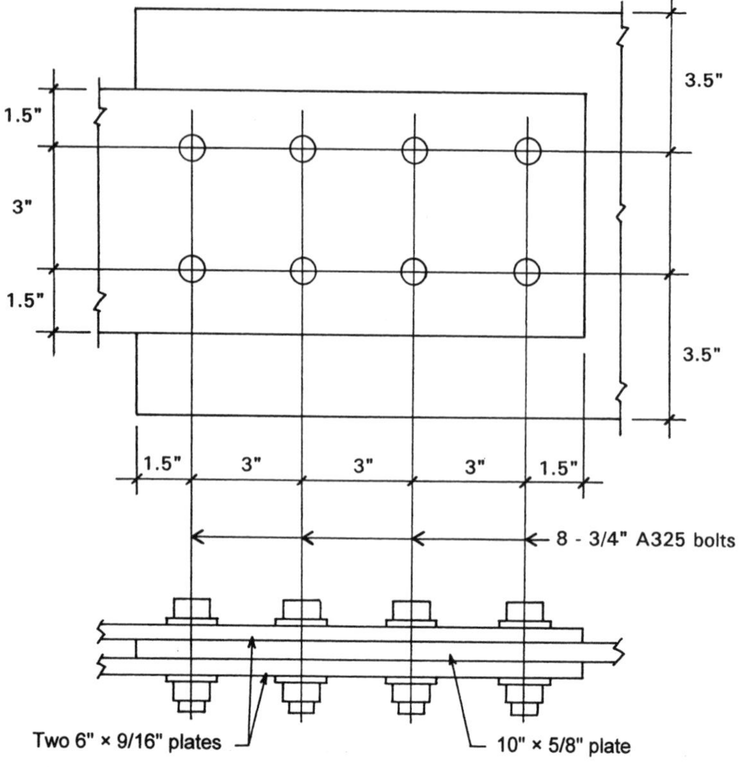

Figure 11.8 Solution for example problem.

The total resistance to tearing is thus

$$T = 34.8 + 49.5 = 114 \text{ kips}$$

Because this is greater than the combined load on the two end bolts (51 kips), the plate is not critical for tearing in block shear.

The solution for the connection is displayed in the top and side views in Figure 11.8. Connections that transfer compression between the joined parts are essentially the same with regard to the bolt stresses and bearing on the parts. Stress on the net section in the joined parts is not likely to be critical since the compression members are likely to be designed for a relatively low stress due to column action.

Problem 11.2.A A bolted connection of the general form shown in Figure 11.6 is to be used to transmit a tension force of 75 kips dead load and 100 kips live load by using 7/8-in. A325 bolts and plates of A992 steel with $F_y = 50$ ksi and $F_u = 58$ ksi. The outer plates are to be 8 in. wide and the center plate is to be 12 in. wide. Find the required thicknesses of the plates and the number of bolts needed if the bolts are placed in two rows. Sketch the final layout of the connection.

Problem 11.2.B Design the connection for the data in Problem 11.2.A, except that the outer plates are 9 in. wide and the bolts are placed in three rows.

11.3 BOLTED FRAMING CONNECTIONS

The joining of structural steel members in a structural system generates a wide variety of situations, depending on the form of the connected parts, the type of connecting device used, and the nature and magnitude of the forces that must be transferred between the members.

Framing connections quite commonly involve the use of welding and bolting in a single connection, as illustrated in the figures. In general, welding is favored for fabrication in the shop and bolting for erection in the field. If this practice is recognized, the connections must be developed with a view to the overall fabrication and erection process and some decision made regarding what is to be done where. With the best of designs, however, the contractor who is awarded the work may have some ideas about these procedures and may suggest alterations in the details.

Development of connection details is particularly critical for structures in which a great number of connections occur. The truss is one such structure.

Framed Beam Connections

The connection shown in Figure 11.9*a* is the type used most frequently in the development of framed structures that consist of I-shaped beams and H-shaped columns. This device is referred to as a *framed beam connection*, for which there are several design considerations:

Type of Fastening. Fastening of the angles to the supported beam and to the support may be accomplished with welds or with any of several types of structural bolt. The most common practice is to weld the angles to the supported beam's web in the fabricating shop and to bolt the angles to the support (column face or supporting beam's web) in the field (the erection site).

Number of Fasteners, If bolts are used, this refers to the number of bolts used on the supported beam web, there being twice this number of bolts in the outstanding legs of the angles. The capacities are matched, however, because the web bolts are in double shear and the others in single shear. For smaller beams, or for light loads in general, angle leg sizes are typically narrow, being just enough to accommodate a single row of bolts, as shown in Figure 11.9*b*. However, for very large beams and for greater loads, a wider leg may be used to accommodate two rows of bolts.

Size of the Angles. Leg width and thickness of the angles depend on the size of fasteners and the magnitude of loads. Width of the outstanding legs may also depend on space available, especially if attachment is to the web of a column.

Length of the Angles. Length must be that required to accommodate the number of bolts. Standard layout of the bolts is that shown in Figure 11.9, with bolts at 3-in. spacing and end distance of 1.25 in. This will accommodate up to 1-in.-diameter bolts. However, the angle length is also limited to the distance available on the beam web; that is, the total length of the flat portion of the beam web (see Figure 11.9*a*).

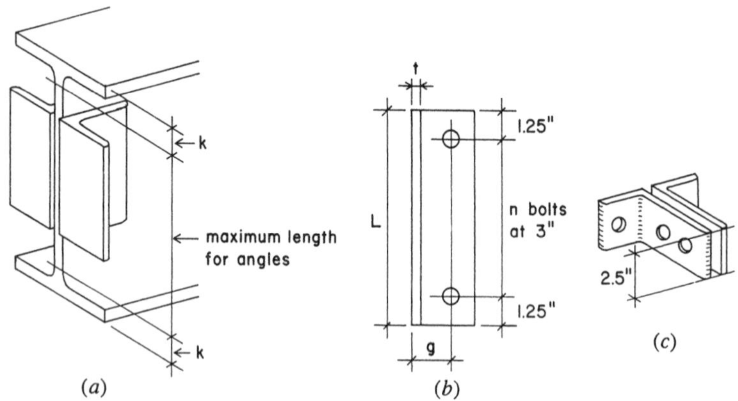

Figure 11.9 Framed beam connections for rolled shapes using intermediate connecting angles.

The AISC *Steel Manual* (Ref. 5) provides considerable information to support the design of these frequently used connecting elements. Data are provided for both bolted and welded fastenings. Predesigned connections are tabulated and can be matched to magnitudes of loadings and to sizes (primarily depths) of the beams (mostly W shapes) that can accommodate them.

Although there is no specified limit for the minimum size of a framed connection to be used with a given beam, a general rule is to use one with the angle length at least one-half of the beam depth. This rule is intended in the most part to ensure some minimum stability against rotational effects at the beam ends. For very shallow beams the special connector shown in Figure 11.9*c* may be used. Regardless of loading, this requires the angle leg at the beam web to accommodate two rows of bolts (one bolt in each row) simply for the stability of the angles.

There are many structural effects to consider for these connections. Of special concern is the inevitable bending in the connection that occurs as shown in Figure 11.2*c*. The bending moment arm for this twisting action is the gage distance of the angle leg, dimension g as shown in Figure 11.9*b*. It is a reason for choosing a relatively narrow angle leg.

If the top flange of the supported beam is cut back—as it commonly is when connection is to another beam—either vertical shear in the net cross section or block shear failure (Figure 11.3*c*) may be critical. Both of these conditions will be aggravated when the supported beam has a very thin web, which is a frequent condition because the most efficient beam shapes are usually the lightest shapes in their nominal size categories.

Another concern for the thin beam web is the possibility of critical bearing stress in the bolted connection. Combine a choice for a large bolt with one for a beam with a thin web, and this is likely to be a problem.

11.4 BOLTED TRUSS CONNECTIONS

A major factor in the design of trusses is the development of the truss joints. Since a single truss typically has several joints, the joints must be relatively easy to produce and economical, especially if there is a large number of trusses of a single type in the building structural system. Considerations involved in the design of connections for the joints include the truss configuration, member shapes and sizes, and the fastening method—usually welding or high-strength bolts.

In most cases the preferred method of fastening for connections made in the fabricating shop is welding. Trusses are usually shop fabricated in the largest units possible, which means the whole truss for modest spans or the maximum-sized unit that can be transported for large trusses. Bolting is mostly used for connections made at the building site. For the small truss, bolting is usually done only for the connections to supports and to supported elements or bracing. For the large truss, bolting may also be done at splice points between shop-fabricated units. All of this is subject to many considerations relating to the nature of the rest of the building structure, the particular location of the site, and the practices of local fabricators and erectors.

Two common forms for light steel trusses are shown in Figure 11.10. In Figure 11.10*a* the truss members consist of pairs of angles and the joints are achieved by using steel gusset plates to which the members are attached. For top and bottom chords the angles are often made continuous through the joint, reducing the number of connectors required and the number of separate cut pieces of the angles. For flat-profiled, parallel-chord trusses of modest size, the chords are sometimes made from tees, with interior members fastened directly to the tee web (Figure 11.10*b*).

Figure 11.11 shows a layout for several joints of a light roof truss, employing the system shown in Figure 11.10*a*. This is a form commonly used in the past for roofs with high slopes, with many short-span trusses fabricated in a single piece in the shop, usually

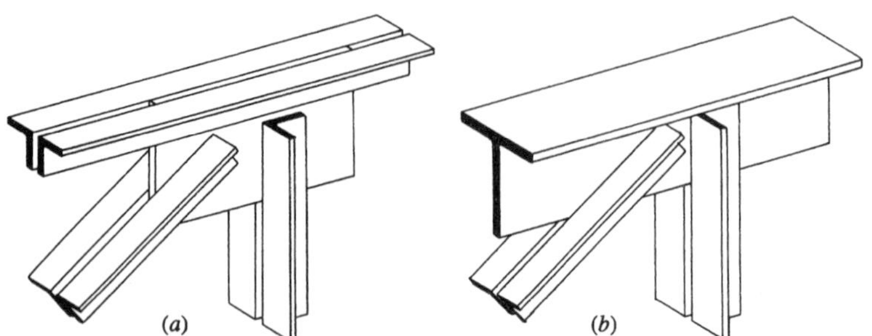

Figure 11.10 Common framing details for light steel trusses.

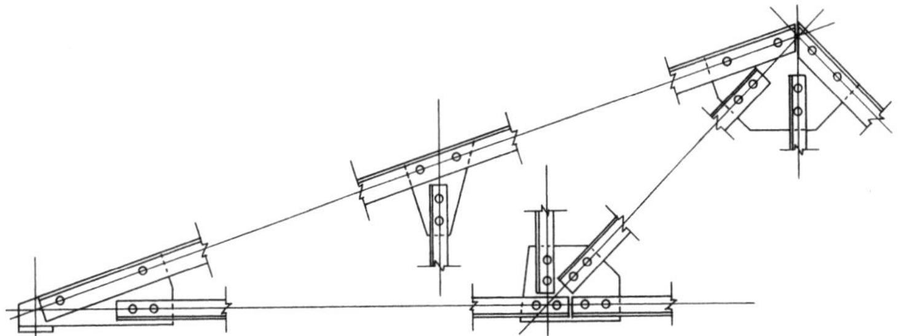

Figure 11.11 Typical form of a light steel truss with double-angle members and bolted connections with gusset plates.

with riveted joints. Trusses of this form are now mostly welded or use high-strength bolts as shown in Figure 11.11.

Development of the joint designs for the truss shown in Figure 11.11 would involve many considerations, including:

Truss Member Size and Load Magnitude. This determines primarily the size and type of connector (bolt) required, based on individual connector capacity.

Angle Leg Size. This relates to the maximum diameter of bolt that can be used, based on angle gages and minimum edge distances. (See Table 11.3.)

Thickness and Profile Size of Gusset Plates. The preference is to have the lightest weight added to the structure (primarily for the cost per pound of the steel), which is achieved by reducing the plates to a minimum thickness and general minimum size.

Layout of Members at Joints. The aim is to have the action lines of the forces (vested in the rows of bolts) all meet at a single point, thus avoiding twisting in the joint.

Many of the points mentioned are determined by data. Minimum edge distances for bolts (Table 11.2) can be matched to usual gage dimensions for angles (Table 11.3). Forces in members can be related to bolt capacities in Table 11.1, the general intent being to keep the number of bolts to a minimum in order to make the required size of the gusset plate smaller.

Other issues involve some judgment or skill in the manipulation of the joint details. For really tight or complex joints, it is often necessary to study the form of the joint with carefully drawn large-scale layouts. Actual dimensions and form of the member ends and gusset plates may be derived from these drawings.

The truss shown in Figure 11.11 has some features that are quite common for small trusses. All member ends are connected by only two bolts, the minimum required by the specifications. This simply indicates that the minimum-sized bolt chosen has sufficient capacity to develop the forces in all members with only two bolts. At the top chord joint between the support and the peak, the top chord member is shown as being continuous (uncut) at the joint. This is quite common where the lengths of members available are greater than the joint-to-joint distances in the truss, a cost savings in member fabrication as well as connection.

If there is only one or a few of the trusses as shown in Figure 11.11 to be used in a building, the fabrication may indeed be as shown in the illustration. However, if there are many such trusses, or the truss is actually a manufactured, standardized product, it is much more likely to be fabricated employing welding for shop work and bolting only for field connections.

12

LIGHT-GAGE FORMED STEEL STRUCTURES

Many structural elements are formed from sheet steel. Elements formed by the rolling process must be heat softened, whereas those produced from sheet steel are ordinarily made without heating the steel: Thus, the common description for these elements is *cold formed*. Because they are typically formed from thin sheet stock, they are also referred to as *light-gage* steel products.

12.1 LIGHT-GAGE STEEL PRODUCTS

Figure 12.1 illustrates the cross sections of some common products formed from sheet steel. Large corrugated or fluted panels are in wide use for wall paneling and for structural decks for roofs and floors (Figure 12.1*a*). These products are made by a number of manufacturers, and information regarding their structural properties may be obtained directly from the manufacturer. General information on structural decks may also be obtained from the Steel Deck Institute (see Ref. 9).

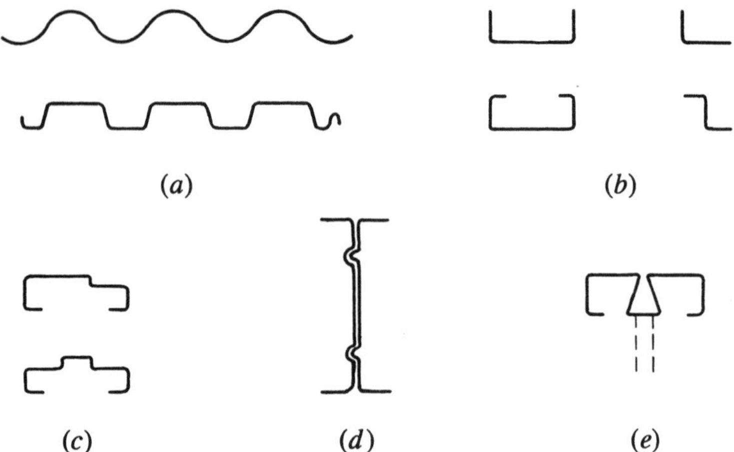

Figure 12.1 Cross-sectional shapes of common cold-formed sheet steel products.

Cold-formed shapes range from the simple L, C, and U shapes (Figure 12.1*b*) to the special forms produced for various construction systems, such as door and window frames (Figure 12.1*c*). Structures for some buildings may be almost entirely comprised of cold-formed products.

While some cold-formed and fabricated elements of sheet steel may be used for parts of structural systems, a major use of these products is for the formation of structural frames for partitions, curtain walls, suspended ceilings, and door and window framing. In large buildings, fire safety requirements usually prevent use of wood for these applications, so the noncombustible steel products are widely chosen.

12.2 LIGHT-GAGE STEEL DECKS

Steel decks consisting of formed sheet steel are produced in a variety of configurations, as shown in Figure 12.2. The simplest is the corrugated sheet, shown in Figure 12.2*a*. This may be used as the single, total surface for walls and roofs of utilitarian buildings (tin shacks). For more demanding usage, it is used mostly as the surfacing of a built-up panel or general sandwich-type construction. As a structural deck, the simple corrugated sheet is used for very short spans, typically with a structural-grade concrete fill that effectively serves as

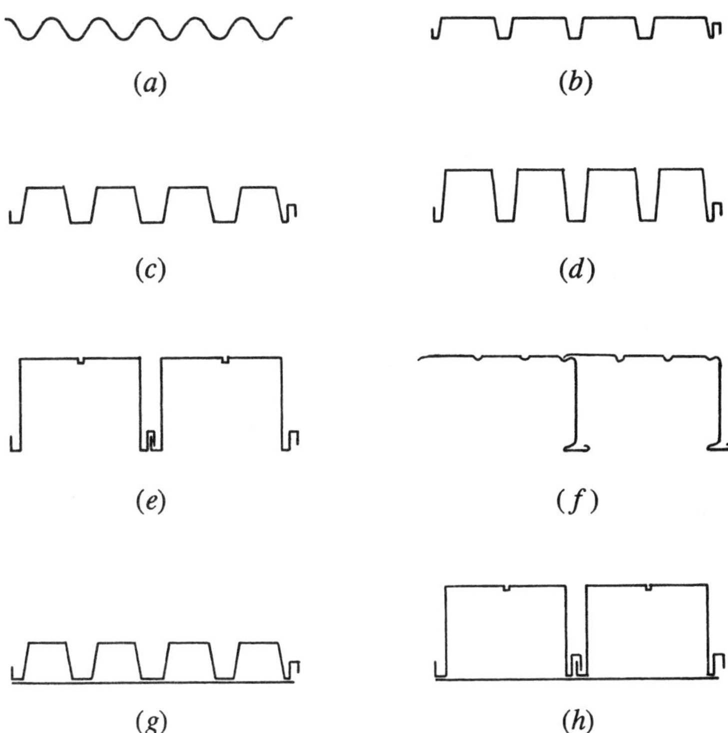

Figure 12.2 Cross-sectional shapes of formed sheet steel deck units.

the spanning deck—the steel sheet serving primarily as forming for the concrete.

A widely used product is that shown in three variations in Figures 12.2*b*–*d*. When used for roof deck, where loads are light, a flat top surface is formed with a very lightweight fill of foamed concrete or gypsum concrete or a rigid sheet material that also serves as insulation. For floors—with heavier loads, need for a relatively hard surface, and concern for bouncing—a structural-grade concrete fill is used, and the deeper ribs of the units shown in Figures 12-2*c* and *d* may be selected to achieve greater spans with wide-spaced beams. Common overall deck heights are 1.5, 3, and 4.5 in.

There are also formed sheet steel decks produced with greater depth, such as those shown in Figures 12.2*e* and *f*. These can achieve considerable span, generally combining the functions of joist and deck in a single unit.

Although used somewhat less now, with the advent of different wiring products and techniques for facilitating the need for frequent and rapid change of building wiring, a possible use for steel deck units is as a conduit for power, signal, or communication wiring. This can be accomplished by closing the deck cells with a flat sheet of steel, as shown in Figures 12.2*g* and *h*. This usually provides for wiring in one direction in a grid, the perpendicular wiring being achieved in conduits buried in the concrete fill.

Decks vary in form (shape of the cross section) and in the thickness (gage) of the steel sheet used to form them. Design choices relate to the desire for a particular form and to the load and span conditions. Units are typically available in lengths of 30 ft or more, which usually permits design for a multiple-span condition; this reduces bending effects only slightly but has a considerable influence on reduction of deflection and bouncing.

Fire protection for floor decks is provided partly by the concrete fill on top of the deck. Protection of the underside is achieved by sprayed-on materials (as also used on beams) or by use of a permanent fire-rated ceiling construction. The latter is no longer favored, however, as many disastrous fires have occurred in the void space between ceilings and overhead floors or roofs.

Other structural uses of the deck must also be considered. The most common use is as a horizontal diaphragm for distribution of lateral forces from wind and earthquakes. Lateral bracing of beams and columns is also often assisted or completely achieved by structural decks.

When structural-grade concrete is used as a fill, there are three possibilities for its relationship to a forming steel deck:

1. The concrete serves strictly as a structurally inert fill, providing a flat surface, fire protection, added acoustic separation, and so on, but no significant structural contribution.
2. The steel deck functions essentially only as a forming system for the concrete fill, the concrete being reinforced and designed as a spanning structural deck.
3. The concrete and sheet steel work together in what is described as *composite structural action*. In effect, the sheet steel on the bottom serves as the reinforcement for midspan bending stresses, leaving a need only for top reinforcement for negative bending moments over the deck supports.

TABLE 12.1 Safe Service Load Capacity of Formed Steel Roof Deck

Deck	Span	Weight^2	Total Safe Service Load (Dead + Live)^c for spans indicated in feet												
Type^a	Condition	(lb/ft)^b	4.0	4.5	5.0	5.5	6.0	6.5	7.0	7.5	8.0	8.5	9.0	9.5	10.0
NR22	Simple	1.6	73	58	47										
NR20		2.0	91	72	58	48	40								
NR18		2.7	125	99	80	66	55	47							
NR22	Two	1.6	80	63	51	42									
NR20		2.0	97	76	62	51	43								
NR18		2.7	128	101	82	68	57	48	42						
NR22	Three +	1.6	100	79	64	53	44								
NR20		2.0	121	96	77	64	54	46							
NR18		2.7	160	126	102	85	71	61	52	45					
IR22	Simple	1.6	84	66	54	44									
IR20		2.0	104	82	67	55	46								
IR18		2.7	142	112	91	75	63	54	46	40					
IR22	Two	1.6	90	71	58	48	40								
IR20		2.0	110	87	70	58	49	41							
IR18		2.7	145	114	93	77	64	55	47	40					
IR22	Three +	1.6	113	89	72	60	50	43							
IR20		2.0	137	108	88	72	61	52	45						
IR18		2.7	181	143	116	96	81	69	59	52	45	40			
WR22	Simple	1.6			90	70	56	46							
WR20		2.0			113	88	70	57	48	40					
WR18		2.7			159	122	96	77	64	54	46	40			
WR22	Two	1.6			96	79	67	57	49	43					
WR20		2.0			123	102	86	73	63	55	48	43			
WR18		2.7			164	136	114	98	84	73	64	57	51	46	41
WR22	Three +	1.6			119	99	83	71	61	53	47	41	36		
WR20		2.0			153	127	107	91	79	68	58	50	43		
WR18		2.7			204	169	142	121	105	91	79	67	58	51	43

Source: Adapted from the *Steel Deck Institute Design Manual for Composite Decks, Form Decks, and Roof Decks* (Ref. 9), with permission of the publisher, the Steel Deck Institute.
[a]Letters refer to rib type (see Figure 12.3). Numbers indicate gage (thickness) of deck sheet steel.
[b]Approximate weight with paint finish; other finishes available.
[c]Total safe allowable service load in lb/ft^2. Loads in parentheses are governed by live-load deflection not in excess of 1/240 of the span, assuming a dead load of 10 lb/ft^2.

Table 12.1 presents data relating to the use of the type of deck unit shown in Figure 12.2*b* for roof structures. These data are adapted from a publication distributed by an industrywide organization referred to in the table footnotes and is adequate for preliminary design work. The reference publication also provides considerable information and standard specifications for usage of the deck. For any final design work for actual construction, structural data for any manufactured products should be obtained directly from the suppliers of the products.

The common usage for roof decks with units as shown in Table 12.1 is that described above as Case 1: structural dependence strictly on the steel deck units. That is the basis for the data in the table given here.

Three different rib configurations are shown for the deck units in Table 12.1, described as *narrow*, *intermediate*, and *wide* rib decks. The configurations for these variations are shown in Figure 12.3. This has some effect on the properties of the deck cross section and thus produces three separate sections in the table. While structural performance may be a factor in choosing the rib width, there are usually other predominating reasons. If the deck is to be welded to its supports (usually required for good diaphragm action), it is done at the bottom of the ribs and the wide rib is required. If a relatively thin topping material is used, the narrow rib is favored.

Rusting is a critical problem for the very thin sheet steel deck. With its top usually protected by other construction, the main problem is the treatment of the underside of the deck. A common practice is to provide the appropriate surfacing of the deck units in the factory. The deck weights in Table 12.1 are based on simple painted surfaces, which is usually the least expensive surface. Surfaces consisting of bonded enamel or galvanizing are also available, adding somewhat to the deck weight.

As described previously, these products are typically available in lengths up to 30 ft or more. Depending on the spacing of supports, therefore, various conditions of continuity of the deck may occur. Recognizing this condition, the table provides three cases for continuity: simple span (one span), two spans, and three or more spans.

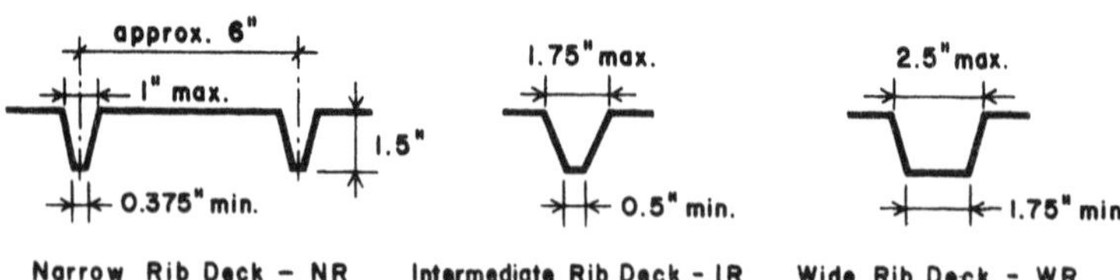

Figure 12.3 Reference for Table 12.1.

Problems 12.2.A–F Using data from Table 12.1, select the lightest steel deck for the following:

A. Simple span of 7 ft, total load of 45 psf
B. Simple span of 5 ft, total load of 50 psf
C. Two-span condition, span of 8.5 ft, total load of 45 psf
D. Two-span condition, span of 6 ft, total load of 50 psf
E. Three-span condition, span of 6 ft, total load of 50 psf
F. Three-span condition, span of 8 ft, total load of 50 psf

12.3 LIGHT-GAGE STEEL SYSTEMS

Proprietary steel structural systems are produced by many manufacturers for various applications. While some systems exist for developing structures for entire buildings, a larger market is that for the systems used for wall framing, ceiling structures, and supports for building service elements. With systems for large structures, rolled shapes or trusses may be used for larger elements, with light-gage elements creating the infilling structure, bracing, and various secondary framing.

The light-gage elements and systems widely developed for partition and ceiling framing for large buildings can be utilized to produce a stud–rafter–joist system that emulates the classic light wood frame with 2-in. nominal lumber elements.

IV

CONCRETE CONSTRUCTION

The term *concrete* covers a variety of products that share a common character: They consist of a mass of loose particles (called the aggregate) that is bound together by some cementing material. Included in this group are asphalt paving and precast shingle tiles, but the material in this part deals primarily with the more familiar material described by the term—that produced with Portland cement as the binder and sand and gravel as the inert mass of loose particles. When used for building structures, provisions are made with concrete construction to compensate for the low tensile strength of the material. Three different methods are currently in use: the addition of fibrous material to the concrete mix, prestressing to add a compressive stress to counteract tension stresses, and insertion of steel reinforcing rods to the cast concrete. When reinforced with steel rods, the construction is described as reinforced concrete, which is the form of construction treated in this part. Work for investigation and design in this part is based on the requirements of *Building Code Requirements for Structural Concrete*, published by the American Concrete Institute (ACI) (Ref. 10), hereinafter referred to as the ACI Code.

13

REINFORCED CONCRETE STRUCTURES

This part deals primarily with concrete formed with the common binding agent of *Portland cement* and a loose mass consisting of sand and gravel. With minor variations, this is the material used mostly for structural concrete—to produce building structures, pavements, and foundations. For common structural uses, the tensile strength weakness of the concrete is modified by the addition of steel reinforcing rods (deformed round bars); when this occurs, the construction is described as *reinforced concrete*.

13.1 GENERAL CONSIDERATIONS

Concrete made from natural materials was used by ancient builders thousands of years ago. Modern concrete, made with industrially produced cement, was developed in the early part of the nineteenth century when the process for producing Portland cement was developed. Because of its lack of tensile strength (in comparison to its compression strength), however, concrete was used principally for crude, massive structures—foundations, bridge piers, and heavy walls.

In the mid- to late-nineteenth century, several builders experimented with the technique of inserting iron or steel rods into relatively thin structures of concrete to enhance their ability to resist tensile forces. This was the beginning of what we now know as reinforced concrete.

From ancient times until now, there has been a steady accumulation of experience derived from experiments, research, and—most recently—intense development of commercial products. As a result, there is now available to the designer an immense variety of products under the general classification of concrete, although the range is somewhat smaller if structural usage is required.

Forms of Concrete Structures

For building structures, concrete is mostly used with one of three basic construction methods. The first is called *site-cast concrete,* in which the wet concrete mix is deposited in formwork at the location where it is to be used. This method is also described as *cast-in-place* or *in situ* construction.

A second method consists of casting portions of the structure at a location away from the desired location of the construction. These elements—described as *precast concrete*—are then moved into position, much as are blocks of stone or parts of steel frames.

Finally, concrete may be used for masonry construction—in one of two ways. Precast units of concrete, called concrete masonry units (CMUs), may be used in a manner similar to bricks or stones. Or, concrete fill may be used to produce solid masonry by being poured into cavities in the masonry construction produced with bricks, stone, or CMUs. The latter technique, combined with the insertion of steel reinforcement into the cavities, is widely used for masonry structures today. The use of concrete-filled masonry, however, is one of the oldest forms of concrete construction—used extensively by the Romans and the builders of early Christian churches.

Concrete is produced in great volume for various forms of construction. Other than for pavements, the widest general use of concrete for building construction is for foundations. Almost every building has a concrete foundation, whether the major above-ground construction is concrete, masonry, wood, steel, aluminum, or fabric. For small buildings with shallow footings and no basement, the total foundation system may be modest, but for large buildings and those with many below-ground levels, there may well be a gigantic underground concrete structure.

For above-ground building construction, concrete is generally used in situations that fully realize the various advantages of the basic material and the common systems that derive from it. For structural applications, this means using the major compressive resistance of the material and in some situations its relatively high stiffness and inertial resistance (major dead weight). However, in many applications, the nonrotting, vermin- and insect-resistive, and fire-resistive properties may be of major significance. And for many uses, its relatively low bulk volume cost is important.

Strength of Concrete

The quality of concrete of greatest significance for structural purposes is its resistance to compressive stress. As such, the common practice is to specify a desired limiting capacity of compressive stress, designated as f'_c, to design a concrete mix to achieve that limit, and to test samples of cast and hardened concrete to verify its true capacity for compression.

For design work, the capacity of concrete for all purposes is established as some percentage of f'_c. Attainment of a quality of concrete to achieve a particular level of compressive resistance generally also serves to certify various other properties, such as hardness, density, and durability. Choice for the desired strength is typically based on the form of construction. For most purposes a strength of 3000–5000 psi for f'_c is usually adequate. However, strengths of 20,000 psi and higher have recently been achieved for lower columns in very tall structures. At the other end of this spectrum is the situation where quality control may be less assured and the designer may assume the attainment of a relatively low strength, basing conservative design computations on strength as low as 2000 psi—while doing all possible to achieve a better concrete.

As it makes up the major bulk of the finished concrete, the aggregate is of primary importance for stress resistance. It must be hard and durable and must be graded in size so that small particles fill the voids between larger ones, producing a dense mass before the cement and water are added. The weight of concrete is usually determined primarily by the density of the aggregate.

The other major factor for concrete strength is the amount of water used for mixing. The basic idea is to use as little water as possible, as an excess will water down the water–cement mixture and produce weak, porous concrete. However, this must be balanced against the need for a wet mix that can be easily placed in forms and finished. A lot of skill and some science are involved in producing an ideal mix.

A final consideration for strength is the control of conditions during the early life of a cast mix. The mobile wet mix hardens relatively quickly but gains its highest potential strength over some period of time. It is important to control the water content and the temperature of the hardened concrete during this critical period if the best quality concrete is to be expected.

Stiffness of Concrete

As with other materials, the stiffness of concrete is measured by the *modulus of elasticity*, designated E. This modulus is established by tests and is the ratio of stress to strain. Since strain has no unit designation (measured as inch/inch, etc.), the unit for E thus becomes the unit for stress, usually lb/in.^{2}

The magnitude of elasticity for concrete, E_c, depends on the weight of the concrete and its strength. For values of unit weight between 90 and 155 lb/ft^3 or pcf, the value of E_c is determined as

$$E_c = w^{1.5} \times 33\sqrt{f'_c}$$

The unit weight for ordinary stone–aggregate concrete is usually assumed to be an average of 145 pcf. Substituting this value for w in the equation, an average concrete modulus of $E_c = 57{,}000\sqrt{f'_c}$.

Distribution of stresses and strains in reinforced concrete is dependent on the concrete modulus, the steel modulus being a constant. In the design of reinforced concrete members the term n is employed. This is the ratio of the modulus of elasticity of steel to that of concrete, or $n = E_s/E_c$, and E_s is taken as 29,000 ksi, a constant.

Creep

When subjected to long-duration stress at a high level, concrete has a tendency to *creep*, a phenomenon in which strain increases over time under constant stress. This has effects on deflections and on the distributions of stresses between the concrete and reinforcing. Some of the implications of this for design are discussed in dealing with design of beams and columns.

Cement

The cement used most extensively in building construction is Portland cement. Of the five types of standard Portland cement generally available in the United States and for which the American Society for Testing and Materials has established specifications, two types account for most of the cement used in buildings. These are a general-purpose cement for use in concrete designed to reach its required strength in about 28 days and a high-early-strength cement for use in concrete that attains its design strength in a period of a week or less.

All Portland cements set and harden by reacting with water, and this hydration process is accompanied by generation of heat. In massive concrete structures such as dams, the resulting temperature rise of the materials becomes a critical factor in both design and construction, but the problem is usually not significant in building construction. A low-heat cement is designed for use where the heat rise during hydration is a critical factor. It is, of course, essential that the cement actually used in construction correspond to that employed in designing the mix, to produce the specified compressive strength of the concrete.

Air-entrained concrete is produced by using special cement or by introducing an additive during mixing of the concrete. In addition to improving workability (mobility of the wet mix), air entrainment permits lower water–cement ratios and significantly improves the durability of the concrete. Air-entraining agents produce billions of microscopic air cells throughout the concrete mass. These minute voids prevent accumulation of water in cracks and other large voids which, on freezing, would permit the water to expand and result in spalling away of the exposed surface of the concrete.

Reinforcement

The steel used in reinforced concrete consists of round bars, mostly of the deformed type, with lugs or projections on their surfaces. The surface deformations help to develop a greater bond between the steel bars and the enclosing concrete mass.

Purpose of Reinforcement. The essential purpose of steel reinforcing is to reduce the failure of the concrete due to tensile stresses. Structural actions are investigated for the development of tension in the structural members and steel reinforcement in the proper amount is placed within the concrete mass to resist the tension. In some situations steel reinforcement may also be used to increase compressive resistance since the ratio of magnitudes of strength of the two materials is quite high; thus the steel displaces a much weaker material and the member gains significant strength.

Tension stress can also be induced by shrinkage of the concrete during its hydration period from the initial wet mix. Temperature variations may also induce tension in many situations. To provide for these latter actions, a minimum amount of reinforcing is used in surface-type members such as walls and paving slabs, even when no structural action is visualized.

Stress–Strain Considerations. The most common grades of steel used for ordinary reinforcing bars are Grade 40 and Grade 60, having yield strengths of 40 and 60 ksi, respectively. The yield strength of the steel is of primary interest for two reasons. Plastic yielding of the steel generally represents the limit of its practical utilization for reinforcing of the concrete since the extensive deformation of the steel in its plastic range results in major cracking of the concrete. Thus, for service load conditions, it is desirable to keep the stress in the steel within its elastic range of behavior where deformation is minimal.

The second reason for the importance of the yield character of the reinforcing is its ability to impart a generally yielding nature (plastic deformation character) to the otherwise typically very brittle concrete structure. This is of particular importance for dynamic loading and is a major consideration in design for earthquake forces. Also of importance is the residual strength of the steel beyond its yield stress limit. The steel continues to resist stress in its plastic range and then gains a second, higher, strength before failure. Thus, the failure induced by yielding is only a first-stage response and a second level of resistance is reserved.

Cover. Ample concrete protection, called *cover*, must be provided for the steel reinforcement. This is important to protect the steel from rusting and to be sure that it is well engaged by the mass of concrete. Cover is measured as the distance from the outside face of the concrete to the edge of the reinforcing bar.

Code minimum requirements for cover are $^3/_4$ in. for walls and slabs and $1\,^1/_2$ in. for beams and columns. Additional distance of cover is required for extra fire protection or for special conditions of exposure of the concrete surface to weather or by contact with the ground.

Spacing of Bars. Where multiple bars are used in concrete members (which is the common situation), there are both upper and lower limits for the spacing of the bars. Lower limits are intended to facilitate the flow of wet concrete during casting and to permit adequate development of the concrete-to-steel stress transfers for individual bars.

Maximum spacing is generally intended to assure that there is some steel that relates to a concrete mass of limited size; that is, there is not too extensive a mass of concrete with no reinforcement. For relatively thin walls and slabs, there is also a concern of scale of spacing related to the thickness of the concrete.

Amount of Reinforcement. For structural members the amount of reinforcement is determined from structural computations as that required for the tension force in the member. This amount (in total cross-sectional area of the steel) is provided by some combination of bars. In various situations, however, there is a minimum amount of reinforcement that is desirable, which may on occasion exceed the amount determined by computation.

Minimum reinforcement may be specified as a minimum number of bars or as a minimum amount, the latter usually based on the amount of the cross-sectional area of the concrete member. These requirements are discussed in the sections that deal with the design of the various types of structural members.

Standard Reinforcing Bars. In early concrete work reinforcing bars took various shapes. A problem that emerged was the proper bonding of the steel bars within the concrete mass, due to the tendency of the bars to slip or pull out of the concrete. This issue is still a critical one and is discussed in Section 13.7.

In order to anchor the bars in the concrete, various methods were used to produce something other than the usual smooth surfaces on bars. After much experimentation and testing, a single set of bars was developed with surface deformations consisting of ridges. These deformed bars were produced in graduated sizes with bars identified by a single number (see Table 13.1).

TABLE 13.1 Properties of Deformed Reinforcing Bars

Bar Size Designation	Nominal Weight lb/ft	Diameter in.	Cross-Sectional Area in.^{2}
No. 3	0.376	0.375	0.11
No. 4	0.668	0.500	0.20
No. 5	1.043	0.625	0.31
No. 6	1.502	0.750	0.44
No. 7	2.044	0.875	0.60
No. 8	2.670	1.000	0.79
No. 9	3.400	1.128	1.00
No. 10	4.303	1.270	1.27
No. 11	5.313	1.410	1.56
No. 14	7.650	1.693	2.25
No. 18	13.600	2.257	4.00

For bars numbered 2–8, the cross-sectional area is equivalent to a round bar having a diameter of as many eighths of an inch as the bar number. Thus, a No. 4 bar is equivalent to a round bar of $^4/_8$ or $^1/_2$ in. diameter. Bars numbered from 9 up lose this identity and are essentially identified by the tabulated properties in a reference document.

The bars in Table 13.1 are developed in U.S. units but can, of course, be used with their properties converted to metric units. However, a new set of bars has been developed, deriving their properties more logically from metric units. The general range of sizes is similar for both sets of bars, and design work can readily be performed with either set. Metric-based bars are obviously more popular outside the United States, but for domestic use (nongovernment) in the United States, the old bars are still in wide use. This is part of a wider conflict over units which is still going on.

The work in this book uses the old inch-based bars, simply because the computational examples are done in U.S. units. In addition, many of the references still in wide use have data presented basically with U.S. units and the old bar sizes.

13.2 GENERAL APPLICATION OF STRENGTH METHODS

Strength design in effect consists of designing members to fail; thus, the ultimate strength of the member at failure (called its design strength) is the only type of resistance considered. The basic procedure of the strength method consists of determining a factored (increased) design load and comparing it to the factored (usually reduced) ultimate resistance of the structural member.

The ACI Code (Ref. 10) provides various combinations of loads that must be considered for design. Each type of load (live, dead, wind, earthquake, snow, etc.) is given an individual factor in these load equations. See discussion in Chapter 1.

The design strength of individual concrete members (i.e., their usable ultimate strength) is determined by the application of assumptions and requirements given in the code and is further modified by the use of a strength reduction factor φ as follows:

$\varphi = 0.90$ for flexure, tension, and combinations of these
$= 0.70$ for columns with spirals
$= 0.65$ for columns with ties
$= 0.75$ for shear and torsion
$= 0.65$ for compressive bearing
$= 0.55$ for flexure in plain (not reinforced) concrete

13.3 BEAMS: ULTIMATE STRENGTH METHOD

The primary concerns for beams relate to their necessary resistance to bending and shear and some limitations on their deflection. For wood or steel beams the usual concern is only for the singular maximum values of bending and shear in a given beam. For concrete beams, on the other hand, it is necessary to provide for the values of bending and shear as they vary along the entire length of a beam, even through multiple spans in the case of continuous beams, which are a common occurrence in concrete structures. For simplification of the work it is necessary to consider the actions of a beam at a specific location, but it should be borne in mind that this action must be integrated with all the other effects on the beam throughout its length.

When a member is subjected to bending, such as the beam shown in Figure 13.1*a*, internal resistances of two basic kinds are generally required. Internal actions are "seen" by visualizing a cut section, such as that taken at *X–X* in Figure 13.1*a*. Removing the portion of the beam to the left of the cut section, its free-body actions are as shown in Figure 13.1*b*. At the cut section, consideration of static equilibrium requires the development of the internal shear force (V in the figure) and the internal resisting moment (represented by the force couple: C and T in the figure).

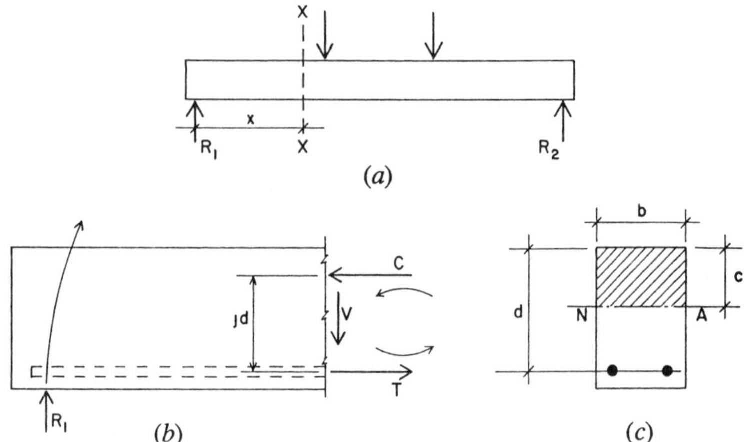

Figure 13.1 Bending action in a reinforced concrete beam.

If a beam consists of a simple rectangular concrete section with tension reinforcement only, as shown in Figure 13.1*c*, the force *C* is considered to be developed by compressive stresses in the concrete—indicated by the shaded area above the neutral axis. The tension force, however, is considered to be developed by the steel alone, ignoring the tensile resistance of the concrete. For low-stress conditions the latter is not true, but at a greater level of stress the tension-weak concrete will indeed crack, virtually leaving the steel unassisted, as assumed.

At moderate levels of stress, the resisting moment is visualized as shown in Figure 13.2*a*, with a linear variation of compressive stress from zero at the neutral axis to a maximum value of f_c at the edge of the section. As stress levels increase, however, the nonlinear stress–strain character of the concrete becomes more significant, and it becomes necessary to acknowledge a more realistic form for the compressive stress variation, such as that shown in Figure 13.2*b*. As stress levels approach the limit of the concrete, the compression becomes vested in an almost constant magnitude of unit stress, concentrated near the top of the section. For strength design, in which the moment capacity is expressed at the ultimate limit, it is common to assume the form of stress distribution shown in Figure 13.2*c*, with the limit for the concrete stress set at 0.85 f'_c. Expressions for the moment capacity derived from this assumed distribution compare reasonably with the response of beams tested to failure in laboratory experiments.

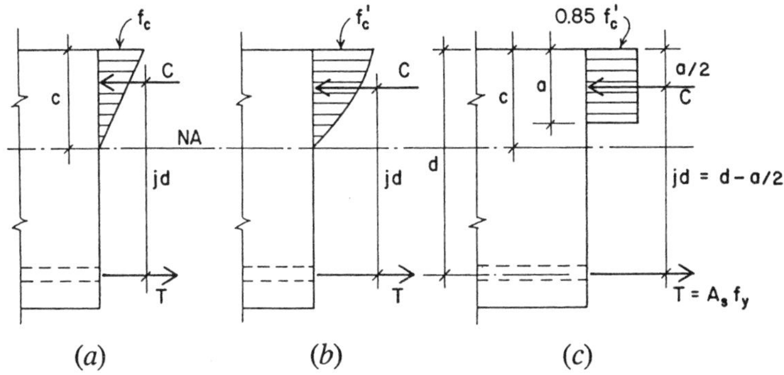

Figure 13.2 Development of bending stress actions in a reinforced concrete beam.

Response of the steel reinforcement is more simply visualized and expressed. Since the steel area in tension is concentrated at a small location with respect to the size of the beam, the stress in the bars is considered to be a constant. Thus, at any level of stress the total value of the internal tension force may be expressed as

$$T = A_s f_s$$

and for the practical limit of T,

$$T = A_s f_y$$

The following is a presentation of the formulas and procedures used in the strength method. The discussion is limited to a rectangular beam section with tension reinforcement only.

Referring to Figure 13.3, the following are defined:

b = width of the concrete compression zone
d = effective depth of the section for stress analysis; from the centroid of the steel to the edge of the compressive zone
h = overall depth (height) of the section
A_s = cross-sectional area of reinforcing bars
ρ = percentage of reinforcement, defined as

$$\rho = \frac{A_s}{bd}$$

Figure 13.2*c* shows the rectangular "stress block" that is used for analysis of the rectangular section with tension reinforcing only by

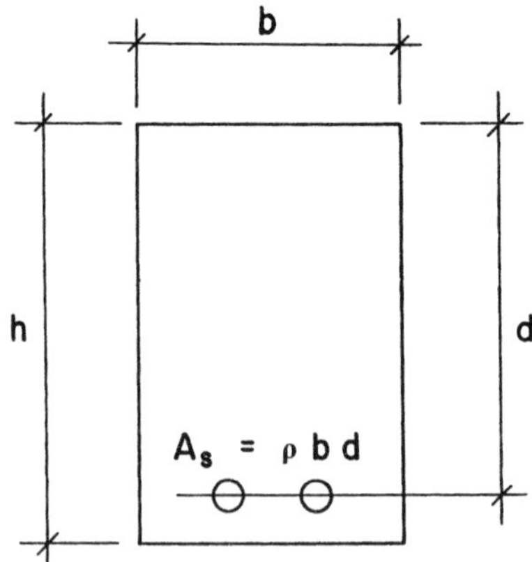

Figure 13.3 Reference notation for a reinforced concrete beam.

the strength method. This is the basis for investigation and design as provided for in the ACI Code (Ref. 10).

The rectangular stress block is based on the assumption that a concrete stress of $0.85f_c'$ is uniformly distributed over the compression zone, which has dimensions equal to the beam width b and the distance a, which locates a line parallel to and above the neutral axis. The value of a is determined from the expression $a = \beta_1 \times c$, where β_1 (beta one) is a factor that varies with the compressive strength of the concrete, and c is the distance from the extreme fiber to the neutral axis (see Figure 13.2*c*). For concrete having $a = \beta_1 \times c$ equal to or less than 4000 psi, the ACI Code (Ref. 10) gives a maximum value for 1 $a = 0.85c$.

With the rectangular stress block, the magnitude of the compressive force is expressed as

$$C = (0.85f_c')(b)(a)$$

and it acts at a distance of $a/2$ from the top of the beam.

The arm of the resisting force couple then becomes $d - (a/2)$, and the developed resisting moment as governed by the concrete is

$$M_c = C\left(d - \frac{a}{2}\right) = 0.85f_c'ba\left(d - \frac{a}{2}\right) \qquad (13.3.1)$$

With T expressed as $A_s \times f_y$, the developed moment as governed by the reinforcement is

$$M_t = T\left(d - \frac{a}{2}\right) = A_s f_y \left(d - \frac{a}{2}\right) \qquad (13.3.2)$$

A formula for the dimension a of the stress block can be derived by equating the compression and tension forces; thus

$$0.85 f_c' ba = A_s f_y \qquad a = \frac{A_s f_y}{0.85 f_c' b} \qquad (13.3.3)$$

By expressing the area of steel in terms of a percentage ρ, the formula for a may be modified as follows:

$$\rho = \frac{A_s}{bd} \quad \rightarrow \quad A_s = \rho bd$$

$$a = \frac{(\rho bd) f_y}{0.85 f_c' b} = \frac{\rho d f_y}{0.85 f_c'} \quad \text{or} \quad \frac{a}{\mathrm{d}} = \frac{\rho f_y}{0.85 f_c'} \qquad (13.3.4)$$

A useful reference is the so-called *balanced section,* which occurs when use of the exact amount of reinforcement results in the simultaneous development of the limiting stresses in the concrete and steel. The balanced section for strength design is visualized in terms of strain rather than stress. The limit for a balanced section is expressed in the form of the percentage of steel required to produce balanced conditions. The formula for this percentage is

$$\rho_b = \frac{0.85 f_c'}{f_y} \times \frac{87}{87 + f_y} \qquad (13.3.5)$$

in which f_c' and f_y are in units of ksi.

Returning to the formula for the developed resisting moment (13.3.2), as expressed in terms of the steel, a useful formula may be derived as follows:

$$\begin{aligned} M_t &= A_s f_y \left(d - \frac{a}{2}\right) \\ &= (\rho bd) f_y \left(d - \frac{a}{2}\right) \\ &= (\rho bd) f_y (d) \left(1 - \frac{a}{2d}\right) \\ &= (bd^2)\left[\rho f_y \left(1 - \frac{a}{2d}\right)\right] \end{aligned}$$

Thus,

$$M_t = Rbd^2 \tag{13.3.6}$$

where

$$R = \rho f_y \left(1 - \frac{a}{2d}\right) \tag{13.3.7}$$

With the reduction factor applied, the design moment for a section is limited to nine-tenths of the theoretical resisting moment.

Values for the balanced section factors (ρ, R, and a/d) are given in Table 13.2 for various combinations of f'_c and f_y. The balanced section is not necessarily a practical one for design. In most cases it is desirable to have the steel reinforcing yield first, therefore creating a ductile failure of the member. This is achieved by using less than the balanced reinforcing for a given concrete section. In special circumstances it may also be possible, or even desirable, to use compressive reinforcing in addition to tension reinforcing. Nevertheless, the balanced section is often a useful reference when design is performed.

Beams with reinforcement less than that required for the balanced moment are called *underbalanced sections* or *underreinforced sections.* If a beam must carry bending moment in excess of the balanced moment for the section, it is necessary to provide some compressive reinforcement, as discussed in Section 13.2. The balanced section is not necessarily a design ideal but is useful in establishing the limits for the section.

TABLE 13.2 Balanced Section Properties for Rectangular Sections with Tension Reinforcement Only

f_y ksi	f'_c ksi	ρ	a/d	R ksi
40	2	0.0291	0.685	0.766
	3	0.0437	0.685	1.149
	4	0.0582	0.685	1.531
	5	0.0728	0.685	1.914
60	2	0.0168	0.592	0.708
	3	0.0252	0.592	1.063
	4	0.0335	0.592	1.417
	5	0.0419	0.592	1.771

In the design of concrete beams, there are two situations that commonly occur. The first occurs when the beam is entirely undetermined; that is, the concrete dimensions and the reinforcement are unknown. The second occurs when the concrete dimensions are given, and the required reinforcement for a specific bending moment must be determined. The following examples illustrate the use of the formulas just developed for each of these problems.

Example 1. The service load bending moments on a beam are 58 kip-ft for dead load and 38 kip-ft for live load. The beam is 10 in. wide, f'_c is 3000 psi, and f_y is 60 ksi. Determine the depth of the beam and the tensile reinforcing required.

Solution: The first step is to determine the required moment using the load factors. Thus,

$$M_u = 1.2(M_{DL}) + 1.6(M_{LL}) = 1.2(58) + 1.6(38) = 130 \text{ kip-ft}$$

With the capacity reduction of 0.90 applied, the desired moment capacity of the section is determined as

$$M_t = \frac{M_u}{0.90} = \frac{130}{0.90} = 145 \text{ kip-ft} \times 12 = 1739 \text{ kip-in.}$$

The reinforcement ratio as given in Table 13.2 is $\rho = 0.0252$. The required area of reinforcement for this section may thus be determined from the relationship

$$A_s = \rho bd$$

While there is nothing especially desirable about a balanced section, it does represent the beam section with the least depth if tension reinforcing only is used. Therefore, proceed to find the required balanced section for this example.

To determine the required effective depth, d, use equation (13.3.6); thus,

$$M_t = Rbd^2$$

With the value $R = 1.063$ ksi from Table 13.2,

$$M_t = 1739 \text{ kip} - \text{in.} = 1.063 \text{ ksi}(10 \text{ in.})(d)^2$$

and

$$d = \sqrt{\frac{M_t}{Rb}} = \sqrt{\frac{1739}{1.063(10)}} = 12.8 \text{ in.}$$

If this value is used for d, the required steel area may be found using the value $p = 0.0252$ from Table 13.2. Thus,

$$A_s = \rho bd = 0.0252(10)(12.8) = 3.23 \text{ in.}^2$$

Although they are not given in this example, there are often some considerations other than flexural behavior alone that influence the choice of specific dimensions for a beam. These may include:

Design for shear

Coordination of the depths of a set of beams in a framing system

Coordination of the beam dimensions and placement of reinforcement in adjacent beam spans

Coordination of beam dimensions with supporting columns

Limiting beam depth to provide overhead clearance beneath the structure

If the beam is of the ordinary form shown in Figure. 13.4, the specified dimension is usually that given as h. Assuming the use of a No. 3 U-stirrup, a cover of 1.5 in., and an average-size reinforcing bar of 1 in. diameter (No. 8 bar), the design dimension d will be less than h

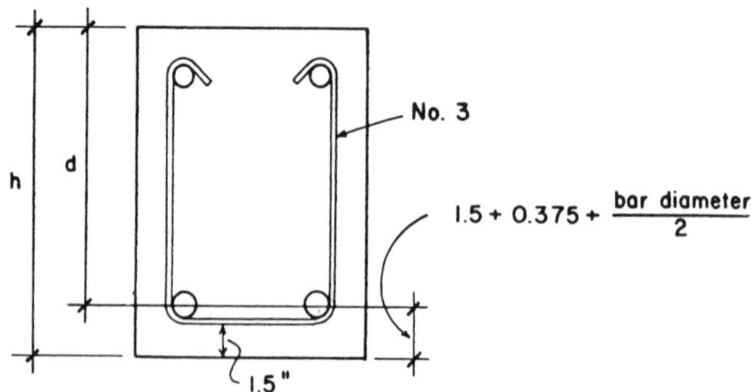

Figure 13.4 Common form of a reinforced concrete beam.

by 2.375 in. Lacking other considerations, the overall required depth of the beam (h) will be 15.175 in.

Next select a set of reinforcing bars to obtain this area. For the purpose of the example, select bars all of a single size (see Table 13.1); the number required will be as follows:

No. 6 bars: 3.23/0.44 = 7.3, or 8

No. 7 bars: 3.23/0.60 = 5.4, or 6

No. 8 bars: 3.23/0.79 = 4.1, or 5

No. 9 bars: 3.23/1.00 = 3.3, or 4

No. 10 bars: 3.23/1.27 = 2.5, or 3

No. 11 bars: 3.23/1.56 = 2.1, or 3

In real design situations there are always various additional considerations that influence the choice of the reinforcing bars. One general desire is that of having the bars in a single layer, as this keeps the centroid of the steel as close as possible to the edge (bottom in this case) of the member, giving the greatest value for d with a given height (h) of a concrete section. With the section as shown in Figure 13.5, a beam width of 10 in. will yield a net width of 6.25 in.

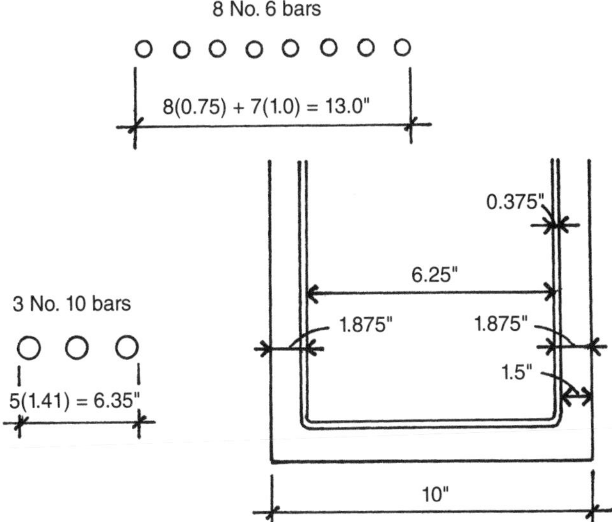

Figure 13.5 Consideration of beam width for proper spacing of a single layer of reinforcement.

inside the No. 3 stirrups, determined as the outside width of 10 in. less 2 (1.5-in. cover) and 2 (0.375-in. stirrup diameter). Applying the code criteria for minimum spacing for this situation, the required width for the various bar combinations can be determined. Minimum space required between bars is one bar diameter or a minimum of 1 in. (See discussion in Section. 13.1.) Two examples for this are shown in Figure 13.5. It will be found that none of the choices will fit this beam width. Thus, the beam width must be increased or two layers of bars must be used.

If there are reasons, as there often are, for not selecting the least deep section with the greatest amount of reinforcing, a slightly different procedure must be used, as illustrated in the following example.

Example 2. Using the same data as in Example 1, find the reinforcement required if the desired beam section has $b = 10$ in. and $d = 18$ in.

Solution: The first two steps in this situation would be the same as in Example 1—to determine M_u and M_t. The next step would be to determine whether the given section is larger than, smaller than, or equal to a balanced section. Since this investigation has already been done in Example 1, observe that the 10×18-in. section is larger than a balanced section. Thus, the actual value of a/d will be less than the balanced section value of 0.592 from Table 13.2. The next step would then be as follows.

Estimate a value for a/d—something smaller than the balanced value. For example, try $a/d = 0.3$. Then

$$a = 0.3d = 0.3(18 \text{ in.}) = 5.4 \text{ in.}$$

With this value for a, use equation (13.3.2) to find a required value for A_s. Referring to Figure 13.2,

$$M_t = T(jd) = (A_s f_y)\left(d - \frac{a}{2}\right)$$

$$A_s = \frac{M_t}{f_y\left(d - \frac{a}{2}\right)} = \frac{1739}{60(15.3)} = 1.89 \text{ in.}$$

Next test to see if the estimate for a/d was close by finding a/d using equation (13.3.4). Thus,

$$\rho = \frac{A_s}{bd} = \frac{1.89}{10(18)} = 0.0105$$

and, from the same equation,

$$\frac{a}{d} = \frac{\rho f_y}{0.85 f_c} = \frac{0.0105 \times 60}{0.85 \times 3} = 0.247$$

Thus,

$$a = 0.247 \times 18 = 4.45 \text{ in.}$$

$$d - \frac{a}{2} = 15.8 \text{ in.}$$

If this value for $d - a/2$ is used to replace that used earlier, the required value of A_s will be slightly reduced. In this example, the correction will be only a few percent. If the first guess of a/d had been way off, it may justify another run through the analysis to get closer to an exact answer.

For beams that are classified as underreinforced (section dimensions larger than the limit for a balanced section), check for the minimum required reinforcement. For a rectangular section, the ACI Code (Ref. 10) specifies that a minimum area be

$$A_s = \frac{3\sqrt{f'_c}}{f_y}(bd)$$

but not less than

$$A_s = \frac{200}{f_y}(bd)$$

On the basis of these requirements, values for minimum reinforcement for rectangular sections with tension reinforcement only are given in Table 13.3 for two grades of steel and three concrete strengths. For the example, with a concrete strength of 3000 psi and f_y of 60 ksi, the minimum area of steel is thus

$$A_s = 0.00333(bd) = 0.00333(10 \times 18) = 0.6 \text{ in.}^2$$

which is clearly not critical in this case.

Problem 13.3.A A rectangular concrete beam has $f'_c = 3000$ psi and steel with $f_y = 40$ ksi. Select the beam dimensions and reinforcement for a balanced section if the beam sustains a moment due to dead load of 60 kip-ft and a moment due to live load of 90 kip-ft.

TABLE 13.3 Minimum Required Tension Reinforcement for Rectangular Sections[a]

f'_c (psi)	f_y = 40 ksi	f_y = 60 ksi
3000	0.0050	0.00333
4000	0.0050	0.00333
5000	0.0053	0.00354

[a]Required A_s equals table value times *bd* of the beam section.

Problem 13.3.B Same as Problem 13.3.A, except f'_c = 4000 psi, f_y = 60 ksi, Moment due to dead load = 36 kip-ft, and Moment due to live load = 65 kip-ft.

Problem 13.3.C Find the area of steel reinforcement required and select the bars for the beam in Problem 13.3.A if the section dimensions are b = 16 in. and d = 32 in.

Problem 13.3.D Find the area of steel reinforcement required and select the bars for the beam in Problem 13.3.B if the section dimensions are b = 14 in. and d = 25 in.

Use of Beam Tables

Compiling tables for design of concrete beams is complicated by the large number of values for concrete strength and steel yield strength. Limiting these values to those most commonly used may reduce the amount of tabulation; however, the number of possible combinations of beam dimensions is also extensive.

Table 13.4 contains a limited number of beam examples with a range of values for beam width and effective depth. The table uses a single combination of strength values: 4 ksi for concrete strength and 60 ksi for steel yield stress. For each size of beam listed, four different choices for reinforcement are shown. The four choices for reinforcement are based on an assumption of a value for the ratio *a*/*d*. Specific combinations of bars mostly do not conform exactly to the *a*/*d* values but approximate them with practical bar choices. The percentage of steel area corresponding to the *a*/*d* values are bracketed between the requirement for minimum reinforcement in Table 13.3 and the upper limit represented by the balanced section values in Table 13.2.

TABLE 13.4 Factored Moment Resistance of Concrete Beams, φM_r

	Approximate Values for a/d			
	0.1	0.2	0.3	0.4
	Approximate Values for ρ			
$b \times d$ (in.)	0.00567	0.01133	0.0170	0.0227
10 × 14	2 #6	2 #8	3 #8	3 #9
	53	90	126	151
10 × 18	3 #5	2 #9	3 #9	(3 #10)
	68	146	207	247
10 × 22	2 #7	3 #8	(3 #10)	(3 #11)
	113	211	321	371
12 × 16	2 #7	3 #8	4 #8	3 #11
	82	154	193	270
12 × 20	2 #8	3 #9	4 #9	(2 #10 + 2 #11)
	135	243	306	407
12 × 24	2 #8	3 #9	(4 #10)	(4 #11)
	162	292	466	539
15 × 20	3 #7	4 #8	5 #9	(4 #11)
	154	256	382	449
15 × 25	3 #8	4 #9	4 #11	(3 #10 + 3 #11)
	253	405	597	764
15 × 30	3 #8	5 #9	(5 #11)	(3 #10 + 4 #11)
	304	607	895	1085
18 × 24	3 #8	5 #9	6 #10	(6 #11)
	243	486	700	809
18 × 30	3 #9	6 #9	(6 #11)	(8 #11)
	384	729	1074	1348
18 × 36	3 #10	6 #10	(7 #11)	(9 #11)
	566	1110	1504	1819
20 × 30	3 #10	7 #9	6 #11	(9 #11)
	489	850	1074	1516
20 × 35	4 #9	5 #11	(7 #11)	(10 #11)
	598	1106	1462	1966
20 × 40	6 #8	6 #11	(9 #11)	(12 #11)
	810	1516	2148	2696
24 × 32	6 #8	7 #10	(8 #11)	(11 #11)
	648	1152	1527	1977
24 × 40	6 #9	7 #11	(10 #11)	(14 #11)
	1026	1769	2387	3145
24 × 48	5 #10	(8 #11)	(13 #11)	(17 #11)
	1303	2426	3723	4583

Note: Table yields values of factored moment resistance in kip-ft with reinforcement indicated. Reinforcement choices shown in parentheses require greater width of beam or use of two stacked layers of bars. $f_c' = 4$ ksi and $f_y = 60$ ksi

A practical consideration for the amount of reinforcement that can be used in a beam is that of the spacing required as related to the beam width. In Table 13.4 the bar combinations that cannot be accommodated in a single layer are indicated in parentheses.

For each combination of concrete dimensions and bar choices Table 13.4 yields a value for the factored moment resistance of the beam. The following example illustrates a possible use for this table.

Example 3. Using Table 13.4, find acceptable choices of beam dimensions and reinforcement for a factored moment of 1000 kip-ft.

Solution: Possible choices from the table are

15 × 30, 3 No. 10 + 4 No. 11 (requires two layers)
18 × 30, 6 No. 11 (requires two layers)
18 × 36, 6 No. 10
20 × 30, 6 No. 11
20 × 35, 5 No. 11
24 × 32, 7 No. 10
24 × 40, 6 No. 9

Using this range of possibilities, together with other design considerations for the beam size, a quick approximation for the beam design can be determined. A review of the work for Example 1 should indicate the practical use of this easily determined information.

Problems 13.3.E, F Using Table 13.4 find acceptable choices for beam dimensions and reinforcement for a factored moment of, E, 400 kip-ft and, F, 1200 kip-ft.

13.4 BEAMS IN SITE-CAST SYSTEMS

In site-cast construction it is common to cast as much of the total structure as possible in a single, continuous pour. The length of the workday, the size of the available work crew, and other factors may affect this decision. Other considerations involve the nature, size, and form of the structure. For example, a convenient single-cast unit for a multistory building may consist of the whole floor structure for one level if it can be cast in a single workday.

Planning of the concrete construction is itself a major design task. The issue in consideration here is that such work typically results in the achievement of continuous beams and slabs versus the common condition of simple spanning elements in wood and steel construction. The design of continuous-span elements involves more complex investigation for behaviors due to the statically indeterminate nature of internal force resolution involving bending moments, shears, and deflections. For concrete structures, additional complexity results from the need to consider conditions all along the beam length, not just at locations of maximum responses.

In the upper part of Figure 13.6 is shown the condition of a simple-span beam subjected to a uniformly distributed loading. The typical bending moment diagram showing the variation of bending

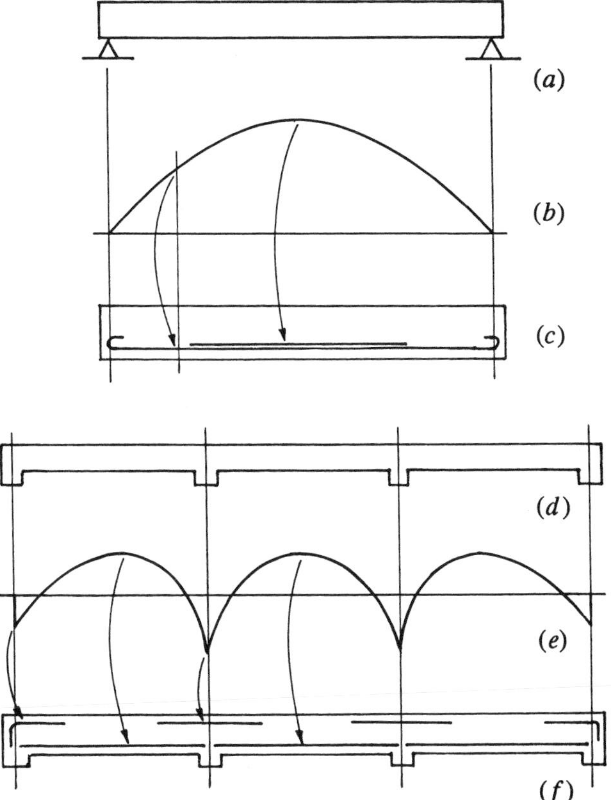

Figure 13.6 Utilization of tension reinforcement in concrete beams.

moment along the beam length takes the form of a parabola, as shown in Figure 13.6*b*. As with a beam of any material, the maximum effort in bending resistance must respond to the maximum value of the bending moment, here occurring at the midspan. For a concrete beam, the concrete section and its reinforcement must be designed for this moment value. However, for a long span and a large beam with a lot of reinforcement, it may be possible to reduce the amount of steel at points nearer to the beam ends. That is, some steel bars may be full length in the beam, while some others are only partial length and occur only in the midspan portion (see Figure 13.6*c*).

Figure 13.6*d* shows the typical situation for a continuous beam in a site-cast slab and beam framing system. For a single uniformly distributed loading the moment diagram takes a form as shown in Figure 13.6*e*, with positive moments near the beam's midspan and negative moments at the supports. Locations for reinforcement that respond to the sign of these moments are shown on the beam elevation in Figure 13.6*f*.

For the continuous beam, separate requirements for the beam's moment resistance must be considered at each of the locations of peak values on the moment diagram. However, there are many additional concerns as well. Principal considerations include the following:

T-Beam Action. At points of positive bending moment (midspan) the beam and slab monolithic construction must be considered to function together, giving a T-shaped form for the portion of the beam section that resists compression.

Use of Compression Reinforcement. If the beam section is designed to resist the maximum bending moment, which occurs at only one point along the beam length, the section will be overstrong for all other locations. For this or other reasons, it may be advisable to use compressive reinforcement to reduce the beam size at the singular locations of maximum bending moment. This commonly occurs at support points and refers to the negative bending moment requiring steel bars in the top of the beam. At these points an easy way to develop compressive reinforcement is to simply extend the bottom steel bars (used primarily for positive moments) through the supports.

Spanning Slabs. Design of site-cast beams must usually be done in conjunction with the design of the slabs that they support. Basic considerations for the slabs are discussed in Section 13.5.

The whole case for the slab–beam site-cast system is discussed in Chapter 20.

Beam Shear. While consideration for bending is a major issue, beams must also be designed for shear effects. This is discussed in Section 13.6. While special reinforcement is typically added for shear resistance, its interaction and coexistence in the beam with the flexural reinforcement must be considered.

Development of Reinforcement. This refers generally to the proper anchorage of the steel bars in the concrete so that their resistance to tension can be developed. At issue are the exact locations of the end cutoffs of the bars and the details such as that for the hooked bars at the discontinuous ends, as shown in Figure 13.6*f*. It also accounts for the hooking of the ends of the bars in the simple beam. The general problems of bar development are discussed in Section 13.7.

T Beams

When a floor slab and its supporting beams are cast at the same time, the result is monolithic construction in which a portion of the slab on each side of the beam serves as the flange of a T beam. The part of the section that projects below the slab is called the web or stem of the T beam. This type of beam is shown in Figure 13.7*a*. For positive moment, the flange is in compression and there is ample concrete to resist compressive stresses, as shown in Figure 13.7*b* or *c*. However, in a continuous beam, there are negative bending moments over the supports, and the flange here is in the tension stress zone with compression in the web. For this situation the beam is assumed to behave essentially as a rectangular section with dimensions b_w and d, as shown in Figure 13.7*d*. This section is also used in determining resistance to shear. The required dimensions of the beam are often determined by the behavior of this rectangular section. What remains for the beam is the determination of the reinforcement required at the midspan where the T-beam action is assumed.

The effective flange width (b_f) to be used in the design of symmetrical T beams is limited to one-fourth the span length of the beam. In addition, the overhanging width of the flange on either side of the web is limited to eight times the thickness of the slab or one-half the clear distance to the next beam.

In monolithic construction with beams and one-way spanning solid slabs, the effective flange area of the T beams is usually quite

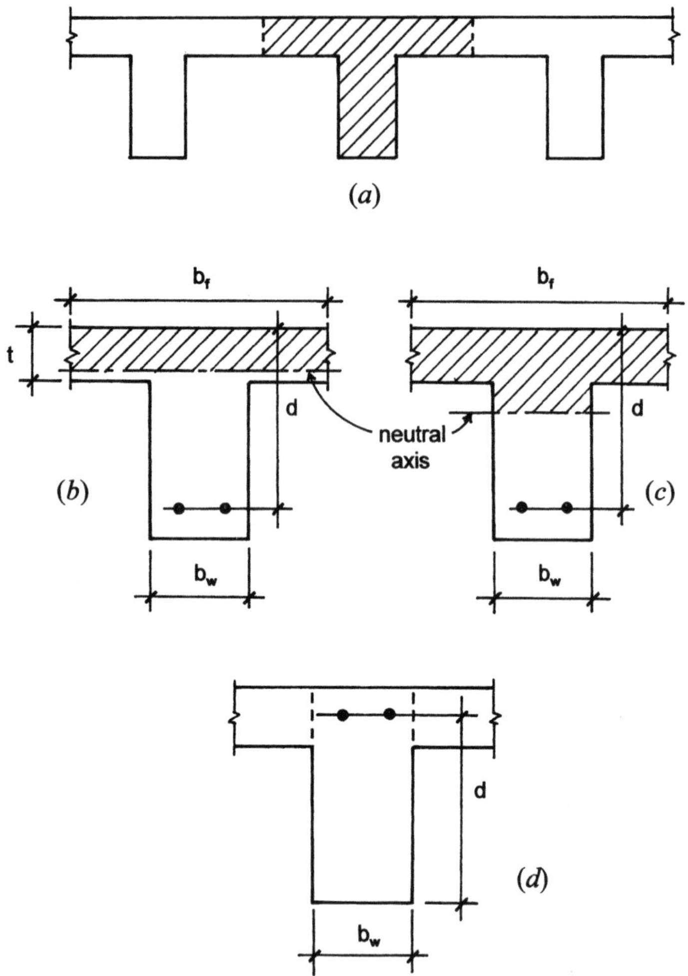

Figure 13.7 Considerations for T beams.

capable of resisting the compressive stresses caused by positive bending moments. With a large flange area, as shown in Figure 13.7*a*, the neutral axis of the section usually occurs quite high in the beam web. If the compression developed in the web is ignored, the net compression force may be considered to be located at the centroid of the trapezoidal stress zone that represents the stress distribution in the flange, and the compression force is located at something less than $t/2$ from the top of the beam.

An approximate analysis of the T section that avoids the need to find the location of the neutral axis and the centroid of the trapezoidal stress zone consists of the following steps:

1. Determine the effective flange width for the T, as previously described.
2. Ignore compression in the web and assume a constant value for compressive stress in the flange (see Figure 13.8). Thus,

$$jd = d - \frac{t}{2}$$

Then, find the required steel area as

$$M_r = \frac{M_u}{0.9} = T(jd) = A_s f_y \left(d - \frac{t}{2}\right)$$

$$A_s = \frac{M_r}{f_y\left(d - \frac{t}{2}\right)}$$

3. Check the compressive stress in the concrete as

$$f_c = \frac{C}{b_f t} \le 0.85 f'_c$$

where

$$C = \frac{M_r}{jd} = \frac{M_r}{d - \frac{t}{2}}$$

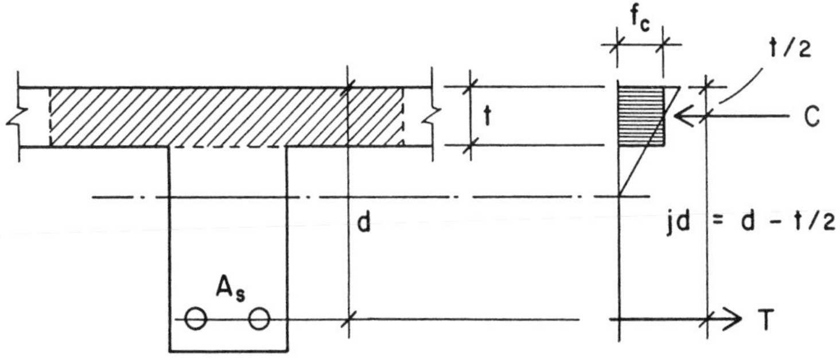

Figure 13.8 Basis for simplified analysis of a T beam.

The value of maximum compressive stress will not be critical if this computed value is significantly less than the limit of $0.85 f'_c$.

4. T beams ordinarily function for positive moments in continuous beams. Since these moments are typically less than those at the beam supports, and the required section is typically derived for the more critical bending at the supports, the T beam is typically considerably underreinforced. This makes it necessary to consider the problem of minimum reinforcement, as discussed for the rectangular section. The ACI Code (Ref. 10) provides special requirements for this for the T beam, for which the minimum area required is defined as the greater value of

$$A_s = \frac{6\sqrt{f'_c}}{F_y}(b_w d)$$

or

$$A_s = \frac{3\sqrt{f'_c}}{F_y}(b_f d)$$

where b_w = width of the beam web
b_f = effective width of the T flange

The following example illustrates the use of this procedure. It assumes a typical design situation in which the dimensions of the section (b_f, b_w, d, and t; see Figure 13.7) are all predetermined by other design considerations and the design of the T section is reduced to the requirement to determine the area of tension reinforcement.

Example 4. A T section is to be used to resist positive moment. The following data are given: beam span = 18 ft, beams are 9 ft center to center, slab thickness is 4 in., beam stem dimensions are $b_w = 15$ in. and $d = 22$ in., $f'_c = 4$ ksi, $f_y = 60$ ksi. Find the required area of steel and select the reinforcing bars for a dead-load moment of 125 kip-ft plus a live-load moment of 100 kip-ft.

Solution: Determine the effective flange width (necessary only for a check on the concrete stress). The maximum value for the flange width is

$$b_f = \frac{\text{span}}{4} = \frac{18 \times 12}{4} = 54 \text{ in.}$$

TABLE 13.5 Options for the T-Beam Reinforcement

Bar Size	No. of Bars	Actual Area Provided (in.^{2})	Width Required[a] (in.)
7	6	3.60	14
8	5	3.95	13
9	4	4.00	12
10	3	3.81	11
11	3	4.68	11

[a]From Table 14.1.

or

$$b_f = \text{center to center beam spacing} = 19 \times 12 = 108 \text{ in.}$$

or

$$b_f = \text{beam stem width} + 16 \times \text{slab thickness} = 15 + (16 \times 4) = 79 \text{ in.}$$

The limiting value is therefore 54 in. Next find the required steel area:

$$M_u = 1.2(125) + 1.6(100) = 310 \text{ kip-ft}$$

$$M_r = \frac{M_u}{0.9} = \frac{310}{0.9} = 344 \text{ kip-ft}$$

$$A_s = \frac{M_r}{f_y\left(d - \frac{t}{2}\right)} = \frac{344 \times 12}{60\left(22 - \frac{4}{2}\right)} = 3.44 \text{ in.}^2$$

Select bars using Table 13.5, which incorporates consideration for the adequacy of the stem width. From the table choose four No. 9 bars, actual $A_s = 4.00$ in.^{2} From Table 13.5, the required width for four No. 9 bars is 12 in., less than the 15 in. provided.

Check the concrete stress:

$$C = \frac{M_r}{jd} = \frac{344 \times 12}{20} = 206.4 \text{ kips}$$

$$f_c = \frac{C}{b_f t} = \frac{206.4}{54 \times 4} = 0.956 \text{ ksi}$$

Compare this to the limiting stress:

$$0.85 f'_c = 0.85(4) = 3.4 \text{ ksi}$$

Thus, compressive stress in the flange is clearly not critical.

Using the beam stem width of 15 in. and the effective flange width of 54 in., the minimum area of reinforcement is determined as the greater of

$$A_s = \frac{6\sqrt{f_c'}}{F_y}(b_w d) = \frac{6\sqrt{4000}}{60{,}000}(15 \times 22) = 2.09 \text{ in.}^2$$

or

$$A_s = \frac{3\sqrt{f_c'}}{F_y}(b_f d) = \frac{3\sqrt{4000}}{60{,}000}(54 \times 22) = 3.76 \text{ in.}^2$$

The minimum area required is thus greater than the computed area of 3.44 in.^{2} The choice of using four No. 9 bars is not affected by this development since its area is 4.00 in.^{2} Some of the other data in Table 13.5 must be adjusted if other bar choices are considered.

The examples in this section illustrate procedures that are reasonably adequate for beams that occur in ordinary beam and slab construction. When special T sections occur with thin flanges (t less than $d/8$ or so), these methods may not be valid. In such cases more accurate investigation should be performed using the requirements of the ACI Code.

Problem 13.4.A Find the area of steel reinforcement required for a concrete T beam for the following data: $f_c' = 3$ ksi, $f_y = 50$ ksi, $d = 28$ in., $t = 6$ in., $b_w = 16$ in., $b_f = 60$ in., and the section sustains a factored bending moment of $M_u = 360$ kip-ft.

Problem 13.4.B Same as Problem 13.4.A, except $f_c' = 4$ ksi, $f_y = 60$ ksi, $d = 32$ in., $t = 5$ in., $b_w = 18$ in., $b_f = 54$ in., $M_u = 500$ kip-ft.

Beams with Compression Reinforcement

There are many situations in which steel reinforcement is used on both sides of the neutral axis in a beam. When this occurs, the steel on one side of the axis will be in tension and that on the other side in compression. Such a beam is referred to as a double-reinforced beam or simply as a beam with compressive reinforcement (it being assumed that there is also tensile reinforcement). Various situations involving such reinforcement have been discussed in the preceding sections. In summary, the most common occasions for such reinforcement include:

1. The desired resisting moment for the beam exceeds that for which the concrete alone is capable of developing the necessary compressive force.
2. Other functions of the section require the use of reinforcement on both sides of the beam. These include the need for bars to support U-stirrups and situations when torsion is a major concern.
3. It is desired to reduce deflections by increasing the stiffness of the compressive side of the beam. This is most significant for reduction of long-term creep deflections.
4. The combination of loading conditions on the structure results in reversal moments on the section at a single location; that is, the section must sometimes resist positive moment and other times resist negative moment.
5. Anchorage requirements (for development of reinforcement) require that the bottom bars in a beam be extended a significant distance into the supports.

The precise investigation and accurate design of doubly reinforced sections, whether performed by the working stress or by strength design methods, are quite complex and are beyond the scope of this book. The following discussion presents an approximation method that is adequate for the preliminary design of a doubly reinforced section. For real design situations, this method may be used to establish a first trial design, which should be precisely investigated using more rigorous methods.

For the beam with double reinforcement, as shown in Figure 13.9*a*, consider the total resisting moment for the section to be the sum of the following two component moments:

M_1 (Figure 13.9*b*) is comprised of a section with tension reinforcement only (A_{s1}). This section is subject to the usual procedures for design and investigation, as discussed in Section 13.3.

M_2 (Figure 13.9*c*) is comprised of two opposed steel areas (A_{s2} and A_s') that function in simple moment couple action, similar to the flanges of a steel beam or the top and bottom chords of a truss.

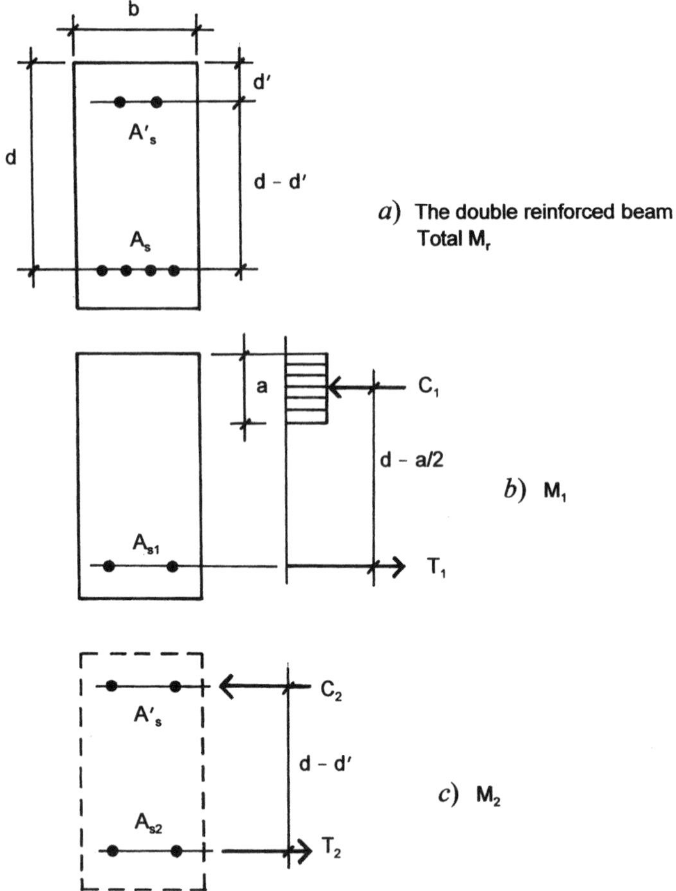

Figure 13.9 Basis for simplified analysis of a double reinforced beam.

The limit for M_1 is the so-called balanced moment, as described in Section 13.3. Given the values for steel yield stress and the specified strength of the concrete, the factors for definition of the properties of this balanced section can be obtained from Table 13.2. Given the dimensions for the concrete section, b and d, the limiting moment resistance for the section can be determined using the value of R from the table. If the capacity of the section as thus determined is less than the required factored moment (M_r), compressive reinforcing is required. If the balanced resistance is larger than the required factored moment, compressive reinforcement is not required. Although not actually required, the compressive reinforcement may be used for any of the reasons previously mentioned.

When M_1 is less than the factored required moment, it may be determined using the balanced value for R in Table 13.2. Then the required value for M_2 may be determined as $M_2 = M_r - M_1$. This is seldom the case in design work, as the amount of tensile reinforcement required to achieve the balanced moment capacity of the section is usually not practical to be fit into the section.

When the potential balanced value for M_1 is greater than the required factored moment, the properties given in Table 13.2 may be used for an approximation of the value for the required steel area defined as A_{s1}. For the true value of the required M_1 the table values may be adjusted as follows:

1. The actual value for a/d will be smaller than the table value.
2. The actual value for ρ will be less than the table value.

The approximation procedure that follows starts with the assumption that the concrete dimensions describe a section with the potential M_r greater than that required. The first step of the procedure is therefore to establish what is actually an arbitrary amount of compressive reinforcement. Using this reinforcement as A_{s2}, a value for M_2 is determined. The value for M_1 is then found by subtracting M_2 from the required moment for the section.

Ordinarily, we expect that $A_{s2} = A'_s$ since the same grade of steel is usually used for both. However, there are two special considerations that must be made. The first involves the fact that A_{s2} is in tension, while A'_s is in compression. Therefore, A'_s must be dealt with in a manner similar to that for column reinforcement. This requires, among other things, that the compressive reinforcement be braced against buckling using ties similar to those in a tied column.

The second consideration for A'_s involves the distribution of stress and strain on the section. Referring to Figure 13.2*a*, it may be observed that, under normal circumstances, A'_s will be closer to the neutral axis than A_{s2}. Thus, the stress in A'_s will be lower than that in A_{s2} if pure elastic conditions are assumed. It is common practice to assume steel to be doubly stiff when sharing stress with concrete in compression due to shrinkage and creep effects. Thus, in translating from linear strain conditions to stress distribution, use the relation $f_c = f_s'/2n$ (where $n = E_s/E_c$).

For the approximate method it is really not necessary to find separate values for A_{s1} and A_{s2}. This is because of an additional approximation assumption that the value for a is $2 \times d'$. Thus, the moment arm for both A_{s1} and A_{s2} is the same, and the value for the total tension reinforcement can be simply determined as

$$A_s = \frac{\text{required } M_r}{f_y(d - d')}$$

This value for A_s can actually be determined as soon as the values for d and d' are established.

With the total tension reinforcement established, the next step involves the determination of A_s' and A_{s2}. Compression reinforcement in beams ordinarily ranges from 0.2 to 0.4 times the total tension reinforcement. For this approximation method we will determine the area to be $0.3 \times A_s$. We will also assume the stress in the compression reinforcement to be one-half of the yield stress. This permits a definition of the resisting moment M_2 as follows:

$$M_2 = A_s'\left(\frac{f_y}{2}\right)(d - d') = 0.3A_s\left(\frac{f_y}{2}\right)(d - d')$$
$$= 0.15A_s f_y(d - d')$$

Using this moment the amount of the tension reinforcement that is required for the development of the steel force couple is twice A_s', and the amount of the tension steel devoted to development of M_2 is found by subtracting this from the total tension reinforcement.

For a final step, the value of A_{s1} may be used to compute a value for the percentage of steel relating to the moment resistance of the section without the compression reinforcement. If this percentage is less than that listed in Table 13.2, the concrete stress will not be critical. However, this situation is predetermined if the total potential balanced resisting moment is determined first and compared to the required factored resisting moment.

For the section that is larger than the balanced section defined by Table 13.2 the procedure can be shortened. The following example will serve to illustrate this procedure.

Example 5. A concrete section with $b = 18$ in. and $d = 21.5$ in. is required to resist service load moments as follows: dead-load moment = 175 kip-ft, live-load moment = 160 kip-ft. Using strength methods, find the required reinforcement. Use $f'_c = 3$ ksi and $f_y = 60$ ksi.

Solution: Using Table 13.2, find $a/d = 0.592$, $\rho = 0.0252$, and $R =$ 1.063 in kip-in.

The factored moment for the section is

$$M_u = 1.2(175) + 1.6(160) = 466 \text{ kip-ft}$$

and the required factored resisting moment is

$$M_r = \frac{M_u}{0.9} = \frac{466}{0.9} = 518 \text{ kip-ft}$$

Using the R value for the balanced section, the maximum resisting moment of the section is

$$M_b = Rbd^2 = \frac{1.063}{12}(18)(21.5)^2 = 737 \text{ kip-ft}$$

As this is considerably larger than the required resisting moment, the section is qualified as "underbalanced," that is, it will be understressed as relates to the compression resistance of the section. It is reasonable, therefore, to use the simplified formula for the tension reinforcing; thus,

$$A_s = \frac{M_r}{f_y(d - d')} = \frac{518 \times 12}{60(21.5 - 2.5)} = 5.45 \text{ in.}^2$$

And a reasonable assumption for the compressive reinforcement is

$$A_s' = 0.3A_s = 0.3(5.45) = 1.63 \text{ in.}^2$$

Bar combinations may next be found for these two steel areas.

If the situation exists in which it is desired to use a given concrete section for a resisting moment that exceeds the balanced limit described by the values in Table 13.2, a different procedure is required. For this case the first two steps are the same as in the preceding example: determining the required resisting moment M_r and the limiting balanced moment M_b. The tension reinforcement required for the balanced moment can be determined with the balanced percentage p from Table 13.2. Then M_b becomes the moment M_1 as shown in Figure 13.9*b*. As shown in Figure 13.9*c*, M_2 is determined as the difference between M_r and M_b. The compression reinforcement and the additional tension reinforcement required for M_2 can then be determined. The following example illustrates this procedure.

Example 6. Find the reinforcement required for the beam in Example 4 if the required resisting moment M_r is 900 kip-ft.

Solution: The first step is the determination of the limiting balanced moment M_b. For this section this value was computed in Example 4 as 737 kip-ft. As the required moment exceeds this value, compression reinforcement is required and the moment for this is determined as

$$M_2 = M_r - M_b = 900 - 737 = 163 \text{ kip-ft}$$

For the balanced moment the required tension reinforcement can be computed using the balanced p from Table 13.2. Thus,

$$A_{s1} = \rho bd = 0.0252 \times 18 \times 21.5 = 9.75 \text{ in.}^2$$

For the determination of the tension reinforcement required for M_2, the procedure involves the use of the moment arm $d - d'$. Thus,

$$A_{s2} = \frac{M_2}{f_y(d - d')} = \frac{163 \times 12}{60(19)} = 1.72 \text{ in.}^2$$

and the total required tension reinforcing is

$$A_s = 9.75 + 1.72 = 11.5 \text{ in.}^2$$

For the compressive reinforcement assume a stress approximately equal to one-half the yield stress. Thus, the area of steel required is twice the value of A_{s2}, $2(1.72) = 3.44 \text{ in.}^2$ This requirement can be met with three No. 10 bars providing an area of 3.81 in.^2

Options for the tension reinforcement are given in Table 13.6. As the table indicates, it is not possible to get the bars into the 18-in.-wide beam by placing them in a single layer. Options are to use two layers of bars or to increase the beam width. Frankly, this is not a good beam

TABLE 13.6 Options for the Tension Reinforcement—Example 5

Bar Size	Area of One Bar (in.^{2})	No. of Bars	Actual Area Provided (in.^{2})	Width Required[a] (in.)
8	0.79	15	11.85	33
9	1.00	12	12.00	30
10	1.27	10	12.70	28
11	1.56	8	12.48	25

[a]Using data from Table 14.1.

design and would most likely not be acceptable unless extreme circumstances force the use of the limited beam size. This is pretty much the typical situation for beams required to develop resisting moments larger than the balanced moment.

Problem 13.4.C A concrete section with $b = 16$ in. and $d = 19.5$ in. is required to develop a bending moment strength of $M_r = 400$ kip-ft. Use of compressive reinforcement is desired. Find the required reinforcement. Use $f'_c = 4$ ksi and $f_y = 60$ ksi.

Problem 13.4.D Same as Problem 13.4.C, except required $M_r = 1000$ kip-ft, $b = 20$ in., $d = 27$ in.

Problem 13.4.E Same as Problem 13.4.C except $M_r = 640$ kip-ft.

Problem 13.4.F Same as Problem 13.4.D except $M_r = 1400$ kip-ft.

13.5 SPANNING SLABS

Concrete slabs are frequently used as spanning roof or floor decks, often occurring in monolithic, cast-in-place slab and beam framing systems. There are generally two basic types of slabs: one-way spanning and two-way spanning. The spanning condition is not so much determined by the slab as by its support conditions. As part of a general framing system, the one-way spanning slab is discussed in Section 14.1. The following discussion relates to the design of one-way solid slabs using procedures developed for the design of rectangular beams.

Solid slabs are usually designed by considering the slab to consist of a series of 12-in.-wide planks. Thus, the procedure consists of simply designing a beam section with a predetermined width of 12 in. Once the depth of the slab is established, the required area of steel is determined, specified as the number of square inches of steel required per foot of slab width.

Reinforcing bars are selected from a limited range of sizes, appropriate to the slab thickness. For thin slabs (4–6-in.-thick) bars may be of a size from No. 3 to No. 6 or so (nominal diameters from 3/8 to 3/4 in.). The bar size selection is related to the bar spacing, the combination resulting in the amount of reinforcing in terms of an average amount of square inches of steel per one foot unit of slab width. Spacing is limited by code regulation to a maximum of three times the slab thickness. There is no minimum spacing, other than that required for proper placing of the concrete; however, a very close spacing indicates a very large number of bars, making for laborious installation.

Every slab must be provided with two-way reinforcement, regardless of its structural functions. This is partly to satisfy requirements for shrinkage and temperature effects. The amount of this minimum reinforcement is specified as a percentage ρ of the gross cross-sectional area of the concrete as follows:

1. For slabs reinforced with grade 40 or grade 50 bars:

$$\rho = \frac{A_s}{bt} = 0.002\ (0.2\%)$$

2. For slabs reinforced with grade 60 bars:

$$p = \frac{A_s}{bt} = 0.0018\ (0.18\%) \quad \rho = \frac{A_s}{bt} = 0.0018\ (0.18\%)$$

Center-to-center spacing of this minimum reinforcement must not be greater than five times the slab thickness or 18 in.

Minimum cover for slab reinforcement is normally 3/4 in., although exposure conditions or need for a high fire rating may require additional cover. For a thin slab reinforced with large bars, there will be a considerable difference between the slab thickness t and the effective depth d, as shown in Figure 13.10. Thus, the practical efficiency of the slab in flexural resistance decreases rapidly as the slab thickness is decreased. For this and other reasons, very thin slabs (less than 4 in. thick) are often reinforced with wire fabric rather than sets of loose bars.

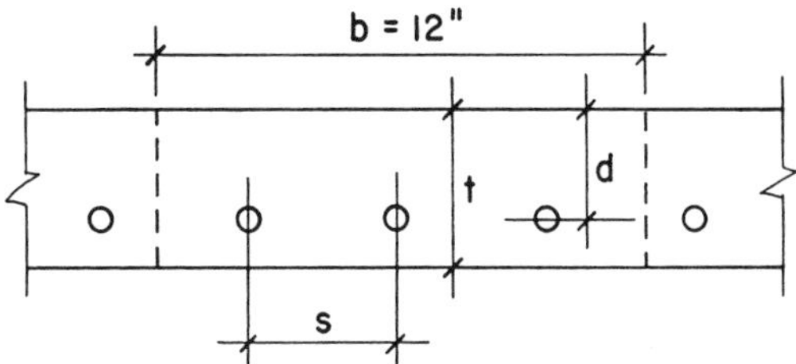

Figure 13.10 Reference for slab design.

Shear reinforcement is seldom used in one-way slabs, and consequently, the maximum unit shear stress in the concrete must be kept within the limit for the concrete without reinforcement. This is usually not a concern, as unit shear is usually low in one-way slabs, except for exceptionally high loadings.

Table 13.7 gives data that are useful in slab design, as demonstrated in the following example. Table values indicate the average amount of

TABLE 13.7 Areas Provided by Spaced Reinforcement

Bar Spacing (in.)	Area Provided (in.^{2}/ft width)									
	No. 2	No. 3	No. 4	No. 5	No. 6	No. 7	No. 8	No. 9	No. 10	No. 11
3	0.20	0.44	0.80	1.24	1.76	2.40	3.16	4.00		
3.5	0.17	0.38	0.69	1.06	1.51	2.06	2.71	3.43	4.35	
4	0.15	0.33	0.60	0.93	1.32	1.80	2.37	3.00	3.81	4.68
4.5	0.13	0.29	0.53	0.83	1.17	1.60	2.11	2.67	3.39	4.16
5	0.12	0.26	0.48	0.74	1.06	1.44	1.89	2.40	3.05	3.74
5.5	0.11	0.24	0.44	0.68	0.96	1.31	1.72	2.18	2.77	3.40
6	0.10	0.22	0.40	0.62	0.88	1.20	1.58	2.00	2.54	3.12
7	0.08	0.19	0.34	0.53	0.75	1.03	1.35	1.71	2.18	2.67
8	0.07	0.16	0.30	0.46	0.66	0.90	1.18	1.50	1.90	2.34
9	0.07	0.15	0.27	0.41	0.59	0.80	1.05	1.33	1.69	2.08
10	0.06	0.13	0.24	0.37	0.53	0.72	0.95	1.20	1.52	1.87
11	0.05	0.12	0.22	0.34	0.48	0.65	0.86	1.09	1.38	1.70
12	0.05	0.11	0.20	0.31	0.44	0.60	0.79	1.00	1.27	1.56
13	0.05	0.10	0.18	0.29	0.40	0.55	0.73	0.92	1.17	1.44
14	0.04	0.09	0.17	0.27	0.38	0.51	0.68	0.86	1.09	1.34
15	0.04	0.09	0.16	0.25	0.35	0.48	0.63	0.80	1.01	1.25
16	0.04	0.08	0.15	0.23	0.33	0.45	0.59	0.75	0.95	1.17
18	0.03	0.07	0.13	0.21	0.29	0.40	0.53	0.67	0.85	1.04
24	0.02	0.05	0.10	0.15	0.22	0.30	0.39	0.50	0.63	0.78

steel area per foot of slab width provided by various combinations of bar size and spacing. Table entries are determined as follows:

$$A_s/\text{ft} = (\text{single bar area})\frac{12}{\text{bar spacing}}$$

Thus, for No. 5 bars at 8-in. centers,

$$A_s/\text{ft} = (0.31)\left(\frac{12}{8}\right) = 0.465 \text{ in.}^2/\text{ft}$$

It may be observed that the table entry for this combination is rounded off to a value of 0.46 in.^{2}/ft.

Example 7. A one-way solid concrete slab is to be used for a simple span of 14 ft. In addition to its own weight, the slab carries a superimposed dead load of 30 psf plus a live load of 100 psf. Using $f'_c = 3$ ksi and $f_y = 40$ ksi, design the slab for minimum overall thickness.

Solution: Using the general procedure for design of a beam with a rectangular section (Section 13.2), we first determine the required slab thickness. Thus, for deflection, from Table 13.11,

$$\text{Minimum } t = \frac{L}{25} = \frac{14 \times 12}{25} = 6.72 \text{ in.}$$

For flexure, first determine the maximum bending moment. The loading must include the weight of the slab, for which the thickness required for deflection may be used as a first estimate. Assuming a 7-in.-thick slab, the slab weight is (7/12) (150 pcf) = 87.5 psf, say 88 psf, and the total dead load is 30 + 88 = 118 psf. The factored load is thus

$$w_u = 1.2(\text{dead load}) + 1.6(\text{live load}) = 1.2(118) + 1.6(100) = 302 \text{ psf}$$

The maximum bending moment on a 12-in.-wide strip of the slab thus becomes

$$M_u = \frac{wl^2}{8} = \frac{302(14)^2}{8} = 7400 \text{ ft-lb}$$

and the required factored resisting moment is

$$M_r = \frac{7400}{0.9} = 8220 \text{ ft-lb}$$

For a minimum slab thickness, we consider the use of a balanced section, for which Table 13.2 yields the following properties: $a/d =$ 0.685 and $R = 1.149$ (in kip and inch units). Then the minimum value for bd^2 is

$$bd^2 = \frac{M_r}{R} = \frac{8220 \times 12}{1.149} = 85.9 \text{ in.}^3$$

and, since b is the 12-in. design strip width,

$$d = \sqrt{\frac{85.9}{12}} = 2.68 \text{ in.}$$

Assuming an average bar size of a No. 6 (3/4 in. nominal diameter) and cover of 3/4 in., the minimum required slab thickness based on flexure becomes

$$t = d + \frac{\text{bar diameter}}{2} + \text{cover} = 2.68 + \frac{1.75}{2} + 0.75 = 3.8 \text{ in.}$$

The deflection limitation thus controls in this situation, and the minimum overall thickness is the 6.72-in. dimension. Staying with the 7-in. overall thickness, the actual effective depth with a No. 6 bar will be

$$d = 7.0 - 1.125 = 5.875 \text{ in.}$$

Since this d is larger than that required for a balanced section, the value for a/d will be slightly smaller than 0.685, as found from Table 13.2. Assume a value of 0.4 for a/d and determine the required area of reinforcement as follows:

$$a = 0.4d = 0.4(5.875) = 2.35 \text{ in.}$$

$$A_s = \frac{M}{f_y\left(d - \frac{a}{2}\right)} = \frac{8.22 \times 12}{40\left(5.875 - \frac{2.35}{2}\right)} = 0.525 \text{ in.}^2$$

Using data from Table 13.7, the optional bar combinations shown in Table 13.8 will satisfy this requirement. Note that for bars larger than the assumed No. 6 bar (0.75 in. diameter), d will be slightly less and the required area of reinforcement slightly higher.

The ACI Code (Ref. 10) permits a maximum center-to-center bar spacing of three times the slab thickness (21 in. in this case) or 18 in.,

TABLE 13.8 Alternatives for the Slab Reinforcement

Bar Size	Spacing of Bars Center to Center (in.)	Average A_s in a 12-in. Width
5	7	0.53
6	10	0.53
7	13	0.55
8	18	0.53

whichever is smaller. Minimum spacing is largely a matter of the designer's judgment. Many designers consider a minimum practical spacing to be one approximately equal to the slab thickness. Within these limits, any of the bar size and spacing combinations listed are adequate.

As described previously, the ACI Code (Ref. 10) requires a minimum reinforcement for shrinkage and temperature effects to be placed in the direction perpendicular to the flexural reinforcement. With the grade 40 bars in this example, the minimum percentage of this steel (ρ) is 0.0020, and the steel area required for a 12-in. strip thus becomes

$$A_s = \rho(bt) = 0.0020(12 \times 7) = 0.168 \text{ in.}^2/\text{ft}$$

From Table 13.7, this requirement can be satisfied with No. 3 bars at 7-in. centers or No. 4 bars at 14-in. centers. Both of these spacings are well below the maximum of five times the slab thickness (35 in.) or 18 in.

Although simply supported single slabs are sometimes encountered, the majority of slabs used in building construction are continuous through multiple spans. An example of the design of such a slab is given in Chapter 20.

Problem 13.5.A A one-way solid concrete slab is to be used for a simple span of 16 ft. In addition to its own weight, the slab carries a superimposed dead load of 40 psf and a live load of 100 psf. Using the strength method with $f'_c = 3$ ksi and $f_y = 40$ ksi, design the slab for minimum overall thickness.

Problem 13.5.B Same as Problem 13.5.A, except span = 18 ft, superimposed dead load = 50 psf, live load = 75 psf, $f'_c = 4$ ksi, $f_y = 60$ ksi.

13.6 SHEAR IN BEAMS

From general consideration of shear effects, as developed in the science of mechanics of materials, the following observations can be made:

1. Shear is an ever-present phenomenon, produced directly by slicing actions, by lateral loading in beams, and on oblique sections in tension and compression members.
2. Shear forces produce shear stress in the plane of the force and equal unit shear stresses in planes that are perpendicular to the shear force.
3. Diagonal stresses of tension and compression, having magnitudes equal to that of the shear stress, are produced in directions of 45° from the plane of the shear force.
4. Direct slicing shear force produces a constant magnitude shear stress on affected sections, but beam shear action produces shear stress that varies on the affected sections, having magnitude of zero at the edges of the section and a maximum value at the centroidal neutral axis of the section.

In the discussions that follow it is assumed that the reader has a general familiarity with these relationships.

Consider the case of a simple beam with uniformly distributed load and end supports that provide only vertical resistance (no moment restraint). The distribution of internal shear and bending moment are as shown in Figure 13.11*a*. For flexural resistance, it is necessary to provide longitudinal reinforcing bars near the bottom of the beam. These bars are oriented for primary effectiveness in resistance to tension stresses that develop on a vertical (90°) plane (which is the case at the center of the span, where the bending moment is maximum and the shear approaches zero).

Under the combined effects of shear and bending, the beam tends to develop tension cracks as shown in Figure 13.11*b*. Near the center of the span, where the bending is predominant and the shear approaches zero, these cracks approach 90°. Near the support, however, where the shear predominates and bending approaches zero, the critical tension stress plane approaches 45°, and the horizontal bars are only partly effective in resisting the cracking.

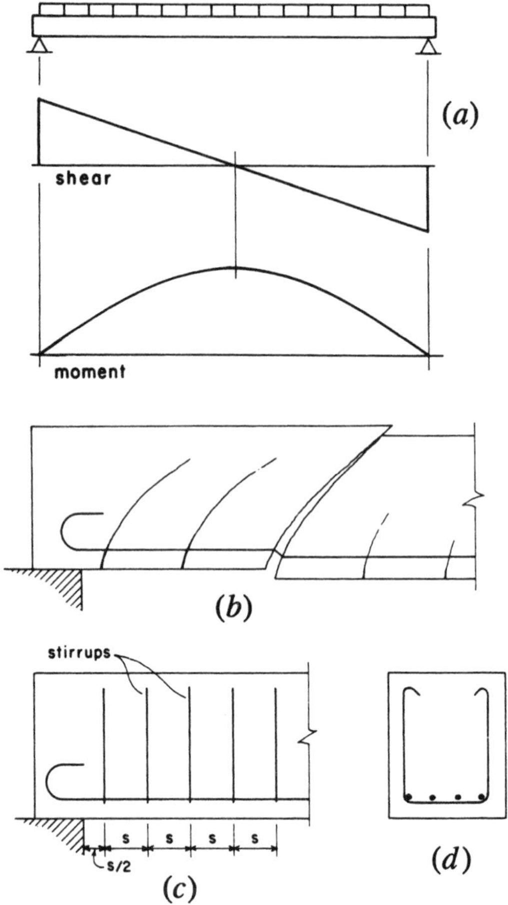

Figure 13.11 Considerations for shear in concrete beams.

Shear Reinforcement for Beams

For beams, the most common form of added shear reinforcement consists of a series of U-shaped bent bars (Figure 13.11*d*) placed vertically and spaced along the beam span, as shown in Figure 13.11*c*. These bars, called *stirrups*, are intended to provide a vertical component of resistance, working in conjunction with the horizontal resistance provided by the flexural reinforcement. In order to develop flexural tension near the support face, the horizontal bars must be bonded to the concrete beyond the point where the stress is developed. Where the ends of simple beams extend only a short distance over the support (a common situation), it is often necessary to bend or hook the bars as shown in Figure 13.11*c*.

The simple span beam and the rectangular section shown in Figure 13.11*d* occur only infrequently in building structures. The most common case is that of the beam section shown in Figure 13.12, which occurs when a beam is cast continuously with a supported concrete slab. In addition, these beams normally occur in continuous spans with negative moments at the supports. Thus, the stress in the beam near the support is as shown in Figure 13.12*a*, with the negative moment producing compressive flexural stress in the bottom of the beam stem. This is substantially different from the case of the simple beam, where the moment approaches zero near the support.

For the purpose of shear resistance, the continuous, T-shaped beam is considered to consist of the section indicated in Figure 13.12*b*. The effect of the slab is ignored, and the section is considered to be a simple rectangular one. Thus, for shear design, there is little difference between the simple span beam and the continuous beam, except for the effect of the continuity on the distribution of shear along the beam span. It is important, however, to understand the relationships between shear and moment in the continuous beam.

Figure 13.13 illustrates the typical condition for an interior span of a continuous beam with uniformly distributed load. Referring to the portions of the beam span numbered 1, 2, and 3, note the following:

Zone 1: In this zone the high negative moment requires major flexural reinforcement consisting of horizontal bars near the top of the beam.

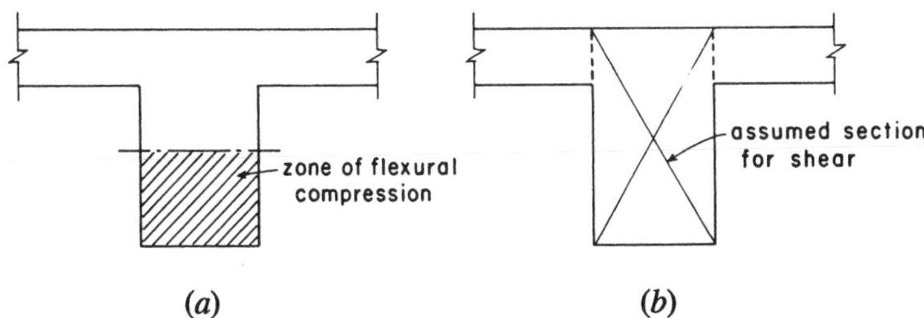

Figure 13.12 Development of negative bending and shear in concrete T beams.

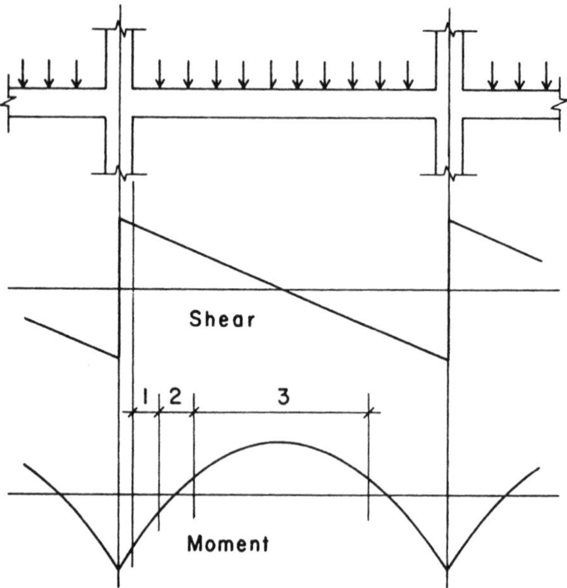

Figure 13.13 Shear and bending in continuous beams.

Zone 2: In this zone, the moment reverses sign; moment magnitudes are low; and, if shear stress is high, the design for shear is a predominant concern.

Zone 3: In this zone, shear consideration is minor and the predominant concern is for positive moment requiring major flexural reinforcement in the bottom of the beam.

Vertical U-shaped stirrups, similar to those shown in Figure 13.14*a*, may be used in the T-shaped beam. An alternate detail for the U-shaped stirrup is shown in Figure 13.14*b*, in which the top hooks are turned outward; this makes it possible to spread the negative-moment reinforcing bars to make placing of the concrete somewhat easier. Figures 13.14*c* and *d* show possibilities for stirrups in L-shaped beams that occur at the edges of large openings or at the outside edge of the structure. This form of stirrup is used to enhance the torsional resistance of the section and also assists in developing the negative-moment resistance in the slab at the edge of the beam.

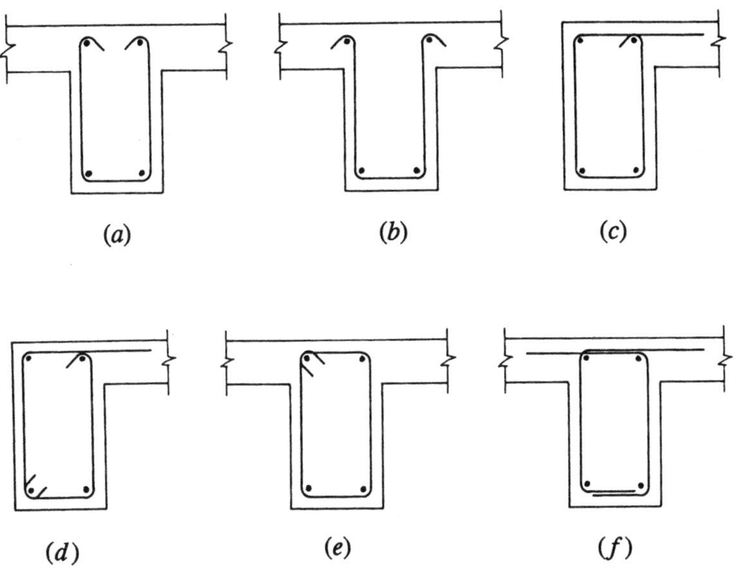

Figure 13.14 Forms for vertical stirrups.

So-called *closed stirrups*, similar to ties in columns, are sometimes used for T- and L-shaped beams, as shown in Figures 13.14*c–f*. These are generally used to improve the torsional resistance of the beam section.

Stirrup forms are often modified by designers or by the reinforcing fabricator's detailers to simplify the fabrication and/or the field installation. The stirrups shown in Figures 13.14*d* and *f* are two such modifications of the basic details in Figures 13.14*c* and *e*, respectively.

The following are considerations and code requirements that apply to design for beam shear:

Concrete Capacity. Whereas the tensile strength of the concrete is ignored in design for flexure, the concrete is assumed to take some portion of the shear in beams. If the capacity of the concrete is not exceeded—as is sometimes the case for lightly loaded beams—there may be no need for reinforcement. The typical case, however, is as shown in Figure 13.15, where the maximum shear V exceeds the capacity of the concrete alone (V_c), and the steel reinforcement is required to absorb the excess, indicated as the shaded portion in the shear diagram.

Minimum Shear Reinforcement. Even when the maximum computed shear stress falls below the capacity of the concrete, the present code requires the use of some minimum amount of shear reinforcement. Exceptions are made in some situations, such as for slabs and very shallow beams. The objective is essentially to toughen the structure with a small investment in additional reinforcement.

Type of Stirrup. The most common stirrups are the simple U-shape or closed forms shown in Figure 13.14, placed in a vertical position at intervals along the beam. It is also possible to place stirrups at an incline (usually 45°), which makes them somewhat more effective in direct resistance to the potential shear cracking near the beam ends (see Figure 13.11*b*). In large beams with high unit shear stress, both vertical and inclined stirrups are sometimes used at the location of the greatest shear.

Size of Stirrups. For beams of moderate size, the most common size for U-stirrups is a No. 3 bar. These bars can be bent relatively tightly at the corners (small radius of bend) in order to fit within the beam section. For larger beams, a No. 4 bar is sometimes used, its strength (as a function of its cross-sectional area) being almost twice that of a No. 3 bar.

Spacing of Stirrups. Stirrup spacings are computed (as discussed in the following sections) on the basis of the amount of reinforcing required for the unit shear stress at the location of the stirrups. A maximum spacing of $d/2$ (i.e., one-half the effective beam depth d) is specified in order to assure that at least one stirrup occurs at the location of any potential diagonal crack (see Figure 13.11*b*). When shear stress is excessive, the maximum spacing is limited to $d/4$.

Critical Maximum Design Shear. Although the actual maximum shear value occurs at the end of the beam, the code permits the use of the shear stress at a distance of d (effective beam depth) from the beam end as the critical maximum for stirrup design. Thus, as shown in Figure 13.16, the shear requiring reinforcement is slightly different from that shown in Figure 13.15.

Total Length for Shear Reinforcement. On the basis of computed shear forces, reinforcement must be provided along the beam length for the distance defined by the shaded portion of the

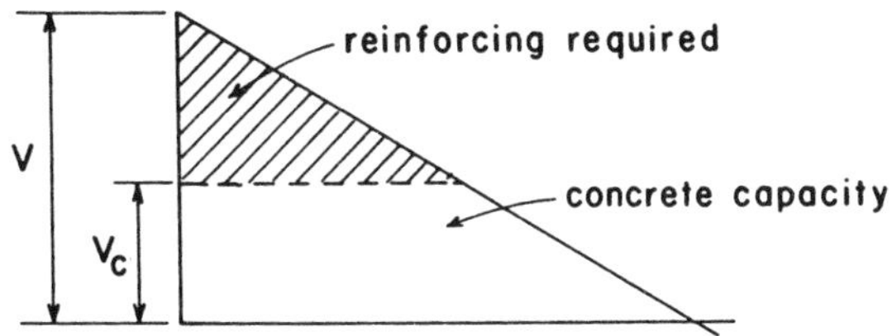

Figure 13.15 Sharing of shear resistance in reinforced concrete beams.

shear stress diagram shown in Figure 13.16. For the center portion of the span, the concrete is theoretically capable of the necessary shear resistance without the assistance of reinforcement. However, the code requires that some shear reinforcement be provided for a distance beyond this computed cutoff point. Earlier codes required that stirrups be provided for a distance equal to the effective depth of the beam beyond the computed cutoff point. Currently, codes require that minimum shear reinforcement be provided as long as the computed shear force exceeds one-half of the capacity of the concrete ($\phi \times V_c/2$). However it is established, the total extended range over which reinforcement must be provided is indicated as *R* on Figure 13.16.

Design for Beam Shear

The following is a description of a procedure for design of shear reinforcement for beams that are experiencing flexural and shear stresses exclusively.

The ultimate shear force (V_u) at any cross section along a given beam due to factored loading must be less than the reduced shear capacity at the section. Mathematically, this is represented as

$$V_u \leq \phi_v \times (V_c + V_s)$$

where V_u = ultimate shear force at the section
φ_v = 0.75 for shear
V_c = shear capacity of the concrete
V_s = shear capacity of the steel reinforcement

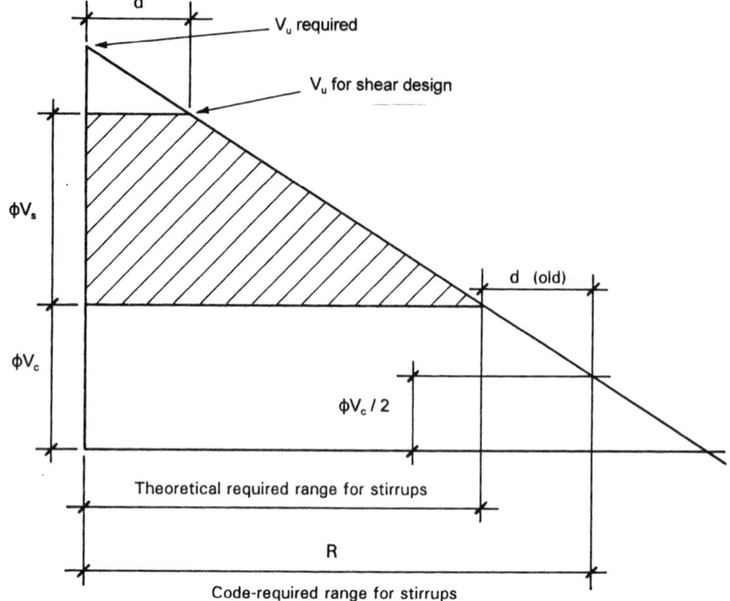

Figure 13.16 Shear stress analysis: ACI Code (Ref. 10) requirements.

For beams of normal-weight concrete, subjected only to flexure and shear, shear force in the concrete is limited to

$$V_c = 2\sqrt{f_c'}bd$$

where f_c' = specified strength of the concrete in psi
b = width of the cross section
d = effective depth of the cross section

When V_u exceeds the limit for (φ_v V_c), reinforcing must be provided, complying with the general requirements discussed previously. Thus,

$$V_s \geq \frac{V_u}{\phi_v} - V_c$$

Required spacing of shear reinforcement is determined as follows. Referring to Figure 13.17, note that the capacity in tensile resistance

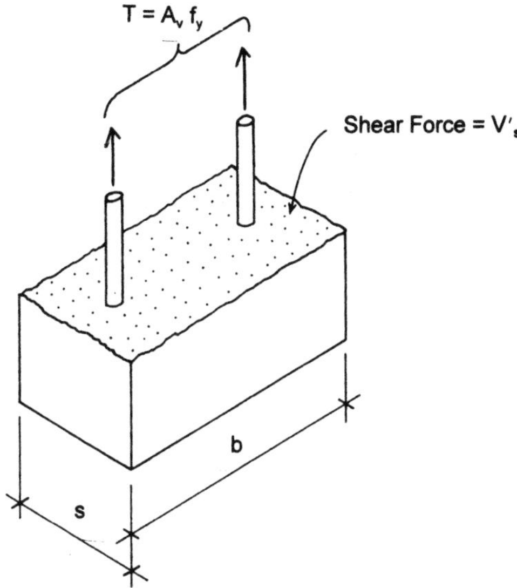

Figure 13.17 Consideration for spacing of a single stirrup.

of a single, two-legged stirrup is equal to the product of the total steel cross-sectional area, A_v, times the yield steel stress. Thus,

$$T = A_v f_y$$

This resisting force opposes part of the steel shear force required at the location of the stirrup, which we will refer to as V_s'. Equating the stirrup tension to this force, an equilibrium equation for one stirrup is obtained:

$$A_v f_y = V_s'$$

The total shear force capacity of the beam in excess of the concrete is determined by the number of stirrups encountered by the shear force acting at a 45° angle through the beam. The number of stirrups will be d/s. Thus, the equilibrium equation for the beam is

$$\left(\frac{d}{s}\right) A_v f_y = V_s'$$

From this equation, an expression for the required spacing can be derived; thus,

$$s \leq \frac{A_v f_y d}{V_s'}$$

The following example illustrates the design procedure for a simple beam.

Example 8. Design the required shear reinforcement for the simple beam shown in Figure 13.18. Use $f_c' = 3$ ksi and $f_y = 40$ ksi and single U-shaped stirrups.

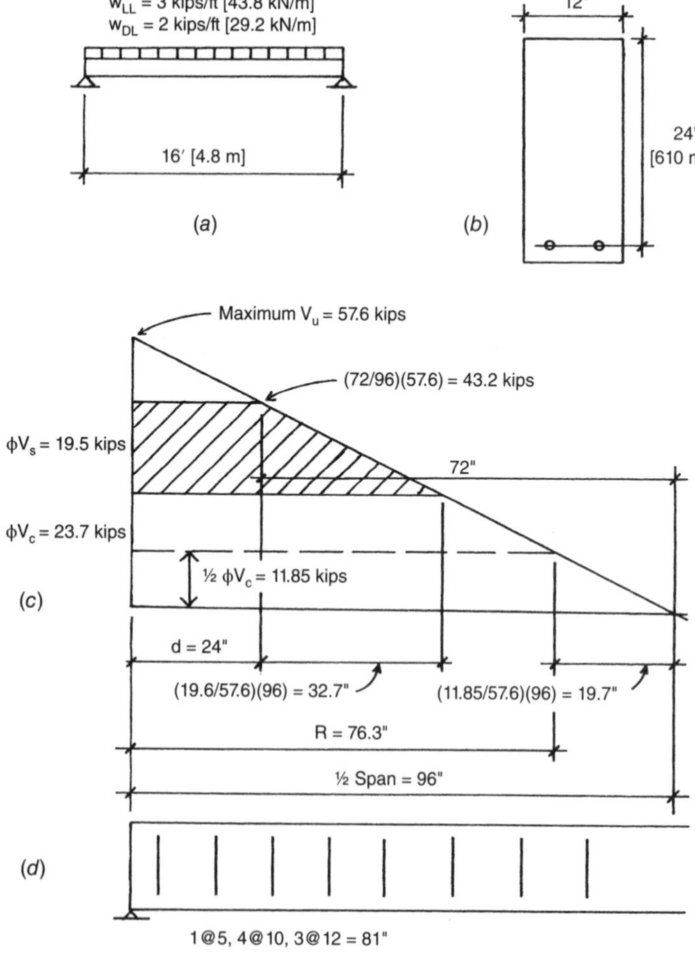

Figure 13.18 Stirrup design: Example 8.

Solution: First, the loading must be factored in order to determine the ultimate shear force:

$$w_u = 1.2(\text{dead load}) + 1.6(\text{live load}) = 1.2(2) + 1.6(3) = 7.2 \text{ kips/ft}$$

The maximum value for the shear is 57.6 kips.

Now construct the ultimate shear force diagram (V_u) for one-half of the beam, as shown in Figure 13.18*c*. For the shear design, the critical shear force is at 24 in. (the effective depth of the beam) from the support. Using proportionate triangles, this value is

$$V_u = \left(\frac{72}{96}\right)(57.6) = 43.2 \text{ kips}$$

The shear capacity of the concrete without reinforcing is given as

$$\phi V_u = (0.75)2\sqrt{f_c{}'}(b)(d) = (0.75)2\sqrt{3000}(12)(24) = 23{,}662 \text{ lb}$$
$$= 23.7 \text{ kips}$$

At the point of critical force, therefore, there is an excess shear force of 43.2 − 23.7 = 19.5 kips that must be carried by reinforcement. Next, complete the construction of the diagram in Figure 13.18*c* to define the shaded portion, which indicates the extent of the required reinforcement. Observe that the excess shear condition extends to 56.7 in. from the support.

In order to satisfy the requirements of the ACI Code (Ref. 10), shear reinforcement must be used wherever the shear force (V_u) exceeds one-half of φV_c. As shown in Figure 13.18*c*, this is a distance of 76.3 in. from the support. The code further stipulates that the minimum cross-sectional area of this reinforcing is

$$A_v = 50\left(\frac{b \times s_{max}}{f_y}\right)$$

With $f_y = 40$ ksi and the maximum allowable spacing of one-half the effective depth, the required area is

$$A_v = 50\left(\frac{12 \times 12}{40000}\right) = 0.18 \text{ in.}^2$$

which is less than the area of $2 \times 0.11 = 0.22$ in.^{2} provided by the two legs of the No. 3 stirrup.

For the maximum V_s value of 19.5 kips, the maximum spacing permitted at the critical point 24 in. from the support is determined as

$$s \leq \frac{A_v f_y d}{\phi V_s} = \frac{(0.22)(40)(24)}{19.5} = 10.8 \text{ in.}$$

Since this is less than the maximum allowable of one-half the depth or 12 in., it is best to calculate one more spacing at a short distance beyond the critical point. For example, at 36 in. from the support the shear force is

$$V_u = \left(\frac{60}{96}\right)(57.6) = 36.0 \text{ kips}$$

and the value of f_c' at this point is 36.0 – 23.7 kips = 12.3 kips. The spacing required at this point is thus

$$s \leq \frac{A_v f_y d}{\phi V_s} = \frac{(0.22)(40)(24)}{12.3} = 17.2 \text{ in.}$$

which indicates that the required spacing drops to the maximum allowed at less than 12 in. from the critical point.

A possible choice for the stirrup spacings is shown in Figure 13.18*d*, with a total of 8 stirrups that extend over a range of 81 in. from the support. There are thus a total of 16 stirrups in the beam, 8 at each end. Note that the first stirrup is placed at 5 in. from the support, which is one-half the computed required spacing; this is a common practice with designers.

Example 9. Determine the required number and spacings for No. 3 U-stirrups for the beam shown in Figure 13.19. Use $f_c' = 3$ ksi and $f_y = 40$ ksi.

Solution: As in Example 7, the shear values are determined, and the diagram in Figure 13.19*c* is constructed. In this case, the maximum critical shear force of 28.5 kips and a shear capacity of concrete (φV_c) of 16.4 kips result in a maximum φV_s value to 12.1 kips, for which the required spacing is

$$s \leq \frac{A_v f_y d}{\phi V_s} = \frac{(0.22)(40)(20)}{12.1} = 14.5 \text{ in.}$$

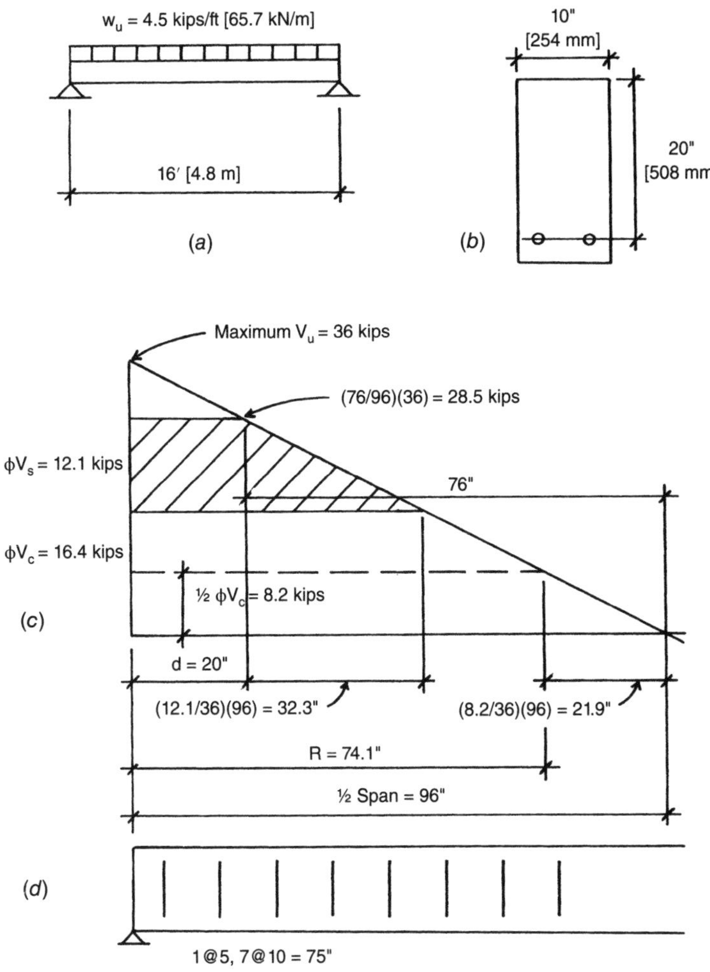

Figure 13.19 Stirrup design: Example 9.

Since this value exceeds the maximum limit of $d/2 = 10$ in., the stirrups may all be placed at the limited spacing, and a possible arrangement is as shown in Figure 13.10*d.* As in Example 7, note that the first stirrup is placed at one-half the required distance from the support.

Example 10. Determine the required number and spacings for No. 3 U-stirrups for the beam shown in Figure 13.20. Use $f_c' = 3$ ksi and $f_y = 40$ ksi.

Solution: In this case, the maximum critical design shear force is found to be less than V_c (16.4 kips from Example 8, Figure 13.19), which in theory indicates that reinforcement is not required. To comply with the code requirement for minimum reinforcement, however, provide stirrups at the maximum permitted spacing out to the point where the shear stress drops to 8.2 kips (one-half of φV_c). To verify

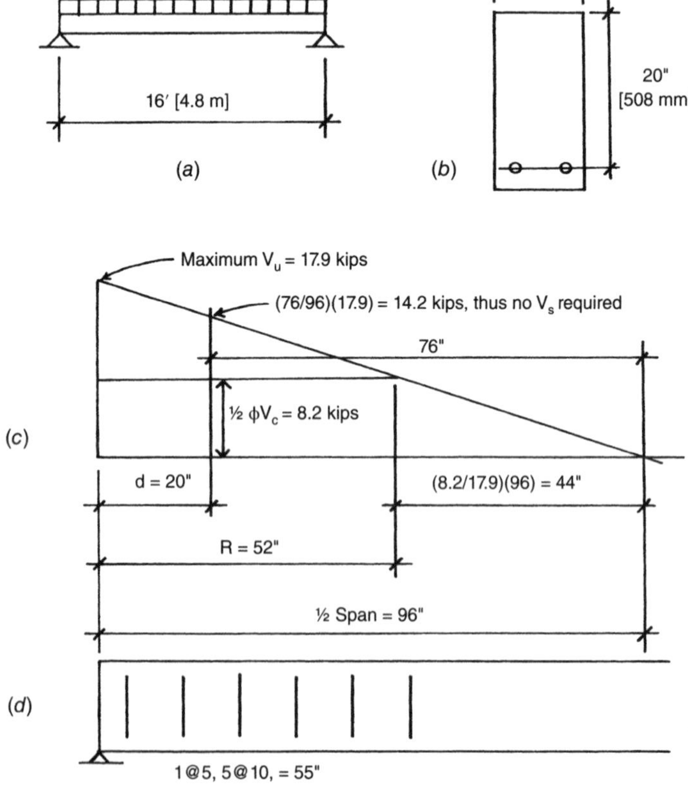

Figure 13.20 Stirrup design: Example 10.

that the No. 3 stirrup is adequate, compute

$$A_v = 50\left(\frac{10 \text{ in.} \times 10 \text{ in.}}{40{,}000 \text{ psi}}\right) = 0.125 \text{ in.}^2$$

which is less than the area of 0.22 in. provided, so the No. 3 stirrup at 10 in. is adequate.

Examples 7–9 have illustrated what is generally the simplest case for beam shear design—that of a beam with uniformly distributed load and with sections subjected only to flexure and shear. When concentrated loads or unsymmetrical loadings produce other forms for the shear diagram, these must be used for design of the shear reinforcement. In addition, where axial forces of tension or compression exist in the concrete frame, consideration must be given to the combined effects when designing for shear.

When torsional moments exist (twisting moments at right angles to the beam), their effects must be combined with beam shear.

Problem 13.6.A A concrete beam similar to that shown in Figure 13.18 sustains a uniform live load of 1.5 klf and a uniform dead load of 1 klf on a span of 24 ft. Determine the layout for a set of No. 3 U-stirrups using the stress method with $f_y = 40$ ksi and $f'_c = 3000$ psi. The beam section dimensions are $b = 12$ in. and $d = 26$ in.

Problem 13.6.B Same as Problem 13.6.A, except the span is 20 ft, $b = 10$ in., $d = 23$ in.

Problem 13.6.C Determine the layout for a set of No. 3 U-stirrups for a beam with the same data as Problem 13.6.A, except the uniform live load is 0.75 klf and the uniform dead load is 0.5 klf.

Problem 13.6.D Determine the layout for a set of No. 3 U-stirrups for a beam with the same data as Problem 13.6.B, except the uniform live load is 1.875 klf and the uniform dead load is 1.25 klf.

13.7 DEVELOPMENT LENGTH FOR REINFORCEMENT

The ACI Code (Ref. 10) defines *development length* as the length of embedment required to develop the design strength of the reinforcement at a critical section. For beams, critical sections occur at points of maximum stress and at points within the span where some of the reinforcement terminates or is bent up or down. For a uniformly loaded simple span beam, the bending moment is a maximum at midspan. The tensile reinforcement required for flexure at this point must extend on both sides a sufficient distance to develop the stress in the bars; however, except for very short spans with large bars, the bar lengths will ordinarily be more than sufficient.

In the simple beam, the bottom reinforcement required for the maximum moment at midspan is not entirely required as the moment decreases toward the end of the span. It is thus sometimes the practice to make only part of the midspan reinforcement continuous for the whole beam length. In this case it may be necessary to assure that the bars that are of partial length are extended sufficiently from the midspan point and that the bars remaining beyond the cutoff point can develop the stress required at that point.

When beams are continuous through the supports, top reinforcement is required for the negative moments at the supports. These top bars must be investigated for the development lengths in terms of the distance they extend from the supports.

For tension reinforcement consisting of bars of No. 11 size and smaller, the code specifies a minimum length for development (L_d) as follows:

For No. 6 bars and smaller:

$$L_d = \frac{f_y d_b}{25\sqrt{f'_c}} \quad \text{but not less than 12 in.}$$

For No. 7 bars and larger:

$$L_d = \frac{f_y d_b}{20\sqrt{f'_c}}$$

In these formulas d_b is the bar diameter.

Modification factors for L_d are given for various situations, as follows:

For top bars in horizontal members with at least 12 in. of concrete below the bars: increase by 1.3.

For flexural reinforcement that is provided in excess of that required by computations: decrease by a ratio of required A_s/provided A_s.

Additional modification factors are given for lightweight concrete, for bars coated with epoxy, for bars encased in spirals, and for bars with f_y in excess of 60 ksi. The maximum value to be used for f_c' is 100 psi.

Table 13.9 gives values for minimum development lengths for tensile reinforcement, based on the requirements of the ACI Code (Ref. 10). The values listed under "Other Bars" are the unmodified length requirements; those listed under "Top Bars" are increased by the modification factor for this situation. Values are given for two concrete strengths and for the two most commonly used grades of tensile reinforcement.

The ACI Code (Ref. 10) makes no provision for a reduction factor for development lengths. As presented, the formulas for development length relate only to bar size, concrete strength, and steel yield strength. They are thus equally applicable for the stress method or the strength method with no further adjustment, except for the conditions previously described.

Example 11. The negative moment in the short cantilever shown in Figure 13.21 is resisted by the steel bars in the top of the beam. Determine whether the development of the reinforcement is adequate without hooked ends on the No. 6 bars if $L_1 = 48$ in. and $L_2 = 36$ in. Use $f_c' = 3$ ksi and $f_y = 60$ ksi.

TABLE 13.9 Minimum Development Length for Tensile Reinforcement (in.)[a]

	$f_y = 40$ ksi				$f_y = 60$ ksi			
	$f_c' = 3$ ksi		$f_c' = 4$ ksi		$f_c' = 3$ ksi		$f_c' = 4$ ksi	
Bar Size	Top Bars[b]	Other Bars	Top Bars[b]	Other Bars	Top Bars[b]	Other Bars	Top Bars[b]	Other Bars
3	15	12	13	12	22	17	19	15
4	19	15	17	13	29	22	25	19
5	24	19	21	16	36	28	31	24
6	29	22	25	19	43	33	37	29
7	42	32	36	28	63	48	54	42
8	48	37	42	32	72	55	62	48
9	54	42	47	36	81	62	70	54
10	61	47	53	41	91	70	79	61
11	67	52	58	45	101	78	87	67

[a]Lengths are based on requirements of the ACI Code (Ref. 10).
[b]Horizontal bars with more than 12 in. of concrete cast below them in the member.

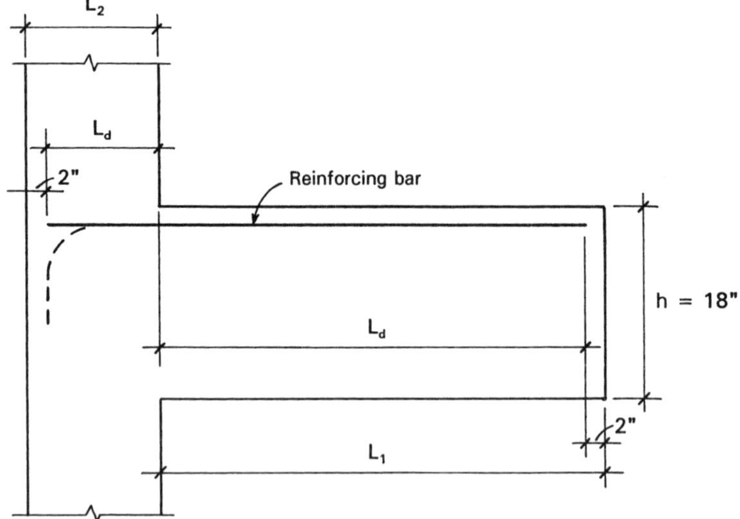

Figure 13.21 Reference for Example 10.

Solution: At the face of the support, anchorage for development must be achieved on both sides: within the support and in the top of the beam. In the top of the beam the condition is one of "Top Bars," as previously defined. Thus, from Table 13.9, a length of 43 in. is required for L_d, which is adequately provided, if cover is minimum on the outside end of the bars.

Within the support, the condition is one of "Other Bars" in the table reference. For this the required length for L_d is 33 in., which is also adequately provided.

Hooked ends are thus not required on either end of the bars, although most designers would probably hook the bars in the support just for the security of the additional anchorage.

Problem 13.7.A A short cantilever is developed as shown in Figure 13.21. Determine whether adequate development is achieved without hooked ends on the bars if L_1 is 36 in., L_2 is 24 in., overall beam height is 16 in., bar size is No. 4, $\sqrt{f_c'} = 4$ ksi, and $f_y = 40$ ksi.

Problem 13.7.B Same as Problem 13.7.A, except $L_1 = 40$ in., $L_2 = 30$ in., No. 5 bar.

Hooks

When details of the construction restrict the ability to extend bars sufficiently to produce required development lengths, development can sometimes be assisted by use of a hooked end on the bar. So-called *standard hooks* may be evaluated in terms of a required development length, L_{dh}. Bar ends may be bent at 90°, 135°, or 180° to produce a hook. The 135° bend is used only for ties and stirrups, which normally consist of relatively small diameter bars.

Table 13.10 gives values for development length with standard hooks using the same variables for f'_c and f_y that are used in Table 13.9. The table values given are in terms of the required development length as shown in Figure 13.22. Note that the table values are for 180° hooks and values may be reduced by 30% for 90° hooks. The following example illustrates the use of the data from Table 13.10 for a simple situation.

Example 12. For the bars in Figure 13.21, determine the length L_{dh} required for development of the bars with a 90° hooked end in the support. Use the same data as in Example 10.

Solution: From Table 13.10, the required length for the data given is 17 in. (No. 6 bar, $f'_c = 3$ ksi, $f_y = 60$ ksi). This may be reduced for the 90° hook to

$$L = 0.70(17) = 11.9 \text{ in.}$$

Problem 13.7.C Find the development length required for the bars in Problem 13.7.A if the bar ends in the support are provided with 90° hooks.

TABLE 13.10 Required Development Length L_{dh} for Hooked Bars (in.)[a]

	$f_y = 40$ ksi		$f_y = 60$ ksi	
Bar Size	$f'_c = 3$ ksi	$f'_c = 4$ ksi	$f'_c = 3$ ksi	$f'_c = 4$ ksi
3	6	6	9	8
4	8	7	11	10
5	10	8	14	12
6	11	10	17	15
7	13	12	20	17
8	15	13	22	19
9	17	15	25	22
10	19	16	28	24
11	21	18	31	27

[a]See Figure 13.22. Table values are for a 180° hook; values may be reduced by 30% for a 90° hook.

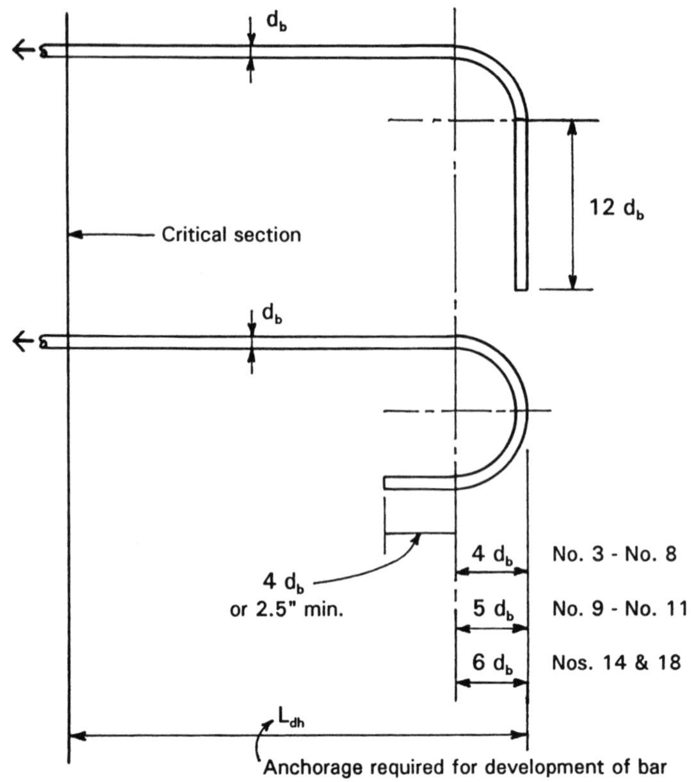

Figure 13.22 Requirements for standard hooks for use of values in Table 13.10.

Problem 13.7.D Find the development length required for the bars in Problem 13.7.B if the bar ends in the support are provided with 90° hooks.

Bar Development in Continuous Beams

Development length is the length of embedded reinforcement required to develop the design strength of the reinforcement at a critical section. Critical sections occur at points of maximum stress and at points within the span at which adjacent reinforcement terminates or is bent up into the top of the beam. For a uniformly loaded simple beam, one critical section is at midspan where the bending moment is maximum. This is a point of maximum tensile stress in the reinforcement (peak bar stress), and some length of bar is required over which the stress can be developed. Other critical sections occur between midspan and the reactions at points where some bars are cut off because they are no longer needed to resist the bending moment; such terminations create peak stress in the remaining bars that extend the full length of the beam.

When beams are continuous through their supports, the negative moments at the supports will require that bars be placed in the top of the beams. Within the span, bars will be required in the bottom of the beam for positive moments. While the positive moment will go to zero at some distance from the supports, the codes require that some of the positive-moment reinforcement be extended for the full length of the span and a short distance into the support.

Figure 13.23 shows a possible layout for reinforcement in a beam with continuous spans and a cantilevered end at the first support. Referring to the notation in the illustration, note the following:

1. Bars *a* and *b* are provided for the maximum moment of positive sign that occurs somewhere near the beam midspan. If all these bars are made full length (as shown for bars *a*), the length L_1 must be sufficient for development (this situation is seldom critical). If bars *b* are partial length as shown in the illustration, then length L_2 must be sufficient to develop bars *b* and length L_3 must be sufficient to develop bars *a*. As was discussed for the simple beam, the partial-length bars must actually extend beyond the theoretical cutoff point (*B* in the illustration) and the true length must include the dashed portions indicated for bars *b*.

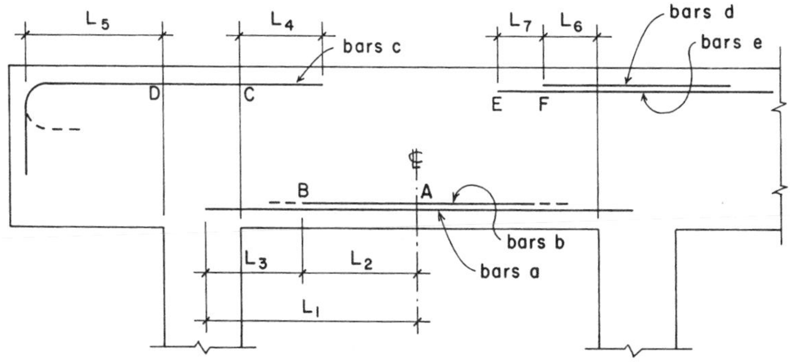

Figure 13.23 Development lengths in continuous beams.

2. For the bars at the cantilevered end, the distances L_4 and L_5 must be sufficient for development of bars c. Distance L_4 is required to extend beyond the actual cutoff point of the negative moment by the extra length described for the partial-length bottom bars. If L_5 is not adequate, the bar ends may be bent into the 90° hook as shown or the 180° hook shown by the dashed line.
3. If the combination of bars shown in the illustration is used at the interior support, L_6 must be adequate for the development of bars d and L_7 adequate for the development of bars e.

For a single loading condition on a continuous beam, it is possible to determine specific values of moment and their location along the span, including the locations of points of zero moment. In practice, however, most continuous beams are designed for more than a single loading condition, which further complicates the problems of determining development lengths required.

Splices in Reinforcement

In various situations in reinforced concrete structures it becomes necessary to transfer stress between steel bars in the same direction. Continuity of force in the bars is achieved by splicing, which may be accomplished by welding, by mechanical means, or by the lapped splice. Figure 13.24 illustrates the concept of the lapped splice, which consists essentially of the development of both bars within the concrete. Because a lapped splice is usually made with the two bars in contact, the lapped length must usually be somewhat greater than the simple development length required in Table 13.10.

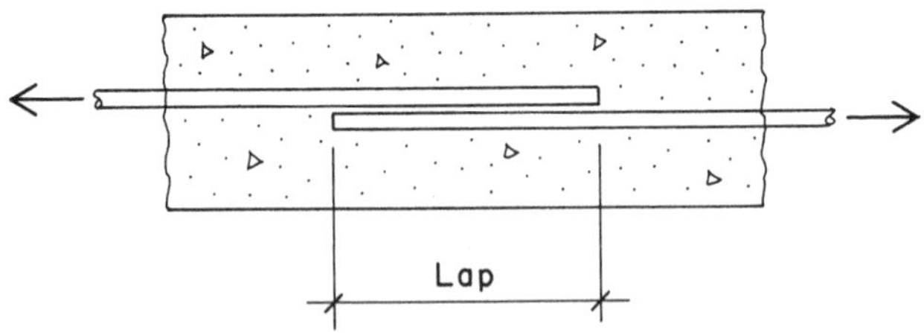

Figure 13.24 Lapped splice for steel reinforcing bars.

For a simple tension lap splice, the full development of the bars usually requires a lap length of 1.3 times that required for simple development of the bars. Lap splices are generally limited to bars of No. 11 size or smaller.

For pure tension members, lapped splicing is not permitted, and splicing must be achieved by welding the bars or by some other mechanical connection. End-to-end butt welding of bars is usually limited to compression splicing of large-diameter bars with high f_y for which lapping is not feasible.

When members have several reinforcement bars that must be spliced, the splicing must be staggered. Splicing is generally not desirable and is to be avoided when possible, but because bars are obtainable only in limited lengths, some situations unavoidably involve splicing. Horizontal reinforcement in walls is one such case. For members with computed stress, splicing should not be located at points of maximum stress, for example, at points of maximum bending. Splicing of compression reinforcement for columns is discussed in the next section.

Development of Compressive Reinforcement

Development length in compression is a factor in column design and in the design of beams reinforced for compression. The absence of flexural tension cracks in the portions of beams where compression reinforcement is employed, plus the beneficial effect of the end bearing of the bars on the concrete, permit shorter developmental lengths in compression than in tension. The ACI Code (Ref. 10) prescribes that L_d for bars in compression shall be computed by the formula

$$L_d = \frac{0.02 f_y d_b}{\sqrt{f'_c}}$$

but shall not be less than $0.0003 f_y d_b$ or 8 in., whichever is greater. Table 13.11 lists compression bar development lengths for a few combinations of specification data.

In reinforced columns both the concrete and the steel bars share the compression force. Ordinary construction practices require the consideration of various situations for development of the stress in the reinforcing bars. Figure 13.25 shows a multistory concrete column

TABLE 13.11 Minimum Development Length for Compressive Reinforcement (in.)

Bar Size	$f_y = 40$ ksi		$f_y = 60$ ksi		
	$f'_c = 3$ ksi	$f'_c = 4$ ksi	$f'_c = 3$ ksi	$f'_c = 4$ ksi	$f'_c = 5$ ksi
3	8	8	8	8	7
4	8	8	11	10	9
5	10	8	14	12	11
6	11	10	17	15	13
7	13	12	20	17	15
8	15	13	22	19	17
9	17	15	25	22	20
10	19	17	28	25	22
11	21	18	31	27	24
14			38	33	29
18			50	43	39

with its base supported on a concrete footing. With reference to the illustration, note the following:

1. The concrete construction is ordinarily produced in multiple, separate pours, with construction joints between the separate pours occurring as shown in the illustration.
2. In the lower column, the load from the concrete is transferred to the footing in direct compressive bearing at the joint between the column and footing. The load from the reinforcing must be developed by extension of the reinforcing into the footing: distance L_1 in the illustration. Although it may be possible to place the column bars in position during casting of the footing to achieve this, the common practice is to use dowels, as shown in the illustration. These dowels must be developed on both sides of the joint: L_1 in the footing and L_2 in the column. If the f'_c value for both the footing and the column are the same, these two required lengths will be the same.
3. The lower column will ordinarily be cast together with the supported concrete framing above it, with a construction joint occurring at the top level of the framing (bottom of the upper column), as shown in the illustration. The distance L_3 is that required to develop the reinforcing in the lower column—bars *a* in the illustration. As for the condition at the top of the footing, the distance L_4 is required to develop the reinforcing

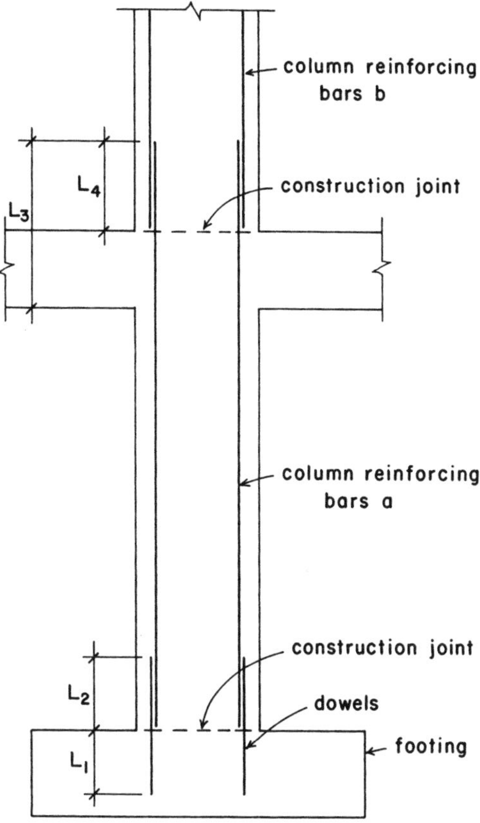

Figure 13.25 Considerations for bar development in concrete columns.

in bars *b* in the upper column. Distance L_4 is more likely to be the critical consideration for the determination of the extension required for bars *a*.

13.8 DEFLECTION CONTROL

Deflection of spanning slabs and beams of cast-in-place concrete is controlled primarily by using a recommended minimum thickness (overall height) expressed as a percentage of the span. Table 13.12 is adapted from a similar table given in the ACI Code (Ref. 10) and yields minimum thickness as a fraction of the span. Table values apply

TABLE 13.12 Minimum Thickness of Slabs or Beams Unless Deflections Are Computed[a]

Type of Member	End Conditions of Span	Minimum Thickness of Slab or Height of Beam	
		f_y = 40 ksi	f_y = 60 ksi
Solid one-way slabs	Simple support	L/25	L/20
	One end continuous	L/30	L/24
	Both ends continuous	L/35	L/28
	Cantilever	L/12.5	L/10
Beams or joists	Simple support	L/20	L/16
	One end continuous	L/23	L/18.5
	Both ends continuous	L/26	L/21
	Cantilever	L/10	L/8

Source: Adapted from material in *Building Code Requirements for Structural Concrete (ACI 318–08)* (Ref. 10), with permission of the publisher, American Concrete Institute.

[a]Refers to overall vertical dimension of concrete section. For normal-weight concrete (145 pcf) only; code provides adjustment for other weights. Valid only for members not supporting or attached rigidly to partitions or other construction likely to be damaged by large deflections.

only for concrete of normal weight (made with ordinary sand and gravel) and for reinforcement with f_y of 40 and 60 ksi. The ACI Code supplies correction factors for other concrete weights and reinforcing grades. The ACI Code further stipulates that these recommendations apply only where beam deflections are not critical for other elements of the building construction, such as supported partitions subject to cracking caused by beam deflections.

Deflection of concrete structures presents a number of special problems. For concrete with ordinary reinforcement (not prestressed), flexural action normally results in some tension cracking of the concrete at points of maximum bending. Thus, the presence of cracks in the bottom of a beam at midspan points and in the top over supports is to be expected. In general, the size (and visibility) of these cracks will be proportional to the amount of beam curvature produced by deflection. Crack size will also be greater for long spans and for deep beams. If visible cracking is considered objectionable, more conservative depth-to-span ratios should be used, especially for spans over 30 ft and beam depths over 30 in.

Creep of concrete results in additional deflections over time. This is caused by the sustained loads—essentially the dead load of the construction. Deflection controls reflect concern for this as well as for the instantaneous deflection under live load, the latter being the major concern in structures of wood and steel.

In beams, deflections, especially creep deflections, may be reduced by the use of some compressive reinforcement. Where deflections are of concern, or where depth-to-span ratios are pushed to their limits, it is advisable to use some compressive reinforcement, consisting of continuous top bars.

When, for whatever reasons, deflections are deemed to be critical, computations of actual values of deflection may be necessary. The ACI Code provides directions for such computations; they are quite complex in most cases and beyond the scope of this work. In actual design work, however, they are required very infrequently.

14

FLAT-SPANNING CONCRETE SYSTEMS

There are many different systems than can be used to achieve flat spans. These are used most often for floor structures, which typically require a dead flat form. However, in buildings with an all-concrete structure, they may also be used for roofs. Site-cast systems generally consist of one of the following basic types:

1. One-way solid slab and beam
2. Two-way solid slab and beam
3. One-way joist construction
4. Two-way flat slab or flat plate without beams
5. Two-way joist construction, called waffle construction

Each system has its own distinct advantages and limits and some range of logical use, depending on required spans, general layout of supports, magnitude of loads, required fire ratings, and cost limits for design and construction.

The floor plan of a building and its intended usage determine loading conditions and the layout of supports. Also of concern are requirements for openings for stairs, elevators, large ducts, skylights, and so on, as these result in discontinuities in the otherwise commonly continuous systems. Whenever possible, columns and bearing walls should be aligned in rows and spaced at regular intervals in order to simplify design and construction and lower costs. However, the fluid concrete can be molded in forms not possible for wood or steel, and many very innovative, sculptural systems have been developed as takeoffs on these basic ones.

14.1 SLAB-AND-BEAM SYSTEMS

The most widely used and most adaptable cast-in-place concrete floor system is that which utilizes one-way solid slabs supported by one-way spanning beams. This system may be used for single spans but occurs more frequently with multiple-span slabs and beams in a system such as that shown in Figure 14.1. In the example shown, the continuous slabs are supported by a series of beams that are spaced at 10 ft center to center. The beams, in turn, are supported by a

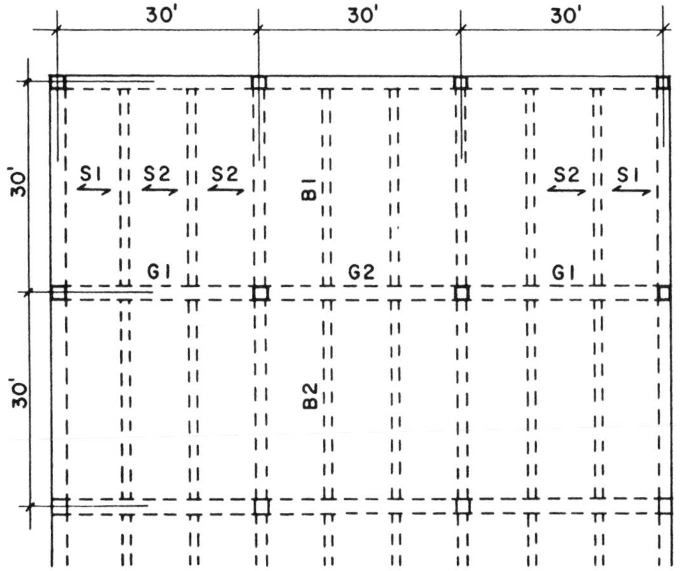

Figure 14.1 Framing layout for a typical slab-and-beam system.

girder and column system with columns at 30-ft centers, every third beam being supported directly by the columns and the remaining beams being supported by the girders.

Because of the regularity and symmetry of the system shown in Figure 14.1, there are relatively few different elements in the basic system, each being repeated several times. While special members must be designed for conditions that occur at the outside edge of the system and at the location of any openings for stairs, elevators, and so on, the general interior portions of the structure may be determined by designing only six basic elements: S1, S2, B1, B2, G1, and G2, as shown in the framing plan (Figure 14.1).

In computations for reinforced concrete, the span length of freely supported beams (simple beams) is generally taken as the distance between centers of supports or bearing areas; it should not exceed the clear span plus the depth of beam or slab. The span length for continuous or restrained beams is taken as the clear distance between faces of supports.

In continuous beams, negative bending moments are developed at the supports and positive moments at or near midspan. This may be readily observed from the exaggerated deformation curve of Figure 14.2*a*. The exact values of the bending moments depend on several factors, but in the case of approximately equal spans supporting uniform loads, when the live load does not exceed three times the dead load, the bending moment values given in Figure 14.2 may be used for design.

The values given in Figure 14.2 are in general agreement with those given in Chapter 8 of the ACI Code (Ref. 10). These values have been adjusted to account for partial live loading of multiple-span beams. Note that these values apply only to uniformly loaded beams. The ACI Code also gives some factors for end-support conditions other than the simple supports shown in Figure 14.2.

Design moments for continuous-span slabs are given in Figure 14.3. With large beams and short-slab spans, the torsional stiffness of the beam tends to minimize the continuity effect in adjacent slab spans. Thus, most slab spans in the slab-and-beam systems tend to function much like individual spans with fixed ends.

Design of a One-Way Continuous Slab

The general design procedure for a one-way solid slab was illustrated in Section 13.5. The example given there is for a simple span slab.

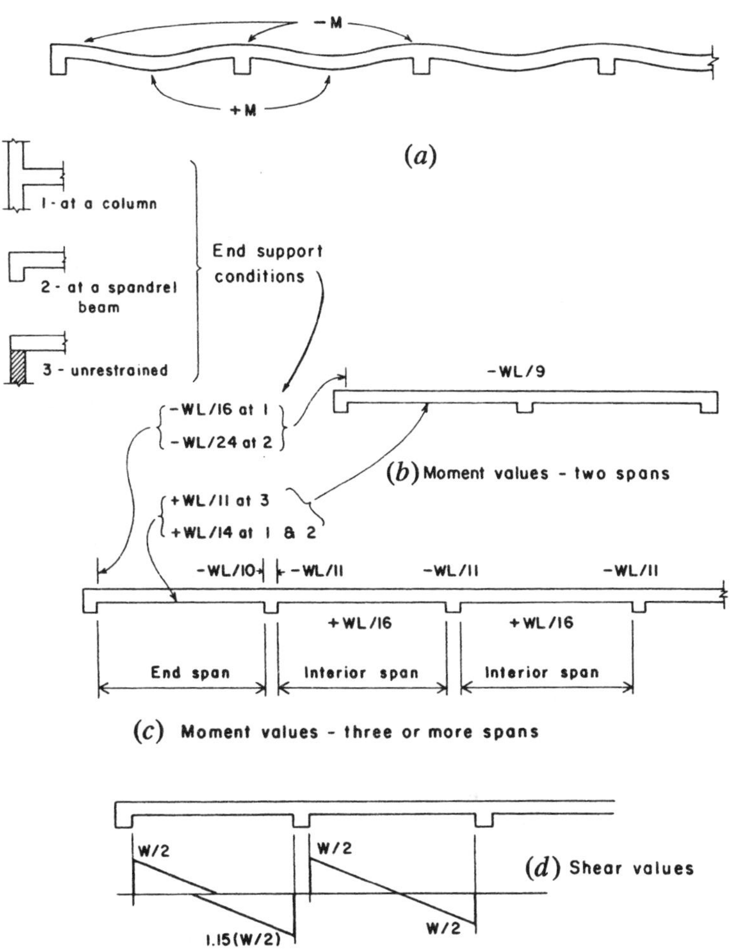

Figure 14.2 Approximate design factors for concrete beams.

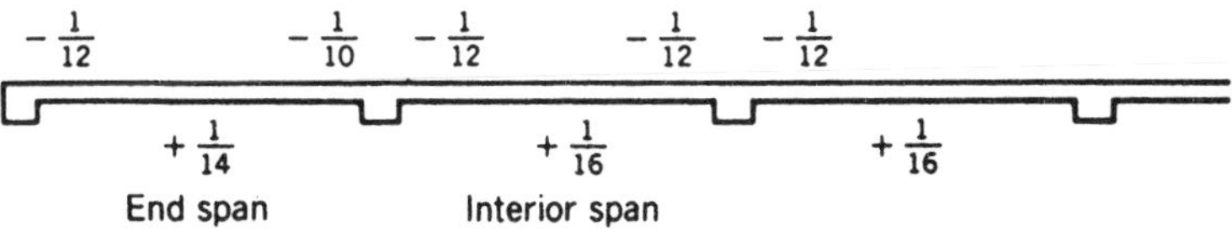

Figure 14.3 Approximate design factors for continuous slabs with spans of 10 continuous slabs.

The following example illustrates the procedure for the design of a continuous solid one-way slab.

Example 1. A solid one-way slab is to be used for a framing system similar to that shown in Figure 14.1. Column spacing is 30 ft with evenly spaced beams occurring at 10 ft center to center. Superimposed loads on the structure (floor live load plus other construction dead load) are a dead load of 38 psf and a live load of 100 psf. Use $f_c' =$ 3 ksi and $f_y = 40$ ksi. Determine the thickness for the slab and select its reinforcement.

Solution: To find the slab thickness, consider three factors: the minimum thickness for deflection, the minimum effective depth for the maximum moment, and the minimum effective depth for the maximum shear. For design purposes the span of the slab is taken as the clear span, which is the dimension from face to face of the supporting beams. With the beams at 10-ft centers, this dimension is 10 ft, less the width of one beam. Since the beams are not given, a dimension must be assumed for them. For this example assume a beam width of 12 in., yielding a clear span of 9 ft.

Consider first the minimum thickness required for deflection. If the slabs in all spans have the same thickness (which is the most common practice), the critical slab is the end span, since there is no continuity of the slab beyond the end beam. While the beam will offer some restraint, it is best to consider this as a simple support; thus, the appropriate factor is $L/30$ from Table 13.12, and

$$\text{Minimum } t = \frac{L}{30} = \frac{9 \times 12}{30} = 3.6 \text{ in.}$$

Assume here that fire-resistive requirements make it desirable to have a relatively heavy slab of 5 in. overall thickness, for which the dead weight of the slab is

$$w = \frac{4}{12} \times 150 = 62 \text{ psf}$$

The total dead load is thus 62 + 38 = 100 psf and the factored total load is

$$w_u = 1.2(100) + 1.6(100) = 280 \text{ psf}$$

Next consider the maximum bending moment. Inspection of the moment values given in Figure 14.3 shows the maximum moment to be

$$M = \frac{1}{10}wL^2$$

With the clear span and the loading as determined, the maximum moment is thus

$$M = \frac{1}{10}(wL^2) = \frac{1}{10}(280 \times (9)^2) = 2268 \text{ ft-lb}$$

and the required resisting moment for the slab is

$$M_R = \frac{2268}{0.9} = 2520 \text{ ft-lb}$$

This moment value should now be compared to the balanced capacity of the design section using the relationships discussed for rectangular beams in Section 13.3. For this computation an effective depth for the design section must be assumed. This dimension will be the slab thickness minus the concrete cover and one-half the bar diameter. With the bars not yet determined, assume an approximate effective depth to be the slab thickness minus 1.0 in.; this will be exactly true with the usual minimum cover of 3/4 in. and a No. 4 bar. Then using the balanced moment R factor from Table 13.2, the maximum resisting moment for the 12-in.-wide design section is

$$M_R = Rbd^2 = (1.149)(12)(4)^2 = 221 \text{ kip-in.}$$

or

$$M_R = 221 \times \frac{1000}{12} = 18400 \text{ ft-lb}$$

As this value is in excess of the required resisting moment of 2520 ft-lb, the slab is adequate for concrete flexural stress.

It is not practical to use shear reinforcement in one-way slabs, and consequently the maximum unit shear stress must be kept within the limit for the concrete alone. The usual procedure is to check the shear stress with the effective depth determined for bending before proceeding to find A_s. Except for very short span slabs with excessively heavy loadings, shear stress is seldom critical.

For interior spans, the maximum shear will be $wL/2$, but for the end span it is usual practice to consider some unbalanced condition for the shear due to the discontinuous end. Use a maximum shear of $1.15(wL/2)$, or an increase of 15% over the simple beam shear value. Thus,

$$\text{Maximum shear} = V_u = 1.15 \times \frac{wL}{2} = 1.15 \times \frac{280 \times 9}{2} = 1449 \text{ lb}$$

and

$$V_R = \frac{1449}{0.75} = 1932 \text{ lb}$$

For the slab section with $b = 12$ in. and $d = 4$ in.

$$V_c = 2\sqrt{f_c{'}}(bd) = 2\sqrt{3000}(12 \times 4) = 5258 \text{ lb}$$

This is considerably greater than the required shear resistance, so the assumed slab thickness is not critical for shear stress.

Having thus verified the choice for the slab thickness, we may now proceed with the design of the reinforcement. For a balanced section, Table 13.2 yields a value of 0.685 for the a/d factor. However, since all sections will be classified as underreinforced (actual moment less than the balanced limit), use an approximate value of 0.4 for a/d. Once the reinforcement for a section is determined, the true value of a/d can be verified using the procedures developed in Section 13.3.

For the slab in this example the following is computed:

$$\frac{a}{d} = 0.4 \quad \text{and} \quad a = 0.4d = 0.4(4) = 1.6 \text{ in.}$$

For the computation of required reinforcement use

$$d - \frac{a}{2} = 4 - \frac{1.6}{2} = 3.2 \text{ in.}$$

Referring to Figure 14.3, note that there are five critical locations for which a moment must be determined and the required steel area computed. Reinforcement required in the top of the slab must be computed for the negative moments at the end support, at the first interior beam, and at the typical interior beam. Reinforcement required in the bottom of the slab must be computed for the positive moments at midspan locations in the first span and in the typical interior spans.

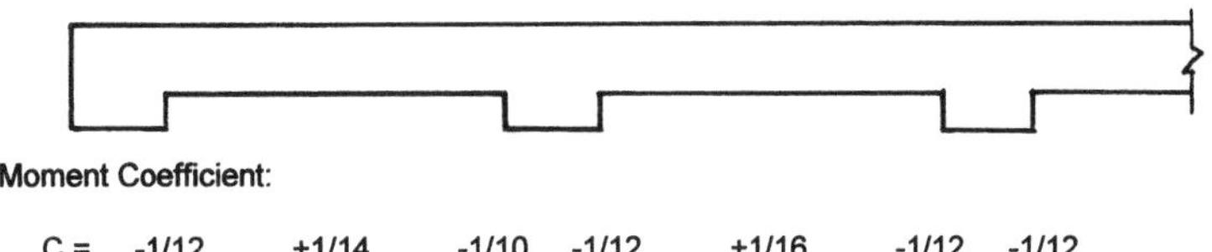

Moment Coefficient:

C = -1/12 +1/14 -1/10 -1/12 +1/16 -1/12 -1/12

Required A_s = 2.36 C (in in.^{2} per ft of slab width)

A_s = 0.197 0.169 0.236 0.148 0.197

Required center-to-center spacing of bars in in. (maximum = 3t = 15 in.):

No. 3 at	6.5	8.5	5.5	8	6.5
No. 4 at	12	14	10	16	12
No. 5 at	18	19	15.5	24	18

Choice:

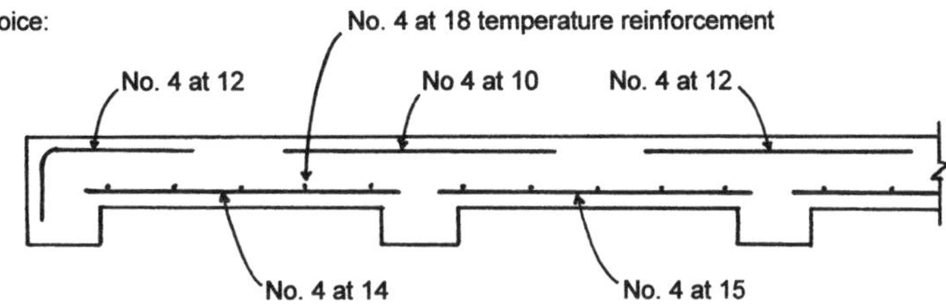

Figure 14.4 Summary of design for the continuous slab.

The design for these conditions is summarized in Figure 14.4. For the data displayed in the figure note the following:

Maximum spacing of reinforcement:

$$s = 3t = 3(5) = 15 \text{ in.}$$

Maximum required bending moment:

$$M_R = (\text{moment factor } C)\left(\frac{wL^2}{0.9}\right)$$

$$= C\left[\frac{280 \times (9)^2}{0.9}\right] \times 12 = 302{,}400C$$

Note that the use of the factor 12 makes this value for the moment in inch-pound units.

Required area of reinforcement:

$$A_s = \frac{M}{f_y(d - a/2)} = \frac{302{,}400 \times C}{40{,}000(3.2)} = 2.36C$$

Using data from Table 13.6, Figure 14.4 shows required spacing for No. 3, 4, and 5 bars. A possible choice for the slab reinforcement, using straight bars, is shown at the bottom of Figure 14.4.

For required temperature reinforcement

$$A_s = 0.002bt = 0.002(12 \times 5) = 0.12 \text{ in.}^2/\text{ft of slab width}$$

Using data from Table 13.7, possible choices are No. 3 at 11 in. or No. 4 at 18 in.

Problem 14.1.A A solid one-way slab is to be used for a framing system similar to that shown in Figure 14.1. Column spacing is 36 ft, with regularly spaced beams occurring at 12 ft center to center. Superimposed dead load on the structure is 40 psf and live load is 80 psf. Use $f'_c = 4$ ksi and $f_y =$ 60 ksi. Determine the thickness for the slab and select the size and spacing for the bars.

Problem 14.1.B Same as Problem 14.1.A, except column spacing is 33 ft, beams are at 11 ft centers, superimposed dead load is 50 psf, and live load is 75 psf.

14.2 GENERAL CONSIDERATIONS FOR BEAMS

The design of a single beam involves a large number of pieces of data, most of which are established for the system as a whole, rather than individually for each beam. Systemwide decisions usually include those for the type of concrete and its design strength (f'_c), the type of reinforcing steel (f_y), the cover required for exposure conditions and the necessary fire rating, and various generally used details for forming of the concrete and placing the reinforcement. Most beams occur in conjunction with solid slabs that are cast monolithically with the beams. Slab thickness is established by the structural requirements of the spanning action between beams and by various concerns, such as those for fire rating, acoustic separation, type of reinforcement, and so on. Design of a single beam is usually limited to determination of the following:

1. Choice of shape and dimensions of the beam cross section
2. Selection of the type, size, and spacing of shear reinforcement
3. Selection of the flexural reinforcement to satisfy requirements based on the variation of moment along the several beam spans

The following are some factors that must be considered in effecting these decisions.

Beam Shape

Figure 14.5 shows the most common shapes used for beams in site-cast construction. The single, simple rectangular section is actually uncommon but does occur in some situations. Design of the

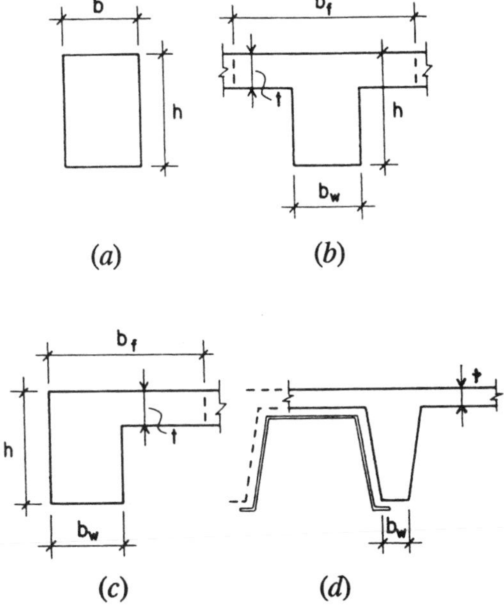

Figure 14.5 Common shapes for beams.

concrete section consists of selecting the two dimensions: the width b and the overall height or depth h.

As mentioned previously, beams occur most often in conjunction with monolithic slabs, resulting in the typical T shape shown in Figure 14.5*b* or the L shape shown in Figure 14.5*c*. The full T shape occurs at the interior portions of the system, while the L shape occurs at the outside of the system or at the side of large openings. As shown in the illustration, there are four basic dimensions for the T and L shapes that must be established in order to fully define the beam section:

t = slab thickness, ordinarily established on its own, rather than as a part of the single beam design

h = overall beam stem depth, corresponding to the same dimension for the rectangular section

b_w = beam stem width, which is critical for consideration of shear and for problems of fitting reinforcing into the section

b_f = so-called effective width of the flange, which is the portion of the slab assumed to work with the beam

A special beam shape is that shown in Figure 14.5*d*. This occurs in concrete joist and waffle construction when "pans" of steel or reinforced plastic are used to form the concrete, the taper of the beam stem being required for easy removal of the forms. The smallest width dimension of the beam stem is ordinarily used for the beam design in this situation.

Beam Width

The width of a beam will affect its resistance to bending. Consideration of the flexure formulas given in Section 13.3 shows that the width dimension affects the bending resistance in a linear relationship (double the width and you double the resisting moment, etc.). On the other hand, the resisting moment is affected by the *square* of the effective beam depth. Thus efficiency—in terms of beam weight or concrete volume—will be obtained by striving for deep, narrow beams instead of shallow, wide ones (just as a 2×8 joist is more efficient than a 4×4 joist in wood).

Beam width also relates to various other factors, however, and these are often critical in establishing the minimum width for a given beam. The formula for shear capacity indicates that the beam width

TABLE 14.1 Minimum Beam Widths[a]

Number of Bars	Bar Size 3	4	5	6	7	8	9	10	11
2	10	10	10	10	10	10	10	10	10
3	10	10	10	10	10	10	10	10	11
4	10	10	10	10	11	11	12	13	14
5	10	11	11	12	12	13	14	15	17
6	11	12	13	14	14	15	17	18	19
7	13	14	15	15	16	17	19	20	22
8	14	15	16	17	18	19	21	23	25
9	16	17	18	19	20	21	23	25	28
10	17	18	19	21	22	23	26	28	30

[a]Minimum width in inches for beams with 1.5-in. cover, No. 3 U-stirrups, clear spacing between bars of one bar diameter or minimum of 1 in. General minimum practical width for any beam with No. 3 U-stirrups is 10 in.

is equally as effective as the depth in shear resistance. Placement of reinforcing bars is sometimes a problem in narrow beams. Table 14.1 gives minimum beam widths required for various bar combinations, based on considerations of bar spacing, minimum concrete cover of 1.5 in., placement of the bars in a single layer, and use of a No. 3 stirrup. Situations requiring additional concrete cover, use of larger stirrups, or the intersection of beams with columns may necessitate widths greater than those given in Table 14.1.

While selection of beam depth is partly a matter of satisfying structural requirements, it is typically constrained by other considerations in the building design. Figure 14.6 shows a section through a typical building floor/ceiling with a concrete slab-and-beam structure. In this situation the critical depth from a general building design point of view is the overall thickness of the construction, shown as H in the illustration. In addition to the concrete structure, this includes allowances for the floor finish, the ceiling construction, and the passage of an insulated air duct. The net usable portion of H for the structure is shown as the dimension h, with the effective structural depth d being something less than h. Since the space defined by H is not highly usable for the building occupancy, there is a tendency to constrain it, which works to limit any extravagant use of d.

Most concrete beams tend to fall within a limited range in terms of the ratio of width to depth. The typical range is for a width-to-depth ratio between 1 : 1.5 and 1 : 2.5, with an average of 1 : 2. This is not a

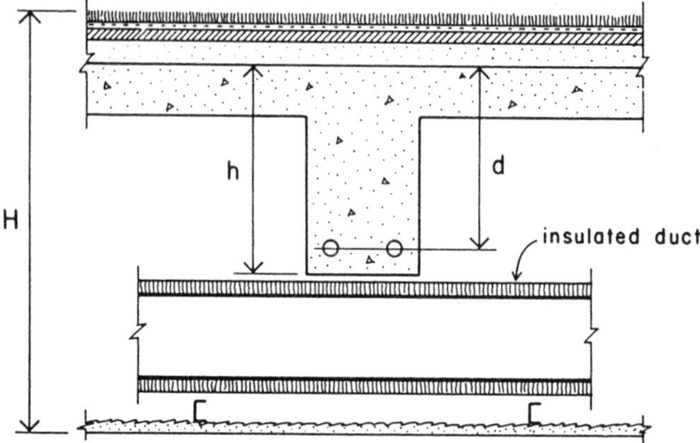

Figure 14.6 Concrete beam in typical multistory construction. Dimension H is critical for architectural planning; dimension d is critical for structural design.

code requirement or a magic rule; it is merely the result of satisfying typical requirements for flexure, shear, bar spacing, economy of use of steel, and deflection.

Deflection Control

Deflection of spanning slabs and beams must be controlled for a variety of reasons. This topic is discussed in Section 13.8. Typically, the most critical decision factor relating to deflection is the overall vertical thickness or height of the spanning member. The ratio of this height dimension to the span length is the most direct indication of the degree of concern for deflection. Recommended minimum beam heights are given in Table 13.12.

15

CONCRETE COLUMNS AND COMPRESSION MEMBERS

In view of the ability of concrete to resist compressive stress and its weakness in tension, it would seem to be apparent that its most logical use is for structural members whose primary task is the resistance of compression. This observation ignores the use of reinforcement to a degree but is nevertheless not without some note. And, indeed, major use is made of concrete for columns, piers, pedestals, posts, and bearing walls—all basically compression members. This chapter presents discussions of the use of reinforced concrete for such structural purposes, with emphasis on the development of columns for building structures. Concrete columns often exist in combination with concrete beam systems, forming rigid frames with vertical planar bents; this subject is addressed as part of the discussion of the concrete example structure in Chapter 20.

15.1 EFFECTS OF COMPRESSION FORCE

When concrete is subjected to a direct compressive force, the most obvious stress response in the material is one of compressive stress,

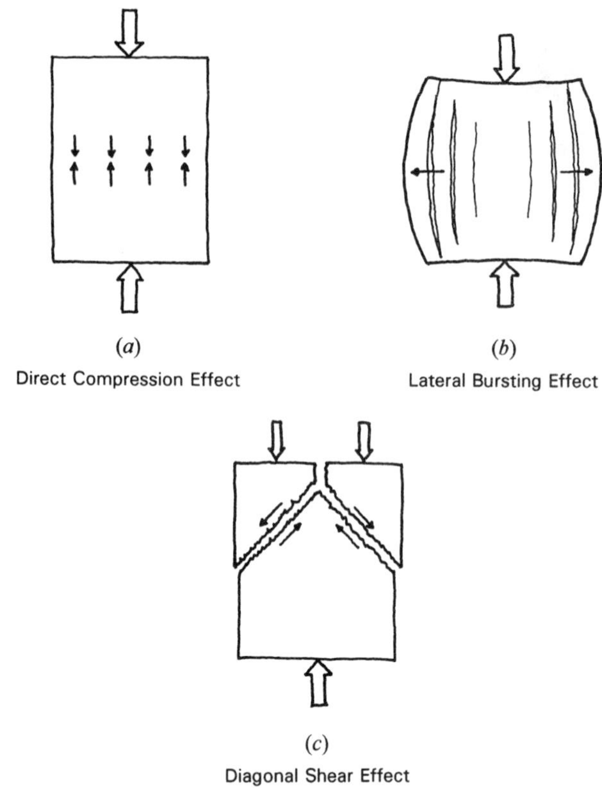

Figure 15.1 Fundamental failure modes of the tension-weak concrete.

as shown in Figure 15.1*a*. This response may be the essential one of concern, as it would be in a wall composed of flat, precast concrete bricks stacked on top of each other. Direct compressive stress in the individual bricks and in the mortar joints between bricks would be a primary situation for investigation.

However, if the concrete member being compressed has some dimension in the direction of the compressive force—as in the case of a column or pier—there are other internal stress conditions that may well be the source of structural failure under the compressive force. Direct compressive force produces a three-dimensional deformation that includes a pushing out of the material at right angles to the force, actually producing tension stress in that direction, as shown in Figure 15.1*b*. In a tension-weak material, this tension action may produce a lateral bursting effect.

Since concrete as a material is also weak in shear, another possibility for failure is along the internal planes where maximum shear stress is developed. This occurs at a 45° angle with respect to the direction of the applied force, as shown in Figure 15.1*c*.

In concrete compression members, other than flat bricks, it is generally necessary to provide for all three stress responses shown in Figure 15.1. In fact, additional conditions can occur if the structural member is also subjected to bending or torsional twisting. Each case must be investigated individually for all the individual actions and the combinations in which they can occur. Design for the concrete member and its reinforcement will typically respond to several considerations of behavior, and the same member and reinforcement must function for all responses. The following discussions focus on the primary function of resistance to compression, but other concerns will also be mentioned. The basic consideration for combined compression and bending is discussed here since present codes require that all columns be designed for this condition.

Reinforcement for Columns

Column reinforcement takes various forms and serves various purposes, the essential consideration being to enhance the structural performance of the column. Considering the three basic forms of column stress failure shown in Figure 15.1, it is possible to visualize basic forms of reinforcement for each condition. This is done in the illustrations in Figures 15.2*a–c*.

To assist the basic compression function steel bars are added with their linear orientation in the direction of the compression force. This is the fundamental purpose of the vertical reinforcing bars in a column. While the steel bars displace some concrete, their superior strength and stiffness make them a significant improvement.

To assist in resistance to lateral bursting (Figure 15.2*b*), a critical function is to hold the concrete from moving out laterally, which may be achieved by so-called *containment* of the concrete mass, similar to the action of a piston chamber containing air or hydraulic fluid. If compression resistance can be obtained from air that is contained, surely it can be more significantly obtained from contained concrete. This is a basic reason for the traditional extra strength of the spiral column and one reason for now favoring very closely spaced ties in

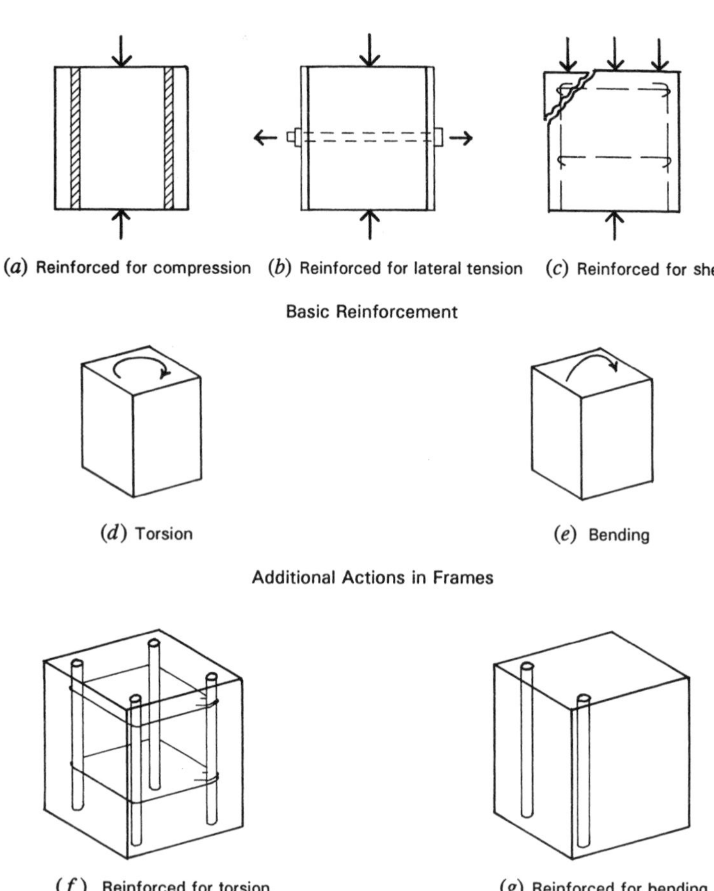

Figure 15.2 Forms and functions of column reinforcement.

tied columns. In retrofitting columns for improved seismic resistance a technique sometimes used is to actually provide a confining, exterior jacket of steel or fiber strand, essentially functioning as illustrated in Figure 15.2*b*.

Natural shear resistance is obtained from the combination of the vertical bars and the lateral ties or spiral, as shown in Figure 15.2*c*. If this is a critical concern, improvements can be obtained by using closer-spaced ties and a larger number of vertical bars that spread out around the column perimeter.

When used as parts of concrete frameworks, columns are also typically subjected to torsion and bending, as shown in Figures 15.2*d* and *e*. Torsional twisting tends to produce a combination of longitudinal tension and lateral shear; thus, the combination of full perimeter ties or spirals and the perimeter vertical bars provide for this in most cases.

Bending, if viewed independently, requires tension reinforcement, just as in an ordinary beam. In the column the ordinary section is actually a doubly reinforced one, with both tension and compression reinforcement for beam action. This function, combined with the basic axial compression, is discussed more fully in later sections of this chapter. An added complexity in many situations is the existence of bending in more than a single direction.

All of these actions can occur in various combinations due to different conditions of loading. Column design is thus a quite complex process if all possible structural functions are considered. A fundamental design principle becomes the need to make multiple usage of the simplest combination of reinforcing elements.

15.2 GENERAL CONSIDERATIONS FOR CONCRETE COLUMNS

Types of Columns

Concrete columns occur most often as the vertical support elements in a structure generally built of cast-in-place concrete (commonly called *site-cast*). This is the situation discussed in this chapter. Very short columns, called *pedestals*, are sometimes used in the support system for columns or other structures. The ordinary pedestal is discussed as a foundation transitional device in Chapter 16. Walls that serve as vertical compression supports are called *bearing walls*.

The site-cast concrete column usually falls into one of the following categories:

1. Square columns with tied reinforcement
2. Oblong columns with tied reinforcement
3. Round columns with tied reinforcement
4. Round columns with spiral-bound reinforcement
5. Square columns with spiral-bound reinforcement
6. Columns of other geometries (L shaped, T shaped, octagonal, etc.) with either tied or spiral-bound reinforcement

Obviously, the choice of column cross-sectional shape is an architectural as well as a structural decision. However, forming methods and costs, arrangement and installation of reinforcement, and relations of the column form and dimensions to other parts of the structural system must also be dealt with.

In tied columns the longitudinal reinforcement is held in place by loop ties made of small-diameter reinforcement bars, commonly No. 3 or No. 4. Such a column is represented by the square section shown in Figure 15.3*a*. This type of reinforcement can quite readily accommodate other geometries as well as the square.

Spiral columns are those in which the longitudinal reinforcing is placed in a circle, with the whole group of bars enclosed by a continuous cylindrical spiral made from steel rod or large-diameter steel

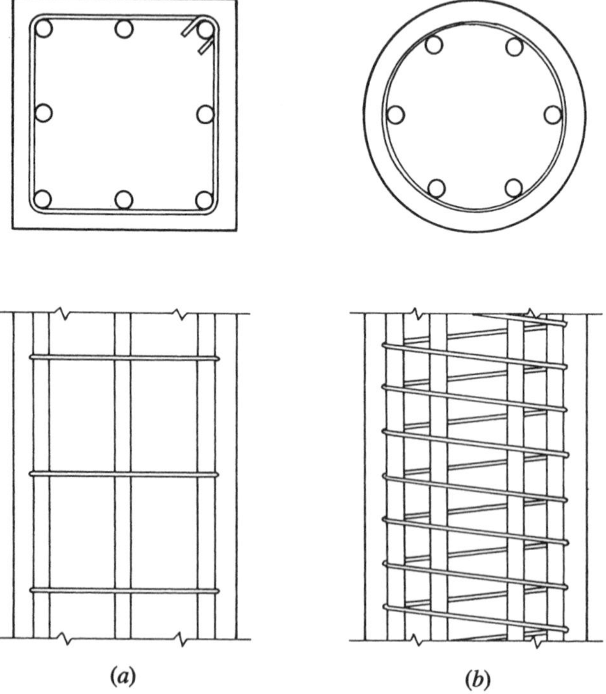

Figure 15.3 Primary forms of column reinforcement: (*a*) rectangular layout of vertical bars with lateral ties and (*b*) circular layout of vertical bars with continuous helix (spiral) wrap.

wire. Although this reinforcing system obviously works best with a round column section, it can be used also with other geometries. A round column of this type is shown in Figure 15.3*b*.

Experience has shown the spiral column to be slightly stronger than an equivalent tied column with the same amount of concrete and reinforcement. For this reason, code provisions have traditionally allowed slightly more load on spiral columns. Spiral reinforcement tends to be expensive, however, and the round bar pattern does not always mesh well with other construction details in buildings. Thus, tied columns are often favored where restrictions on the outer dimensions of the sections are not severe.

A recent development is the use of tied columns with very closely spaced ties. A basic purpose for this is the emulation of a spiral column for achieving additional strength, although many forms of gain are actually obtained simultaneously, as discussed in regard to the illustrations in Figure 15.2.

General Requirements for Columns

Code provisions and practical construction considerations place a number of restrictions on column dimensions and choice of reinforcement.

Column Size. The current code does not contain limits for column dimensions. For practical reasons, the following limits are recommended. Rectangular tied columns should be limited to a minimum area of 100 in.^{2} and a minimum side dimension of 10 in. if square and 8 in. if oblong. Spiral columns should be limited to a minimum size of 12 in. if either round or square.

Reinforcement. Minimum bar size is No. 5. The minimum number of bars is four for tied columns, five for spiral columns. The minimum amount of area of steel is 1% of the gross column area. A maximum area of steel of 8% of the gross area is permitted, but bar spacing limitations makes this difficult to achieve; 4% is a more practical limit. The ACI Code stipulates that for a compression member with a larger cross section than required by considerations of loading, a reduced effective area not less than one-half the total area may be used to determine minimum reinforcement and design strength.

Ties. Ties should be at least No. 3 for bars No. 10 and smaller. No. 4 ties should be used for bars that are No. 11 and larger. Vertical spacing of ties should be not more than 16 times the vertical bar diameter, 48 times the tie diameter, or the least dimension of the column. Ties should be arranged so that every corner and alternate longitudinal bar is held by the corner of a tie with an included angle of not greater than 135°, and no bar should be farther than 6 in. clear from such a supported bar. Complete circular ties may be used for bars placed in a circular pattern.

Concrete Cover. A minimum of 1.5 in. of cover is needed when the column surface is not exposed to weather and is not in contact with the ground. Cover of 2 in. should be used for formed surfaces exposed to the weather or in contact with ground. Cover of 3 in. should be used if the concrete is cast directly against earth without constructed forming, such as occurs on the bottoms of footings.

Spacing of Bars. Clear distance between bars should not be less than 1.5 times the bar diameter, 1.33 times the maximum specified size for the coarse aggregate, or 1.5 in.

Combined Compression and Bending

Due to the nature of most concrete structures, design practices generally do not consider the possibility of a concrete column with axial compression alone. This is to say, the existence of some bending moment is always considered together with the axial force.

Figure 15.4 illustrates the nature of the so-called *interaction response* for a concrete column, with a range of combinations of axial load plus bending moment. In general, there are three basic ranges of this behavior, as follows (see the dashed lines in Figure 15.4):

1. *Large Axial Force, Minor Moment.* For this case the moment has little effect, and the resistance to pure axial force is only negligibly reduced.
2. *Significant Values for Both Axial Force and Moment.* For this case the analysis for design must include the full combined force effects, that is, the interaction of the axial force and the bending moment.
3. *Large Bending Moment, Minor Axial Force.* For this case the column behaves essentially as a doubly reinforced (tension and compression reinforced) member, with its capacity for moment resistance affected only slightly by the axial force.

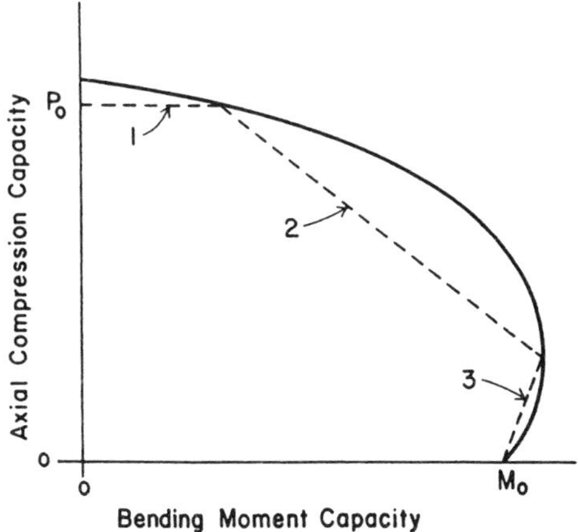

Figure 15.4 Interaction of axial compression (*P*) and bending moment (*M*) in a reinforced concrete column.

In Figure 15.4 the solid line on the graph represents the true response of the column—a form of behavior verified by many laboratory tests. The dashed line represents the generalization of the three types of response just described.

The terminal points of the interaction response—pure axial compression or pure bending moment—may be reasonably easily determined (P_0 and M_0 in Figure 15.4). The interaction responses between these two limits require complex analyses beyond the scope of this book.

Considerations for Column Shape

Usually, a number of possible combinations of reinforcing bars may be assembled to satisfy the steel area requirement for a given column. Aside from providing for the required cross-sectional area, the number of bars must also work reasonably in the layout of the column. Figure 15.5 shows a number of columns with various numbers of bars. When a square tied column is small, the preferred choice is usually that of the simple four-bar layout, with one bar in each corner and a single perimeter tie. As the column gets larger, the distance between the corner bars gets larger, and it is best to use more bars so that the reinforcement is spread out around the column periphery. For a symmetrical layout and the simplest of tie layouts,

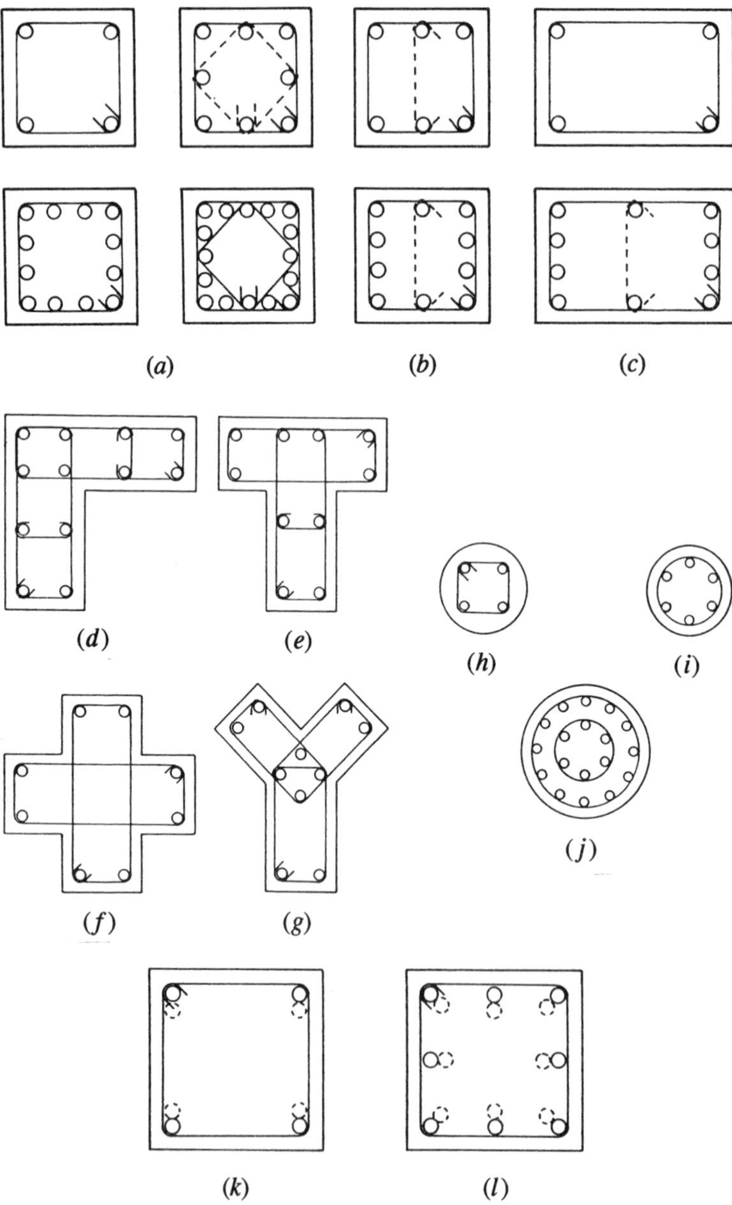

Figure 15.5 Considerations for bar layouts and tie patterns in tied columns.

the best choice is for numbers that are multiples of four, as shown in Figure 15.5*a*. The number of additional ties required for these layouts depends on the size of the column and the considerations discussed in Section 15.4.

An unsymmetrical bar arrangement (Figure 15.5*b*) is not necessarily bad, even though the column and its construction details are otherwise not oriented differently on the two axes. In situations where moments may be greater on one axis, the unsymmetrical layout is actually preferred; in fact, the column shape will also be more effective if it is unsymmetrical, as shown for the oblong shapes in Figure 15.5*c*.

Figures 15.5*d*–*g* show some special column shapes developed as tied columns. Although spirals could be used in some cases for such shapes, the use of ties allows greater flexibility and simplicity of construction. A reason for using ties may be the column dimensions, there being a practical lower limit of about 12 in. in width for a spiral-bound column.

Round columns are frequently formed as shown in Figure 15.5*h*, if built as tied columns. This allows for a minimum reinforcement with four bars. If a round pattern is used (as it must be for a spiral-bound column), the usual minimum number recommended is six bars, as shown in Figure 15.5*i*. Spacing of bars is much more critical in spiral-bound circular arrangements, making it very difficult to use high percentages of steel in the column section. For very large diameter columns it is possible to use sets of concentric spirals, as shown in Figure 15.5*j*.

For cast-in-place columns, vertical splicing of the steel bars is a concern. Two places where this commonly occurs are at the top of the foundation and at floors where a multistory column continues upward. At these points there are three ways to achieve the vertical continuity (splicing) of the steel bars, any of which may be appropriate for a given situation:

1. Bars may be lapped the required distance for development of the compression splice. For bars of smaller dimension and lower yield strengths, this is usually the desired method.
2. Bars may have milled square-cut ends butted together with a grasping device to prevent separation in a horizontal direction.
3. Bars may be welded with full-penetration butt welds or by welding of the grasping device described for method 2.

The choice of splicing methods is basically a matter of cost comparison but is also affected by the size of the bars, the degree of concern for bar spacing in the column arrangement, and possibly a need for some development of tension through the splice if uplift or high magnitudes of moments exist. If lapped splicing is used, a problem that must be considered is the bar layout at the location of the splice, at which point there will be twice the usual number of bars. The lapped bars may be adjacent to each other, but the usual considerations for space between bars must be made. If spacing is not critical, the arrangement shown in Figure 15.5*k* is usually chosen, with the spliced sets of bars next to each other at the tie perimeter. If spacing limits prevent the arrangement in Figure 15.5*k*, that shown in Figure 15.5*l* may be used, with the lapped sets in concentric patterns. The latter arrangement is used for spiral columns, where spacing is often critical.

Bending of steel bars involves the development of yield stress to achieve plastic deformation (the residual bend). As bars get larger in diameter, they are more difficult—and less feasible—to bend. Also, as the yield stress increases, the bending effort gets larger. It is questionable to try to bend bars as large as No. 14 or No. 18 in any grade, and it is also not advised to bend any bars with yield stress greater than 75 ksi. Bar fabricators should be consulted for real limits of this nature.

Columns in Site-Cast Frames

Reinforced concrete columns seldom occur as single, pin-ended members, as opposed to most wood columns and many steel columns. This condition may exist for some precast concrete columns, but almost all site-cast columns occur as members in frames, with interaction of the frame members in the manner of a so-called *rigid frame*.

Rigid frames derive their name from the joints between members, which are assumed to be *moment resistive* (rotationally rigid) and thus capable of transmitting bending moments between the ends of the connected members. This condition may be visualized by considering the entire frame as being cut from a single piece of material, as shown in Figure 15.6*a*. The site-cast concrete frame and the all-welded steel frame most fully approximate this condition.

When the horizontal-spanning frame members (beams) are subjected to vertical gravity loads, the inclination of their ends to rotate transmits bending to the columns connected to their ends, as

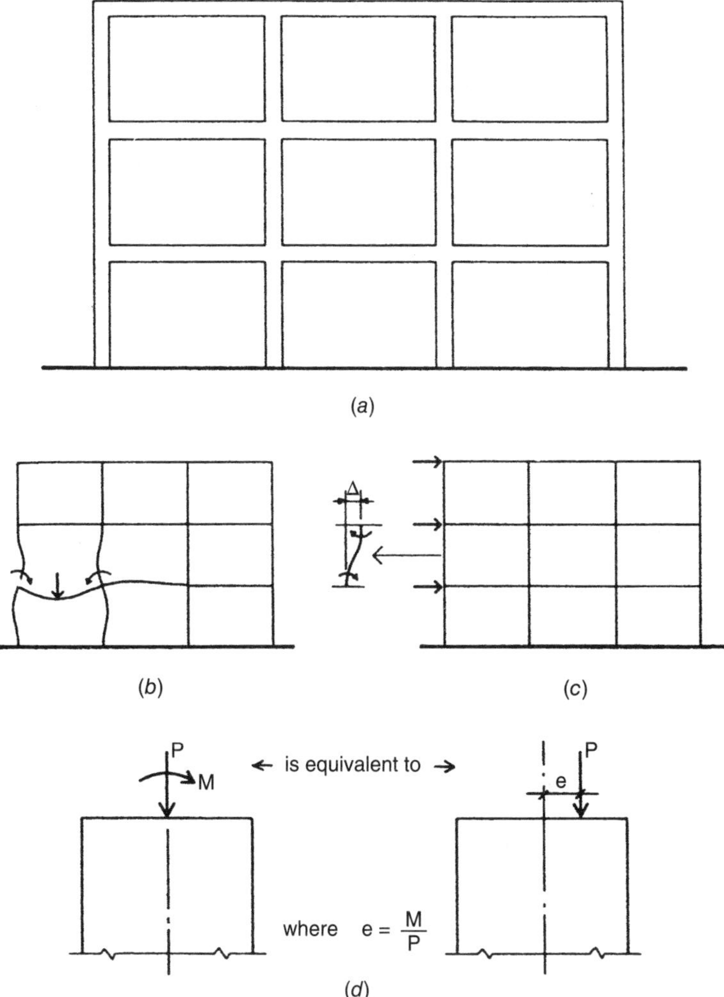

Figure 15.6 Columns in rigid frames.

shown in Figure 15.6*b*. If the frame is subjected to lateral loads (often the case, as rigid frames are frequently used for lateral bracing), the relative horizontal displacement of the column tops and bottoms (called lateral drift) transmits bending to the members connected to the columns (see Figure 15.6*c*). The combination of these loads result in the general case of combined axial loads plus bending in all the members of a rigid frame.

Figure 15.6*d* shows the case for the effect on the cross section of a column in a frame: a condition of axial compression plus bending. For some purposes, it is useful to visualize this as an analogous eccentric compressive force, with the bending produced by the product of the compression times its distance of eccentricity (dimension e in Figure 15.6*d*). It is thus possible to consider the column to have a maximum capacity for compression (with $e = 0$), which is steadily reduced as the eccentricity is increased. This is the concept of the interaction graph (see Figure 15.4).

Multistory Columns

Concrete columns occur frequently in multistory structures. In the site-cast structure, separate stories are typically cast in separate pours, with a cold joint (construction joint) between the successive pours. While this makes for a form of discontinuity, it does not significantly reduce the effective monolithic nature of the framed structure. Compression is continuous by the simple stacking of the levels of the heavy structure, and splicing of the reinforcement develops a form of tension continuity, permitting development of bending moments.

The typical arrangement of reinforcement in multistory columns is shown in Figure 13.25, which illustrates the form of bar development required to achieve the splicing of the reinforcement. This is essentially compressive reinforcement, so its development is viewed in those terms. However, an important practical function of the column bars is simply to tie the structure together through the discontinuous construction joints.

Load conditions change in successive stories of the multistory structure. It is, therefore, common to change both the column size and reinforcement. Design considerations for this are discussed in the examples in Chapter 20.

In very tall structures the magnitude of compression in lower stories requires columns with very high resistance. There is often some practical limit to column sizes, so that all efforts are made to obtain strength increases other than by simply increasing the mass of concrete. The three basic means of achieving this are:

1. Increase the amount of reinforcement, packing columns with the maximum amount that is feasible and allowable by codes.

2. Increase the yield strength of the steel, using as much as twice the strength for ordinary bars.
3. Increase the strength of the concrete.

The superstrength column is a clear case for use of the highest achievable concrete strengths and is indeed the application that has resulted recently in spiraling high values for design strength. Strengths exceeding 20,000 psi have been achieved.

15.3 DESIGN METHODS AND AIDS FOR CONCRETE COLUMNS

At the present, design of concrete columns is mostly achieved by using either tabulations from handbooks or computer-aided procedures. Using the code formulas and requirements to design by "hand operation" with both axial compression and bending present at all times is prohibitively laborious. The number of variables present (column shape and size, f_c', f_y, number and size of bars, arrangement of bars, etc.) adds to the usual problems of column design to make for a situation much more complex than those for wood or steel columns.

The large number of variables also works against the efficiency of handbook tables. Even if a single concrete strength (f_c') and a single steel yield strength (f_y) are used, tables would be very extensive if all sizes, shapes, and types (tied and spiral) of columns were included. Even with a very limited range of variables, handbook tables are much larger than those for wood or steel columns, They are, nevertheless, often quite useful for preliminary design estimation of column sizes. The obvious preference when relationships are complex, requirements are tedious and extensive, and there are a large number of variables, is for a computer-aided system. It is hard to imagine a professional design office that is turning out designs of concrete structures on a regular basis at the present without computer-aided methods. The reader should be aware that the software required for this work is readily available.

As in other situations, the common practices at any given time tend to narrow down to a limited usage of any type of construction, even though the potential for variation is extensive. It is thus possible to use some very limited but easy-to-use design aids to make early selections for design. These approximations may be adequate for preliminary building planning, cost estimates, and some preliminary structural analyses.

Approximate Design of Tied Columns

Tied columns are much preferred due to the relative simplicity and usually lower cost of their construction, plus their adaptability to various column shapes (square, round, oblong, T shape, L shape, etc.). Round columns—most naturally formed with spiral-bound reinforcing—are often made with ties instead, when the structural demands are modest.

The column with moment is often designed using the equivalent eccentric load method. The method consists of translating a compression plus bending situation into an equivalent one with an eccentric load, the moment becoming the product of the load and the eccentricity (see Figure 15.6*d*). This method is often used in the presentation of tabular data for column capacities.

Figures 15.7–15.10 yield safe ultimate factored capacities for a selected number of sizes of square tied columns with varying percentages of reinforcement. Allowable axial compression loads are given for various degrees of eccentricity, which is a means for handling axial load and bending moment combinations. The computed moment on the column is translated into an equivalent eccentric loading. Data for the curves were computed by strength design methods, as currently required by the ACI Code (Ref. 10).

When bending moments are relatively high in comparison to axial loads, round or square column shapes are not the most efficient, just as they are not for spanning beams. Figures 15.11 and 15.12 yield safe ultimate factored capacities for columns with rectangular cross sections. To further emphasize the importance of major bending resistance, all the reinforcement is assumed to be placed on the narrow sides, thus utilizing it for its maximum bending resistance effect.

The following examples illustrate the use of Figures 15.7–15.12 for the design of square and rectangular tied columns.

Example 1. A square tied column with $f'_c = 5$ ksi and steel with $f_y = 60$ ksi sustains an axial compression load of 150 kips dead load and 250 kips live load with no computed bending moment. Find the minimum practical column size if reinforcement is a maximum of 4% and the maximum size if reinforcement is a minimum of 1%.

Solution: As in all problems, this one begins with the determination of the factored ultimate axial load P_u:

$$P_u = 1.2(P_{\text{dead}}) + 1.6(P_{\text{live}}) = 1.2(150) + 1.6(250) = 580 \text{ kips}$$

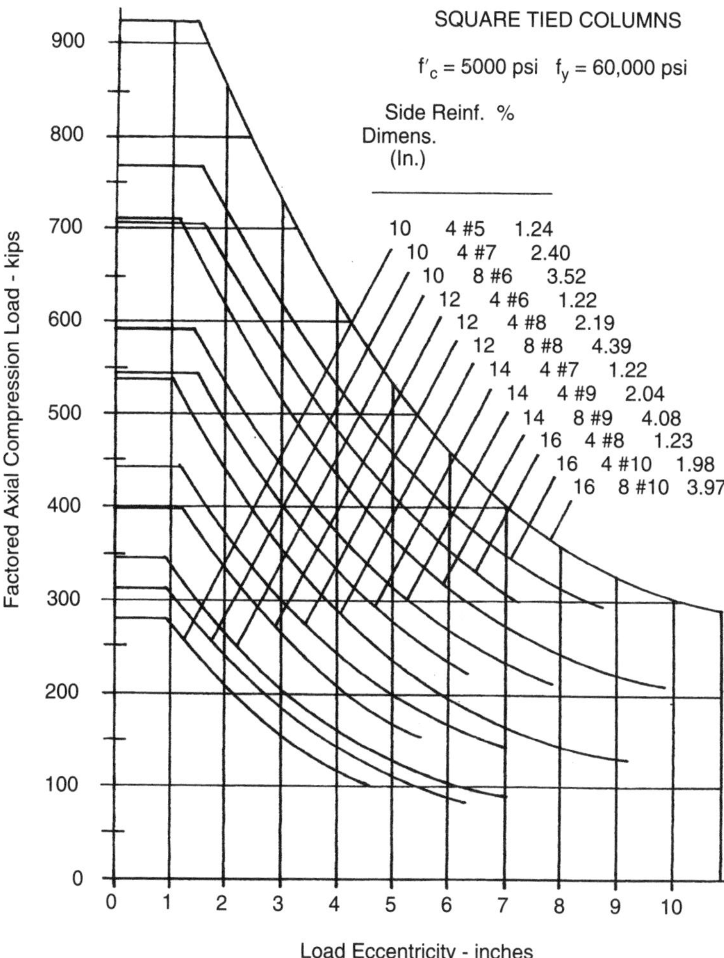

Figure 15.7 Maximum factored axial compression capacity for selected square tied columns.

With no real consideration for bending, the maximum axial load capacity may be determined from the graphs by simply reading up the left edge of the figure. The curved lines actually end some distance from this edge since the code requires a minimum bending for all columns.

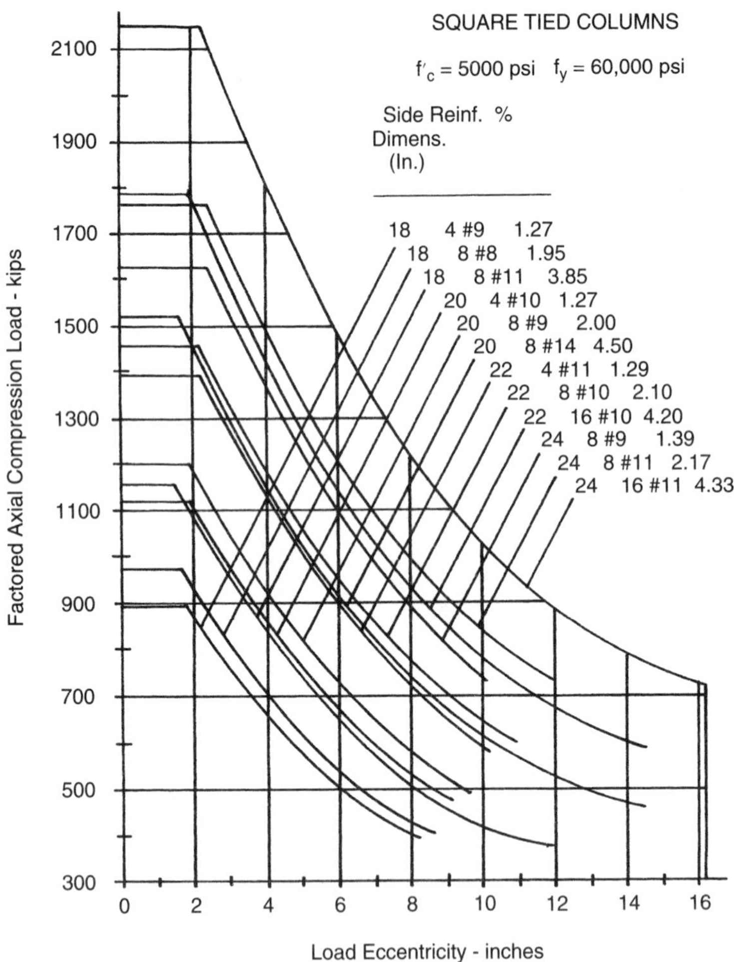

Figure 15.8 Maximum factored axial compression capacity for selected square tied columns.

Using Figure 15.7, the minimum size is a 14-in. square column with four No. 9 bars, for which the graph yields a maximum capacity of approximately 590 kips. Note that this column has a steel percentage of 2.04%.

What constitutes the maximum size is subject to some judgment. Any column with a curve above that of the chosen minimum column

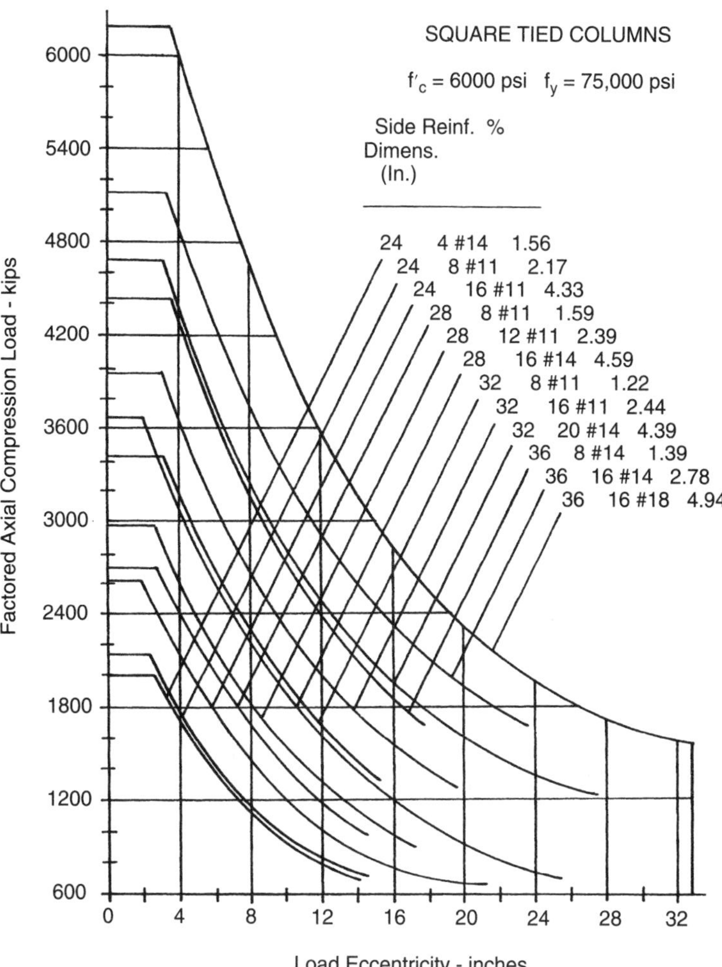

Figure 15.9 Maximum factored axial compression capacity for selected square tied columns.

will work. It becomes a matter of increasing redundancy of capacity. However, there are often other design considerations involved in developing a whole structural system, so these examples are quite academic. See the discussion for the example building in Chapter 20.

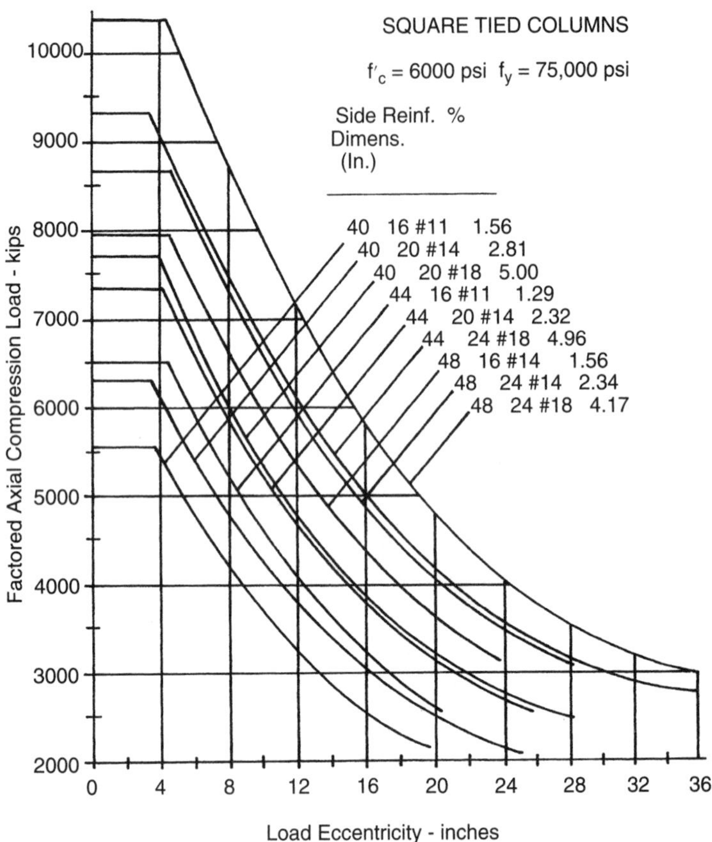

Figure 15.10 Maximum factored axial compression capacity for selected square tied columns.

For this example, it may be observed that the minimum-size choice is for a 14-in. square column. Thus, going up to a 15- or 16-in. size will reduce the reinforcement. We may thus note from the limited choices in Figure 15.7 that a maximum size is 16 in. square with four No. 8 bars, capacity is 705 kips, and $\rho = 1.23\%$. Since this is close to the usual recommended minimum reinforcement percentage (1%), columns of larger size will be increasingly redundant in strength, that is, structurally oversized in designer's lingo.

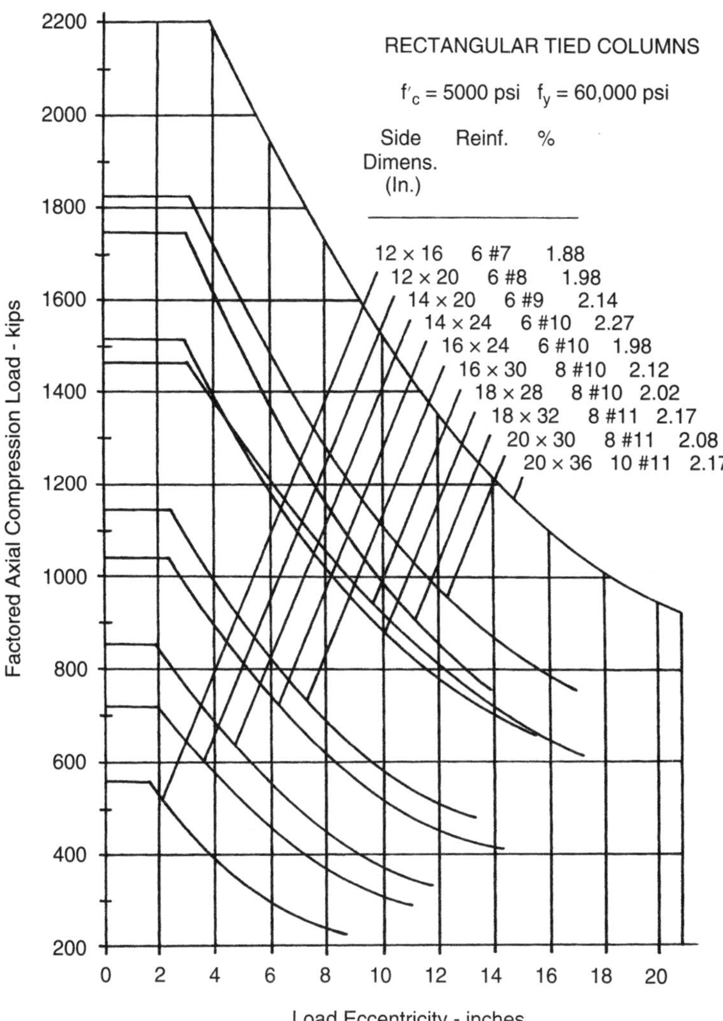

Figure 15.11 Maximum factored axial compression capacity for selected rectangular tied columns. Bending moment capacity determined for the major axis with reinforcement equally divided on the short sides of the column section.

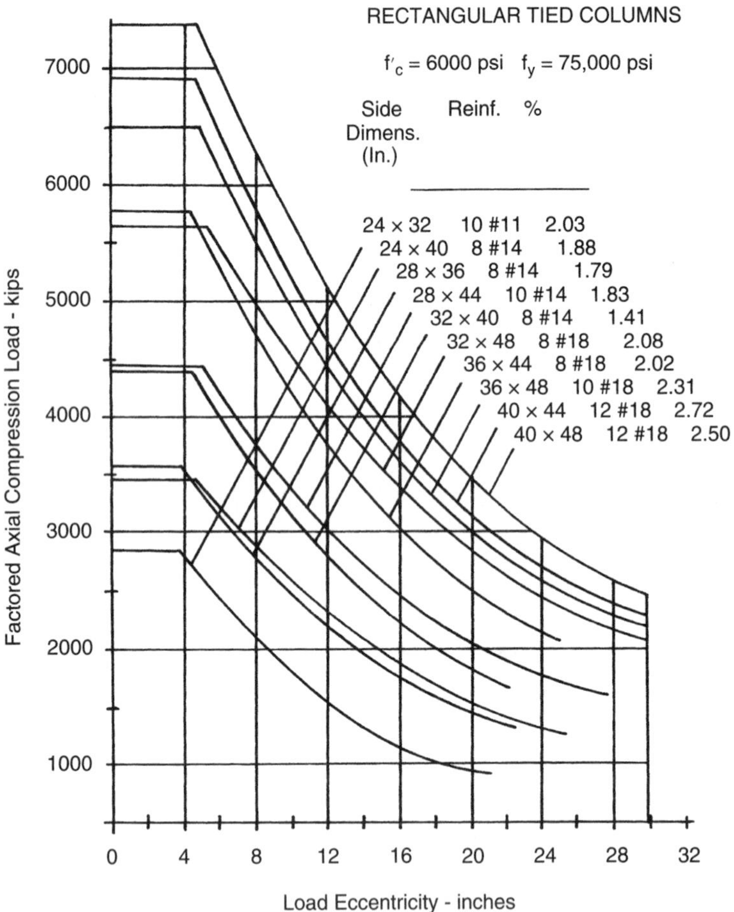

Figure 15.12 Maximum factored axial compression capacity for selected rectangular tied columns. Bending moment capacity determined for the major axis with reinforcement equally divided on the short sides of the column section.

Example 2. A square tied column with $f_c' = 5$ ksi and steel with $f_y =$ 60 ksi sustains an axial load of 150 kips dead load and 250 kips live load and a bending moment of 75 kip-ft dead load and 125 kip-ft live load. Determine the minimum-size column and its reinforcement.

Solution: First determine the ultimate axial load and ultimate bending moment. From Example 1, $P_u = 580$ kips:

$$M_u = 1.2(M_{\text{dead}}) + 1.6(M_{\text{live}}) = 1.2(75) + 1.6(125) = 290 \text{ kip-ft}$$

Next determine the equivalent eccentricity. Thus,

$$e = \frac{M_u}{P_u} = \frac{290 \times 12}{580} = 6 \text{ in.}$$

Then, from Figure 15.8, minimum size is 18 in. square with eight No. 11 bars, capacity at 6-in. eccentricity is approximately 650 kips. Note that the steel percentage is 3.85%. If this is considered to be too high, use a 20 × 20-in. column with four No. 10 bars with a capacity of approximately 675 kips, or a 22 × 22-in. column with four No. 11 bars with a capacity of approximately 900 kips.

Example 3. Select the minimum-size rectangular column for the same data as in Example 2.

Solution: With the factored axial load of 580 kips and the eccentricity of 6 in., Figure 15.11 yields the following: 14 × 24-in. column, six No. 10 bars, with a capacity of approximately 730 kips.

Problems 15.3.A–C Using Figures 15.7–15.10, select the minimum-size square tied column and its reinforcement for the following data:

	Concrete Strength (psi)	Axial Compressive Load (kips)		Bending Moment (kip-ft)	
		Live	Dead	Live	Dead
A	5000	80	100	30	25
B	5000	100	140	40	60
C	5000	150	200	100	100

Problems 15.3.D–F From Figures 15.11 and 15.12, determine minimum sizes for rectangular columns for the same data as in Problems 15.3.A–C.

Round Columns

Round columns, as discussed previously, may be designed and built as spiral columns, or they may be developed as tied columns with bars in a rectangular layout or with the bars placed in a circle and held by a series of round circumferential ties. Because of the cost of spirals, it is usually more economical to use the tied column, so it is often used unless the additional compressive strength or other behavioral characteristics of the spiral column are required.

Figure 15.13 gives safe loads for round columns that are designed as tied columns. As for the square and rectangular columns in

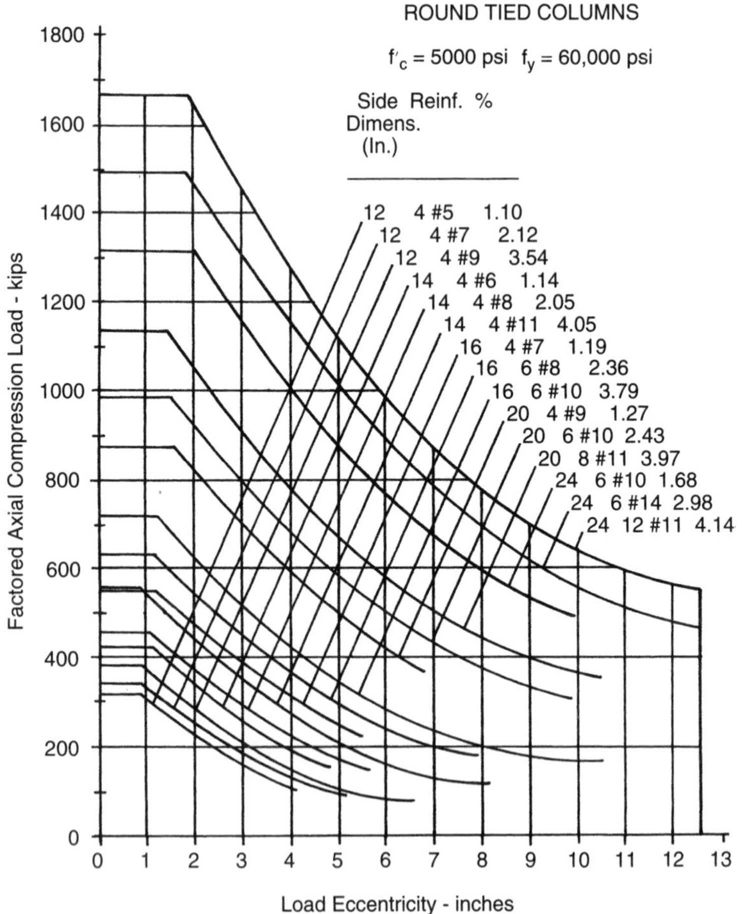

Figure 15.13 Maximum factored axial compression capacity for selected round tied columns.

Figures 15.7–15.12, load values have been adapted from values determined by strength design methods, and use is similar to that demonstrated in the preceding examples.

Problems 15.3.G–I Using Figure 15.13, pick the minimum-size round column and its reinforcing for the load and moment combinations in Problems 15.3.A–C.

15.4 SPECIAL CONSIDERATIONS FOR CONCRETE COLUMNS

Slenderness

Cast-in-place concrete columns tend to be stout in profile, so that slenderness related to buckling failure is much less often a critical concern than with columns of wood or steel. Earlier editions of the ACI Code (Ref. 10) provided for consideration of slenderness but permitted the issue to be ignored when the L/r of the column fell below a controlled value. For rectangular columns this meant that the effect was ignored when the ratio of unsupported height to side dimension was less than about 12. This is roughly analogous to the case for the wood column with L/d less than 11.

Slenderness effects must also be related to the conditions of bending for the column. Since bending is usually induced at the column ends, the two typical cases are those shown in Figure 15.14.

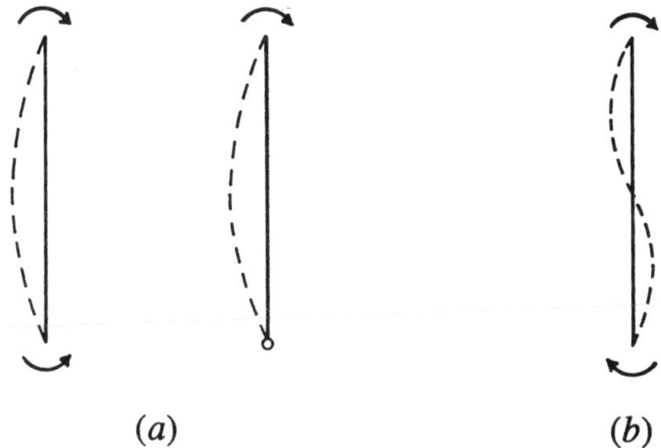

Figure 15.14 Assumed conditions of bending moments at column ends for consideration of column slenderness.

If a single end moment exists or two equal end moments exist, as shown in Figure 15.14*a*, the buckling effect is magnified and the *P*–delta effect is maximum. The condition in Figure 15.14*a* is not the common case, however, the more typical condition in framed structures being that shown in Figure 15.14*b*, for which the code treats the problem as one of moment magnification.

When slenderness must be considered, the ACI Code provides procedures for a reduction of column axial load capacity. One should be aware, however, that reduction for slenderness is not considered in design aids such as tables or graphs.

Development of Compressive Reinforcement

In multistory buildings, it is usually necessary to splice the vertical reinforcement in columns. Steel bars are available in limited lengths, making it necessary to do some splicing between stories of the structure. Thus, the vertical load transfer from an upper column to the column below it is achieved in two parts: from concrete to concrete by direct bearing and from steel to steel by splicing. This transfer must also occur at the joint between the lowest column and its supporting foundation. These development problems are treated in Section 13.7.

Vertical Concrete Compression Elements

There are several types of construction elements used to resist vertical compression for building structures. Dimensions of elements are used to differentiate between the defined elements. Figure 15.15 shows four such elements, described as follows:

Wall. Walls of one or more story height are often used as bearing walls, especially in concrete and masonry construction. Walls may be quite extensive in length but are also sometimes built in relatively short segments.

Pier. When a segment of wall has a length that is less than six times the wall thickness, it is called a pier or sometimes a wall pier.

Column. Columns come in many shapes but generally have some extent of height in relation to dimensions of the cross section. The usual limit for consideration as a column is a minimum height of three times the column diameter (e.g., side dimension). A wall pier may serve as a column, so the name distinction gets somewhat ambiguous.

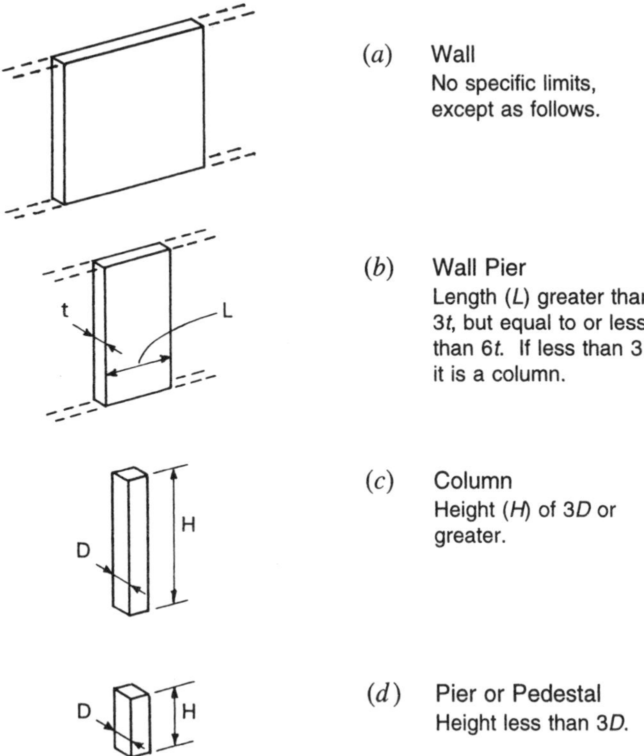

Figure 15.15 Classification of concrete compression elements.

Pedestal. A pedestal is really a short column; that is, a column with height not greater than three times its thickness. This element is also frequently called a pier, adding to the confusion of names.

To add more confusion, most large, relatively stout and massive concrete support elements are typically also called piers. These may be used to support bridges, long-span roof structures, or any other extremely heavy load. Identity in this case is more a matter of overall size rather than any specific proportions of dimensions. Bridge supports and supports for arch-type structures are also sometimes called *abutments*.

One more use of the word *pier* is for description of a type of foundation element, which is also sometimes called a *caisson*. This consists essentially of a concrete column cast in a vertical shaft that is dug in the ground.

Walls, piers, columns, and pedestals may also be formed from concrete masonry units (CMUs). The pedestal, as used with foundation systems, is discussed in Chapter 16.

Concrete Masonry Columns and Piers

Structural columns may be formed with CMU construction for use as entities or as part of a general CMU structure. In light construction, column pedestals are commonly formed with CMU construction, especially if no other site-cast concrete is being used, other than for ground slabs and foundations.

Figure 15.16 shows several forms of CMU columns, commonly used with construction that is generally reinforced to qualify as structural masonry. Figure 15.16*a* shows the minimum column, formed with two block units in plan. The positions of the two blocks are ordinarily rotated 90° in alternating courses of the construction, as shown in the figure. This column is totally filled with concrete and ordinarily reinforced with a vertical rod in each block cavity. Horizontal ties—necessary for full qualification as a structural column—must be placed in the mortar joints, which must usually be at least 0.5 in. thick to accommodate the ties.

Figure 15.16*b* shows the four-unit column, forming in this case a small void area in the center of the column. Capacity of this column may be varied, with the minimum column having concrete fill and steel rods in only the corner voids. Additional rods and fill may be placed in the other voids for a stronger column. Finally, the center void may also be filled.

Even larger columns may be formed with a perimeter of CMU construction and an increasing center void. These may be constituted as hollow masonry shell structures or may have significantly large concrete columns or piers cast into their voids.

It is also possible to form a reinforced concrete column by casting the concrete inside a boxlike shell made from CMU pieces that define a considerable void. The simplest form for this is shown in Figure 15.16*c*, using two U-shaped units for each course of the masonry. Columns as small as 8 in. wide could be made this way, but the usual smallest size is one with a 12-in.-wide side, and the most common size is one with a 16-in. side, producing an exterior that exactly resembles the column in Figure 15.16*a*. With 16-in. units, the

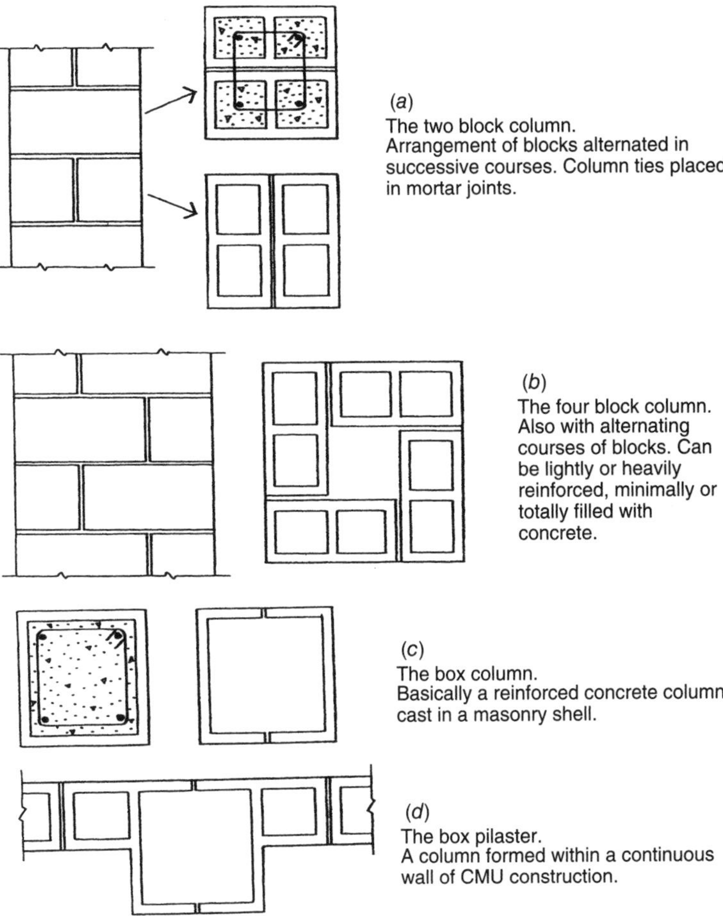

Figure 15.16 Forms of CMU columns.

net size of the concrete column on the inside is about 13.5 in., which is a significant concrete column.

The form of column in Figure 15.16*c* is also frequently used to produce pilasters in continuous walls of CMU construction. This is typically done by using alternating courses of units, with one course being as shown in Figure 15.16*c* and the alternating course being as shown in Figure 15.16*d*.

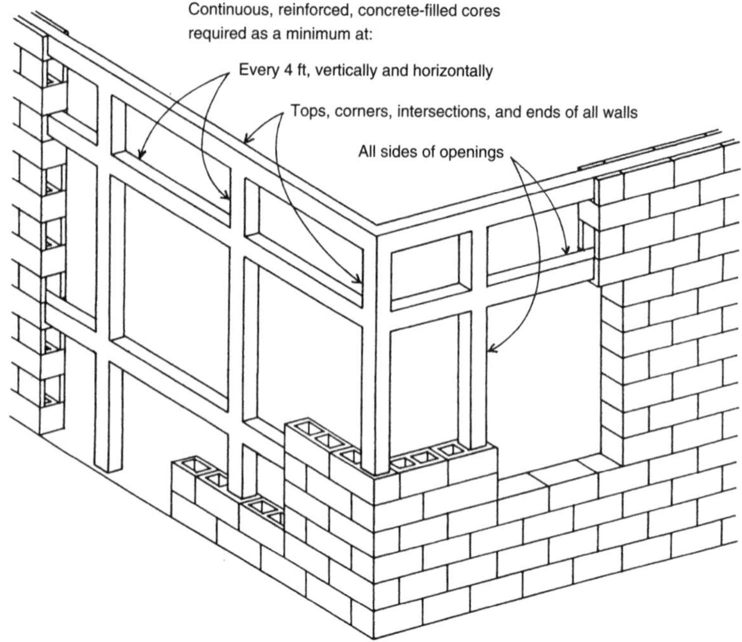

Figure 15.17 Common form of reinforced masonry construction with CMUs.

Shown in Figure 15.17 is a common form of structural masonry with CMUs, called *reinforced masonry*. In this type of construction, concrete is used in two ways:

1. For precast units that are laid up with mortar in the time-honored fashion
2. To fill selected vertically aligned voids and special horizontal courses after steel rods have been inserted

The result of using the concrete fill and steel reinforcement produces a reinforced concrete rigid frame inside the CMU construction. This type of construction is most popular in western and southern states in the United States.

16

FOUNDATIONS

Almost every building has a foundation built into the ground as its base. In times past, stone or masonry construction was the usual form of this structure. Today, however, site-cast concrete is the choice. This is one of the most common and extensive uses for concrete in building construction.

16.1 SHALLOW BEARING FOUNDATIONS

The most common foundation consists of pads of concrete placed beneath the building. Because most buildings make a relatively shallow penetration into the ground, these pads—called footings—are generally classified as *shallow bearing foundations.* For simple economic reasons, shallow foundations are generally preferred. However, when adequate soil does not exist at a shallow location, driven piles or excavated piers (caissons)—which extend some distance below the building—must be used; these are called *deep foundations.*

The two common footings are the wall footing and the column footing. Wall footings occur in strip form, usually placed symmetrically beneath the supported wall. Column footings are most often simple square pads supporting a single column. When columns are very close together or at the very edge of the building site, special footings that carry more than a single column may be used.

Two other basic construction elements that occur frequently with foundation systems are foundation walls and pedestals. Foundation walls may be used as basement walls or merely to provide a transition between more deeply placed footings and the above-ground building construction. Foundation walls are common with above-ground construction of wood or steel, as these constructions must be kept from contact with the ground.

Pedestals are actually short columns used as transitions between the building columns and their bearing footings. These may also be used to keep wood or steel columns above ground or they may serve a structural purpose to facilitate the transfer of a highly concentrated force from a column to a widely spread footing.

16.2 WALL FOOTINGS

Wall footings consist of concrete strips placed under walls. The most common type is that shown in Figure 16.1, consisting of a strip with a rectangular cross section placed in a symmetrical position with respect to the wall and projecting an equal distance as a cantilever from both faces of the wall. For soil pressure the critical dimension of the footing is its width as measured perpendicular to the wall.

Footings ordinarily serve as construction platforms for the walls they support. Thus, a minimum width is established by the wall thickness plus a few inches on each side. The extra width is necessary

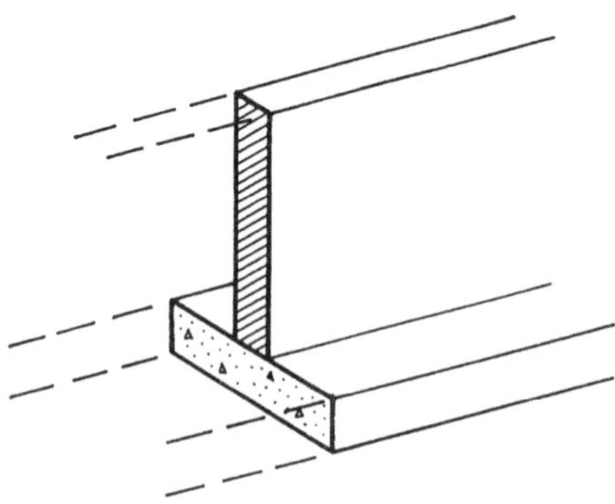

Figure 16.1 Typical form of a strip wall footing.

because of the crude form of foundation construction but also may be required for support of forms for concrete walls. A minimum projection of 2 in. is recommended for masonry walls and 3 in. for concrete walls.

With relatively light vertical loads, the minimum construction width may be adequate for soil bearing. Walls ordinarily extend some distance below grade, and allowable bearing will usually be somewhat higher than for very shallow footings. With the minimum recommended footing width, the cantilever bending and shear will be negligible, so no transverse (perpendicular to the wall) reinforcement is used. However, some longitudinal reinforcement is recommended.

As the wall load increases and a wider footing is required, the transverse bending and shear require some reinforcement. At some point the increased width also determines a required thickness. Otherwise recommended minimum thickness is 8 in. for nonreinforced footings and 10 in. for reinforced footings.

Determination of the Footing Width

Footing width is determined by soil pressure, assuming that the minimum width required for construction is not adequate for bearing. Since footing weight is part of the total load on the soil, the required width cannot be precisely determined until the footing thickness is known. A common procedure is to assume a footing thickness, design for the total load, verify the structural adequacy of the thickness, and—if necessary—modify the width once the final thickness is determined. The current ACI Code (Ref. 10) calls for using the unfactored loading (service load) when determining footing width by soil pressure. The example demonstrates this procedure.

Determination of the Footing Thickness

If the footing has no transverse reinforcement, the required thickness is determined by the tension stress limit of the concrete, in either flexural stress or diagonal stress due to shear. Transverse reinforcement is not required until the footing width exceeds the wall thickness by some significant amount, usually 2 ft or so. A good rule of thumb is to provide transverse reinforcement only if the cantilever edge distance for the footing (from the wall face to the footing edge) exceeds the footing thickness. For average conditions, this means for footings of about 3 ft wide or greater.

If transverse reinforcement is used, the critical concerns become for shear in the concrete and tension stress in the reinforcing.

Thicknesses determined by shear will usually assure a low bending stress in the concrete, so the cantilever beam action will involve a very low percentage of steel. This is in keeping with the general rule for economy in foundation construction, which is to reduce the amount of reinforcement to a minimum.

Minimum footing thicknesses are a matter of design judgment, unless limited by building codes. The ACI Code (Ref. 10) recommends limits of 8 in. for unreinforced footings and 10 in. for footings with transverse reinforcement. Another possible consideration for the minimum footing thickness is the necessity for placing dowels for wall reinforcement.

Selection of Reinforcement

Transverse reinforcement is determined on the basis of flexural tension and development length due to the cantilever action. Longitudinal reinforcement is usually selected on the basis of providing minimum shrinkage reinforcement. A reasonable value for the latter is a minimum of 0.0015 times the gross concrete area (area of the cross section of the footing). Cover requirements are for 2 in. from formed edges and 3 in. from surfaces generated without forming (such as the footing bottom). For practical purposes, it may be desirable to coordinate the spacing of the footing transverse reinforcement with that of any dowels for wall reinforcement. Reinforcement for shear is required by code only when the ultimate shear capacity due to factored loading (V_u) is greater than the factored shear capacity of the concrete (φV_c).

Example 1 illustrates the design procedure for a reinforced wall footing. Data for predesigned footings are given in Table 16.1. Figure 16.2 provides an explanation of the table entries. The use of unreinforced footings is not recommended for footings greater than 3 ft in width. Note: In using the strip method, a strip width of 12 in. is an obvious choice when using U.S. units but not with metric units. To save space the computations are performed with U.S. units only, but some metric equivalents are given for key data and answers.

Example 1. Design a wall footing with transverse reinforcement for the following data:

Footing design load = 3750 lb/ft dead load and 5000 lb/ft live load of wall length.

TABLE 16.1 Allowable Loads on Wall Footings (see Figure 16.2)

Maximum Soil Pressure (lb/ft²)	Minimum Wall Thickness, *t* (in.)		Allowable Load on Footing[a] (lb/ft)	Footing Dimensions (in.)		Reinforcement	
	Concrete	Masonry		*h*	*w*	Long Direction	Short Direction
1000	4	8	2,625	10	36	3 No. 4	No. 3 at 17
	4	8	3,062	10	42	2 No. 5	No. 3 at 12
	6	12	3,500	10	48	4 No. 4	No. 4 at 18
	6	12	3,938	10	54	3 No. 5	No. 4 at 13
	6	12	4,375	10	60	3 No. 5	No. 4 at 10
	6	12	4,812	10	66	5 No. 4	No. 5 at 13
	6	12	5,250	10	72	4 No. 5	No. 5 at 11
1500	4	8	4,125	10	36	3 No. 4	No. 3 at 11
	4	8	4,812	10	42	2 No. 5	No. 4 at 14
	6	12	5,500	10	48	4 No. 4	No. 4 at 11
	6	12	6,131	11	54	3 No. 5	No. 5 at 16
	6	12	6,812	11	60	5 No. 4	No. 5 at 12
	6	12	7,425	12	66	4 No. 5	No. 5 at 11
	8	16	8,100	12	72	5 No. 5	No. 5 at 10
2000	4	8	5,625	10	36	3 No. 4	No. 4 at 15
	6	12	6,562	10	42	2 No. 5	No. 4 at 12
	6	12	7,500	10	48	4 No. 4	No. 5 at 13
	6	12	8,381	11	54	3 No. 5	No. 5 at 12
	6	12	9,520	12	60	4 No. 5	No. 5 at 10
	8	16	10,106	13	66	4 No. 5	No. 5 at 10
	8	16	10,875	15	72	6 No. 5	No. 5 at 10
3000	6	12	8,625	10	36	3 No. 4	No. 4 at 11
	6	12	10,019	11	42	4 No. 4	No. 5 at 14
	6	12	11,400	12	48	3 No. 5	No. 5 at 11
	6	12	12,712	14	54	6 No. 4	No. 5 at 11
	8	16	14,062	15	60	5 No. 5	No. 5 at 10
	8	16	15,400	16	66	5 No. 5	No. 6 at 13
	8	16	16,725	17	72	6 No. 5	No. 6 at 11

[a] Allowable loads do not include the weight of the footing, which has been deducted from the total bearing capacity. Criteria: $f'_c = 2000$ psi, Grade 40 bars.

Wall thickness for design = 6 in.
Maximum soil pressure = 2000 psf
Concrete design strength = 2000 psi
Steel yield stress = 40,000 psi

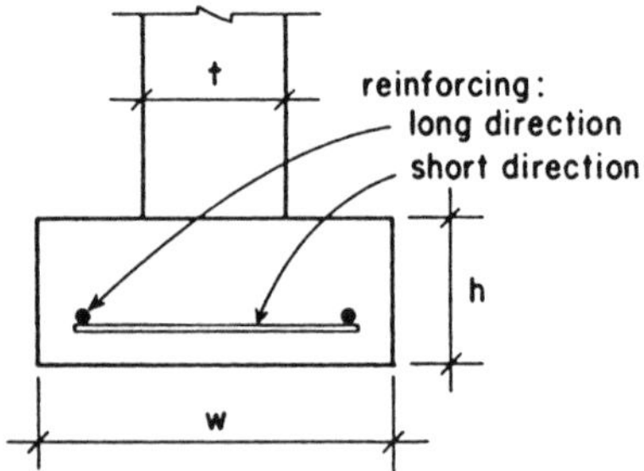

Figure 16.2 Reference for Table 16.1.

Solution: For the reinforced footing the only concrete stress of concern is that of shear. Concrete flexural stress will be low because of the low percentage of reinforcement. As with the unreinforced footing, the usual design procedure consists of making a guess for the footing thickness, determining the required width for soil pressure, and then checking the footing stress.

Try $h = 12$ in. Then, footing weight = 150 psf, and the net usable soil pressure is 2000 − 150 = 1850 lb/ft^2.

The footing design load is unfactored when determining footing width; therefore, the load is 8750 lb/ft. Required footing width is 8750/1850 = 4.73 ft, or 4.73(12) = 56.8 in., say 57 in., or 4 ft 9 in., or 4.75 ft. With this width, the design soil pressure for stress is 8750/4.75 = 1842 psf.

For the reinforced footing it is necessary to determine the effective depth, that is, the distance from the top of the footing to the center of the steel bars. For a precise determination this requires a second guess: the steel bar diameter (D). For the example a guess is made of a No. 6 bar with a diameter of 0.75 in. With the cover of 3 in., this produces an effective depth of $d = h - 3 - (D/2) = 12 - 3 - (0.75/2) = 8.625$ in.

Concern for precision is academic in footing design, however, considering the crude nature of the construction. The footing bottom is formed by a hand-dug soil surface, unavoidably roughed up during placing of the reinforcement and casting of the concrete. The value of d will, therefore, be taken as 8.6 in.

Next, we need to determine how much the soil is pushing back up on the footing using the factored loads. This load is

$$w_u = 1.2(3750) + 1.6(5000) = 12500 \text{ lb/ft}$$

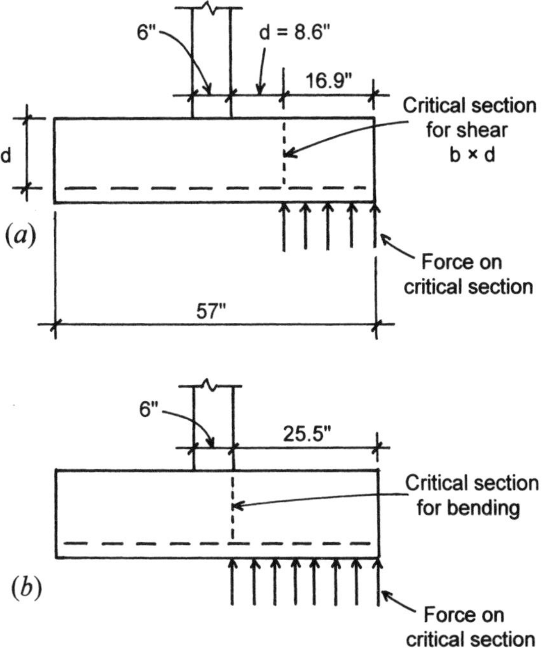

Figure 16.3 Considerations for shear and bending in the wall footing.

With a width of 4.75 ft, the factored design soil pressure is

$$p_d = \frac{12500}{4.75 \times 1} = 2630 \text{ psf}$$

The critical section for shear stress is taken at a distance of *d* from the face of the wall. As shown in Figure 16.3*a*, this places the shear section at a distance of 16.9 in. from the footing edge. At this location the shear force is determined as

$$V_u = 2630 \times \frac{16.9}{12} = 3700 \text{ lb}$$

and the shear capacity of the concrete is

$$\phi V_c = 0.75\left(2\sqrt{f_c'}\right)(bd) = 0.75\left(2\sqrt{2000}\right)(12 \times 8.6) = 6920 \text{ lb}$$

It is possible, therefore, to reduce the footing thickness. However, cost-effectiveness is usually achieved by reducing the steel reinforcement to a minimum. Low-grade concrete dumped into a hole in the ground is relatively inexpensive, compared to the cost of steel bars. Selection of footing thickness, therefore, becomes a matter of design judgment, unless the footing width becomes as much as five times or so the wall thickness, at which point stress limits may become significant.

If a thickness of 11 in. is chosen for this example, the shear capacity decreases only slightly, and the required footing width will remain effectively the same. A new effective depth of 7.6 in. will be used, but the design soil pressure for stresses will remain the same as it relates only to the width of the footing.

The bending moment to be used for determination of the steel bars is computed as follows (see Figure 16.3*b*). The force on the cantilevered edge of the footing is

$$F = \frac{25.5}{12} \times 2630 = 5589 \text{ lb}$$

and the cantilever bending moment at the wall face is thus

$$M_u = 5589 \times \frac{25.5}{2} = 71260 \text{ in.-lb}$$

$$M_t = \frac{M_u}{\phi} = \frac{71260}{0.9} = 79178 \text{ in.-lb}$$

Assuming an underreinforced section with *a*/*d* approximately 0.2, the required steel area per foot of wall length is

$$A_s = \frac{M_t}{f_y(d - a/d)} = \frac{79178}{40000(7.6 - (0.2 \times 7.6)/2)} = 0.29 \text{ in.}^2$$

The spacing required for a given bar size to satisfy this requirement can be derived as follows:

$$\text{Required spacing} = (\text{area of bar}) \times \frac{12}{\text{required area/ft}}$$

Thus, for a No. 3 bar

$$s = 0.11 \times \frac{12}{0.290} = 4.6 \text{ in.}$$

TABLE 16.2 Options for the Reinforcement for Example 1

Bar Size	Area of Bar (in.^{2})	Area Required for Bending (in.^{2})	Bar Spacing Required (in.)	Bar Spacing Selected (in.)
3	0.11	0.290	4.6	4.5
4	0.20	0.290	8.3	8.0
5	0.31	0.290	12.8	12.5
6	0.44	0.290	18.2	18.0

Using this procedure, the required spacings for bar sizes 3–7 are shown in the fourth column of Table 16.2. Bar sizes and spacings can be most easily selected using handbook tables that yield the average steel areas for various combinations of bar size and spacing. One such table is Table 13.6, from which the spacings shown in the last column of Table 16.2 were selected, indicating a range of choices for the footing transverse reinforcement. Selection of the actual bar size and spacing is a matter of design judgment, for which some considerations are the following:

1. Maximum recommended spacing is 18 in.
2. Minimum recommended spacing is 6 in., to reduce the number of bars and make placing of concrete easier.
3. Preference is for smaller bars as long as spacing is not too close.
4. A practical spacing may be that of the spacing of vertical reinforcement in the supported wall, for which footing dowels are required (or some full number multiple or division of the wall bar spacing).

With these considerations in mind, a choice may be made for either the No. 5 bars at 12.5-in. spacing or the No. 4 bars at 8-in. spacing. The No. 6 bars at 24-in. spacing would not be a good choice due to the large distance between bars; thus, the practice is to use a maximum spacing of 18 in. for any spaced set of bars. Another consideration that must be made for the choice of reinforcement is that regarding the required development length for anchorage. With 2 in. of edge cover, the bars will extend 23.5 in. from the critical bending section at the wall face (see Figure 16.3*b*). Inspection of Table 13.9 will show that this is an adequate length for all the bar sizes used in Table 16.2. Note that the placement of the bars in the footing falls in the classification of "other" bars in Table 13.9.

For the longitudinal reinforcement, the minimum steel area is

$$A_s = (0.0015)(11)(57) = 0.94 \text{ in.}^2$$

Using three No. 5 bars yields

$$A_s = (3)(0.31) = 0.93 \text{ in.}^2$$

Table 16.1 gives values for wall footings for four different soil pressures. Table data were derived using the procedures illustrated in the example. Figure 16.2 shows the dimensions referred to in the table.

Problem 16.2.A Using concrete with a design strength of 2000 psi and grade 40 bars with a yield strength of 40 ksi, design a wall footing for the following data: wall thickness = 10 in.; dead load on footing = 5000 lb/ft and live load = 7000 lb/ft; and maximum soil pressure = 2000 psf.

Problem 16.2.B Same as Problem 16.2.A, except wall is 15 in. thick, dead load is 6000 lb/ft and live load is 8000 lb/ft, and maximum soil pressure is 3000 psf.

16.3 COLUMN FOOTINGS

The great majority of independent or isolated column footings are square in plan, with reinforcement consisting of two equal sets of bars at right angles to each other. The column may be placed directly on the footing or it may be supported by a pedestal, consisting of a short column that is wider than the supported column. The pedestal helps to reduce the so-called *punching shear* effect in the footing; it also slightly reduces the edge cantilever distance and thus the magnitude of bending in the footing. The pedestal thus allows for a thinner footing and slightly less footing reinforcement. However, another reason for using a pedestal may be to raise the bottom of the supported column above the ground, which is important for columns of wood and steel.

The design of a column footing is based on the following considerations:

Maximum Soil Pressure. The sum of the unfactored superimposed load on the footing and the unfactored weight of the footing must not exceed the limit for bearing pressure on the supporting soil material. The required total plan area of the footing is derived on this basis.

Design Soil Pressure. By itself, simply resting on the soil, the footing does not generate shear or bending stresses. These are developed only by the superimposed load. Thus, the soil pressure to be used for designing the footing is determined as the factored superimposed load divided by the actual chosen plan area of the footing.

Control of Settlement. Where buildings rest on highly compressible soil, it may be necessary to select footing areas that assure a uniform settlement of all the building foundation supports. For some soils, long-term settlement under dead load only may be more critical in this regard and must be considered as well as maximum soil pressure limits.

Size of the Column. The larger the column, the less will be the shear and bending stresses in the footing since these are developed by the cantilever effect of the footing projection beyond the edges of the column.

Shear Capacity Limit for the Concrete. For square-plan footings this is usually the only critical stress in the concrete. To achieve an economical design, the footing thickness is usually chosen to reduce the need for reinforcement. Although small in volume, the steel reinforcement is a major cost factor in reinforced concrete construction. This generally rules against any concerns for flexural compression stress in the concrete. As with wall footings, the factored load is used when determining footing thickness and any required reinforcement.

Flexural Tension Stress and Development Length for the Bars. These are the main concerns for the steel bars, on the basis of the cantilever bending action. It is also desired to control the spacing of the bars between some limits.

Footing Thickness for Development of Column Bars. When a footing supports a reinforced concrete or masonry column, the compressive force in the column bars must be transferred to the footing by development action (called *doweling*), as discussed in Chapter 13. The thickness of the footing must be adequate for this purpose.

Example 2 illustrates the design process for a simple, square column footing.

Example 2. Design a square column footing for the following data:

Column load = 200 kips dead load and 300 kips live load
Column size = 15 in. square
Maximum allowable soil pressure = 4000 axim
Concrete design strength = 3000 psi
Yield stress of steel reinforcement = 40 ield

Solution: A quick guess for the footing size is to divide the load by the maximum allowable soil pressure. Thus,

$$A = \frac{500}{4} = 125 \text{ ft}^2 \quad w = \sqrt{125} = 11.2 \text{ ft}$$

This does not allow for the footing weight, so the actual size required will be slightly larger. However, it gets the guessing quickly into the approximate range.

For a footing this large the first guess for the footing thickness is a real shot in the dark. However, any available references to other footings designed for this range of data will provide some reasonable first guess.

Try $h = 31$ in.. Then, footing weight = (31/12)(150) = 388 psf. Net usable soil pressure = 4000 − 388 = 3612 psf. The required plan area of the footing is thus

$$A = \frac{500000}{3612} = 138.4 \text{ ft}^2$$

and the required width for a square footing is

$$w = \sqrt{138.4} = 11.76 \text{ ft}$$

Try $w = 11$ ft 9 in., or 11.75 ft. Then, design soil pressure = $500{,}000/(11.75)^2 = 3622$ psf.

For determining reinforcement and footing thickness, a factored soil pressure is needed:

$$P_u = 1.2(P_{\text{dead}}) + 1.6(P_{\text{live}}) = 1.2(200) + 1.6(300) = 720 \text{ kips}$$

and, for the design soil pressure,

$$p_d = \frac{720}{(11.75)^2} = 5.22 \text{ ksf} \Rightarrow 5220 \text{ psf}$$

Determination of the bending force and moment are as follows (see Figure 16.4):

Bending force:

$$F = 5220 \times \frac{63}{12} \times 11.75 = 322{,}000 \text{ lb}$$

Bending moment:

$$M_u = 322{,}000 \times \frac{63}{12} \times \frac{1}{2} = 845{,}000 \text{ ft-lb}$$

For design

$$M_t = \frac{M_u}{\phi} = \frac{845{,}000}{0.9} = 939{,}000 \text{ ft-lb}$$

This bending moment is assumed to operate in both directions on the footing and is provided for with similar reinforcement in each direction. However, it is necessary to place one set of bars on top of the perpendicular set, as shown in Figure 16.5, and there are thus different effective depths in each direction. A practical procedure is to use the average of these two depths, that is, a depth equal to the

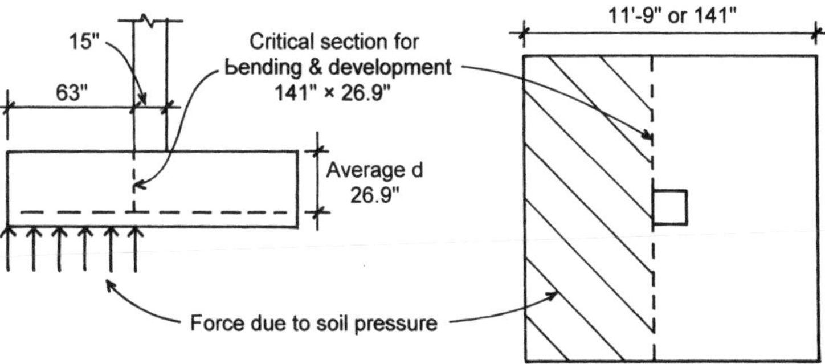

Figure 16.4 Considerations for bending and bar development in the column footing.

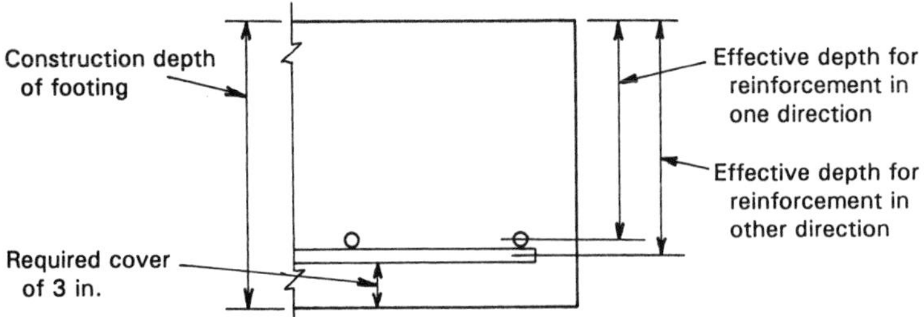

Figure 16.5 Considerations for effective depth of the column footing with two-way reinforcement.

footing thickness minus the 3-in. cover and one bar diameter. This will theoretically result in a minor overstress in one direction, which is compensated for by a minor understress in the other direction.

It is also necessary to assume a size for the reinforcing bar in order to determine the effective depth. As with the footing thickness, this must be a guess, unless some reference is used for approximation. Assuming a No. 9 bar for this footing, the effective depth thus becomes

$$d = h - \text{cover} - \left(\frac{\text{bar diameter}}{2}\right) = 31 - 3 - \left(\frac{1.13}{2}\right) = 27.4 \text{ in.}$$

The section resisting the bending moment is one that is 141 in. wide and has a depth of 26.9 in. Using a resistance factor for a balanced section from Table 13.2, the balanced moment capacity of this section is determined as follows:

$$M_R = Rbd^2 = \frac{1149 \times 141 \times (27.4)^2}{12} = 10{,}100{,}000 \text{ ft-lb}$$

which is almost 10 times the required moment.

From this analysis it may be seen that the compressive bending stress in the concrete is not critical. Furthermore, the section may be classified as considerably underreinforced, and a conservative value can be used for *jd* (or *d–a*/2) in determining the required reinforcement.

The critical stress condition in the concrete is that of shear, either in beam-type action or in punching action. Referring to Figure 16.6, the investigation for these two conditions is as follows:

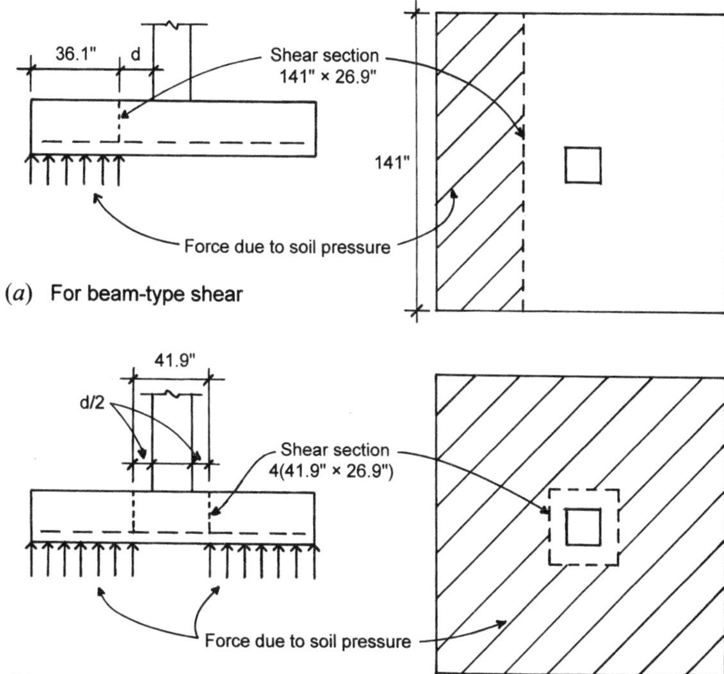

Figure 16.6 Considerations for shear in the column footing.

For beam-type shear (Figure 16.6*a*):

$$V_u = 5220 \times 11.75 \times \frac{36.1}{12} = 185{,}000 \text{ lb}$$

For the shear capacity of the concrete:

$$V_c = 2\sqrt{f_c'}(bd) = 2\sqrt{3000}(141 \times 26.9) = 415{,}000 \text{ lb}$$

$$\phi V_c = 0.75(415{,}000) = 311{,}000 \text{ lb}$$

For punching shear (Figure 16.6*b*):

$$V_u = 5220\left[(11.75)^2 - \left(\frac{41.9}{12}\right)^2\right] = 657{,}000 \text{ lb}$$

Shear capacity of the concrete is

$$V_c = 4\sqrt{f_c'}(bd) = 4\sqrt{3000}(4 \times 41.9 \times 26.9) = 988{,}000 \text{ lb}$$

$$\phi V_c = 0.75(988{,}000) = 741{,}000 \text{ lb}$$

Although the beam shear force is low, the punching shear force is close to the limit, so the 31-in. thickness is probably the minimum allowable dimension.

Using an assumed value of 0.9 for j, the area of steel required is determined as

$$A_s = \frac{M}{f_y jd} = \frac{939{,}000 \times 12}{40{,}000 \times 0.9 \times 27.4} = 11.4 \text{ in.}$$

There are several bar size-and-number combinations that may be selected to satisfy this area requirement. A range of possible choices is shown in Table 16.3. Also displayed in the table are data relating to two other considerations for the bar choice: the center-to-center spacing of the bars and the development lengths required. Spacings given in the table assume the first bar to be centered at 4 in. from the footing edge. Maximum spacing should be limited to 18 in. and minimum to about 6 in.

Required development lengths are taken from Table 13.9. The development length available is a maximum of the distance from the column face to the footing edge minus a 2-in. cover, in this case, a distance of 61 in.

TABLE 16.3 Options for the Reinforcement for the Column Footing

Number and Size of Bars	Area of Steel Provided in.^{2}	Required Development Length[a] in.	Center-to-Center Spacing in.
20 No. 7	12.0	32	7.0
15 No. 8	11.85	37	9.5
12 No. 9	12.0	42	12.1
10 No. 10	12.7	47	14.7
8 No. 11	12.48	52	19.0

[a]From Table 13.9, values for "Other Bars," $f_y = 40$ ksi, $f_c' = 3$ ksi.

Inspection of Table 16.3 reveals that all the combinations given are acceptable. In most cases, designers prefer to use the largest possible bar in the fewest number since handling of the bars is simplified with fewer bars, which is usually a savings of labor time and cost.

Although the computations have established that the 31-in. dimension is the least possible thickness, it may be more economical to use a thicker footing with less reinforcement, assuming the usual ratio of costs of concrete and steel. In fact, if construction cost is the major determinant, the ideal footing is the one with the lowest combined cost for excavation, forming, concrete, and steel.

One possible limitation for the footing reinforcement is the total percentage of steel. If this is excessively low, the section is hardly being reinforced. The ACI Code stipulates that the minimum reinforcement be the same as that for temperature reinforcement in slabs, a percentage of 0.002 A_g for grade 40 bars and 0.0015A_g for grade 60 bars. For this footing cross section of 141 in. × 31 in. with grade 40 bars, this means an area of

$$A_s = 0.002(141 \times 31) = 8.74 \text{ in.}^2$$

There are a number of other considerations that may affect the selection of footing dimensions, such as the following:

Restricted Thickness. Footing thickness may be restricted by excavation problems, water conditions, or presence of undesirable soil materials at lower strata. Thickness may be reduced by use of pedestals as discussed in Section 16.4.

Need for Dowels. When the footing supports a reinforced concrete or masonry column, dowels must be provided for the vertical column reinforcement, with sufficient extension into the footing for development of the bars. This problem is discussed in Section 13.48.

Restricted Footing Width. Proximity of other construction or close spacing of columns sometimes makes it impossible to use the required square footing. For a single column a possible solution is the use of an oblong (called a rectangular) footing. For multiple columns a combined footing is sometimes used. A special footing is the cantilever footing, used when footings cannot extend beyond the building face. An extreme case occurs when the entire building footprint must be used in a single large footing, called a mat foundation.

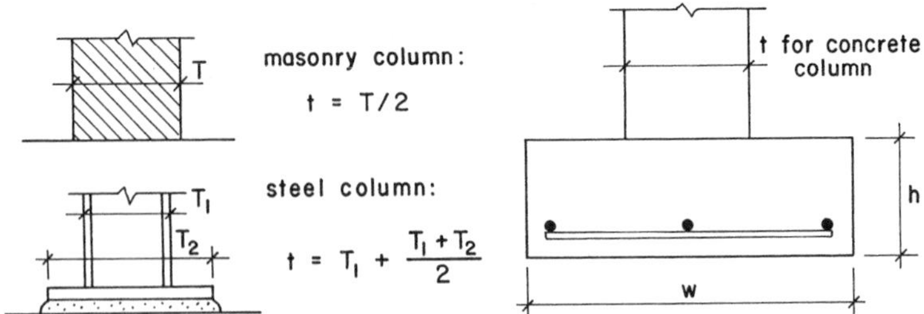

Figure 16.7 Reference for Table 16.4.

Table 16.4 yields the allowable superimposed load for a range of predesigned footings and soil pressures. This material has been adapted from more extensive data in *Simplified Design of Building Foundations* (Ref. 11). Figure 16.7 indicates the symbols used for dimensions in Table 16.4.

Problem 16.3.A Design a square footing for a 14-in. square column and a superimposed dead load of 100 kips and a live load of 100 kips. The maximum permissible soil pressure is 3000 psf. Use concrete with a design strength of 3 ksi and grade 40 reinforcing bars with yield strength of 40 ksi.

Problem 16.3.B Same as Problem 16.3.A, except column is 18 in., dead load is 200 kips and live load is 300 kips, and permissible soil pressure is 4000 psf.

16.4 PEDESTALS

A pedestal (also called a pier) is defined by the ACI Code as a short compression member whose height does not exceed three times its width. Pedestals are frequently used as transitional elements between columns and the bearing footings that support them. Figure 16.8 shows the use of pedestals with both steel and reinforced concrete columns. The most common reasons for use of pedestals are:

1. To spread the load on top of the footing. This may relieve the intensity of direct bearing pressure on the footing or may simply permit a thinner footing with less reinforcement due to the wider column.
2. To permit the column to terminate at a higher elevation where footings must be placed at depths considerably below the lowest

TABLE 16.4 Safe Loads for Square Column Footings[a] (see Figure 16.7)

Maximum Soil Pressure (psf)	Minimum Column Width, t (in.)	Service Load on Footing (kips)	Dimensions h (in.)	Dimensions w (ft)	Reinforcement Each Way
1000	8	7	10	3	3 No. 2
	8	10	10	3.5	3 No. 3
	8	14	10	4	4 No. 3
	8	17	10	4.5	4 No. 4
	8	21	10	5	4 No. 5
	8	31	10	6	4 No. 6
	8	42	11	7	6 No. 6
1500	8	12	10	3	3 No. 3
	8	16	10	3.5	3 No. 4
	8	22	10	4	4 No. 4
	8	27	10	4.5	4 No. 5
	8	34	10	5	5 No. 5
	8	49	12	6	5 No. 6
	8	65	13	7	5 No. 7
	8	84	15	8	7 No. 7
	8	105	17	9	8 No. 7
2000	8	16	10	3	3 No. 3
	8	23	10	3.5	3 No. 4
	8	30	10	4	5 No. 4
	8	38	10	4.5	5 No. 5
	8	46	11	5	4 No. 6
	8	66	13	6	6 No. 6
	8	89	15	7	6 No. 7
	8	114	17	8	8 No. 7
	8	143	19	9	7 No. 8
	10	175	20	10	9 No. 8
3000	8	25	10	3	3 No. 4
	8	35	10	3.5	3 No. 5
	8	45	11	4	4 No. 5
	8	57	12	4.5	4 No. 6
	8	71	13	5	5 No. 6
	8	101	15	6	7 No. 6
	10	136	17	7	7 No. 7
	10	177	20	8	7 No. 8
	12	222	21	9	9 No. 8
	12	272	24	10	9 No. 9
	12	324	26	11	10 No. 9
	14	383	28	12	10 No. 10

(*continued*)

TABLE 16.4 *(Continued)*

Maximum Soil Pressure (psf)	Minimum Column Width, t (in.)	Service Load on Footing (kips)	Dimensions h (in.)	w (ft)	Reinforcement Each Way
4000	8	34	10	3	4 No. 4
	8	47	11	3.5	4 No. 5
	8	61	12	4	5 No. 5
	8	77	13	4.5	5 No. 6
	8	95	15	5	5 No. 6
	8	136	18	6	6 No. 7
	10	184	20	7	8 No. 7
	10	238	23	8	8 No. 8
	12	300	25	9	8 No. 9
	12	367	27	10	10 No. 9
	14	441	29	11	10 No. 10
	14	522	32	12	11 No. 10
	16	608	34	13	13 No. 10
	16	698	37	14	13 No. 11
	18	796	39	15	14 No. 11

[a]Service loads do not include the weight of the footing, which has been deducted from the total bearing capacity. Service load is considered 40% dead load and 60% live load. Grade 40 reinforcement. $f'_c = 3$ ksi.

parts of the building. This is generally most significant for steel columns.

3. To provide for the required development length of reinforcing in reinforced concrete columns, where footing thickness is not adequate for development within the footing.
4. To effect a transition between a column with very high concrete strength and a footing with only moderate concrete strength.

Figure 16.8*d* illustrates the third situation described. Referring to Table 13.11, we may observe that a considerable development length is required for large-diameter bars made from high grades of steel. If the minimum required footing does not have a thickness that permits this development, a pedestal may offer a reasonable solution. However, there are many other considerations to be made in the decision, and the column reinforcing problem is not the only factor in this situation.

If a pedestal is quite short with respect to its width (see Figure 16.8*e*), it may function essentially the same as a column

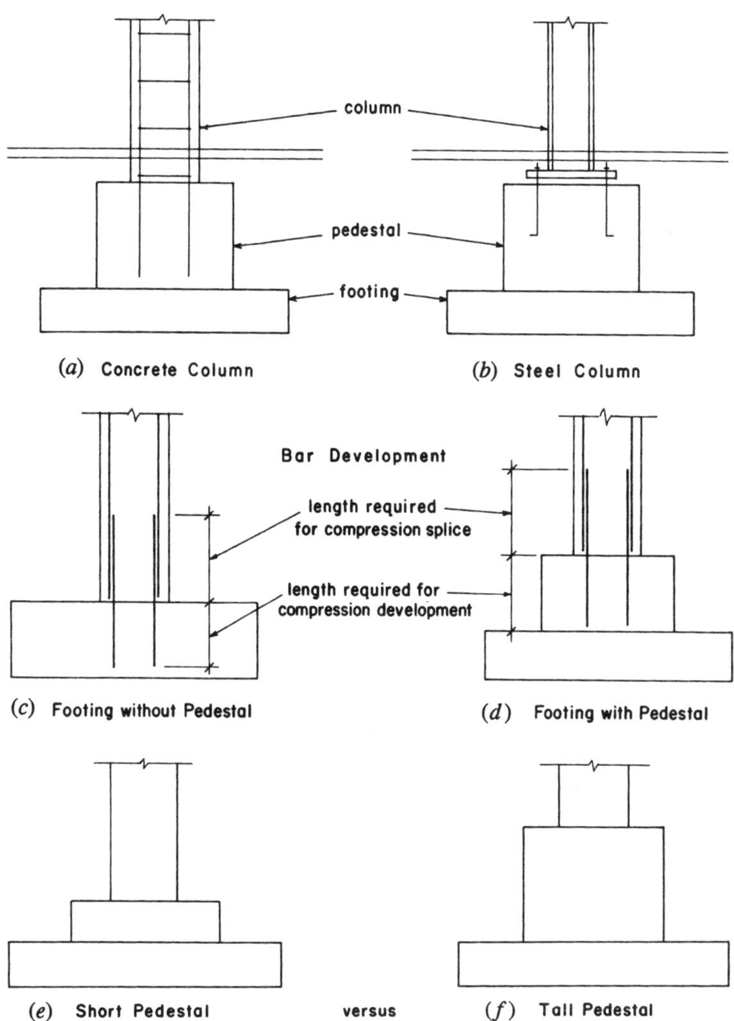

Figure 16.8 Considerations for use of pedestals.

footing and may develop significant values for shear and bending stresses. This condition is likely to occur if the pedestal width exceeds twice the column width and the pedestal height is less than one-half of the pedestal width. In such cases, the pedestal must be designed by the same procedures used for an ordinary column footing.

V

STRUCTURAL SYSTEMS FOR BUILDINGS

This part contains examples of the design of structural systems for buildings. The buildings selected for design are not intended as examples of good architectural design but rather have been selected to create a range of common situations in order to be able to demonstrate the use of various structural components. Design of individual elements of the structural systems is based on the materials presented in earlier chapters. The purpose here is to show a broader context of design work by dealing with the whole structure and with the building in general.

17

GENERAL CONSIDERATIONS FOR BUILDING STRUCTURES

This chapter contains discussions of some general issues relating to design of building structures. These concerns have mostly not been addressed in the presentations in earlier chapters but require consideration when dealing with whole building design situations. Application of this material is illustrated in the design examples in Chapters 18–20.

17.1 CHOICE OF BUILDING CONSTRUCTION

Materials, methods, and details of building construction vary considerably on a regional basis. There are many factors that affect this situation, including the effects of response to climate and regional availability of construction materials and products. Even in a single region, differences occur between individual buildings, based on styles of architectural design and techniques of builders. Nevertheless, at any given time there are usually a few predominant, popular methods of construction that are employed for most buildings of a given type and size. The construction methods and details shown here are reasonable, but in no way are they intended to illustrate a singular, superior style of building.

It is not possible to choose the materials and forms for a building structure without considering the integration of the structure with the general building construction. In some cases it may also be necessary to consider the elements required for various building services, such as those for piping, electrical service, lighting, communication, roof drainage, and the HVAC (heating, ventilating, and air conditioning) systems.

For multistory buildings it is necessary to accommodate the placement of stairs, elevators, and the vertical elements for various building services—particularly for air ducts between building levels. A major consideration for multistory buildings is the planning of the various levels so that they work when superimposed on top of each other. Bearing walls and columns must be supported from below.

Choice of the general structural system as well as the various individual elements of the system is typically highly dependent on the general architectural design of the building. Ideally, the two issues—structural planning and architectural planning—are dealt with simultaneously from preliminary design to final construction drawings.

17.2 STRUCTURAL DESIGN STANDARDS

Use of methods, procedures, and reference data for structural design is subject to the judgment of the designer. Many guides exist, but some individual selection is often required. Strong influences on choices include:

- Building code requirements from the enforceable statutes relating to the location of the building
- Acceptable design standards as published by professional groups, such as the reference from the American Society of Civil Engineers (ASCE) referred to frequently in this book (Ref. 1)
- Recommended design standards from industry organizations, such as the American Institute of Steel Construction (AISC) and the American Concrete Institute (ACI)
- The body of work from current texts and references produced by respected authors

Some reference is made to these sources in this book. However, much of the work is also simply presented in a manner familiar to the authors, based on their own experiences. If study of this subject is pursued by readers, they are sure to encounter styles and opinions that differ from those presented here. Making one's own choices in the face of those conflicts is part of the progress of professional growth.

17.3 STRUCTURAL DESIGN PROCESS

The accomplishment of the structural design work for a building project involves the planning of the structure and the application of a design process for use of standard resources and methods. Design work tends to be highly repetitive from one project to the next, so individual design organizations and individual designers generally develop some standard process to assure thorough completion of the work.

An early task to be done is the establishment of the enforceable building code for the project. Once this is established, specific values can be assigned for design loads, as discussed in Chapter 1. Application of design criteria from the building code may also be spelled out in detail by the code, although this is mostly a task for the designer. Examples of project-related determination of design loads are given in the building design cases in Chapters 18–20.

Standards of Performance for Design

Loads used for structural design must be derived primarily from enforceable building codes. However, the principal concern of codes is public health and safety. Performance of the structure for other concerns may not be adequately represented in the minimum requirements of the building code. Issues sometimes not included in code requirements are:

Effects of deflection of spanning structures on nonstructural elements of the construction

Sensations of bounciness of floors by building occupants

Protection of structural elements from damage due to weather or normal usage

Damage to nonstructural construction and building services due to movements of the structure during windstorms or earthquakes

It is quite common for professional structural designers to have situations where they use their own judgment in assigning design loads. This ordinarily means using increased loads, as the minimum loads required by codes must always be recognized.

17.4 DEVELOPMENT OF STRUCTURAL SYSTEMS

Structural Planning

Planning a structure requires the ability to perform two major tasks. The first is the logical arranging of the structure itself, regarding its geometric form, its actual dimensions and proportions, and the ordering of the elements for basic stability and reasonable interaction. All of these issues must be faced, whether the building is simple or complex, small or large, of ordinary construction or unique. Spanning beams must be supported and have depths adequate for the spans; horizontal thrusts of arches must be resolved; columns above should be centered over columns below; and so on.

The second major task in structural planning is the development of the relationships between the structure and the building in general. The building plan must be "seen" as a structural plan. The two may not be quite the same, but they must fit together. "Seeing" the structural plan (or possibly alternative plans) inherent in a particular architectural plan is a major task for designers of building structures.

Hopefully, architectural planning and structural planning are done interactively, not one after the other. The more the architect knows about the structural problems and the structural designer knows about architectural problems, the more likely it is possible that an interactive design development may occur.

Although each individual building offers a unique situation if all of the variables are considered, the majority of building design problems are highly repetitious. The problems usually have many alternative solutions, each with its own set of pluses and minuses in terms of various points of comparison. Choice of the final design involves the comparative evaluation of known alternatives and the eventual selection of one.

The word *selection* may seem to imply that all the possible solutions are known in advance, not allowing for the possibility of a new solution. The more common the problem, the more this may be true. However, the continual advance of science and technology and the fertile imagination of designers make new solutions an ever-present possibility, even for the most common problems. When the problem is truly a new one in terms of a new building use, a jump in scale, or a new performance situation, there is a real need for innovation. Usually, however, when new solutions to old problems are presented, their merits must be compared to established previous solutions in order to justify them. In its broadest context the selection process includes the consideration of all possible alternatives: those well known, those new and unproven, and those only imagined.

Building Systems Integration

Good structural design requires integration of the structure into the whole physical system of the building. It is necessary to realize the potential influences of structural design decisions on the general architectural design and on the development of the systems for power, lighting, thermal control, ventilation, water supply, waste handling, vertical transportation, firefighting, and so on. The most popular structural systems have become so in many cases largely because of their ability to accommodate the other subsystems of the building and to facilitate popular architectural forms and details.

Economics

Dealing with dollar cost is a very difficult, but necessary, part of structural design. For the structure itself, the bottom-line cost is the delivered cost of the finished structure, usually measured in units of dollars per square foot of the building. For individual components, such as a single wall, units may be used in other forms. The individual cost factors or components, such as cost of materials, labor, transportation, installation, testing, and inspection, must be aggregated to produce a single unit cost for the entire structure.

Designing for control of the cost of the structure is only one aspect of the cost problem, however. The more meaningful cost is that for the entire building construction. It is possible that certain cost-saving efforts applied to the structure may result in increases of cost of other parts of the construction. A common example is that of the floor structure for multistory buildings. Efficiency of floor beams occurs with the generous provision of beam depth in proportion to the span. However, adding inches to beam depths with the unchanging need for dimensions required for floor and ceiling construction and installation of ducts and lighting elements means increasing the floor-to-floor distance and the overall height of the building. The resulting increases in cost for the added building skin, interior walls, elevators, piping, ducts, stairs, and so on may well offset the small savings in cost of the beams. The really effective cost-reducing structure is often one that produces major savings of nonstructural costs, in some cases at the expense of less structural efficiency.

Real costs can only be determined by those who deliver the completed construction. Estimates of cost are most reliable in the form of actual offers or bids for the construction work. The farther the cost estimator is from the actual requirement to deliver the goods, the more speculative the estimate. Designers, unless they are in the actual employ of the builder, must base any cost estimates on educated guesswork deriving from some comparison with similar work recently done in the same region. This kind of estimating must be adjusted for the most recent developments in terms of the local markets, competitiveness of builders and suppliers, and the general state of the economy. Then the four best guesses are placed in a hat and one is drawn out.

Serious cost estimating requires training and experience and a source of reliable, timely information. For major projects various sources are available in the form of publications or computer databases.

The following are some general rules for efforts that can be made in the structural design work in order to have an overall general cost-saving attitude.

1. Reduction of material volume is usually a means of reducing cost. However, unit prices for different grades must be noted. Higher grades of concrete or wood may be proportionally more expensive than the higher stress values they represent; more volume of cheaper material may be less expensive.
2. Use of standard, commonly stocked products is usually a cost savings, as special sizes or shapes may be premium priced. Wood 2 × 3 studs are generally higher in price than 2 × 4 studs since the 2 × 4 is so widely used and bought in large quantities.
3. Reduction in the complexity of systems is usually a cost savings. Simplicity in purchasing, handling, managing of inventory, and so on, will be reflected in lower bids as builders anticipate simpler tasks. Use of the fewest number of different

grades of materials, sizes of fasteners, and other such variables is as important as the fewest number of different parts. This is especially true for any assemblage done on the building site; large inventories may not be a problem in a factory but usually are on a restricted site.

4. Cost reduction is usually achieved when materials, products, and construction methods are familiar to local builders and construction workers. If real alternatives exist, choice of the "usual" one is the best course.
5. Do not guess at cost factors; use real experience, yours or others. Costs vary locally, by job size and over time. Keep up to date with cost information.
6. In general, labor cost is greater than material cost. Labor for building forms, installing reinforcement, pouring, and finishing concrete surfaces is *the* major cost factor for site-poured concrete. Savings in these areas are much more significant than saving of material volume.
7. For buildings of an investment nature, time is money. Speed of construction may be a major advantage. However, getting the structure up fast is not a true advantage unless the other aspects of the construction can take advantage of the time gained. Steel frames often go up quickly, only to stand around while the rest of the work catches up.

18

BUILDING ONE

The building in this chapter consists of a simple, single-story, box-shaped building. Lateral bracing is developed in response to wind load. Several alternatives are considered for the building structure.

18.1 GENERAL CONSIDERATIONS

Figure 18.1 shows the general form, the construction of the basic building shell, and the form of the wind-bracing shear walls for Building One. The drawings show a building profile with a generally flat roof (with minimal slope for drainage) and a short parapet at the roof edge. This structure is generally described as a *light wood frame* and is the first alternative to be considered for Building One. The following data are used for design:

Roof live load = 20 psf (reducible)
Wind load as determined from the ASCE standard (Ref. 1)
Wood framing lumber of Douglas fir–larch

18.2 DESIGN OF THE WOOD STRUCTURE FOR GRAVITY LOADS

With the construction as shown in Figure 18.1*f*, the roof dead load is determined as follows:

Three-ply felt and gravel roofing	5.5 psf
Glass fiber insulation batts	0.5 psf
½-in.-thick plywood roof deck	1.5 psf
Wood rafters and blocking (estimate)	2.0 psf
Ceiling framing	1.0 psf
½-in.-thick drywall ceiling	2.5 psf
Ducts, lights, and so on	3.0 psf
Total roof dead load for design	16.0 psf

Assuming a partitioning of the interior as shown in Figure 18.2*a*, various possibilities exist for the development of the spanning roof and ceiling framing systems and their supports. Interior walls may be used for supports, but a more desirable situation in commercial uses is sometimes obtained by using interior columns that allow for rearrangement of interior spaces. The roof framing system shown in Figure 18.2*b* is developed with two rows of interior columns placed at the location of the corridor walls. If the partitioning shown in Figure 18.2*a* is used, these columns may be totally out of view (and not intrusive in the building plan) if they are incorporated in the wall construction. Figure 18.2*c* shows a second possibility for the roof framing using the same column layout as in Figure 18.2*b*. There may be various reasons for favoring one of these framing schemes over the other. Problems of installation of ducts, lighting, wiring, roof drains, and fire sprinklers may influence this structural design decision. For this example, the scheme shown in Figure 18.2*b* is arbitrarily selected for illustration of the design of the elements of the structure.

Installation of membrane-type roofing ordinarily requires at least a ½-in.-thick roof deck. Such a deck is capable of up to 32-in. spans in a direction parallel to the face ply grain (the long direction of ordinary 4 × 8-ft panels). If rafters are not over 24 in. on center—as they are likely to be for the schemes shown in Figures 18.2*b* and *c*—the panels may be placed so that the plywood span is across the face grain. An advantage in the latter arrangement is the reduction in the amount of blocking between rafters that is required at panel edges not falling on

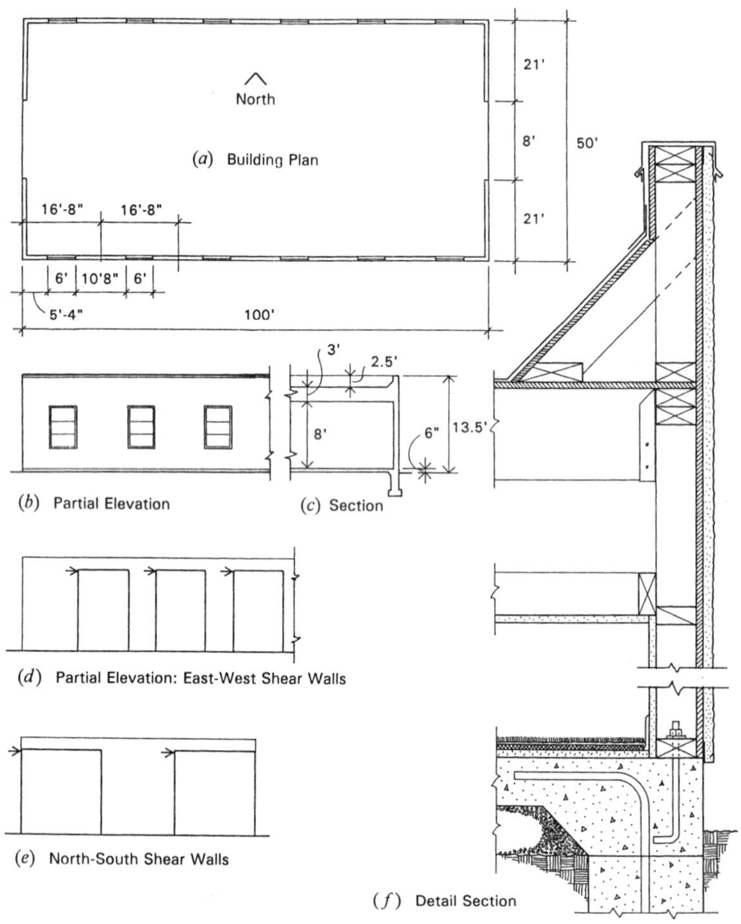

Figure 18.1 Building One: general form.

a rafter. The reader is referred to the discussion of plywood decks in Section 5.11.

A common choice of grade for rafters is No. 2, for which Table 5.1*a* yields an allowable bending stress of 900 psi × 1.15 = 1035 psi (repetitive use) and a modulus of elasticity of 1,600,000 psi. Since the data for this case fall approximately within the criteria for Table 5.11, possible choices from the table are for either 2 × 10 s at 12-in. centers or 2 × 12 s at 16-in. centers. It should be noted that the parameters for

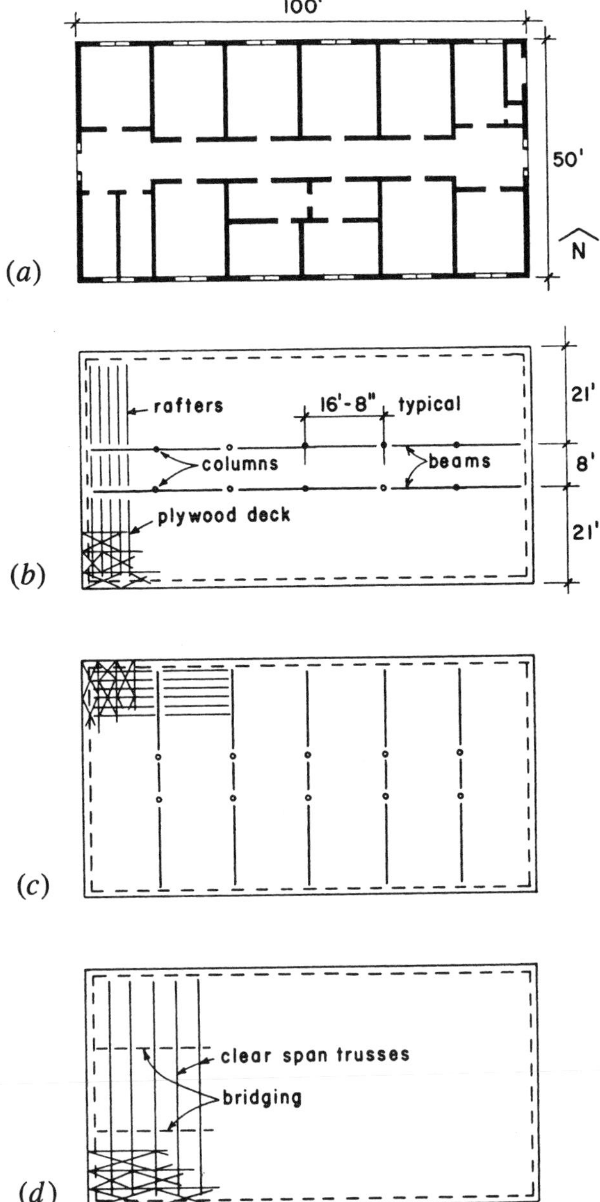

Figure 18.2 Developed plan for the interior partitioning and alternatives for the roof framing.

Table 5.11 are a live load of 20 psf and a dead load of 10 psf or a live load of 20 psf and a dead load of 20 psf. The latter was used as it will yield a more conservative approximation. Interpolating between the two sections of the table is a possibility but is usually not considered worth the effort.

A ceiling may be developed by direct attachment to the underside of the rafters. However, the construction as shown here indicates a ceiling at some distance below the rafters, allowing for various service elements to be incorporated above the ceiling. Such a ceiling might be framed independently for short spans (such as at a corridor) but is more often developed as a *suspended ceiling*, with hanger elements from the overhead structure used to shorten the span of ceiling framing.

The wood beams as shown in Figure 18.2*b* are continuous through two spans, with a total length of 33 ft 4 in. and beam spans of 16 ft 8 in. For the two-span beam the maximum bending moment is the same as for a simple span, the principal advantage being a reduction in deflection. The total roof load area for one beam span is

$$A = \left(\frac{21+8}{2}\right) \times 16.67 = 242 \text{ ft}^2$$

This permits the use of a reduced live load of 16 psf. Thus, the unit of the uniformly distributed load on the beam is found as

$$w = (16 \text{ psf LL} + 16 \text{ psf DL}) \times \frac{21+8}{2} = 464 \text{ lb/ft}$$

Adding a bit for the beam weight, a design for 480 lb/ft is reasonable, for which the maximum bending moment is

$$M = \frac{wL^2}{8} = \frac{480 \times (16.67)^2}{8} = 16{,}673 \text{ ft} - \text{lb}$$

A common minimum grade for beams is No. 1. The allowable bending stress depends on the beam size and the load duration. Assuming a 15% increase for load duration, Table 5.1*a* yields the following:

$$\text{For a } 4\times \text{member:} \quad F_b = 1.15(1000) = 1150 \text{ psi}$$

$$\text{For a } 5\times \text{or larger:} \quad F_b = 1.15(1350) = 1552 \text{ psi}$$

Then, for a 4 ×

$$\text{Required } S = \frac{M}{F_b} = \frac{16{,}673 \times 12}{1150} = 174 \text{ in}^3$$

From Table A.8, the largest 4-in.-thick member is a 4 × 16 with $S = 135.7$ in.^{3}, which is not adequate. (*Note:* deeper 4 × members are available but are quite laterally unstable and thus not recommended.) For a thicker member the required S may be determined as

$$S = \frac{1150}{1552} \times 174 = 129 \text{ in.}^3$$

for which possibilities include a 6 × 14 with $S = 167$ in.^{3} or an 8 × 12 with $S = 165$ in.^{3}

Although the 6 × 14 has the least cross-sectional area and ostensibly the lower cost, various considerations of the development of construction details may affect the beam selection. This beam could also be formed as a built-up member from a number of 2 × members. Where deflection and long-term sag are critical, a wise choice might be to use a glued-laminated section, a laminated veneer lumber section, or even a steel-rolled section. This may also be a consideration if shear is critical, as is often the case with heavily loaded beams.

A minimum slope of the roof surface for drainage is usually 2%, or approximately 1/4 in./ft. If drainage is achieved as shown in Figure 18.3*a*, this requires a total slope of $1/4 \times 25 = 6.25$ in. from the center to the edge of the roof. There are various ways of achieving this sloped surface, including the simple tilting of the rafters.

Figure 18.3*b* shows some possibilities for the details of the construction at the center of the building. As shown here the rafters are kept flat and the roof profile is achieved by attaching cut 2 × members to the tops of the long rafters and using a short profiled rafter at the corridor. Ceiling joists for the corridor are supported directly by the corridor walls. Other ceiling joists are supported at their ends by the walls and at intermediate points by suspension from the rafters.

The typical column at the corridor supports a load approximately equal to the entire spanning load for one beam, or

$$P = 480 \times 16.67 = 8000 \text{lb}$$

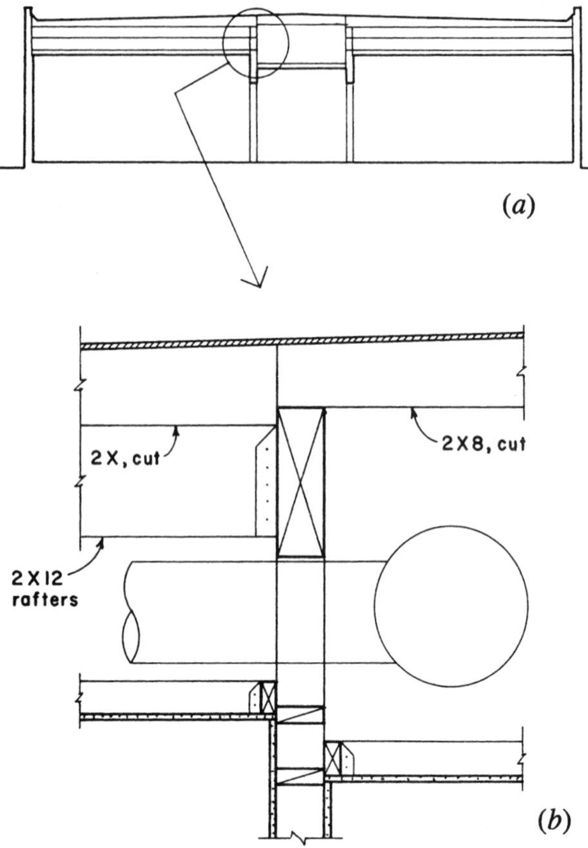

Figure 18.3 Construction details.

This is a light load, but the column height (greater than 10 ft) requires something larger than a 4 × size. (See Table 6.1.) If a 6 × 6 is not objectionable, it is adequate in the lower stress grades. However, it is common to use a steel pipe or tubular section, either of which can probably be accommodated in a stud partition wall.

18.3 DESIGN FOR LATERAL LOADS

Design of the building structure for wind includes consideration for the following:

1. Inward and outward pressure on exterior building surfaces, causing bending of the wall studs and an addition to the gravity loads on roofs
2. Total lateral (horizontal) force on the building, requiring bracing by the roof diaphragm and the shear walls
3. Uplift on the roof, requiring anchorage of the roof structure to its supports
4. Total effect of uplift and lateral forces, possibly resulting in overturn (toppling) of the entire building

Uplift on the roof depends on the roof shape and the height above ground. For this low, flat-roofed building, the ASCE standard (Ref. 1) requires an uplift pressure of 10.7 psf. In this case the uplift pressure does not exceed the roof dead weight of 16 psf, so special anchorage of the roof construction is not required. However, common use of metal framing devices for light wood frame construction provides an anchorage with considerable resistance.

Overturning of the building is not likely critical for a building with this squat profile (50 ft wide by only 13.5 ft high). Even if the overturning moment caused by wind exceeds the restoring moment due to the building dead weight, the sill anchor bolts will undoubtedly hold the building down in this case. Overturn of the whole building is usually more critical for towerlike building forms or for extremely light construction. Of separate concern is the overturn of individual bracing elements, in this case the individual shear walls, which will be investigated later.

Wind Force on the Bracing System

The building's bracing system must be investigated for horizontal force in the two principal orientations, east–west and north–south,

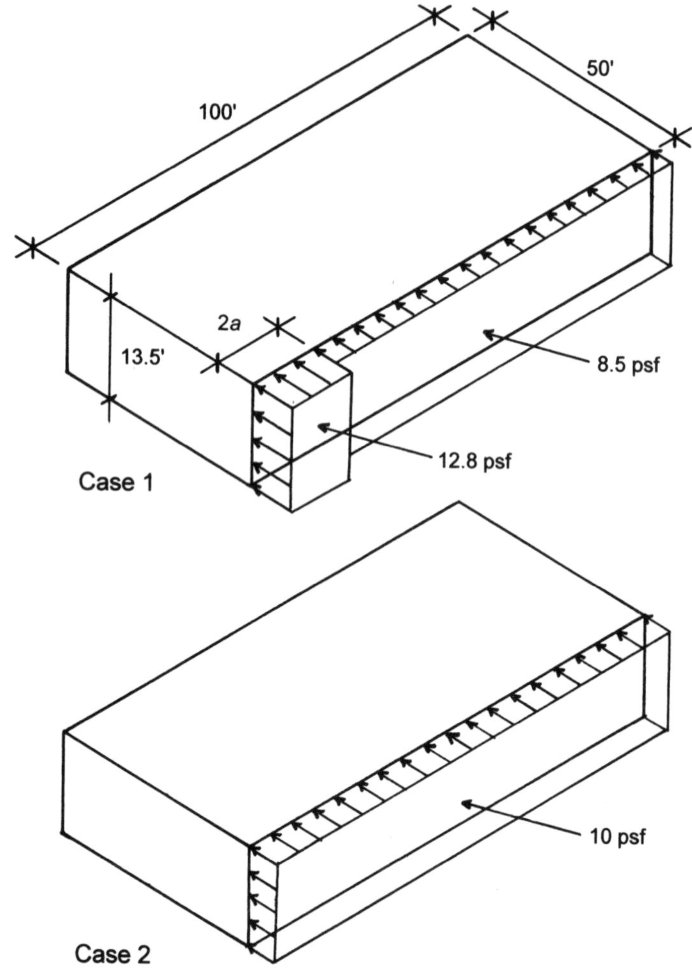

Figure 18.4 Wind pressure on the south wall.

and, if the building is not symmetrical, in each direction on each building axis: east, west, north, and south.

The horizontal wind force on the north and south walls of the building is shown in Figure 18.4. This force is generated by a combination of positive (direct, inward) pressure on the windward side and negative (suction, outward) on the leeward side of the building. The pressures shown as Case 1 in Figure 18.4 are obtained from data in the

ASCE standard (Ref. 1). (See discussion of wind loads in Chapter 1.) The single pressures shown in the figure are intended to account for the combination of positive and suction pressures. The ASCE standard provides for two zones of pressure—a general one and a small special increased area of pressure at one end. The values shown in Figure 18.4 for these pressures are derived by considering a critical wind velocity of 90 mph and an exposure condition B, as described in the standard.

The range for the increased pressure in Case 1 is defined by the dimension a and the height of the windward wall. The value of a is established as 10% of the least plan dimension of the building or 40% of the wall height, whichever is smaller, but not less than 3 ft. For this example a is determined as 10% of 50 ft, or 5 ft. The distance for the pressure of 12.8 psf in Case 1 is thus $2(a) = 10$ ft.

The design standard also requires that the bracing system be designed for a minimum pressure of 10 psf on the entire area of the wall. This sets up two cases (Case 1 and Case 2 in Figure 18.4) that must be considered. Since the concern for the design is the generation of maximum effect on the roof diaphragm and the end shear walls, the critical conditions may be determined by considering the development of end reaction forces and maximum shear for an analogous beam subjected to the two loadings. This analysis is shown in Figure 18.5, from which it is apparent that the critical concern for the end shear walls and the maximum effect in the roof diaphragm is derived from Case 2 in Figure 18.4.

The actions of the horizontal wind force resisting system in this regard are illustrated in Figure 18.6. The initial force comes from wind pressure on the building's vertical sides. The wall studs span vertically to resist this uniformly distributed load, as shown in Figure 18.6a. Assuming the wall function to be as shown in Figure 18.6a, the north–south wind force delivered to the roof edge is determined as

$$\text{Total } W = (10 \text{ psf})(100 \times 13.5) = 13{,}500 \text{ lb}$$

$$\text{Roof edge } W = 13{,}500 \times \frac{6.75}{11} = 8284 \text{ lb}$$

In resisting this load, the roof functions as a spanning member supported by the shear walls at the east and west ends of the building. The investigation of the diaphragm as a 100-ft simple span beam with

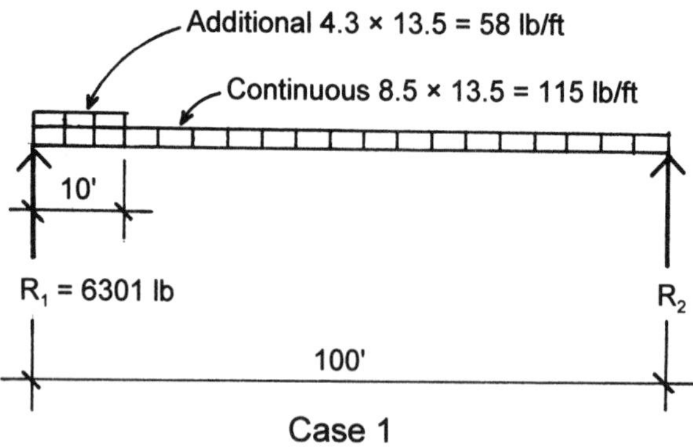

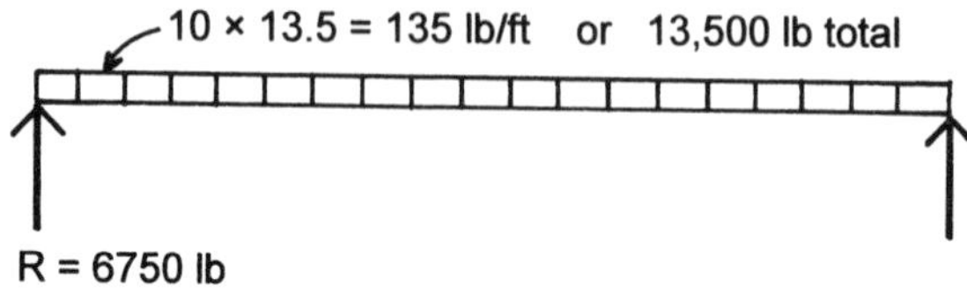

Figure 18.5 Resultant wind forces on the end shear walls.

uniformly distributed loading is shown in Figure 18.7. The end reaction and maximum diaphragm shear force is found as

$$R = V = \frac{8284}{2} = 4142 \text{ lb}$$

which produces a maximum unit shear in the 50-ft-wide diaphragm of

$$v = \frac{\text{shear force}}{\text{roof width}} = \frac{4142}{50} = 82.8 \text{lb/ft}$$

From Table 18.1, a variety of selections are possible. Variables include the class of the plywood, the panel thickness, the width of supporting rafters, the nail size and spacing, the use of blocking, and the layout pattern of the plywood panels. Assuming a minimum

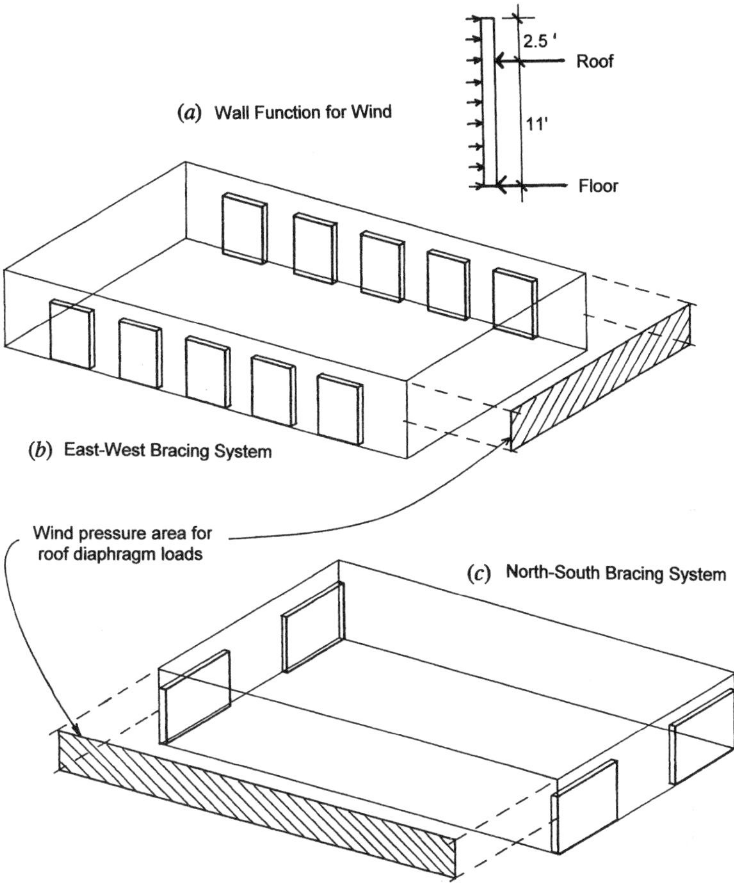

Figure 18.6 Wall functions and wind pressure development.

plywood thickness for the flat roof at $^1/_2$ in. (given as $^{15}/_{32}$ in the table), a possible choice is as follows:

APA rated sheathing, $^{15}/_{32}$ in. thick, 2 × rafters, 8d nails at 6 in. at all panel edges, blocked diaphragm

For these criteria the table yields a capacity of 270 lb/ft.

TABLE 18.1 Load Values for Plywood Diaphragms

PANEL GRADE	COMMON NAIL SIZE	MINIMUM NAIL PENETRATION IN FRAMING (inches)	MINIMUM NOMINAL PANEL THICKNESS (inches)	MINIMUM NOMINAL WIDTH OF FRAMING MEMBER (inches)	BLOCKED DIAPHRAGMS: Nail spacing (in.) at diaphragm boundaries (all cases), at continuous panel edges parallel to load (Cases 3 and 4) and at all panel edges (Cases 5 and 6) × 25.4 for mm: 6; Nail spacing (in.) at other panel edges × 25.4 for mm: 6	4; 6	$2^1/_2$ [2]; 4	2 [2]; 3	UNBLOCKED DIAPHRAGMS: Nails spaced 6″ (152 mm) max. at supported edges: Case 1 (No unblocked edges or continuous joints parallel to load)	All other configurations (Cases 2, 3, 4, 5 and 6)
		× 25.4 for mm			× 0.0146 for N/mm					
Structural 1	6d	$1^1/_4$	$^5/_{16}$	2	185	250	375	420	165	125
				3	210	280	420	475	185	140
	8d	$1^1/_2$	$^3/_8$	2	270	360	530	600	240	180
				3	300	400	600	675	265	200
	10d[3]	$1^5/_8$	$^{15}/_{32}$	2	320	425	640	730	285	215
				3	360	480	720	820	320	240
C-D, C-C, Sheathing, and other grades covered in UBC Standard 23-2 or 23-3	6d	$1^1/_4$	$^5/_{16}$	2	170	225	335	380	150	110
				3	190	250	380	430	170	125
			$^3/_8$	2	185	250	375	420	165	125
				3	210	280	420	475	185	140
	8d	$1^1/_2$	$^3/_8$	2	240	320	480	545	215	160
				3	270	360	540	610	240	180
			$^7/_{16}$	2	255	340	505	575	230	170
				3	285	380	570	645	255	190
			$^{15}/_{32}$	2	270	360	530	600	240	180
				3	300	400	600	675	265	200
	10d[3]	$1^5/_8$	$^{15}/_{32}$	2	290	385	575	655	255	190
				3	325	430	650	735	290	215
			$^{19}/_{32}$	2	320	425	640	730	285	215
				3	360	480	720	820	320	240

[1]These values are for short-time loads due to wind or earthquake and must be reduced 25 percent for normal loading. Space nails 12 inches (305 mm) on center along intermediate framing members.

Allowable shear values for nails in framing members of other species set forth in Division III, Part III, shall be calculated for all other grades by multiplying the shear capacities for nails in Structural I by the following factors: 0.82 for species with specific gravity greater than or equal to 0.42 but less than 0.49, and 0.65 for species with a specific gravity less than 0.42.

[2]Framing at adjoining panel edges shall be 3-inch (76 mm) nominal or wider and nails shall be staggered where nails are spaced 2 inches (51 mm) or $2^1/_2$ inches (64 mm) on center.

[3]Framing at adjoining panel edges shall be 3-inch (76 mm) nominal or wider and nails shall be staggered where 10d nails having penetration into framing of more than $1^5/_8$ inches (41 mm) are spaced 3 inches (76 mm) or less on center.

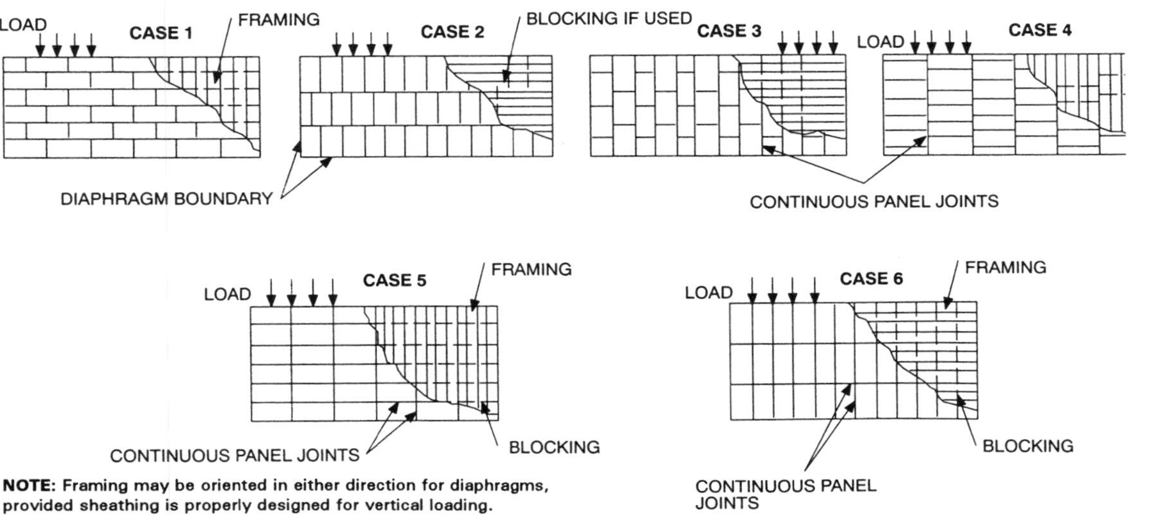

NOTE: Framing may be oriented in either direction for diaphragms, provided sheathing is properly designed for vertical loading.

Source: Reproduced from *Introduction to Lateral Design* (Ref. 12), with permission of the publisher, APA—The Engineered Wood Association.

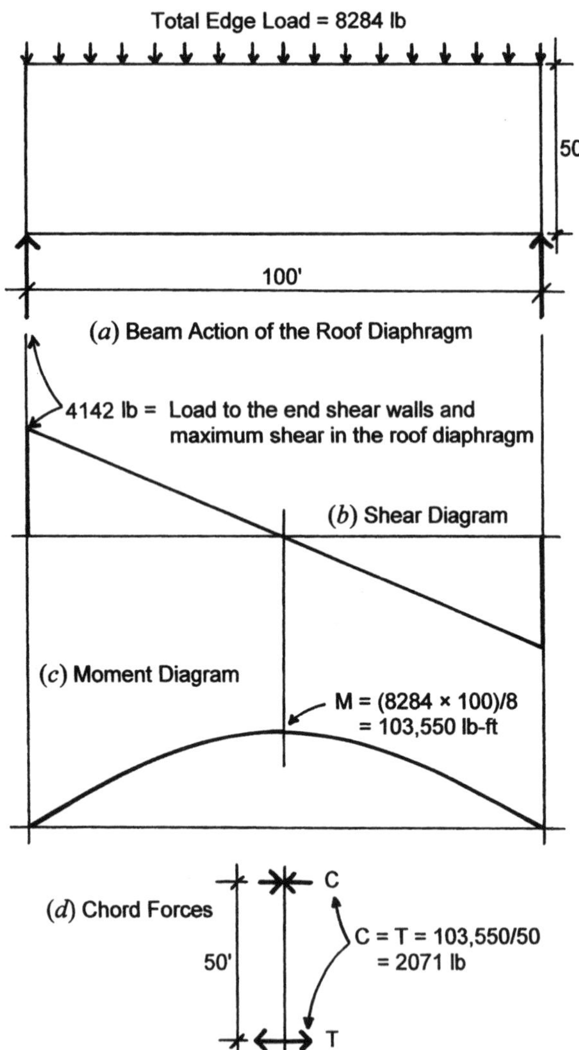

Figure 18.7 Spanning function of the roof diaphragm.

In this example, if the need for the minimum-thickness plywood is accepted, it turns out that the minimum construction is more than adequate for the required lateral force resistance. Had this not been the case and the required capacity had resulted in considerable nailing beyond the minimum, it would be possible to graduate the nailing spacing from that required at the building ends to minimal nailing in the center portion of the roof. (See the form of the shear variation across the roof width.)

The moment diagram shown in Figure 18.7 indicates a maximum value of 104 kip-ft at the center of the span. This moment is used to determine the maximum force in the diaphragm chord at the roof edges. The force must be developed in both compression and tension as the wind direction reverses. With the construction as shown in Figure 18.1*f*, the top plate of the stud wall is the most likely element to be utilized for this function. In this case the chord force of 2071 lb, as shown in Figure 18.7, is quite small, and the doubled 2 × member should be capable of resisting the force. However, the building length requires the use of several pieces to create this continuous plate, so the splices for the double member should be investigated.

The end reaction force for the roof diaphragm, as shown in Figure 18.7, must be developed by the end shear walls. As shown in Figure 18.1, there are two walls at each end, both 21 ft long in plan. Thus, the total shear force is resisted by a total of 42 ft of shear wall and the unit shear in the wall is

$$v = \frac{4142}{42} = 98.6 \text{ lb/ft}$$

As with the roof, there are various considerations for the selection of the wall construction. Various materials may be used on both the exterior and interior surfaces of the stud wall. The common form of construction shown in Figure 18.1*f* indicates gypsum drywall on the inside and a combination of plywood and stucco (cement plaster) on the outside of this wall. All three of these materials have rated resistances to shear wall stresses. However, with a combination of materials, it is common practice to consider only the strongest of the materials to be the resisting element. In this case that means the plywood sheathing on the exterior of the wall. From Table 18.2, a possible choice is

APA rated sheathing, 3/8 in. thick, with 6d nails at 6-in. spacing at all panel edges

For this criteria the table allows a unit shear of 200 lb-ft. Again, this is minimal construction. For higher loadings a greater resistance can be obtained by using better plywood, thicker panels, larger nails, closer nail spacing, and—sometimes—wider studs. Unfortunately, the nail spacing cannot be graduated—as it may be for the roof—as the unit shear is a constant value throughout the height of the wall.

Figure 18.8*a* shows the loading condition for investigation of the overturn effect on the end shear wall. Overturn is resisted by the so-called restoring moment, due to the dead load on the wall—in this case a combination of the wall weight and the portion of roof dead load supported by the wall. Safety is considered adequate if the restoring moment is at least 1.5 times the overturning moment. A comparison is therefore made between the value of 1.5 times the overturning moment and the restoring moment as follows:

$$\text{Overturning moment} = (2.071)(11)(1.5) = 34.2\text{kip} - \text{ft}$$

$$\text{Restoring moment} = (3 + 6)(21/2) = 94.5\text{kip} - \text{ft}$$

This indicates that no tie-down force is required at the wall ends (force T as shown in Figure 18.7). Details of the construction and other functions of the wall may provide additional resistances to overturn. However, some designers prefer to use end anchorage devices (called tie-down anchors) at the ends of all shear walls, regardless of loading magnitudes.

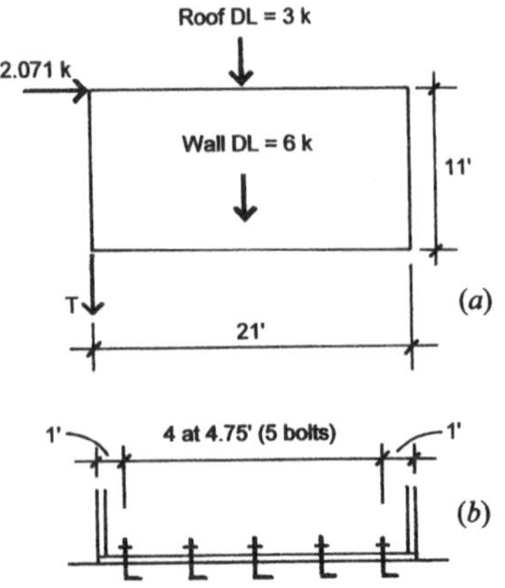

Figure 18.8 Functions of the end shear wall.

TABLE 18.2 Load Values for Plywood Shear Walls

PANEL GRADE	MINIMUM NOMINAL PANEL THICKNESS (inches)	MINIMUM NAIL PENETRATION IN FRAMING (inches)	PANELS APPLIED DIRECTLY TO FRAMING					PANELS APPLIED OVER $^{1}/_{2}$-INCH (13 mm) OR $^{5}/_{8}$-INCH (16 mm) GYPSUM SHEATHING				
			Nail Size (Common or Galvanized Box)[5]	Nail Spacing at Panel Edges (in.) × 25.4 for mm				Nail Size (Common or Galvanized Box)[5]	Nail Spacing at Panel Edges (in.) × 25.4 for mm			
				6	4	3	2		6	4	3	2
	× 25.4 for mm			× 0.0146 for N/mm					× 0.0146 for N/mm			
Structural I	$^{5}/_{16}$	$1^{1}/_{4}$	6d	200	300	390	510	8d	200	300	390	510
	$^{3}/_{8}$	$1^{1}/_{2}$	8d	230[4]	360[4]	460[4]	610[4]	10d	280	430	550	730
	$^{7}/_{16}$			255[4]	395[4]	505[4]	670[4]					
	$^{15}/_{32}$			280	430	550	730					
	$^{15}/_{32}$	$1^{5}/_{8}$	10d	340	510	665	870	—	—	—	—	—
C-D, C-C Sheathing, plywood panel siding and other grades covered in UBC Standard 23-2 or 23-3	$^{5}/_{16}$	$1^{1}/_{4}$	6d	180	270	350	450	8d	180	270	350	450
	$^{3}/_{8}$			200	300	390	510		200	300	390	510
	$^{3}/_{8}$	$1^{1}/_{2}$	8d	220[4]	320[4]	410[4]	530[4]	10d	260	380	490	640
	$^{7}/_{16}$			240[4]	350[4]	450[4]	585[4]					
	$^{15}/_{32}$			260	380	490	640					
	$^{15}/_{32}$	$1^{5}/_{8}$	10d	310	460	600	770	—	—	—	—	—
	$^{19}/_{32}$			340	510	665	870					
			Nail Size (Galvanized Casing)					Nail Size (Galvanized Casing)				
Plywood panel siding in grades covered in UBC Standard 23-2	$^{5}/_{16}$	$1^{1}/_{4}$	6d	140	210	275	360	8d	140	210	275	360
	$^{3}/_{8}$	$1^{1}/_{2}$	8d	160	240	310	410	10d	160	240	310	410

[1] All panel edges backed with 2-inch (51 mm) nominal or wider framing. Panels installed either horizontally or vertically. Space nails at 6 inches (152 mm) on center along intermediate framing members for $^{3}/_{8}$-inch (9.5 mm) and $^{7}/_{16}$-inch (11 mm) panels installed on studs spaced 24 inches (610 mm) on center and 12 inches (305 mm) on center for other conditions and panel thicknesses. These values are for short-time loads due to wind or earthquake and must be reduced 25 percent for normal loading.

Allowable shear values for nails in framing members of other species set forth in Division III, Part III, shall be calculated for all other grades by multiplying the shear capacities for nails in Structural I by the following factors: 0.82 for species with specific gravity greater than or equal to 0.42 but less than 0.49, and 0.65 for species with a specific gravity less than 0.42.

[2] Where panels are applied on both faces of a wall and nail spacing is less than 6 inches (152 mm) on center on either side, panel joints shall be offset to fall on different framing members or framing shall be 3-inch (76 mm) nominal or thicker and nails on each side shall be staggered.

[3] Where allowable shear values exceed 350 pounds per foot (5.11 N/mm), foundation sill plates and all framing members receiving edge nailing from abutting panels shall not be less than a single 3-inch (76 mm) nominal member. Nails shall be staggered.

[4] The values for $^{3}/_{8}$-inch (9.5 mm) and $^{7}/_{16}$-inch (11 mm) panels applied direct to framing may be increased to values shown for $^{15}/_{32}$-inch (12 mm) panels, provided studs are spaced a maximum of 16 inches (406 mm) on center or panels are applied with long dimension across studs.

[5] Galvanized nails shall be hot-dipped or tumbled.

Source: Reproduced from *Introduction to Lateral Design* (Ref. 12), with permission of the publisher, APA—The Engineered Wood Association.

Finally, the walls will be bolted to the foundation with code-required sill bolts, which provide some resistance to uplift and overturn effects. At present most codes do not permit sill bolts to be used for computed resistances to these effects due to the cross-grain bending that is developed in the wood sill members.

The sill bolts are used, however, for resistance to the sliding of the wall. The usual minimum bolting is with $^1/_2$-in. bolts, spaced at a maximum of 6-ft centers, with a bolt not less than 12 in. from the wall ends. This results in a bolting for this wall as shown in Figure 18.8*b*. The five bolts shown should be capable of resisting the lateral force, using the values given by the codes.

For buildings with relatively shallow foundations, the effects of shear wall anchorage forces on the foundation elements should also be investigated. For example, the overturning moment is also exerted on the foundations and may cause undesirable soil stresses or require some structural resistance by foundation elements.

Another area of concern has to do with the transfer of forces from element to element in the whole lateral force resisting structural system. A critical point of transfer in this example is at the roof-to-wall joint. The force delivered to the shear walls by the roof diaphragm must actually be passed through this joint, by the attachments of the construction elements. The precise nature of this construction must be determined and must be investigated for these force actions.

18.4 ALTERNATIVE STEEL AND MASONRY STRUCTURE

Alternative construction for Building One is shown in Figure 18.9. In this case the walls are made of CMU construction, and the roof structure consists of a formed sheet steel deck supported by open-web steel joists (light, prefabricated steel trusses). A plan for this framing is shown in Figure 18.2*d*. The following data are assumed for design:

Roof dead load = 15 psf, not including the weight of the structure

Roof live load = 20 psf, reducible for large supported areas

Construction consists of:

K-series open-web steel joists (see Section 9.10)
Reinforced hollow concrete masonry construction
Formed sheet steel deck (see Table 12.1)
Deck surfaced with lightweight insulating concrete fill

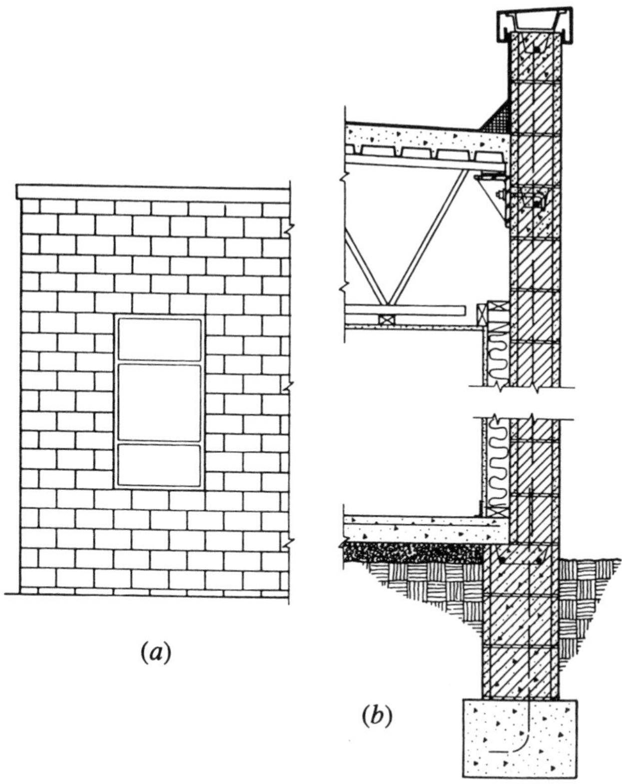

Figure 18.9 Building One: alternative steel and masonry structure.

Multiple-ply, hot-mopped, felt-and-gravel roofing
Suspended ceiling with gypsum drywall

The section in Figure 18.9*b* indicates that the wall continues above the top of the roof to create a parapet and the steel trusses are supported at the wall face. The span of the joists is thus established as approximately 48 ft, which is used for their design.

As the construction section shows, the roof deck is placed directly on top of the trusses and the ceiling is supported by attachment to the bottom of the trusses. For reasonable drainage of the roof surface a slope of at least $^1/_4$ in./ft (2%) must be provided. The roof slope may be provided by tilting the trusses or using a variable-depth truss with the top chord sloped and the bottom chord horizontally flat. The following work assumes a constant depth of the trusses for design purposes.

Design of the Roof Structure

Spacing of the open-web joists must be coordinated with the selection of the roof deck and the details for construction of the ceiling. For a trial design, a spacing of 4 ft is assumed. From Table 12.1, with deck units typically achieving three spans or more, the lightest deck in the table (22 gage) may be used. Choice of the deck configuration (rib width) depends on the type of materials placed on top of the deck and the means used to attach the deck to the supports.

Adding the weight of the deck to the other roof dead load produces a total dead load of 17 psf for the superimposed load on the joists. As illustrated in Section 9.10, the design for a K-series joist is as follows:

Joist dead load = 4(17) = 68 lb/ft (not including joist)
Joist live load = 4(20) = 80 lb/ft
Total factored load = 1.2(68) + 1.6(80) = 82 + 128 = 210 lb/ft + the factored joist weight

For the 48-ft span, the following alternative choices are obtained from Table 9.4:

24K6 at 9.7 plf, total factored load = 1.2(9.7 + 68) + 128 = 96 + 128 = 221 lb/ft (less than the table value of 233 lb/ft)
24K9 at 12.0 plf, total factored load = 1.2(12 + 68) + 128 = 96 + 128 = 224 lb/ft (less than the table value of 313 lb/ft)
26K5 at 10.6 plf, total load = 1.2(68 + 10.6) + 128 = 94 + 128 = 222 lb/ft (less than the table value of 233 lb/ft)

The live-load capacity of the 24K6 is 77 lb/ft, which is less than the 80 lb/ft being placed on the joist. The 24K6 is strong enough but would have too much deflection and therefore should be eliminated from consideration. The live-load capacity for the 24K9 and the 26K5 for *L*/360 deflection exceeds the requirement for both of these choices. Either of these joists would be acceptable choices.

While the 26K5 is the lightest permissible choice, there may be compelling reasons for using a deeper joist. For example, if the ceiling is directly attached to the bottoms of the joists, a deeper joist will provide more space for passage of building service elements. Deflection will also be reduced if a deeper joist is used. Pushing the live-load deflection to the limit means a deflection of $(1/360)(48 \times 12) = 1.9$ in. While this may not be critical for the roof surface, it can present problems for the underside of the structure, involving sag of ceilings or difficulties with nonstructural walls built up to the ceiling.

Choice of a 30K7 at 12.3 plf results in considerably less deflection at a small premium in additional weight.

It should be noted that Table 9.5 is abridged from a larger table in the reference, and there are therefore many more choices for joist sizes. The example here is meant only to indicate the process for use of such references.

Specifications for open-web joists give requirements for end support details and lateral bracing (see Ref. 6). If the 30K7 is used for the 48-ft span, for example, four rows of bridging are required.

Although the masonry walls are not designed for this example, it should be noted that the support indicated for the joists in Figure 18.9*b* results in an eccentric load on the wall. This induces bending in the wall, which may be objectionable. An alternative detail for the roof-to-wall joint is shown in Figure 18.10 in which the joists sit directly on the wall with the joist top chord extending to form a short cantilever. This is a common detail, and the reference supplies data and suggested details for this construction.

Alternative Roof Structure with Interior Columns

If a clear spanning roof structure is not required for this building, it may be possible to use some interior columns and a framing system for the roof with quite modest spans. Figure 18.11*a* shows a framing plan for a system that uses columns at 16 ft 8 in. on center in each direction. While short-span joists may be used with this system, it would also be possible to use a longer span deck, as indicated on the plan. This span exceeds the capability of the deck with 1.5-in. ribs, but decks with deeper ribs are available.

A second possible framing arrangement is shown in Figure 18.11*b* in which the deck spans the other direction and only two rows of beams are used. This arrangement allows for wider column spacing; while that increases the beam spans, a major cost savings is represented by the elimination of 60% of the interior columns and their footings.

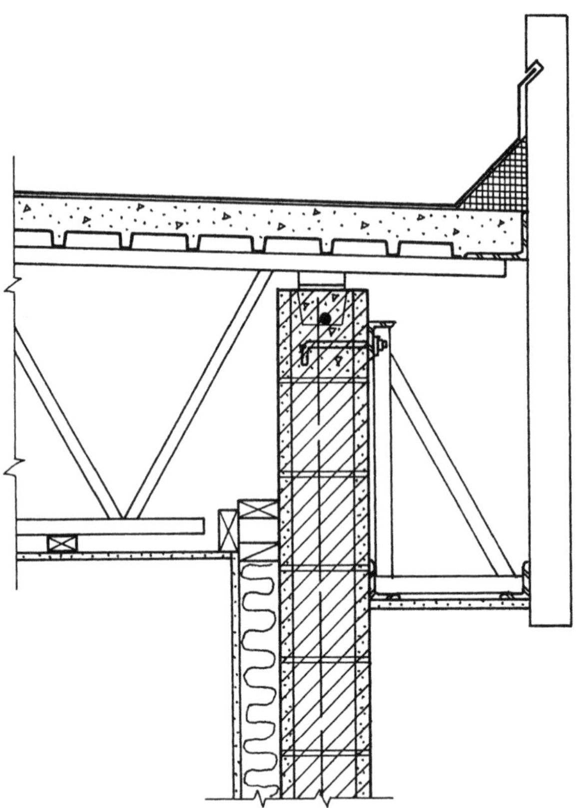

Figure 18.10 Building One: variation of the roof-to-wall joint.

Beams in continuous rows can sometimes be made to simulate a continuous beam action without the need for moment-resistive connections. Use of beam splice joints off the columns, as shown in Figure 18.11*c*, allows for relatively simple connections but some advantages of the continuous beam. A principal gain thus achieved is a reduction in deflections.

For the beam in Figure 18.11*b*, assuming a slightly heavier deck, an approximate dead load of 20 psf will result in a beam factored load of

$$w = 16.67[1.2(20) + 1.6(16)] = 827\ \text{lb/ft} + \text{beam weight} \approx 900\ \text{lb/ft}$$

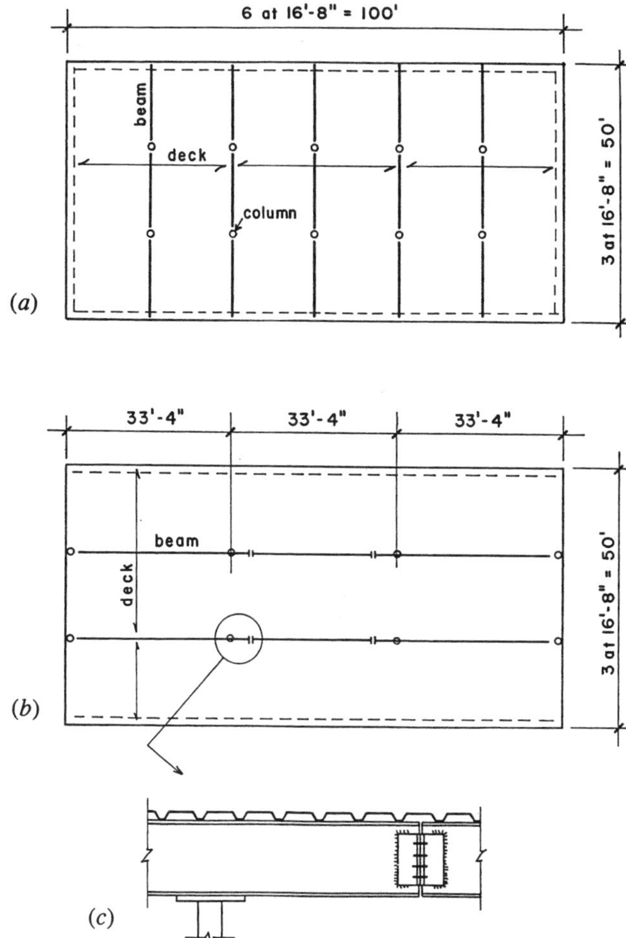

Figure 18.11 Building One: options for the roof framing with interior columns.

(Note that the beam periphery of $33.3 \times 16.67 = 555\ \text{ft}^2$ qualifies the beam for a roof live-load reduction, indicating the use of 16 psf as discussed in Section 1.8.)

The simple beam bending moment for the 33.3-ft span with the factored load is

$$M_u = \frac{wL^2}{8} = \frac{0.900 \times (33.3)^2}{8} = 125 \text{ kip} - \text{ft}$$

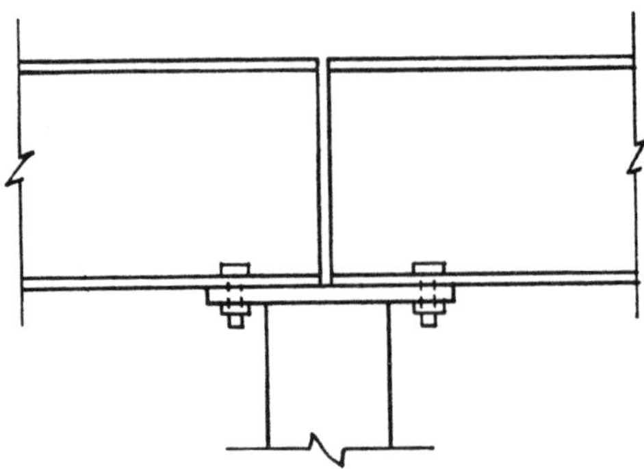

Figure 18.12 Framing detail at the top of the column with simple beam action.

and the required moment resistance of the beam is therefore

$$M_n = \frac{M_u}{\phi_b} = \frac{125}{0.9} = 139 \text{ kip-ft}$$

From Table 9.1, the lightest W-shape beam permitted is a W 12 × 26. It is assumed that the continuous connection of the deck to the beam top flange is adequate to consider the beam to have continuous lateral support, permitting the use of a resisting moment of M_p. If the three-span beam is constructed with three simple-spanning segments, the detail at the top of the column will be as shown in Figure 18.12.

While a W 12 × 26 is strong enough for this loading, its deflection under service (nonfactored) loads must be checked and compared to the standards to make sure they are acceptable. The allowable deflection criteria for roofs are generally that the maximum deflection under live loads only should be less than or equal to *L*/240 and under total service loads must be less than or equal to *L*/180:

$$\Delta_{\text{live load}} \leq 240 = \frac{33.3 \times 12}{240} = 1.67 \text{ in.}$$

$$\Delta_{\text{total load}} \leq \frac{L}{180} = \frac{33.3 \times 12}{180} = 2.22 \text{ in.}$$

The total live service load for this beam is 8.89 kips and its total live plus dead service load is 20 kips. A W 12 × 26 has a moment of inertia of 204 in.[4] Using Figure 3.18, the actual deflections are determined:

$$\Delta_{\text{live load}} = \frac{5WL^3}{384EI} = \frac{5(8.89)(400)^3}{384(29{,}000)(204)} = 1.25 \text{ in.} < 1.67 \text{ in.}$$

$$\Delta_{\text{total load}} = \frac{5WL^3}{384EI} = \frac{5(20)(400)^3}{384(29{,}000)(204)} = 2.8 \text{ in.} > 2.22 \text{ in.}$$

A W 12 × 26 does not meet the criteria for total load deflection; therefore other beams capable of handling the load must be investigated. The next two sections are the W 14 × 26 and the W 16 × 26. The moment of inertia for the W 14 × 26 is 245 in.[4]:

$$\Delta_{\text{live load}} = \frac{5WL^3}{384EI} = \frac{5(8.89)(400)^3}{384(29{,}000)(245)} = 1.04 \text{ in.} < 1.67 \text{ in.}$$

$$\Delta_{\text{total load}} = \frac{5WL^3}{384EI} = \frac{5(20)(400)^3}{384(29{,}000)(245)} = 2.34 \text{ in.} > 2.22 \text{ in.}$$

The moment of inertia for the W 16 × 26 is 301 in.[4]:

$$\Delta_{\text{live load}} = \frac{5WL^3}{384EI} = \frac{5(8.89)(400)^3}{384(29{,}000)(301)} = 0.85 \text{ in.} < 1.67 \text{ in.}$$

$$\Delta_{\text{total load}} = \frac{5WL^3}{384EI} = \frac{5(20)(400)^3}{384(29{,}000)(301)} = 1.9 \text{ in.} < 2.22 \text{ in.}$$

The W 16 × 26 is the first beam that meets all the required criteria. It is also possible to consider the use of the beam framing indicated in Figure 18.11*b* with a continuous beam having pinned connections off the columns. This will reduce both the maximum bending moment and the deflection for the beam and most likely permit a slightly lighter or shallower beam.

The total load on the beam is the approximate load on the column. Thus, the column factored load is 0.900(33.3) = 30 kips. The required design strength of the column is therefore 30(1/0.85) = 35.3 kips. Assuming an unbraced column height of 10 ft, the following choices may be found for the column:

From Table 10.3, a 3-in. pipe (nominal size, standard weight)

From Table 10.4, an HSS 3-in. square tube, with $^3/_{16}$-in.-thick wall

18.5 ALTERNATIVE TRUSS ROOF

If a gabled (double-sloped) roof form is desirable for Building One, a possible roof structure is shown in Figure 18.13. The building profile shown in Figure 18.13*a* is developed with a series of trusses, spaced at plan intervals, as shown for the beam-and-column rows in Figure 18.11*a*. The truss form is shown in Figure 18.13*b*. The complete results of an algebraic analysis for a unit loading on this truss are displayed in Figure 2.14. The true unit loading for the truss is derived from the form of construction and is approximately 10 times the unit load used in the example. This accounts for the values of the internal forces in the members as displayed in Figure 18.13*e*.

The detail in Figure 18.13*d* shows the use of double-angle members with joints developed with gusset plates. The top chord is extended to form the cantilevered edge of the roof. For clarity of the structure, the detail shows only the major structural elements. Additional construction would be required to develop the roofing, ceiling, and soffit.

In trusses of this size, it is common to extend the chords without joints for as long as possible. Available lengths depend on the sizes of members and the usual lengths in stock by local fabricators. Figure 18.13*c* shows a possible layout that creates a two-piece top chord and a two-piece bottom chord. The longer top-chord piece is thus 36 ft plus the overhang, which may be difficult to obtain if the angles are small.

The roof construction illustrated in Figure 18.13*d* shows the use of a long-span steel deck that bears directly on top of the top chord of the trusses. This option simplifies the framing by eliminating the need for intermediate framing between the trusses. For the truss spacing of 16 ft 8 in. as shown in Figure 18.11*a*, the deck will be quite light and this is a feasible system. However, the direct bearing of the deck adds a spanning function to the top chord, and the chords must be considerably heavier to work for this added task.

The loading condition for the truss as shown in Section 2.6 indicates concentrated forces of 1000 lb each at the top-chord joints. [*Note:* It is a typical procedure to assume this form of loading, even though the actual load is distributed along the top chord (roof load) and the bottom chord (ceiling load).] If the total of the live load, roof dead load, ceiling dead load, and truss weight is approximately 60 psf, the single-joint load is

$$P = (60)(10)(16.67) = 10{,}000 \text{ lb}$$

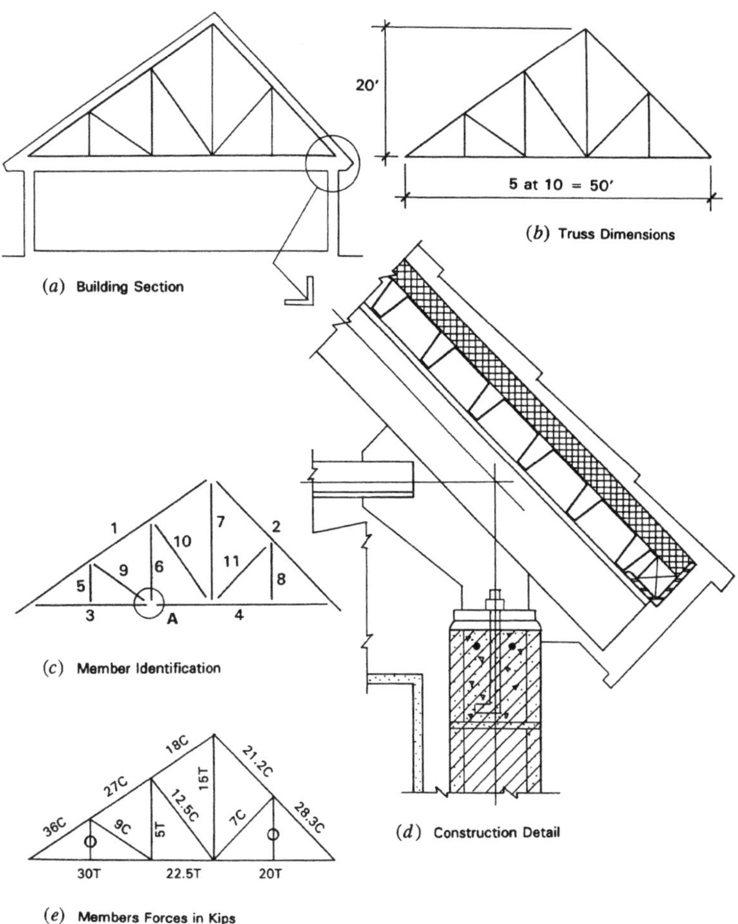

Figure 18.13 Building One: alternative truss roof structure.

This is 10 times the load in the truss in Section 2.6, so the internal forces for the gravity loading will be 10 times those shown in Figure 2.13. These values are shown here in Figure 18.13*e*.

Various forms may be used for the members and the joints of this truss. The loading and span are quite modest here, so the truss members will be quite small and joints will have minimum forces. A common form for this truss would be one using tee shapes for the top and bottom chords and double angles for interior members with the angles welded directly to the tee stems (see Figure 11.10*b*). Bolted connections are possible but probably not practical for this size truss.

18.6 FOUNDATIONS

Foundations for Building One would be quite minimal. For the exterior bearing walls, the construction provided will depend on concerns for frost and the location below ground of suitable bearing material. Options for the wood structure are shown in Figure 18.14.

Where frost is not a problem and suitable bearing can be achieved at a short distance below the finished grade, a common solution is to use the construction shown in Figure 18.14*a*, called a *grade beam*. This is essentially a combined footing and short foundation wall in one. It is typically reinforced with steel bars in the top and bottom to give it some capacity as a continuous beam, capable of spanning over isolated weak spots in the supporting soil.

Where frost *is* a problem, local codes will specify a minimum distance from the finished grade to the bottom of the foundation. To reach this distance, it may be more practical to build a separate footing and foundation wall, as shown in Figure 18.14*b*. This short,

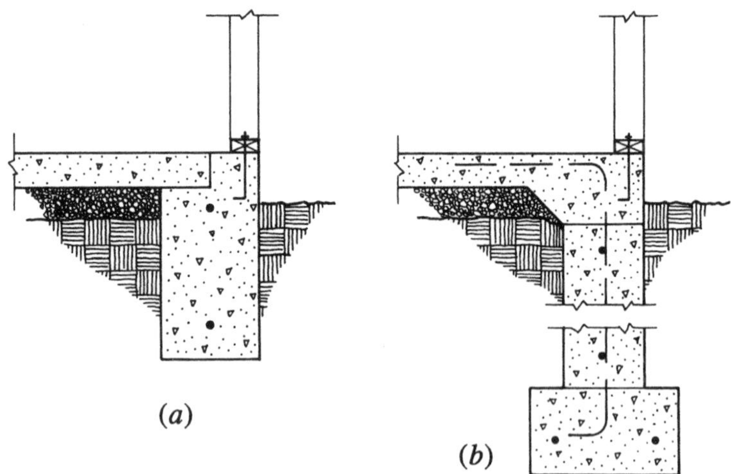

Figure 18.14 Options for the exterior wall foundations for the wood structure.

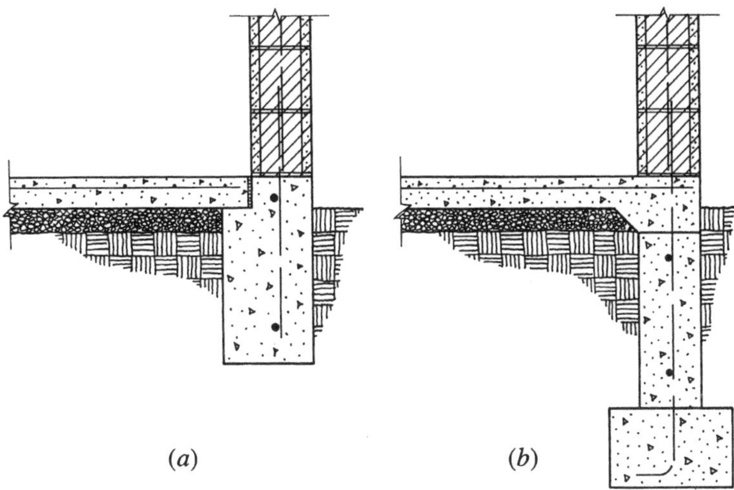

Figure 18.15 Options for the exterior wall foundations for the masonry wall structure.

continuous wall may also be designed for some minimal beamlike action, similar to that for the grade beam.

For either type of foundation, the light loading of the roof and the wood stud wall will require a very minimal width of foundation if the bearing soil material is at all adequate. If bearing is not adequate, then this type of foundation (shallow bearing footings) must be replaced with some form of deep foundation (piles or caissons), which presents a major structural design problem.

Figure 18.15 shows foundation details for the masonry wall structure, similar to those for the wood structure. Here the extra weight of the masonry wall may require some more width for the bearing elements, but the general form of construction will be quite similar. An alternative for the foundation wall in either Figure 18.14*b* or Figure 18.15*b* is to use grout-filled concrete blocks instead of the cast concrete wall.

Footings for any interior columns for Building One would also be minimal, due to the light loading from the roof structure and the low value of live load for the roof.

19

BUILDING TWO

Figure 19.1 shows a building that consists essentially of stacking the plan for Building One to produce a two-story building. The profile section of the building shows that the structure for the second story is developed the same as the roof structure and walls for Building One. Here, for both the roof and the second floor, the framing option chosen is that shown in Figure 18.2*b*.

While the framing layout is similar, the principal difference between the roof and floor structures has to do with the loadings. Both the dead load and live load are greater for the floor. In addition, the deflection of long-spanning floor members is a concern both for the dimension and for the bounciness of the structure.

The two-story building sustains a greater total wind load, although the shear walls for the second story will be basically the same as for Building One. The major effect in this building is the force generated in the first-story shear walls. In addition, there is a second horizontal diaphragm: the second-floor deck.

Some details for the second-floor framing are shown in Figure 19.1*c*. Roof framing details are similar to those shown

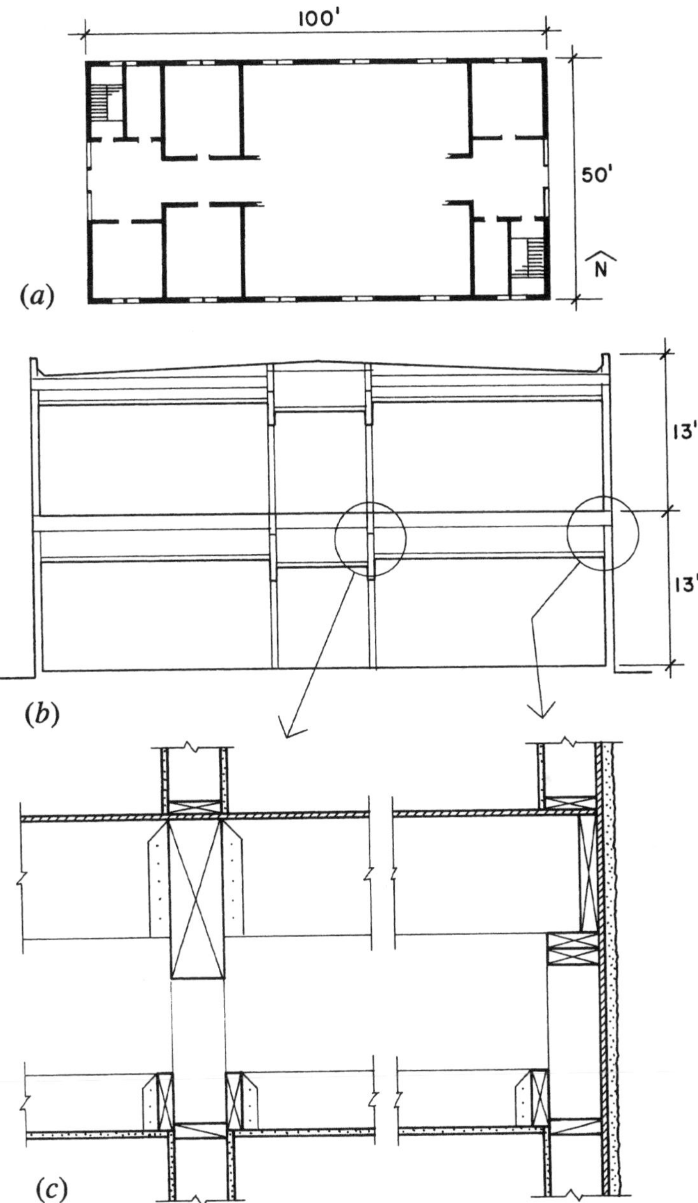

Figure 19.1 Building Two: general form and construction details.

for Building One in Figure 18.3*b*. As with Building One, an option here is to use a clear spanning roof structure—most likely with light trusses—that would eliminate the need for the corridor wall columns in the second floor.

19.1 DESIGN FOR GRAVITY LOADS

For design of the second-floor structure the following construction is assumed. The weight of the ceiling is omitted, assuming it to be supported by the first-story walls.

Carpet and pad	3.0 psf
Fiberboard underlay	3.0 psf
Concrete fill, 1.5 in.	12.0 psf
Plywood deck, 3/4 in.	2.5 psf
Ducts, lights, wiring	3.5 psf
Total, without joists	24.0 psf

Minimum live load for office areas is 50 psf. However, the code requires the inclusion of an extra load to account for possible additional partitioning; usually 25 psf. Thus, the full design live load is 75 psf. At the corridor the live load is 100 psf. Many designers would prefer to design the whole floor for a live load of 100 psf, thereby allowing for other arrangements or occupancies in the future. As the added partition load is not required for this live load, it is only an increase of about 20% in the total load. With this consideration, the total design floor load is 124 psf.

With joists at 16-in. centers, the superimposed uniformly distributed load on a single joist is

$$\text{DL} = \frac{16}{12}(24) = 32 \text{ lb/ft} + \text{the joist, say } 40 \text{ lb/ft}$$

$$\text{LL} = \frac{16}{12}(100) = 133 \text{ lb/ft}$$

and the total load is 173 lb/ft. For the 21-ft-span joists the maximum bending moment is

$$M = \frac{wL^2}{8} = \frac{173 \times (21)^2}{8} = 9537 \text{ ft} - \text{lb}$$

For Douglas fir–larch joists of select structural grade and 2-in. nominal thickness, F_b from Table 5.1*a* is $1500 \times 1.15 = 1725$ psi for repetitive member use. Thus, the required section modulus is

$$S = \frac{M}{F_b} = \frac{9537 \times 12}{1725} = 66.3 \text{in.}^3$$

Inspection of Table A.8 shows that there is no 2 × member with this value for section modulus. A possible choice is for a 3 × 14 with $S = 73.151$ in.^{3} This is not a good design since the 3 × members in select structural grade are very expensive. A better choice for this span and load is probably for one of the proprietary fabricated joists (see discussion in Section 5.14). Another alternative to investigate would be to change the direction of the beams and joist as illustrated in Figure 18.2*c*.

The beams support both the 21-ft joists and the short 8-ft corridor joists. The total load periphery carried by one beam is approximately 240 ft^2, for which a reduction of 7% is allowed for the live load (see discussion in Section 1.8). Using the same loading for corridors and offices, the beam load is determined as

DL = (30)(14.5)	= 435 lb/ft
+ beam weight	= 50 lb/ft
+ wall above	= 150 lb/ft
Total DL	= 635 lb/ft
LL = (0.93)(100)(14.5)	= 1349 lb/ft
Total load on beam	= 1984, say 2000 lb/ft

For the uniformly loaded simple beam with a span of 16.67 ft

Total load = W = (2)(16.67) = 33.4 kips

End reaction = maximum beam shear = $W/2$ = 16.7 kips

Maximum bending moment is

$$M = \frac{WL}{8} = \frac{33.4 \times 16.67}{8} = 69.6 \text{ kip} - \text{ft}$$

For a Douglas fir–larch, dense No. 1 grade beam, Table 5.1*a* yields values of $F_b = 1550$ psi, $F_v = 170$ psi, and $E = 1{,}700{,}000$ psi. To satisfy the flexural requirement, the required section modulus is

$$S = \frac{M}{F_b} = \frac{69.6 \times 12}{1.550} = 539 \text{ in.}^3$$

From Table A.8 the least weight section is a 10 × 20 or a 12 × 18.

If the 20-in.-deep section is used, its effective bending resistance must be reduced (see discussion in Section 5.4). Thus the actual moment capacity of the 10 × 20 is reduced by the size factor from Table 5.2 and is determined as

$$\begin{aligned} M &= C_F \times F_b \times S \\ &= (0.947)(1.550)(602.1)(1/12) = 73.6 \text{ kip} - \text{ft} \end{aligned}$$

As this still exceeds the requirement, the selection is adequate. Similar investigations will show the other size options to also be acceptable.

If the actual beam depth is 19.5 in., the critical shear force may be reduced to that at a distance of the beam depth from the support. Thus, an amount of load equal to the beam depth times the unit load can be subtracted from the maximum shear. The critical shear force is thus

$$\begin{aligned} V &= (\text{actual end shear force}) - (\text{beam depth in feet times unit load}) \\ &= 16.7 - 2.0(19.5/12) = 16.7 - 3.25 = 13.45 \text{ kips} \end{aligned}$$

For the 10 × 20 the maximum shear stress is thus

$$f_v = 1.5\frac{V}{A} = 1.5\frac{13{,}450}{185.3} = 109 \text{ psi}$$

This is less than the limiting stress of 170 psi as given in Table 5.1*a*, so the beam is acceptable for shear resistance. However, this is still a really big piece of lumber, and questionably feasible, unless this building is in the heart of a major timber region. It is probably logical to modify the structure to reduce the beam span or to choose a steel beam, a glued-laminated section, or a laminated veneer lumber section in place of the solid-sawn timber.

Although deflection is often critical for long spans with light loads, it is seldom critical for the short-span, heavily loaded beam. The reader may verify this by investigating the deflection of this beam, but the computation is not shown here (see Section 5.7).

For the interior column at the first story the design load is approximately equal to the total load on the second-floor beam plus the load from the roof structure. As the roof loading is about one-third of that for the floor, the design load is about 50 kips for the 10-ft-high column. Table 6.1 yields possibilities for an 8 × 10 or 10 × 10 section. For various reasons it may be more practical to use a steel member here—a round pipe or a square tubular section—which may actually be accommodated within a relatively thin stud wall at the corridor.

Columns must also be provided at the ends of the beams in the east and west walls. Separate column members may be provided at these locations, but it is also common to simply build up a column within the wall from a number of studs.

19.2 DESIGN FOR LATERAL LOADS

Lateral resistance for the second story of Building Two is essentially the same as for Building One. Design consideration here will be limited to the diaphragm action of the second-floor deck and the two-story end shear walls.

The wind loading condition for the two-story building is shown in Figure 19.2*a*. For the same design conditions assumed for wind in Chapter 18, the pressure used for horizontal force on the building bracing system is 10 psf for the entire height of the exterior wall. At the second-floor level the wind load delivered to the edge of the diaphragm is 120 lb/ft, resulting in the spanning action of the diaphragm as shown in Figure 19.2*b*. Referring to the building plan in Figure 19.1*a*, it may be observed that the opening required for the stairs creates a void in the floor deck at the ends of the diaphragm. The net width of the diaphragm is thus reduced to approximately 35 ft at this point, and the unit stress for maximum shear is

$$v = \frac{6000}{35} = 171 \text{ lb/ft}$$

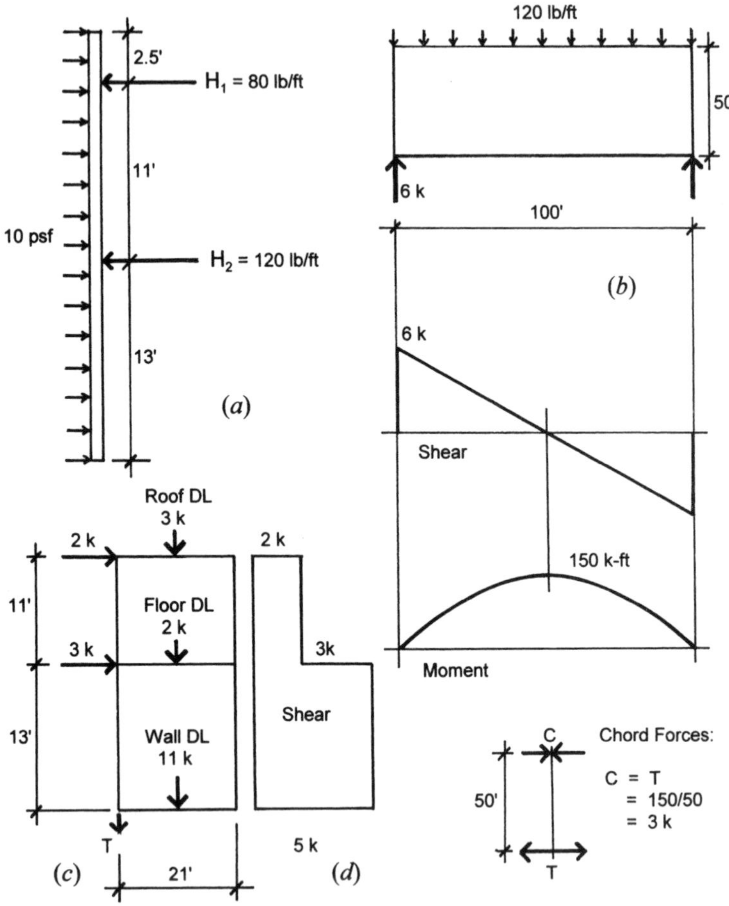

Figure 19.2 Building Two: development of lateral force due to wind.

From Table 18.2 it may be determined that this requires only minimum nailing for a $^{19}/_{32}$-in.-thick plywood deck, which is the usual minimum thickness used for floor decks.

As discussed for the roof diaphragm in Chapter 18, the chord at the edge of the floor diaphragm must be developed by framing members to sustain the computed tension–compression force of 3 kips. Ordinary framing members may be capable of this action if attention is paid to splicing for full continuity of the 100-ft-long edge member.

The construction details for the roof, floor, and exterior walls must be carefully studied to assure that the necessary transfers of force are achieved. These transfers include the following:

1. Transfer of the force from the roof plywood deck (the horizontal diaphragm) to the wall plywood sheathing (the shear walls)
2. Transfer from the second-story shear wall to the first-story shear wall that supports it
3. Transfer from the second-floor deck (horizontal diaphragm) to the first-story wall plywood sheathing (the shear walls)
4. Transfer from the first-story shear wall to the building foundations

In the first-story end shear walls the total lateral load is 5000 lb, as shown in Figure 19.2*d*. For the 21-ft-wide wall the unit shear is

$$v = \frac{5000}{21} = 238\ \text{lb/ft}$$

From Table 18.2 it may be noted that this resistance can be achieved with APA rated sheathing of 3/8-in. thickness, although nail spacing closer than the minimum of 6 in. is required at the panel edges. Choice of APA structural I plywood will permit use of fewer nails.

At the first-floor level, the investigation for overturn of the end shear wall is as follows (see Figure 19.2*c*):

$$\text{Overturning moment} = (2)(24)(1.5) + (3)(13)(1.5)$$

$$\text{Total overturning moment with safety factor}: 130.5\ \text{kip} - \text{ft}$$

$$\text{Restoring moment} = (3 + 2 + 11)(21/2) = 168\ \text{kip} - \text{ft}$$

$$\text{Net overturning effect} = 130.5 - 168 = -37.5\ \text{kip} - \text{ft}$$

As the restoring moment provides a safety factor greater than 1.5, there is no requirement for the anchorage force *T*.

In fact, there are other resisting forces on this wall. At the building corner the end walls are reasonably well attached through the corner framing to the north and south walls, which would need to be lifted to permit overturning. At the sides of the building entrance, with the second-floor framing as described, there is a post in the end of the wall that supports the end of the floor beams. All in all, there is probably no computational basis for requiring an anchor at the ends of the shear walls. Nevertheless, many designers routinely supply such anchors.

19.3 ALTERNATIVE STEEL AND MASONRY STRUCTURE

As with Building One, an alternative construction for this building is one with masonry walls and interior steel framing, with ground floor and roof construction essentially the same as that shown for Building One in Section 18.6. The second floor may be achieved as shown in Figure 19.3. Because of heavier loads, the floor structure here consists of a steel framing system with rolled steel beams supported by steel columns on the interior and by pilasters in the exterior masonry walls.

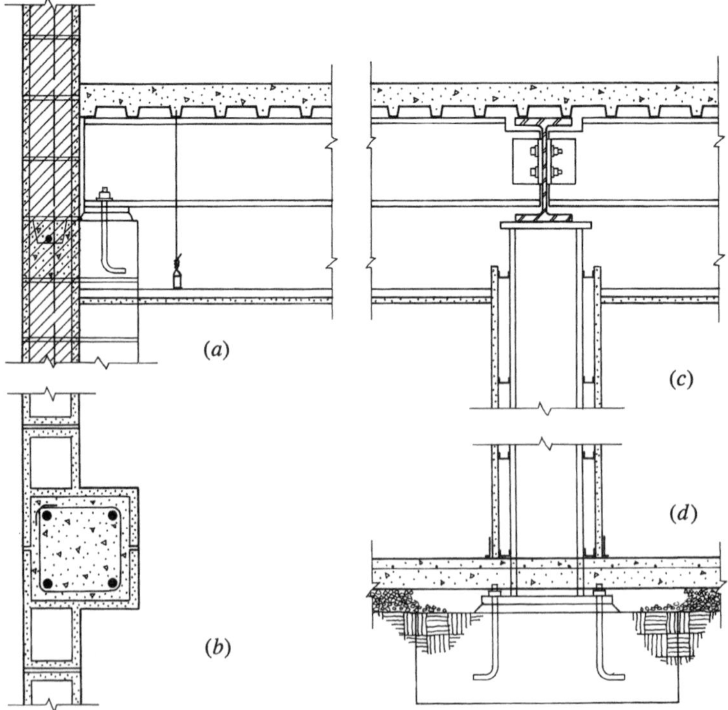

Figure 19.3 Building Two: details for the alternative steel and masonry structure.

The plan detail in Figure 19.3*b* shows the typical pilaster as formed in the CMU construction.

The floor deck consists of formed sheet steel units with a structural-grade concrete fill. This deck spans between steel beams, which are in turn supported by larger beams that are supported directly by the columns. All of the elements of this steel structure can be designed by procedures described in Part III of this book.

Figure 19.4 shows a possible framing plan for the second floor of Building Two, consisting of a variation of the roof framing plan in Figure 18*b*. Here, instead of the long-span roof deck, a shorter span floor deck is used with a series of beams at 6.25-ft spacing supported by the exterior walls and the two interior girders.

Fireproofing details for the steel structure would depend on the local fire zone and the building code requirements. Encasement of steel members in fire-resistive construction may suffice for this small building.

Although the taller masonry walls, carrying both roof and floor loads, have greater stress development than in Building One, it is still possible that minimal code-required construction may be adequate. Provisions for lateral load depend on load magnitudes and code requirements.

Another possibility for Building Two—depending on fire resistance requirements—is to use wood construction for the building

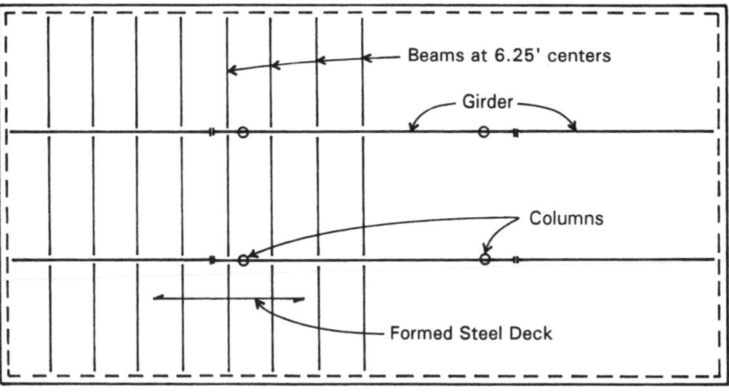

Figure 19.4 Second-floor framing plan with steel beams.

roof and floor systems, together with the masonry walls. Many buildings were built through the nineteenth and early twentieth centuries with masonry walls and interior structures of timber, a form of construction referred to as mill construction. A similar system in present-day construction is one that uses a combination of wood and steel structural elements with exterior masonry walls, as illustrated for Building Three in Section 20.7.

20

BUILDING THREE

Building Three is a modest-size office building, generally qualified as being low rise (see Figure 20.1). In this category there is a considerable range of choice for the construction, although in a particular place, at a particular time, a few popular forms of construction tend to dominate the field.

20.1 GENERAL CONSIDERATIONS

Some modular planning is usually required for this type of building, involving the coordination of dimensions for spacing of columns, window mullions, and interior partitions in the building plan. This modular coordination may also be extended to development of ceiling construction, lighting, ceiling HVAC elements, and the systems for access to electric power, phones, data, and other wiring systems. There is no single magic number for this modular system; all dimensions between 3 and 5 ft have been used and strongly advocated by various designers. Selection of a particular proprietary system for the curtain wall, interior modular partitioning, or an integrated ceiling system may establish a reference dimension.

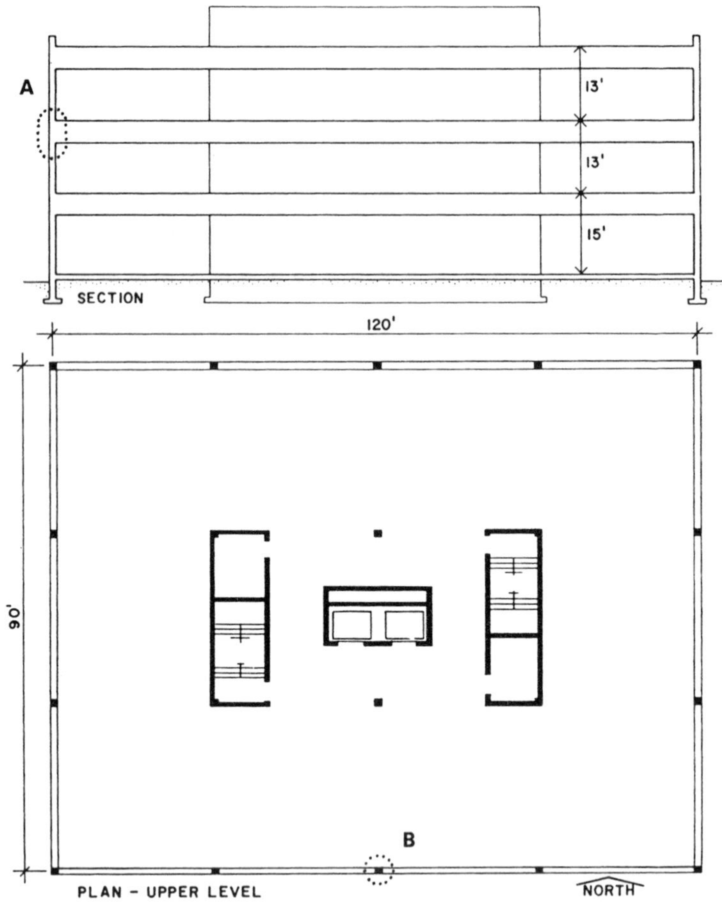

Figure 20.1 Building Three: general form.

For buildings built as investment properties, with speculative occupancies that may vary over the life of the building, it is usually desirable to accommodate future redevelopment of the building interior with some ease. For the basic construction, this means a design with as few permanent structural elements as possible. At a bare minimum, what is usually required is the construction of the major structure (columns, floors, and roof), the exterior walls, and the interior walls that enclose stairs, elevators, rest rooms, and risers for building services. Everything else should be nonstructural or demountable in nature, if possible.

Spacing of columns on the building interior should be as wide as possible, basically to reduce the number of free-standing columns in the rented portion of the building plan. A column-free interior may be possible if the distance from a central core (grouped permanent elements) to the outside walls is not too far for a single span. Spacing of columns at the building perimeter does not affect this issue, so additional columns are sometimes used at this location to reduce their size for gravity loading or to develop a stiffer perimeter rigid-frame system for lateral loads.

The space between the underside of suspended ceilings and the top of floor or roof structures above must typically contain many elements besides those of the basic construction. This usually represents a situation requiring major coordination for the integration of the space needs for the elements of the structural, HVAC, electrical, communication, lighting, and fire-fighting systems. A major design decision that must often be made very early in the design process is that of the overall dimension of the space required for this collection of elements. Depth permitted for the spanning structure and the general level-to-level vertical building height will be established—and not easy to change later if the detailed design of any of the enclosed systems indicates a need for more space.

Generous provision of the space for building elements makes the work of the designers of the various other building subsystems easier, but the overall effects on the building design must be considered. Extra height for the exterior walls, stairs, elevators, and service risers all result in additional cost, making tight control of the level-to-level distance very important.

A major architectural design issue for this building is the choice of a basic form of the construction of the exterior walls. For the column-framed structure, there are two elements that must be integrated: the columns and the nonstructural infill wall. The basic form of the construction as shown in Figure 20.2 involves the incorporation of the columns into the wall, with windows developed in horizontal strips between the columns. With the exterior column and spandrel covers developing a general continuous surface, the window units are thus developed as "punched" holes in the wall.

The windows in this example do not exist as parts of a continuous curtain wall system. They are essentially single individual units, placed in and supported by the general wall system. The curtain wall is developed as a stud-and-surfacing system not unlike the typical light wood stud wall system in character. The studs in this case are

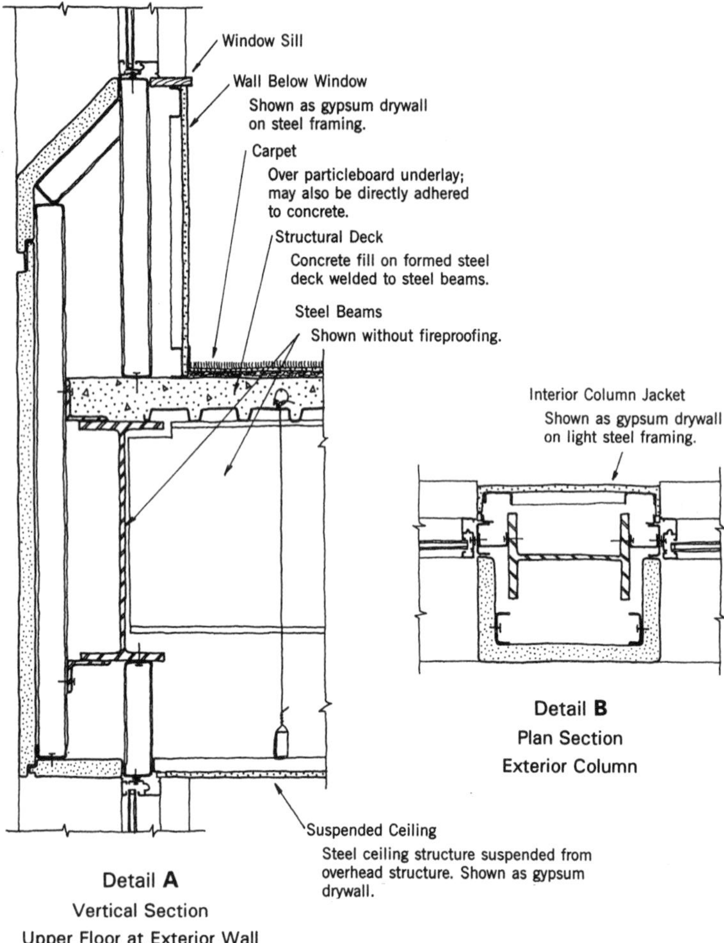

Figure 20.2 Wall, floor, and exterior column construction at the upper floors.

light-gage steel, the exterior covering is a system of metal-faced sandwich panel units, and the interior covering, where required, is gypsum drywall, attached to the metal studs with screws.

Detailing of the wall construction (as shown in detail A of Figure 20.2) results in a considerable interstitial void space. Although taken up partly with insulation materials, this space may easily contain elements for the electrical system or other services. In cold climates, a perimeter hot-water heating system would most likely be used, and it could be incorporated in the wall space shown here.

Design Criteria

The following are used for the design work:

Design codes: ASCE 2010 standard (Ref. 1) and 2012 IBC (Ref. 4)
Live loads:
Roof: 20 psf, reducible as described in Section 1.8.
Floor: from Table 1, 50 psf minimum for office areas, 100 psf for lobbies and corridors, 20 psf for movable partitions
Wind: map speed of 90 mph, exposure B
Assumed construction loads:
 Floor finish: 5 psf
 Ceilings, lights, ducts: 15 psf
 Walls (average surface weight):
  Interior, permanent: 15 psf
  Exterior curtain wall: 25 psf
  Steel for rolled shapes: ASTM A996, $F_y = 50$ ksi

20.2 STRUCTURAL ALTERNATIVES

Structural options for this example are considerable, including possibly the light wood frame if the total floor area and zoning requirements permit its use. Certainly, many steel frame, concrete frame, and masonry bearing wall systems are feasible. Choice of the structural elements will depend mostly on the desired plan form, the form of window arrangements, and the clear spans required for the building interior. At this height and taller, the basic structure must usually be steel, reinforced concrete, or masonry.

Design of the structural system must take into account both gravity and lateral loads. Gravity requires developing horizontal spanning systems for the roof and upper floors and the stacking of vertical supporting elements. The most common choices for the general lateral bracing system are the following (see Figure 20.3):

Core Shear Wall System (Figure 20.3*a*). Use of solid walls around core elements (stairs, elevators, rest rooms, duct shafts) produces a very rigid vertical structure; the rest of the construction may lean on this rigid core.

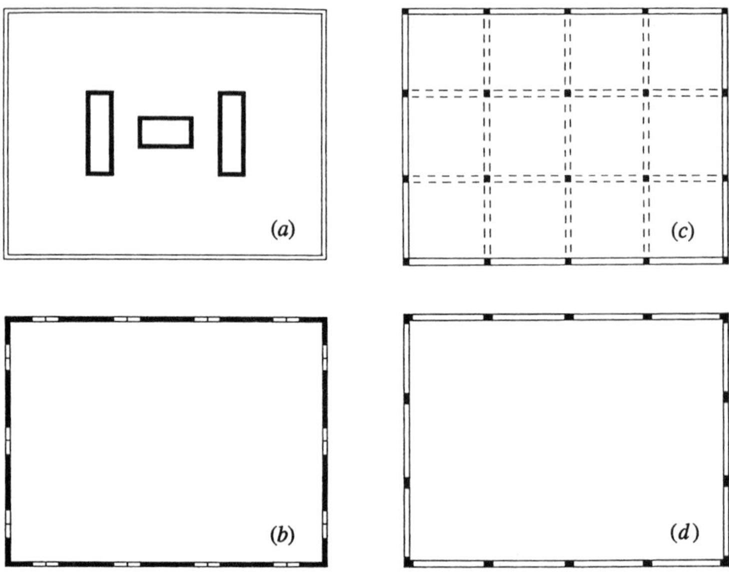

Figure 20.3 Options for the vertical elements of the lateral bracing system.

Truss-Braced Core. Similar to the shear wall core; trussed bents replace solid walls.

Perimeter Shear Walls (Figure 20.3*b*). Turns the building into a tubelike structure; walls may be structurally continuous and pierced by holes for windows and doors or may be built as individual, linked piers between vertical strips of openings.

Mixed Exterior and Interior Shear Walls or Trussed Bents. For some building plans the perimeter or core systems may not be feasible, requiring use of some mixture of walls and/or trussed bents.

Full Rigid-Frame Bent System (aka Moment Frame) (Figure 20.3*c*). Uses all the available bents described by vertical planes of columns and beams.

Perimeter Rigid-Frame Bent System (aka Perimeter Tube) (Figure 20.3*d*). Uses only the columns and spandrel beams in the exterior wall planes, resulting in only two bents in each direction for this building plan.

In the right circumstances, any of these systems may be acceptable for this size building. Each has some advantages and disadvantages from both structural and architectural design points of view.

Presented here are schemes for use of three lateral bracing systems: a truss-braced core, a rigid-frame bent, and multistory shear walls. For the horizontal roof and floor structures, several schemes are also presented.

20.3 DESIGN OF THE STEEL STRUCTURE

Figure 20.4 shows a partial plan of a framing system for the typical upper floor that uses rolled steel beams spaced at a module related to

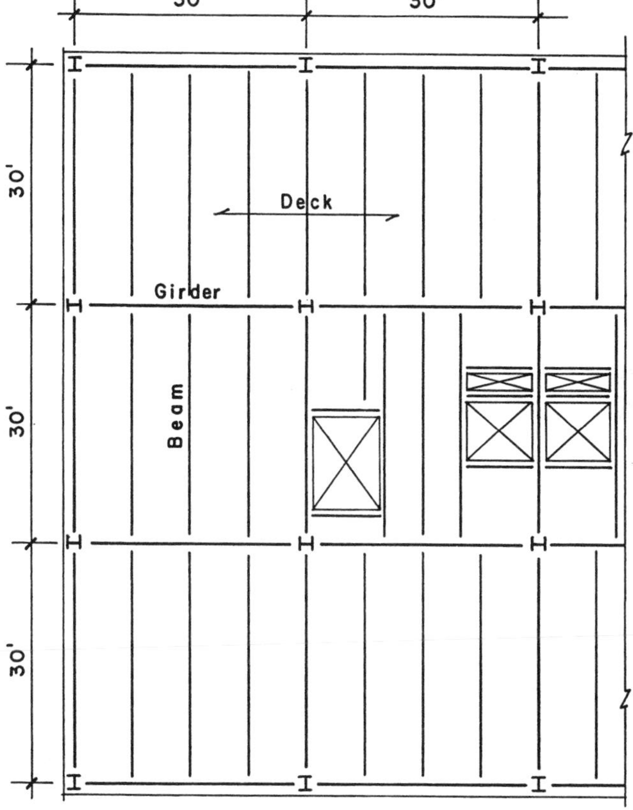

Figure 20.4 Partial framing plan for the steel floor structure for the upper levels.

the column spacing. As shown, the beams are 7.5 ft on center and the beams that are not on the column lines are supported by column line girders. Thus three-fourths of the beams are supported by the girders and the remainder are supported directly by the columns. The beams in turn support a one-way spanning deck.

Within this basic system there are a number of variables:

Beam Spacing. Affects the deck span and the beam loading.

Deck. A variety available, as discussed later.

Beam/Column Relationship in Plan. As shown, permits possible development of vertical bents in both directions.

Column Orientation. The W shape has a strong axis and accommodates framing differently in different directions.

Fire Protection. Various means, as related to codes and general building construction.

These issues and others are treated in the following discussions.

Inspection of the framing plan in Figure 20.4 reveals a few common elements of the system as well as several special beams required at the building core. The discussions that follow are limited to treatments of the common elements, that is, the members labeled "Beam" and "Girder" in Figure 20.4.

For the design of the speculative rental building, it must be assumed that different plan arrangements of the floors are possible. Thus, it is not completely possible to predict where there will be offices and where there will be corridors, each of which requires different live loads. It is thus not uncommon to design for some combinations of loading for the general system that relates to this problem. For the design work here, the following will be used:

For the deck: live load = 100 psf

For the beams: live load = 80 psf, with 20 psf added to dead load for movable partitions

For girders and columns: live load = 50 psf, with 20 psf added to dead load

The Structural Deck

Several options are possible for the floor deck. In addition to structural concerns, which include gravity loading and diaphragm action for lateral loads, consideration must be given to fire protection for the steel, to the accommodation of wiring, piping, and ducts, and to attachment of finish floor, roofing, and ceiling construction. For office buildings there are often networks for electrical power and communication that must be built into the ceiling, wall, and floor constructions.

If the structural floor deck is a concrete slab, either site cast or precast, there is usually a nonstructural fill placed on top of the structural slab; power and communication networks may be buried in this fill. If a steel deck is used, closed cells of the formed sheet steel deck units may be used for some wiring, although this is no longer a common practice.

For this example, the selected deck is a steel deck with 1.5-in.-deep ribs, on top of which is cast a lightweight concrete fill with a minimum depth of 2.5 in. over the steel units. The unit average dead weight of this deck depends on the thickness of the sheet steel, the profile of the deck folds, and the unit density of the concrete fill. For this example, it is assumed that the average weight is 30 psf. Adding to this the assumed weight of the floor finish and suspended items, the total dead load for the deck design is thus 50 psf.

While industry standards exist for these decks (see Ref. 9), data for deck design should be obtained from deck manufacturers.

The Common Beam

As shown in Figure 20.4, this beam spans 30 ft and carries a load strip that is 7.5 ft wide. The total peripheral load support area for the beam is thus $7.5 \times 30 = 225\ \text{ft}^2$. This allows for a reduced live load as follows (see Section 4.2):

$$L = L_0\left(0.25 + \frac{15}{\sqrt{K_{\text{LL}}A_T}}\right) = 80\left(0.25 + \frac{15}{\sqrt{2 \times 225}}\right) = 77\ \text{psf}$$

The beam loading is thus

Live load = 7.5(77) = 578 lb/lineal ft (or plf)

Dead load = 7.5(50 + 20) = 525 plf + beam weight, say 560 plf

Total factored unit load = 1.2(560) + 1.6(578) = 672 + 925 = 1597 plf

Total supported factored load = 1.597(30) = 48 kips

Assuming that the welding of the steel deck to the top flange of the beam provides almost continuous lateral bracing, the beam may be selected on the basis of flexural failure.

For this load and span, Table 9.2 yields the following possible choices: W 16 × 36, W 18 × 40, or W 21 × 44. Actual choice may be affected by various considerations. For example, the table used does not incorporate concerns for deflection or lateral bracing. The deeper shape will obviously produce the least deflection, although in this case the live-load deflection for the 16-in. shape is within the usual limit (see Figure 9.11). This beam becomes the typical member, with other beams being designed for special circumstances, including the column-line beams, the spandrels, and so on.

The Common Girder

Figure 20.5 shows the loading condition for the girder, as generated only by the supported beams. While this ignores the effect of the weight of the girder as a uniformly distributed load, it is reasonable for use in an approximate design since the girder weight is a minor loading.

Note that the girder carries three beams and thus has a total load periphery of 3(225) = 675 ft^2. The reduced live load is thus (see Section 4.3)

$$L = (80 \times 675)\left(0.25 + \frac{15}{\sqrt{2 \times 675}}\right) = 35{,}545 \text{ lb} \quad \text{or} \quad 35.5 \text{ kips}$$

The unit beam load for design of the girder is determined as follows:

Live load = 35.5/3 = 11.8 kips
Dead load = 0.560(30) = 16.8 kips
Factored load = 1.2(16.8) + 1.6(11.8) = 30.0 kips
To account for beam weights use 40 kips
From Figure 20.5, maximum moment is 600 kip-ft

Selection of a member for this situation may be made using various data sources. Since this member is laterally braced at only 7.5-ft intervals, attention must be paid to this point. The maximum moment together with the laterally unbraced length can be used in Table 9.2

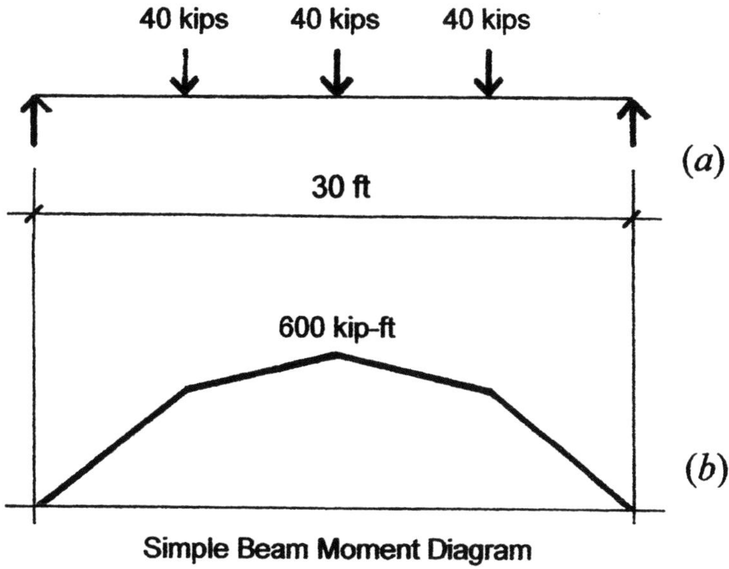

Figure 20.5 Loading condition and moment for the floor girder.

to determine acceptable choices. Possible are W 16 × 77, W 18 × 86, W 21 × 101, W 24 × 117, W 27 × 94, or W 30 × 108. The deeper members will have less deflection and will allow greater room for building service elements in the enclosed floor/ceiling space. However, shallower beams may reduce the required story height, resulting in cost savings.

Computation for deflections may be performed with formulas that recognize the true form of loading. However, approximate deflection values may be found using an equivalent load derived from the maximum moment, as discussed in Section 9.4. For this example, the equivalent uniform load (EUL) is obtained as follows:

$$M = \frac{WL}{8} = 600 \text{ kip-ft}$$

$$W = \frac{8M}{L} = \frac{8 \times 600}{30} = 160 \text{ kips}$$

This hypothetical uniformly distributed load may be used with the formula for deflection of a simple beam (see Figure 3.18) to find an approximate deflection. However, for a quick check, Figure 9.11 indicates that for this span section depths 21 in. or greater will have a total load deflection of less than 1/240 of the span, thus eliminating the W 16 × 77 and W 18 × 86 from consideration. The two girders remaining are the W 21 × 101 (the shallowest) and the W 27 × 94 (the lightest).

While deflection of individual elements should be investigated, there are wider issues regarding deflection, such as the following:

Bounciness of Floors. This involves the stiffness and the fundamental period of spanning elements and may relate to the deck and/or the beams. In general, use of the static deflection limits usually assures a reasonable lack of bounce, but just about anything that increases stiffness improves the situation.

Transfer of Load to Nonstructural Walls. With the building construction completed, live-load deflections of the structure may result in bearing of spanning members on nonstructural construction. Reducing deflections of the structure will help for this, but some special details may be required for attachment between the structure and the nonstructural construction.

Deflection During Construction. The deflection of the girders plus the deflection of the beams adds up to a cumulative deflection at the center of a column bay. This may be critical for live loads but can also create problems during construction. If the steel beams and steel deck are installed dead flat, then construction added later will cause deflection from the flat condition. In this example, that would include the concrete fill, which can cause a considerable deflection at the center of the column bay. One response is to camber (bow upward) the beams by bending them in the fabricating shop so that they deflect to a flat position under the dead load.

Column Design for Gravity Loads

Design of the steel columns must include considerations for both gravity and lateral loads. Gravity loads for individual columns are based on the column's *periphery*, which is usually defined as the area of supported surface on each level supported. Loads are actually delivered to the columns by the beams and girders, but the peripheral area is used for load tabulation and determination of live-load reductions.

If beams are rigidly attached to columns with moment-resistive connections—as is done in development of rigid-frame bents—then gravity loads will also cause bending moments and shears in the columns. Otherwise, the gravity loads are essentially considered only as axial compressive loads.

Involvement of the columns in development of resistance to lateral loads depends on the form of the lateral bracing system. If trussed bents are used, some columns will function as chords in the vertically cantilevered trussed bents, which will add some compressive forces and possibly cause some reversals with net tension in the columns. If columns are parts of rigid-frame bents, the same chord actions will be involved, but the columns will also be subject to bending moments and shears from the rigid-frame lateral actions.

Whatever the lateral force actions may do, the columns must also work for gravity load effects alone. In this part this investigation is made and designs are completed without reference to lateral loads. This yields some reference selections, which can then be modified (but not reduced) when the lateral resistive system is designed. Later discussions in this chapter present designs for both a trussed bent system and a rigid-frame system.

There are several different cases for the columns, due to the framing arrangements and column locations. For a complete design of all columns it would be necessary to tabulate the loading for each different case. For illustration purposes here, tabulation is shown for a hypothetical interior column. The interior column illustrated assumes a general periphery of 900 ft^2 of general roof or floor area. Actually, the floor plan in Figure 20.1 shows that all the interior columns are within the core area, so there is no such column. However, the tabulation yields a column that is general for the interior condition and can be used for approximate selection. As will be shown later, all the interior columns are involved in the lateral force systems, so this also yields a takeoff size selection for the design for lateral forces.

Table 20.1 is a common form of tabulation used to determine the column loads. For the interior columns, the table assumes the existence of a rooftop structure (penthouse) above the core, thus creating a fourth story for these columns.

TABLE 20.1 Service Load Tabulation for the Interior Column

Level Supported	Load Source and Computation	Load Tabulation (lb)	
		Dead Load	Live Load
Penthouse roof	Live load, not reduced = 20 psf × 225		4,500
225 ft^2	Dead load = 40 psf × 225	9,000	
Building roof	Live load, not reduced = 20 psf × 675		13,500
675 ft^2	Dead load = 40 psf × 675	27,000	
Penthouse floor	Live load = 100 psf × 225		22,500
225 ft^2	Dead load = 50 psf × 225	11,250	
	Story loads + loads from above	47,250	40,500
	Reduced live load (50%)		20,250
Third floor	Live load = 50 psf × 900		45,000
900 ft^2	Dead load = 70 psf × 900	63,000	
	Story loads + loads from above	110,250	85,500
	Reduced live load (50%)		42,750
Second floor	Story loads	63,000	45,000
(same as third)	Story loads + loads from above	173,250	130,500
	Reduced live load (50%)		65,250

Table 20.1 is organized to facilitate the following determinations:

1. Dead load on the periphery at each level, determined by multiplying the area by an assumed average dead load per square foot. Loads determined in the process of design of the horizontal structure may be used for this estimate.
2. Live load on the periphery areas.
3. The reduced live load to be used at each story, based on the total supported periphery areas above that story.
4. Other dead loads directly supported, such as the column weight and any permanent walls within the load periphery.
5. The total load collected at each level.
6. A design load for each story, using the total accumulation from all levels supported.

For the entries in Table 20.1, the following assumptions were made:

Roof unit live load = 20 psf (reducible)

Roof dead load = 40 psf (estimated, based on the similar floor construction)

Penthouse floor live load = 100 psf (for equipment, average)
Penthouse floor dead load = 50 psf
Floor live load = 50 psf (reducible)
Floor dead load = 70 psf (including partitions)

Table 20.2 summarizes the design for the four-story column. For the pin-connected frame, a K factor of 1.0 is assumed, and the full story heights are used as the unbraced column lengths. Although column loads in the upper stories are quite low and some small column sizes would be adequate for the loads, a minimum size of 10 in. is maintained for the W shapes for two reasons.

The first consideration involves the form of the horizontal framing members and the type of connections between the columns and the horizontal framing. All the H-shaped columns must usually facilitate framing in both directions, with beams connected both to column flanges and webs. With standard framing connections for field bolting to the columns, minimum beam depths and flange widths are required for practical installation of the connecting angles and bolts.

The second consideration involves the problem of achieving splices in the multistory column. If the building is too tall for a single-piece column, a splice must be used somewhere, and the stacking of one column piece on top of another to achieve a splice is made much easier if the two pieces are of the same nominal size group.

TABLE 20.2 Design of the Interior Column

Design Loads for Each Story (lb)	Possible Choices (Table 10.2)
Penthouse, unbraced height = 13 ft $P_u = \varphi P_n = 1.2(9000) + 1.6(4500)$ $= 10{,}800 + 7200 = 18{,}000$ lb	W 10 × 33
Third story, unbraced height = 13 ft $P_u = \varphi P_n = 1.2(47{,}250) + 1.6(20{,}250)$ $= 58{,}700 + 32{,}400 = 91{,}100$ lb	W 10 × 33
Second story, unbraced height = 13 ft $P_u = \varphi P_n = 1.2(110{,}250) + 1.6(42{,}750)$ $= 132{,}300 + 68{,}400 = 200{,}700$ lb	W 10 × 33, W 12 × 45
First story, unbraced height = 15 ft $P_u = \varphi P_n = 1.2(173{,}250) + 1.6(65{,}250)$ $= 207{,}900 + 104{,}400 = 312{,}300$ lb	W 10 × 54, W 12 × 53, W 14 × 54

Add to this a possible additional concern relating to the problem of handling long pieces of steel during transportation to the site and erection of the frame. The smaller the member's cross section, the shorter the piece that is feasible to handle.

For all of these reasons, a minimum column is often considered to be the W 10 × 33, which is the lightest shape in the group that has an 8-in.-wide flange. It is assumed that a splice occurs at 3 ft above the second-floor level (a convenient, waist-high distance for the erection crew), making two column pieces approximately 18 and 23 ft long. These lengths are readily available and quite easy to handle with the 10-in. nominal shape. On the basis of these assumptions, a possible choice would be for a W 10 × 54 for the ground floor and a W 10 × 33 for columns above the first story.

20.4 ALTERNATIVE FLOOR CONSTRUCTION WITH TRUSSES

A framing plan for the upper floor of Building Three is shown in Figure 20.6, indicating the use of open-web steel joists and joist girders. Although this construction might be extended to the core and the exterior spandrels, it is also possible to retain the use of rolled shapes for these purposes. Although somewhat more applicable to longer spans and lighter loads, this system is reasonably applicable to this situation as well.

One potential advantage of using the all-truss framing for the horizontal structure is the higher degree of freedom of passage of building service elements within the enclosed space between ceilings and the supported structure above. A disadvantage is the usual necessity for greater depth of the structure, adding to building height—a problem that increases with the number of stories.

Design of the Open-Web Joists

General concerns and basic design for open-web joists are presented in Section 9.10. Using the data for this example, a joist design is as follows:

Joists at 3 ft on center, span of 30 ft

Dead load = 3(70) = 210 lb/ft not including joists

Live load = 3(100) = 300 lb/ft not reduced

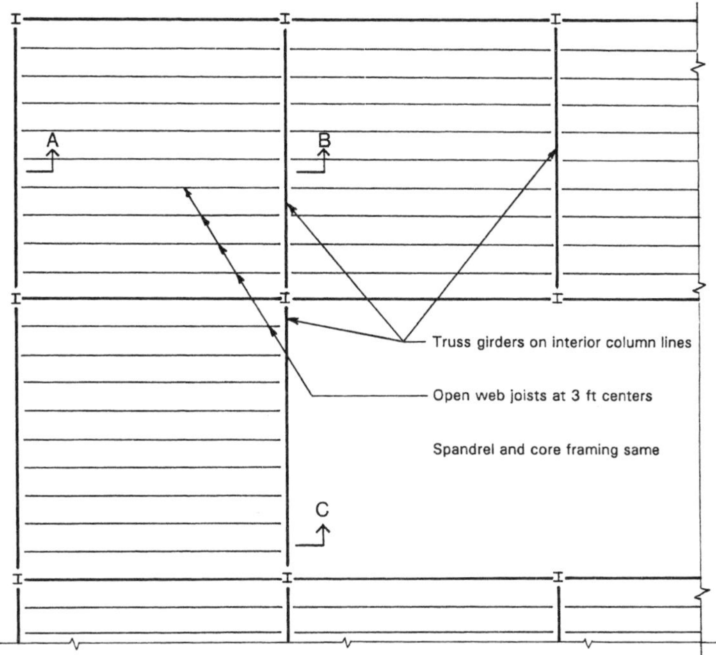

Figure 20.6 Building Three: partial framing plan at the upper floor, using open-web joists and joist girders.

This is a high live load, but it permits location of a corridor anywhere on the plan and also reduces deflection and bounciness.

Total factored load = 1.2(210) + 1.6(300) = 252 + 480 = 732 lb/ft

Referring to Table 9.4, choices may be considered for any joist that will carry the total load of 732 lb/ft, and a live load of 300 lb/ft on a span of 30 ft. The following choices are possible:

24 K9 at 12 lb/ft, permitted load = 807 − 1.2(12) = 794 lb/ft
26 K9, stronger than 24 K9 and only 0.2 lb/ft heavier
28 K8, stronger than 24 K9 and only 0.7 lb/ft heavier
30 K7, stronger than 24 K9 and only 0.3 lb/ft heavier

All of the joists listed are economically equivalent. Choice would be made considering details of the general building construction. A shallower joist depth means a shorter story height and less overall building height. A deeper joist yields more open space in the floor construction for ducts, wiring, piping, and the like and also means less deflection and less floor bounce.

Design of the Joist Girders

Joist girders are also discussed in Section 9.10. Both the joists and the girders are likely to be supplied and erected by a single contractor. Although there are industry standards (see Ref. 6), the specific manufacturer should be consulted for data regarding design and construction details for these products.

The pattern of the joist girder members is somewhat fixed and relates to the spacing of the supported joists. To achieve a reasonable proportion for the panel units of the truss, the dimension for the depth of the girder should be approximately the same as that for the joist spacing.

Considerations for design for the truss girder are as follows:

The assumed depth of the girder is 3 ft, which should be considered a minimum depth for this span ($L/10$). Any additional depth possible will reduce the amount of steel and also improve deflection responses. However, for floor construction in multistory buildings, any increase of this dimension is hard to bargain for.

Use a live-load reduction of 40% (maximum) with a live load of 50 psf. Thus, live load from one joist = $(3 \times 30)(0.6 \times 50) =$ 2700 lb or 2.7 kips.

For dead load add a partition load of 20 psf to the construction load of 40 psf. Thus, dead load = $(3 \times 30)(60) = 5400$ lb, + joist weight of 12 lb/ft $\times$ 30 = 360 lb, total = 5400 + 360 = 5760 lb or 5.76 kips.

Total factored load = $1.2(5.76) + 1.6(2.7) = 6.91 + 4.32 =$ 11.23 kips.

Figure 20.7 shows a possible form for the joist girder. For this form and the computed data the joist specification is as follows:

36G = girder depth of 3 ft
10N = 10 spaces between the joists

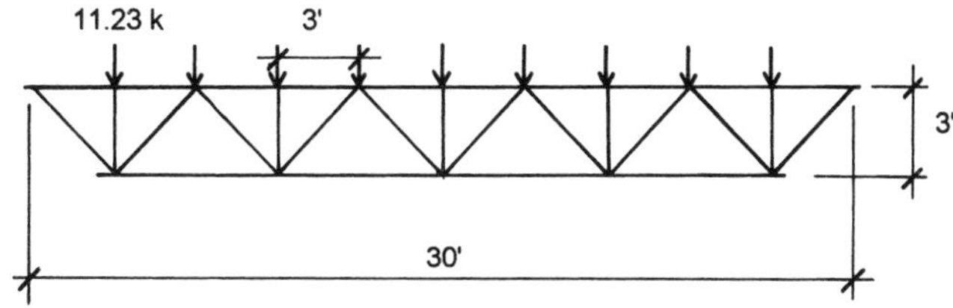

Figure 20.7 Form and data for the joist girder.

11.23K = design factored joist load

Complete specification is thus 36G10N11.23K.

Construction Details for the Truss Structure

Figure 20.8 shows some details for construction of the trussed system. The deck shown here is the same as that for the scheme with W-shape

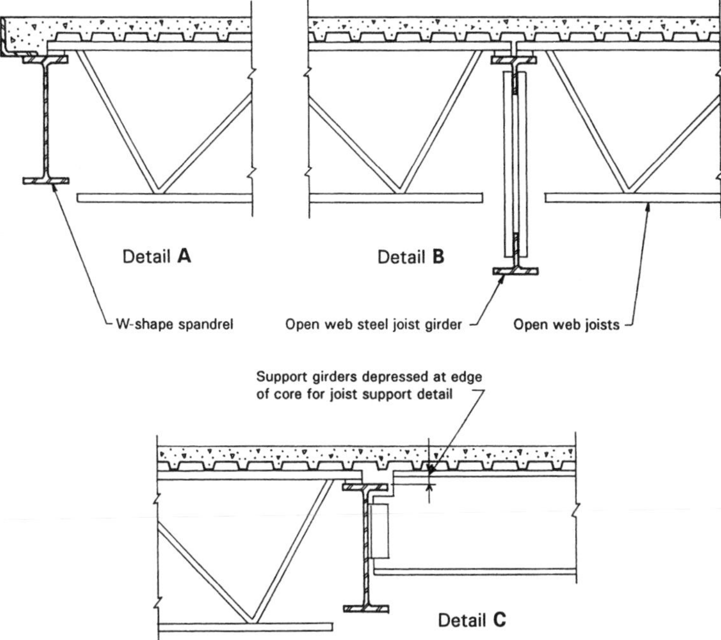

Figure 20.8 Details for the floor system with open-web joists and joist girders. For location of the details see the framing plan in Figure 20.6.

framing, although the shorter span may allow use of a lighter sheet steel deck. However, the deck must also be used for diaphragm action, which may limit the reduction.

Adding to the problem of overall height for this structure is the detail at the joist support, in which the joists must sit on top of the supporting members, whereas in the all-W-shape system the beams and girders have their tops level.

With the closely spaced joists, ceiling construction may be directly supported by the bottom chords of the joists. This may be a reason for selection of the joist depth. However, it is also possible to suspend the ceiling from the deck, as is generally required for the all W-shape structure with widely spaced beams.

Another issue here is the usual necessity to use a fire-resistive ceiling construction, as it is not feasible to encase the joists or girders in fireproofing material.

20.5 DESIGN OF THE TRUSSED BENT FOR WIND

Figure 20.9 shows a partial framing plan for the core area, indicating the placement of some additional columns off the 30-ft grid. These columns are used together with the regular columns and some of the horizontal framing to define a series of vertical bents for the development of the trussed bracing system shown in Figure 20.10. With relatively slender diagonal members, it is assumed that the X bracing behaves as if the tension diagonals function alone. There are thus considered to be four vertical, cantilevered, determinate trusses that brace the building in each direction.

With the symmetrical building exterior form and the symmetrically placed core bracing, this is a reasonable system for use in conjunction with the horizontal roof and upper floor structures to develop resistance to horizontal forces due to wind. The work that follows illustrates the design process using criteria for wind loading from ASCE 2010 (Ref. 4).

For the total wind force on the building, we will assume a base pressure of 15 psf, adjusted for height as described in the ASCE standard (Ref. 1). The design pressures and their zones of application are shown in Figure 20.11.

For investigation of the lateral bracing system, the design wind pressures on the outside wall surface are distributed as edge loadings to the roof and floor diaphragms. These are shown as the forces H_1, H_2, and H_3 in Figure 20.11. The horizontal forces are next shown as loadings to one of the vertical truss bents in Figure 20.12*a*. For the bent loads the total force per bent is determined by multiplying

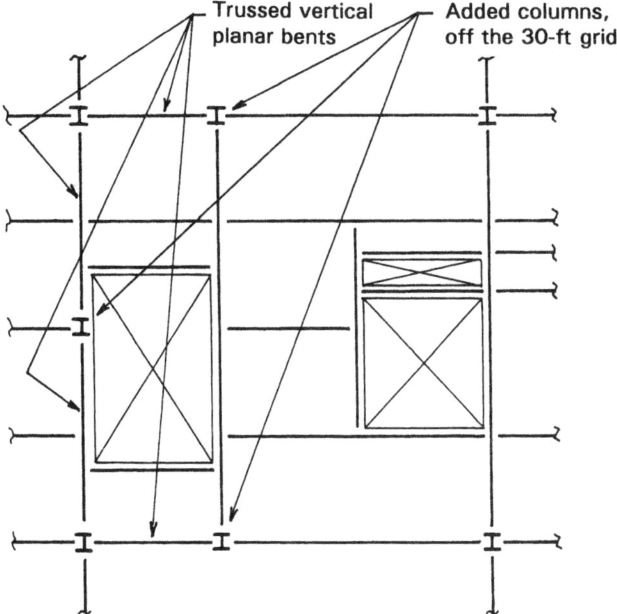

Figure 20.9 Modified framing plan for development of the trussed bents at the building core.

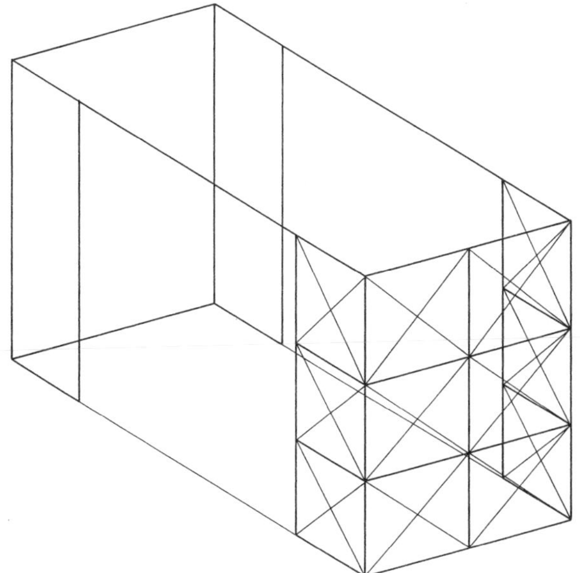

Figure 20.10 General form of the trussed bent bracing system.

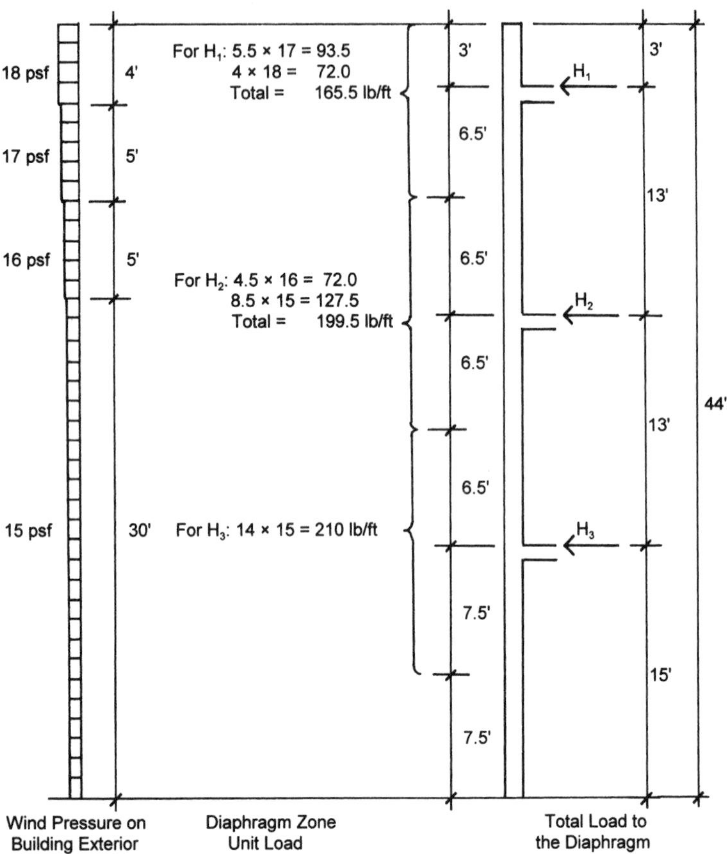

Figure 20.11 Building Three: development of the wind loads to the horizontal diaphragms.

the unit edge diaphragm load by the building width and dividing by the number of bracing bents for load in that direction. The bent loads are thus

$$H_1 = (165.5)(92)/4 = 3807 \text{ lb}$$

$$H_2 = (199.5)(92)/4 = 4589 \text{ lb}$$

$$H_3 = (210)(92)/4 = 4830 \text{ lb}$$

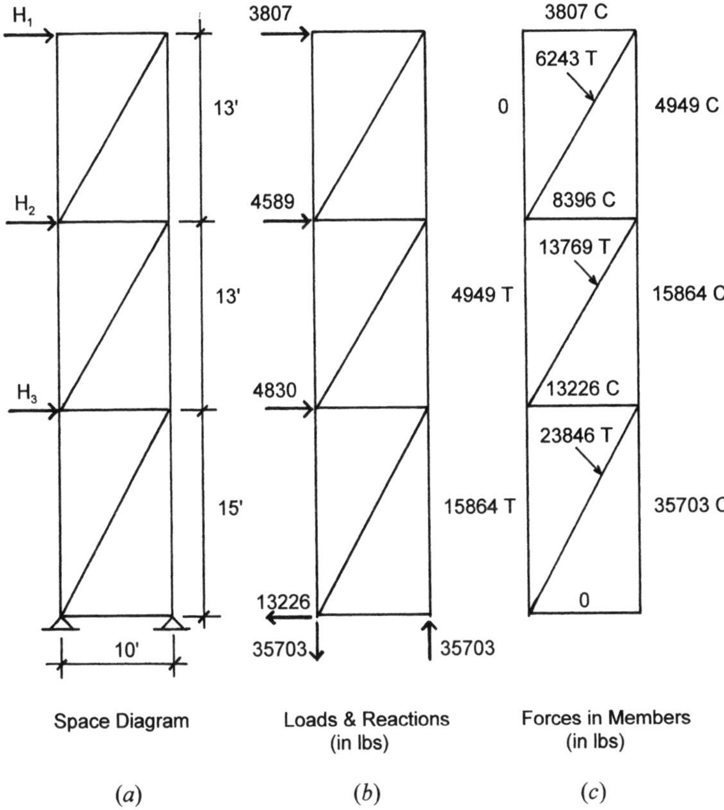

Figure 20.12 Investigation of one of the east–west bents.

The truss loading, together with the reaction forces at the supports, are shown in Figure 20.12*b*. The internal forces in the truss members resulting from this loading are shown in Figure 20.12*c*, with force values in pounds and sense indicated by *C* for compression and *T* for tension.

The forces in the diagonals may be used to design tension members using the factored load combination that includes wind (see Section 1.9). The compression forces in the columns may be added to the gravity loads to see if this load combination is critical for the column design. The uplift tension force at the column should be compared with the dead load to see if the column base needs to be designed for a tension anchorage force.

The horizontal forces should be added to the beams in the core framing and an investigation should be done for the combined bending and compression. Since beams are often weak on their minor axis (y axis), it may be practical to add some framing members at right angles to these beams to brace them against lateral buckling.

Design of the diagonals and their connections to the beam and column frame must be developed with consideration of the form of the elements and some consideration for the wall construction in which they are imbedded. Figure 20.13 shows some possible details for the diagonals and the connections. A detail problem that must be solved is that of the crossing of the two diagonals at the middle of the bent. If double angles are used for the diagonals (a common truss form),

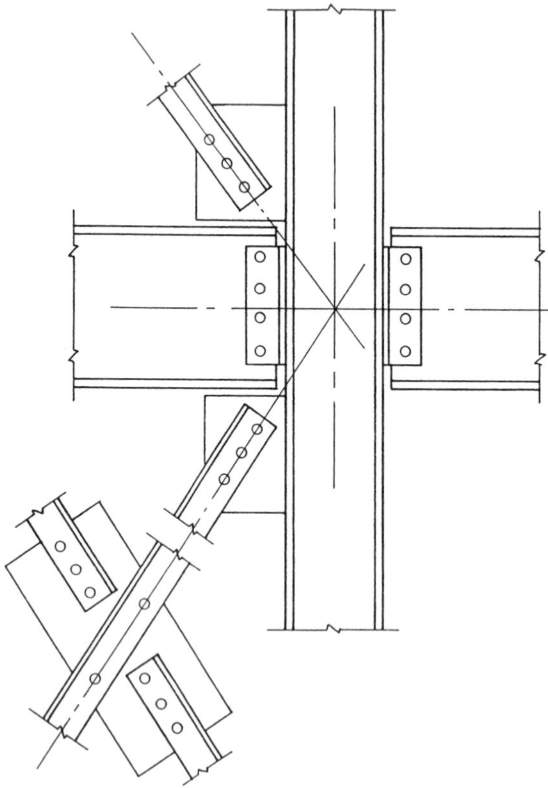

Figure 20.13 Details of the bent construction with bolted joints.

the splice joint shown in Figure 20.13 is necessary. An option is to use either single angles or channel shapes for the diagonals, allowing the members to pass each other back to back at the center. The latter choice, however, involves some degree of eccentricity in the members and connections and a single shear load on the bolts, so it is not advisable if load magnitudes are high. For the tension member, a recommended minimum slenderness is represented by an L/r ratio of 300.

20.6 CONSIDERATIONS FOR A STEEL RIGID FRAME

The general nature of rigid frames is discussed in Section 3.10. A critical concern for multistory, multiple-bay frames is the lateral strength and stiffness of columns. As the building must be developed to resist lateral forces in all directions, it becomes necessary in many cases to consider the shear and bending resistance of columns in two directions (e.g., north–south and east–west). This presents a problem for W-shape columns, as they have considerably greater resistance on their major (x–x) axis versus their minor (y–y) axis. Orientation of W-shape columns in plan thus sometimes becomes a major consideration in structural planning.

Figure 20.14*a* shows a possible plan arrangement for column orientation for Building Three, relating to the development of two major bracing bents in the east–west direction and five shorter and less stiff bents in the north–south direction. The two stiff bents may well be approximately equal in resistance to the five shorter bents, giving the building a reasonably symmetrical response in the two directions.

Figure 20.14*b* shows a plan arrangement for columns designed to produce approximately symmetrical bents on the building perimeter. The form of such perimeter bracing is shown in Figure 20.15.

One advantage of perimeter bracing is the potential for using deeper (and thus stiffer) spandrel beams, as the restriction on depth that applies for interior beams does not exist at the exterior wall plane. Another possibility is to increase the number of columns at the exterior, as shown in Figure 20.14*c*, a possibility that does not compromise the building interior space. With deeper spandrels and closely spaced exterior columns, a very stiff perimeter bent is possible. In fact, such a bent may have very little flexing in the members, and its behavior approaches that of a pierced wall, rather than a flexible frame.

At the expense of requiring much stronger (and heavier and/or larger) columns and expensive moment-resistive connections, the

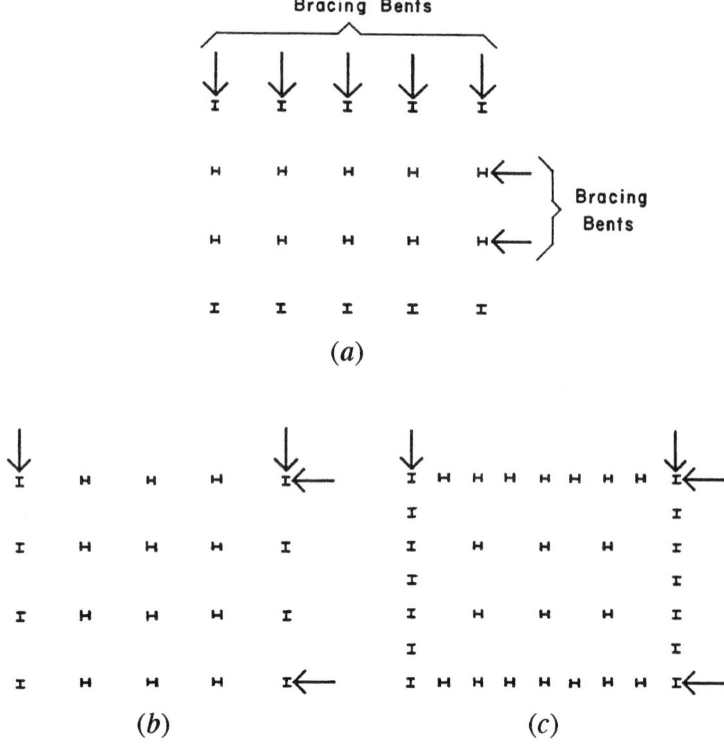

Figure 20.14 Building Three: optional arrangements for the steel W-shape columns for development of the rigid-frame bents.

rigid-frame bracing offers architectural planning advantages with the elimination of solid shear walls or truss diagonals in the walls. However, the lateral deflection (drift) of the frames must be carefully controlled, especially with regard to damage to nonstructural parts of the construction.

20.7 CONSIDERATIONS FOR A MASONRY WALL STRUCTURE

An option for the construction of Building Three involves the use of structural masonry for development of the exterior walls. The walls are used for both vertical bearing loads and lateral shear

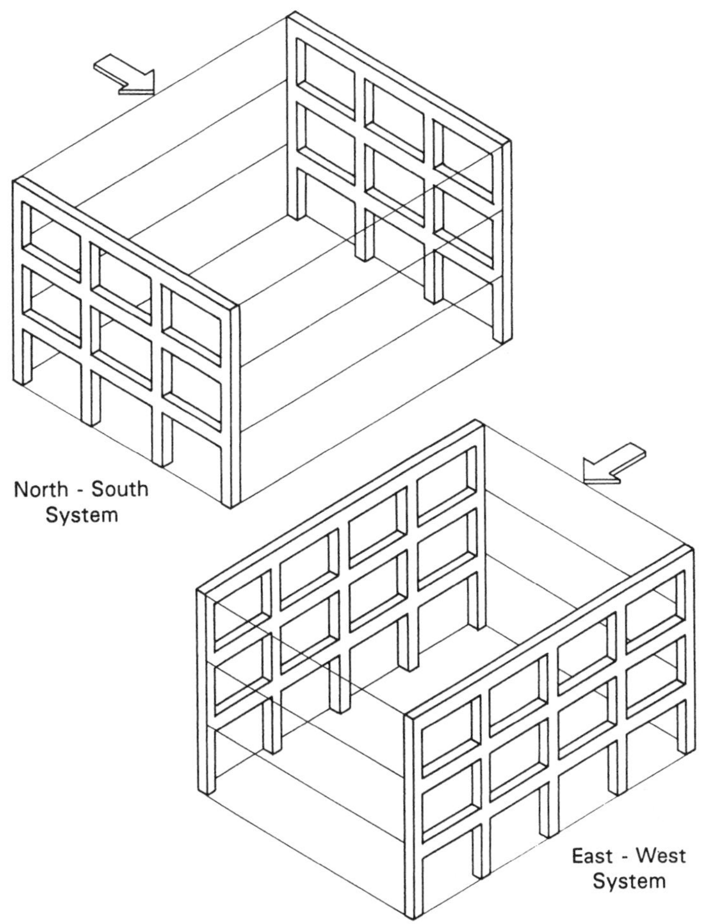

Figure 20.15 Form of the perimeter bent bracing system.

wall functions. The choice of forms of masonry and details for the construction depend very much on regional considerations (climate, codes, local construction practices, etc.) and on the general architectural design. Major differences occur due to variations in the range in outdoor temperature extremes and the specific critical concerns for lateral forces.

General Considerations

Figure 20.16 shows a partial elevation of the masonry wall structure and a partial framing plan of the upper floors. The wood construction shown here is questionably acceptable for fire codes. The example is presented only to demonstrate the general form of the construction.

Plan dimensions for structures using CMUs (concrete blocks) must be developed so as to relate to the modular sizes of typical

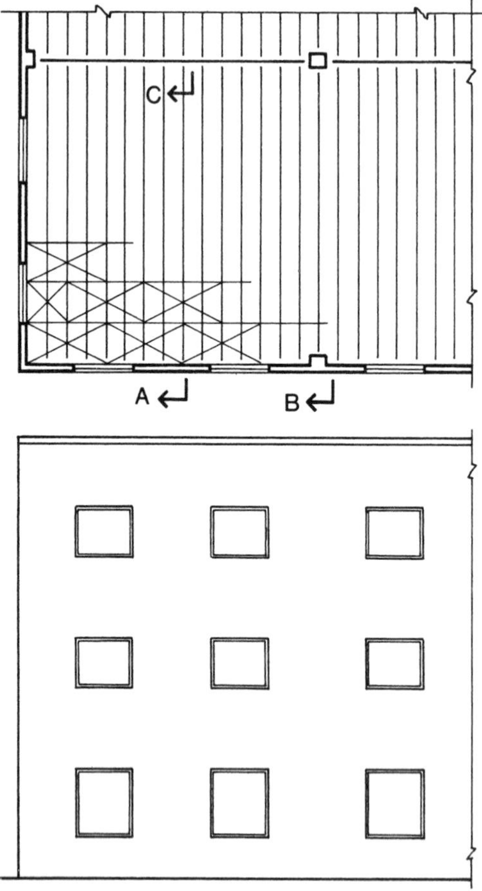

Figure 20.16 Building Three: partial framing plan for the upper floor and partial elevation for the masonry wall structure.

CMUs. There are a few standard sizes widely used, but individual manufacturers often have some special units or will accommodate requests for special shapes or sizes. However, while solid brick or stone units can be cut to produce precise, nonmodular dimensions, the hollow CMUs generally cannot. Thus, the dimensions for the CMU structure itself must be carefully developed to have wall intersections, corners, ends, tops, and openings for windows and doors fall on the fixed modules. (See Figure 18.9*a*.)

There are various forms of CMU construction. The one shown here is that widely used where either windstorm or earthquake risk is high. This is described as *reinforced masonry* and is produced to generally emulate reinforced concrete construction, with tensile forces resisted by steel reinforcement that is grouted into the hollow voids in the block construction. This construction takes the general form shown in Figure 15.17.

Another consideration to be made for the general construction is that involving the relation of the structural masonry to the complete architectural development of the construction, regarding interior and exterior finishes, insulation, incorporation of wiring, and so on.

The Typical Floor

The floor framing system here uses column-line girders that support fabricated joists and a plywood deck. The girders could be glued-laminated timber but are shown here as rolled steel shapes. Supports for the steel girders consist of steel columns on the interior.

As shown in the details in Figure 20.17, the exterior masonry walls are used for direct support of the deck and the joists, through ledgers bolted and anchored to the interior wall face. With the plywood deck also serving as a horizontal diaphragm for lateral loads, the load transfers for both gravity and lateral forces must be carefully developed in the details for this construction.

Since the masonry structure in this scheme is used only for the exterior walls, the construction at the building core is free to be developed by any of the general methods shown for other schemes. If the steel girders and steel columns are used here, it is likely that a general steel framing system might be used for most of the core framing.

Attached only to its top, the supported construction does not provide very good lateral support for the steel girder in resistance to torsional buckling. It is advisable, therefore, to use a steel shape that

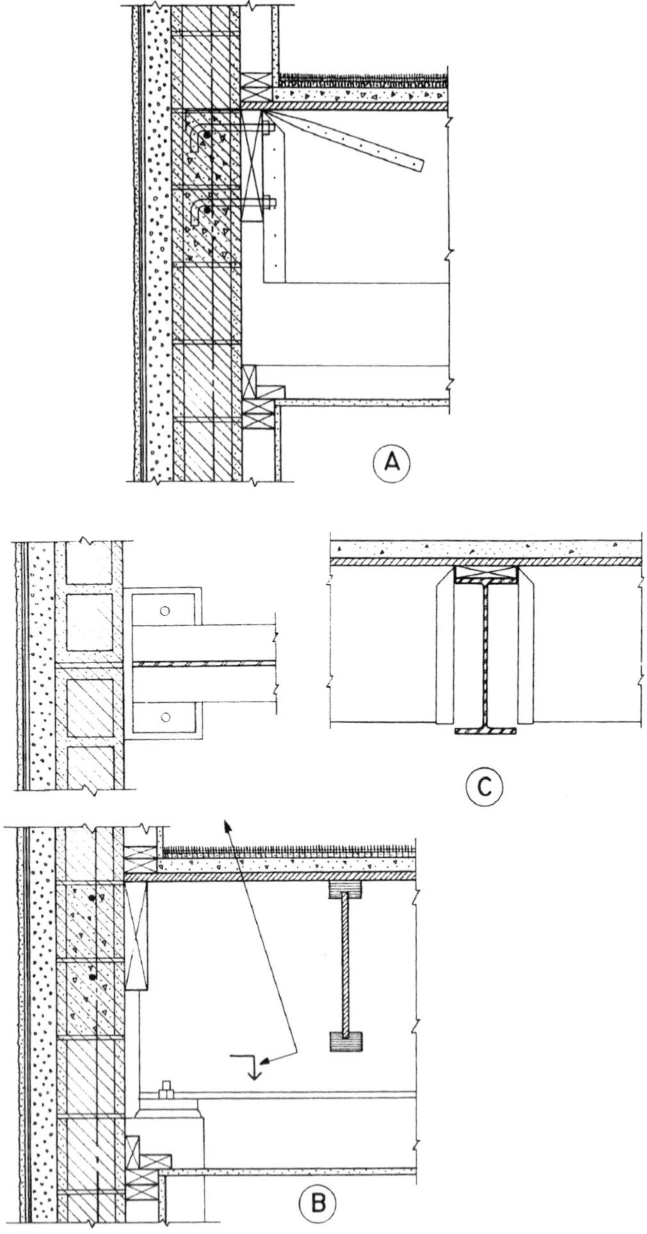

Figure 20.17 Details of the upper floor and the exterior masonry wall.

is not too weak on its *Y* axis, generally indicating a critical concern for lateral unsupported length.

The Masonry Walls

Buildings much taller than this have been achieved with structural masonry, so the feasibility of the system is well demonstrated. The vertical loads increase in lower stories, so it is expected that some increases in structural capacity will be achieved in lower portions of the walls. The two general means for increasing wall strength are to use thicker CMUs or to increase the amount of core grouting and reinforcement.

It is possible that the usual minimum structure—relating to code minimum requirements for the CMU construction—may be sufficient for the top-story walls, with increases made in steps for lower walls. Without increasing the CMU size, there is considerable range between the minimum and the feasible maximum potential for a wall.

It is common to use fully grouted walls (all cores filled) for CMU shear walls. This scheme would technically involve using fully grouted construction for *all* of the exterior walls, which might likely rule against the economic feasibility of the scheme. Adding this to the concerns for thermal movements in the long walls might indicate the wisdom of using some control joints to define individual wall segments.

Design for Lateral Forces

A common solution for lateral bracing is the use of an entire masonry wall as a shear wall, with openings considered as producing the effect of a very stiff rigid frame. As for gravity loads, the total lateral shear force increases in lower stories. Thus, it is also possible to consider the use of the potential range for a wall from minimum construction (defining a minimum structural capacity) to the maximum possible strength with all voids grouted and some feasible upper limit for reinforcement.

The basic approach here is to design the required wall for each story using the total shear at that story. In the end, however, the individual story designs must be coordinated for the continuity of the construction. However, it is also possible that the construction itself could be significantly altered in each story if it fits with architectural design considerations.

Construction Details

There are many concerns for the proper detailing of the masonry construction to fulfill the shear wall functions. There are also many concerns for proper detailing to achieve the force transfers between the horizontal framing and the walls. Some construction details for the building are shown in Figure 20.17. The general framing plan for the upper floor is shown in Figure 20.16 and the locations of the details discussed here are indicated by the section marks on that plan.

Detail A. This shows the general exterior wall construction and the framing of the floor joists at the exterior wall. The wood ledger is used for vertical support of the joists, which are hung from steel framing devices fastened to the ledger. The plywood deck is nailed directly to the ledger to transfer its horizontal diaphragm loads to the wall. Outward forces on the wall must be resisted by anchorage directly between the wall and the joists. Ordinary hardware elements can be used for this, although the exact details depend on the type of forces (wind or seismic), their magnitude, the details of the joists, and the details of the wall construction. The anchor shown in the detail is really only symbolic. General development of the construction here shows the use of a concrete fill on top of the floor deck, furred out wall surfacing with batt insulation on the interior wall side, and a ceiling suspended from the joists.

Detail B. This shows the use of the steel beam for support of the joists and the deck. After the wood lumber piece is bolted to the top of the steel beam, the attachment of the joists and the deck become essentially the same as they would be with a timber girder.

Detail C. This shows the section of the exterior wall at the location of the girder support. The girder is shown with its end resting on top of a steel column, which may be a W shape, a pipe, or a tube. If this detail is used, the column in the story above must rest on top of the girder end. This is indeed possible, if the girder web is braced for the high compression force. However, the girder could also be framed conventionally into the side of the column, with the column continuous through the joint.

20.8 THE CONCRETE STRUCTURE

A structural framing plan for the upper floors in Building Three is presented in Figure 20.18, showing the use of a site-cast concrete slab-and-beam system. Support for the spanning structure is provided by concrete columns. The system for lateral bracing is that shown in Figure 20.15, which uses the exterior columns and spandrel beams as rigid-frame bents at the building perimeter. This is a highly indeterminate structure for both gravity and lateral loads, and its precise engineering design would undoubtedly be done with a computer-aided design process. The presentation here treats the major issues and illustrates an approximate design using highly simplified methods.

Design of the Slab-and-Beam Floor Structure

For the floor structure use $f'_c = 3$ ksi and $f_y = 40$ ksi. As shown in Figure 20.18, the floor framing system consists of a series of parallel

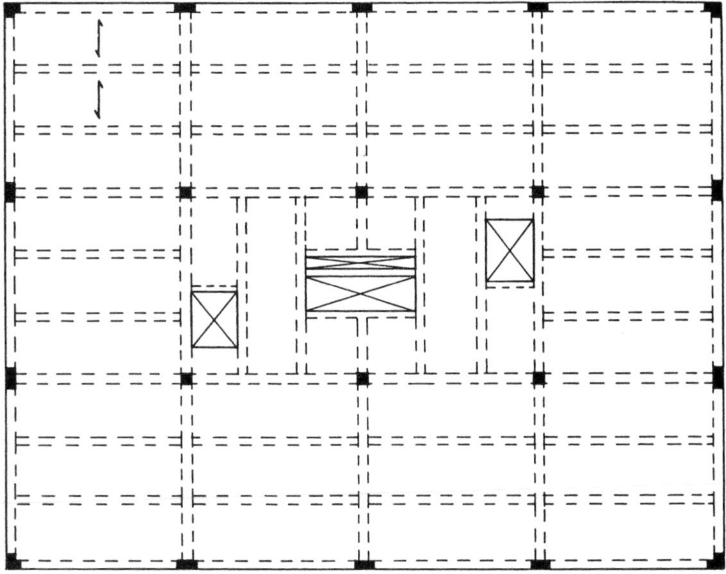

Figure 20.18 Building Three: framing plan for the concrete structure for the upper floor.

beams at 10-ft centers that support a continuous, one-way spanning slab and are in turn supported by column-line girders or directly by columns. Although there are special beams required for the core framing, the system is made up largely of repeated elements. The discussion here will focus on three of these elements: the continuous slab, the four-span interior beam, and the three-span spandrel girder.

Using the approximation method described in Section 14.1, the critical conditions for the slab, beam, and girder are shown in Figure 20.19. Use of these coefficients is reasonable for the slab and beam that support uniformly distributed loads. For the girder, however, the presence of major concentrated loads makes the use of the coefficients somewhat questionable. An adjusted method is thus described later for use with the girder. The coefficients shown in Figure 20.19 for the girder are for uniformly distributed load only (e.g., the weight of the girder itself).

Figure 20.20 shows a section of the exterior wall that illustrates the general form of the construction. The exterior columns and the spandrel beams are exposed to view. Use of the full available depth of the spandrel beams results in a much stiffened bent on the building exterior. As will be shown later, this is combined with the use of oblong-shaped (where one plan dimension is greater than the other) columns at the exterior to create perimeter bents that will indeed absorb most of the lateral force on the structure.

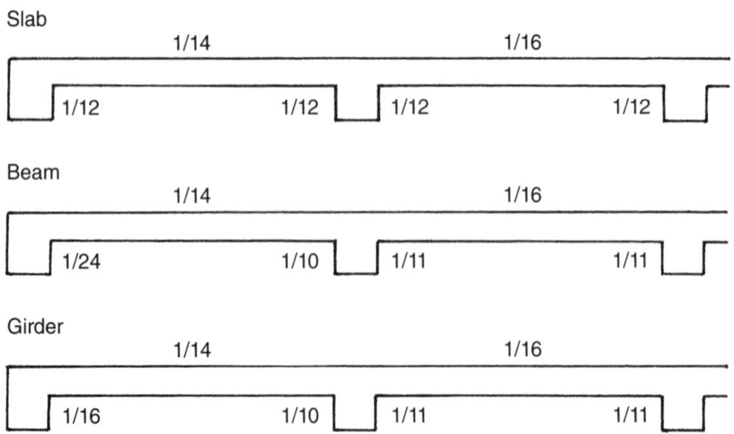

Figure 20.19 Approximate design factors for moments in the slab and beam structure.

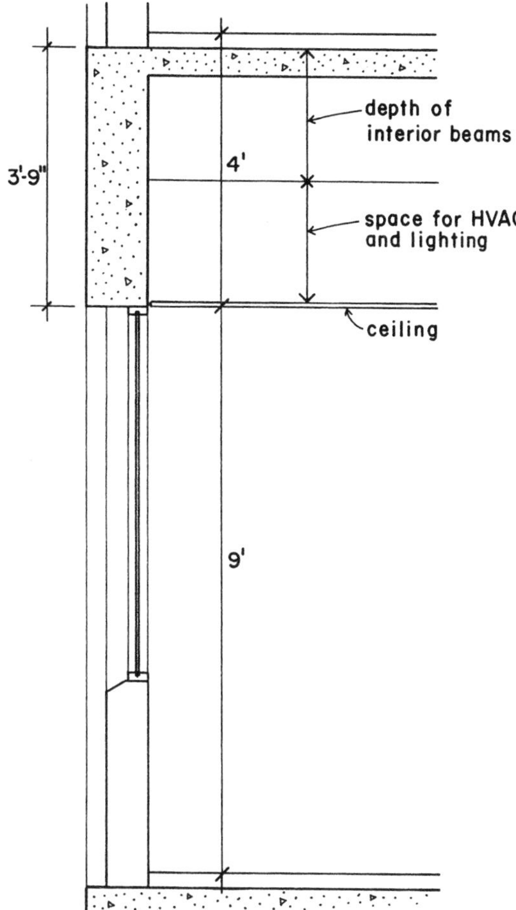

Figure 20.20 Section at the exterior wall with the concrete structure.

The design of the continuous slab is presented as the example in Section 13.5. The use of the 5-in. slab is based on assumed minimum requirements for fire protection. If a thinner slab is possible, the 9-ft clear span would not require this thickness based on limiting bending or shear conditions or recommendations for deflection control. If the 5-in. slab is used, however, the result will be a slab with a low percentage of steel bar weight per square foot—a situation usually resulting in lower cost for the structure.

The unit loads used for the slab design are determined as follows:

Floor live load: 100 psf (at the corridor)
Floor dead load (see Table 1.2):
Carpet and pad at 5 psf
Ceiling, lights, and ducts at 15 psf
2-in.-lightweight concrete fill at 18 psf
5-in.-thick slab at 62 psf
Total dead load: 100 psf

With the slab determined, it is now possible to consider the design of one of the typical interior beams, loaded by a 10-ft-wide strip of slab, as shown in Figure 20.18. The supports for these beams are 30 ft on center. If the beams and columns are assumed to be a minimum of 12 in. wide, the clear span for the beam becomes 29 ft and its load periphery is 29 × 10 = 290 ft^2. Using the ASCE standard (Ref. 4) provisions for reduction of live load (see Section 4.3),

$$L = L_0\left(0.25 + \frac{15}{\sqrt{K_{LL}A_T}}\right) = 100\left(0.25 + \frac{15}{\sqrt{2 \times 290}}\right) = 87 \text{ psf}$$

The beam loading as a per-foot unit load is determined as follows:

$$\text{Live load} = (87 \text{ psf})(10 \text{ ft}) = 870 \text{ lb/ft}$$

Dead load without the beam stem extending below the slab:

$$(100 \text{ psf})(10 \text{ ft}) = 1000 \text{ lb/ft}$$

Estimating a 12-in.-wide × 20.in.-deep beam stem extending below the bottom of the slab, the additional dead load becomes

$$\frac{12 \times 20}{144} \times 150 \text{ lb/ft}^3 = 250 \text{ lb/ft}$$

The total dead load for the beam is thus 1000 + 250 = 1250 lb/ft, and the total uniformly distributed factored load for the beam is

$$\begin{aligned} w_u &= 1.2(1250) + 1.6(870) = 1500 + 1392 \\ &= 2892 \text{ lb/ft}, \quad \text{or} \quad 2.89 \text{ kips/ft} \end{aligned}$$

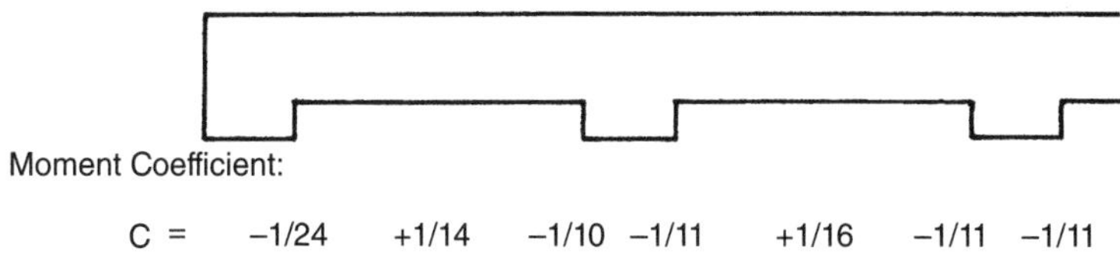

Moment Coefficient:

C = −1/24 +1/14 −1/10 −1/11 +1/16 −1/11 −1/11

Required Reinforcement (in square inches)

Top, A_s = 47.66 C
= 1.99 4.77 4.33

Bottom, A_s = 42.64 C
= 3.05 2.67

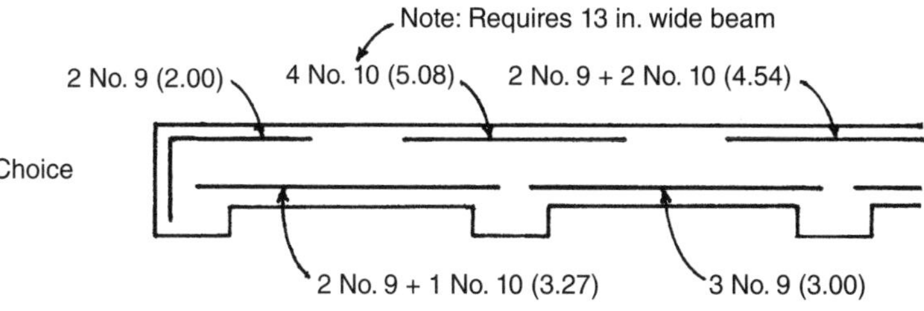

Figure 20.21 Summary of design for the four-span floor beam.

Consider now the four-span continuous beam that is supported by the north–south column-line beams that are referred to as the girders. The approximation factors for design moments for this beam are given in Figure 20.19, and a summary of design data is given in Figure 20.21. Note that the design provides for tension reinforcement only, which is based on an assumption that the beam concrete section is adequate to prevent a critical bending compressive stress in the concrete. Using the strength method (see Section 13.3), the basis for this is as follows.

From Figure 20.19 the maximum bending moment in the beam is

$$M_u = \frac{wL^2}{10} = \frac{2.89 \times (29)^2}{10} = 243 \text{ kip-ft}$$

$$M_r = \frac{M_u}{\phi} = \frac{243}{0.9} = 270 \text{ kip-ft}$$

Then, using factors from Table 13.2 for a balanced section, the required value for bd^2 is determined as

$$bd^2 = \frac{M}{R} = \frac{270 \times 12}{1.149} = 2820$$

With the unit values as used for M and R, this quantity is in units of cubic inches. Various combinations of b and d may now be derived from this relationship, as demonstrated in Section 13.3. For this example, assuming a beam width of 12 in.,

$$d = \sqrt{\frac{2820}{12}} = 15.3 \text{ in.}$$

With minimum cover of 1.5 in., No. 3 stirrups, and moderate-size bars for the tension reinforcement, an overall required beam dimension is obtained by adding approximately 2.5 in. to this derived value for the effective depth. Thus, any dimension selected that is at least 17.8 in. or more will assure a lack of critical bending stress in the concrete. In most cases the specified dimension is rounded off to the nearest full inch, in which case the overall beam height would be specified as 18 in. As discussed in Section 13.3, the balanced section is useful only for establishing a tension failure for the beam (yielding of the reinforcement).

Another consideration for choice of the beam depth is deflection control, as discussed in Section 13.8. From Table 13.12, a minimum overall height of $L/23$ is recommended for the end span of a continuous beam. This yields a minimum overall height of

$$h = \frac{29 \times 12}{23} = 15 \text{ in.}$$

Pushing these depth limits to their minimum is likely to result in high shear stress, a high percentage of reinforcement, and possibly some excessive creep deflection. We will therefore consider the use of an overall height of 24 in., resulting in an approximate value of 24 − 2.5 = 21.5 in. for the effective depth d. Since this is quite close to the size assumed for dead load, no adjustment is made of the previously computed loading for the beam.

For the beams the flexural reinforcement that is required in the top at the supports must pass either over or under the bars in the top of the girders. Figure 20.22 shows a section through the beam with

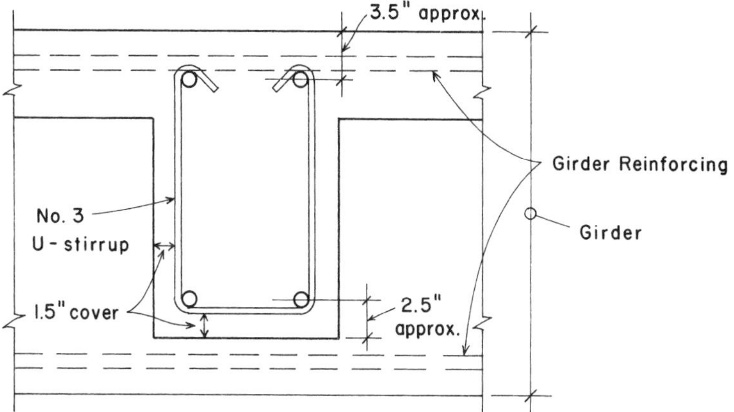

Figure 20.22 Layout of the reinforcement for the intersecting beam and girder.

an elevation of the girder in the background. It is assumed that the much heavier-loaded girder will be deeper than the beams, so the bar intersection problem does not exist in the bottoms of the intersecting members. At the top, however, the beam bars are run under the girder bars, favoring the heavier-loaded girder. For an approximate consideration, an adjusted dimension of 3.5–4 in. should thus be subtracted from the overall beam height to obtain an effective depth for design of the beam. For the remainder of the computations a value of 20 in. is used for the beam effective depth.

The beam cross section must also resist shear, and the beam dimensions should be verified to be adequate for this task before proceeding with design of the flexural reinforcement. Referring to Figure 14.2, the maximum shear force is approximated as 1.15 times the simple span shear of $wL/2$. For the beam this produces a maximum shear of

$$V_u = 1.15\frac{wL}{2} = 1.15 \times \frac{2.89 \times 29}{2} = 48.2 \text{ kips}$$

As discussed in Section 13.6, this value may be reduced by the shear between the support and the distance of the beam effective depth from the support; thus,

$$\text{Design } V = 48.2 - \left(\frac{20}{12} \times 2.89\right) = 43.4 \text{ kips}$$

and the required maximum shear capacity is

$$V = \frac{43.4}{0.75} = 57.9 \text{ kips}$$

Using a d of 20 in. and b of 12 in., the critical shear capacity of the concrete alone is

$$V_c = 2\sqrt{f_c'}bd = 2\sqrt{3000}(12 \times 20)$$

$$= 26{,}290 \text{ lb} \quad \text{or} \quad 26.3 \text{ kips}$$

This leaves a shear force to be developed by the steel equal to

$$V_s' = 57.9 - 26.3 = 31.6 \text{ kips}$$

and the closest stirrup spacing at the beam end is

$$s = \frac{A_v f_y d}{V_s'} = \frac{0.22 \times 40 \times 20}{31.6} = 5.6 \text{ in.}$$

which is not an unreasonable spacing.

For the approximate design shown in Figure 20.21, the required area of steel at the points of support is determined as follows.

Assume a of 6 in., $jd = d - a/2 = 17$ in. Then, using $M_u = C \times w \times L^2$,

$$A_s = \frac{M}{\phi f_y jd} = \frac{C \times 2.89 \times (29)^2 \times 12}{0.9(40 \times 17)} = 47.66C$$

At midspan points, the positive bending moments will be resisted by the slab and beam acting in T-beam action (see Section 13.5). For this condition, an approximate internal moment arm consists of $d - t/2$ and the required steel areas are approximated as

$$A_s = \frac{M}{\phi f_y (d - t2)} = \frac{C \times 2.89 \times (29)^2 \times 12}{0.9[40 \times (21.5 - 2.5)]} = 42.64C$$

Inspection of the framing plan in Figure 20.18 reveals that the girders on the north–south column lines carry the ends of the beams as concentrated loads at their third points (10 ft from each support).

The spandrel girders at the building ends carry the outer ends of the beams plus their own dead weight. In addition, all the spandrel beams support the weight of the exterior curtain walls. The form of the spandrels and the wall construction is shown in Figure 20.20.

The framing plan also indicates the use of widened columns at the exterior walls. Assuming a minimum width of 2 ft, the clear span of the spandrels thus becomes 28 ft. This much stiffened bent, with very deep spandrel beams and widened columns, is used for lateral bracing, as discussed later in this section.

The spandrel beams carry a combination of uniformly distributed loads (spandrel weight plus wall) and concentrated loads (the beam ends). These loadings are determined as follows:

For reduction of the live load, the portion of floor loading carried is 2 times one-half the beam load, or approximately the same as one full beam: 290 ft^2. The design live load for the spandrel girders is thus reduced the same amount as it was for the beams. From the beam loading, therefore:

The total factored load from the beam is

$$P = 2.89 \text{ kips/ft} \times 30/2 \text{ ft} = 43.4 \quad \text{say 44 kips}$$

The uniformly distributed load is basically all dead load, determined as:

Spandrel weight: [(12)(45)/144)](150 pcf) = 563 lb/ft
Wall weight: (25 psf average)(9 ft high) = 225 lb/ft
Total distributed load: 563 + 225 = 788 lb, say 0.8 kips/ft
And the factored load is

$$w_u = 1.2(0.8) = 0.96 \text{ say } 1.0 \text{ kips/ft}$$

For the distributed load, approximate design moments may be determined using the moment coefficients, as was done for the slab and beam. Values for this procedure are given in Figure 20.19. Thus,

$$M_u = C(w \times L^2) = C(1.0 \times 28^2) = 784C$$

The ACI Code does not permit use of coefficients for concentrated loads, but for an approximate design some adjusted coefficients may

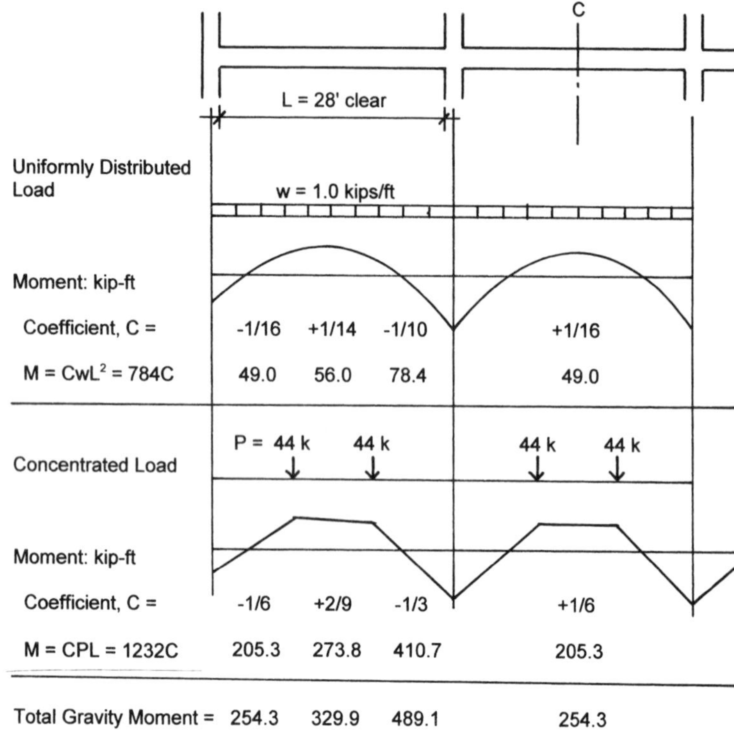

Figure 20.23 Factored gravity load effect on the spandrel girder.

be derived from tabulated loadings for beams with third-point load placement. Using these coefficients, the moments are

$$M_u = C(P \times L) = C(44 \times 28) = 1232C$$

Figure 20.23 presents a summary of the approximation of moments for the spandrel girder. This is, of course, only the gravity loading, which must be combined with effects of lateral loads for complete design of the bents. The design of the spandrel girder is therefore deferred until after the discussion of lateral loads later in this section.

Figure 20.24 Relations between the columns and the floor framing.

Design of the Concrete Columns

The general cases for the concrete columns are as follows (see Figure 20.24):

1. The interior column, carrying primarily only gravity loads due to the stiffened perimeter bents
2. The corner columns, carrying the ends of the spandrel beams and functioning as the ends of the perimeter bents in both directions
3. The intermediate columns on the north and south sides, carrying the ends of the interior girders and functioning as members of the perimeter bents
4. The intermediate columns on the east and west sides, carrying the ends of the column-line beams and functioning as members of the perimeter bents

Summations of the design loads for the columns may be done from the data given previously. As all columns will be subjected to combinations of axial load and bending moments, these gravity loads represent only the axial compression action. Bending moments will be relatively low in magnitude on interior columns, since they are framed into by beams on all sides. As discussed in Chapter 15, all columns are designed for a minimum amount of bending, so routine design, even when done for axial load alone, provides for some residual moment capacity. For an approximate design, therefore, it is reasonable to consider the interior columns for axial gravity loads only.

Figure 20.25 presents a summary of design for an interior column, using loads determined from a column load summation with the data given previously in this section. Note that a single size of 20 in.[2] is used for all three stories, a common practice permitting reuse of

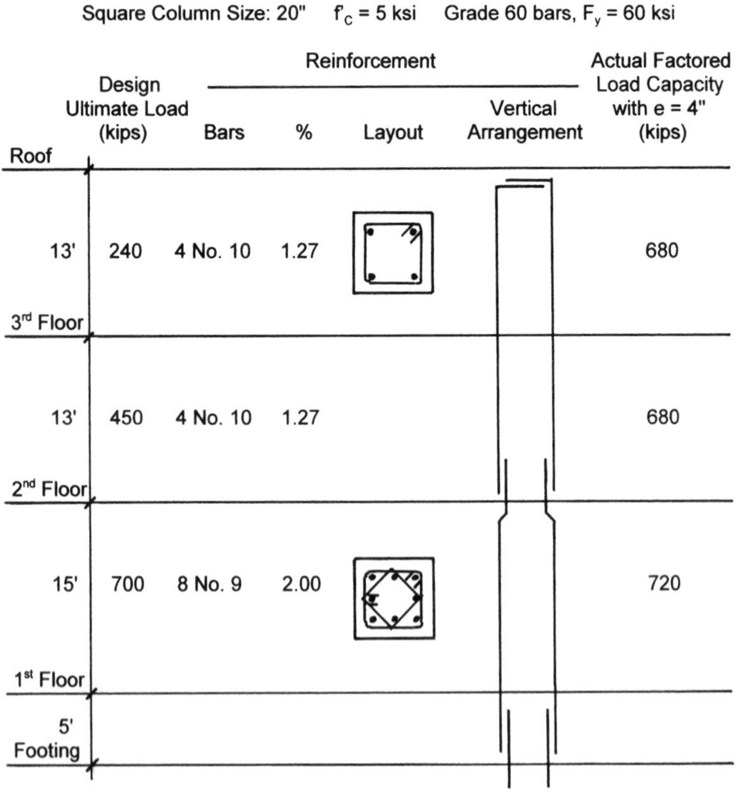

Square Column Size: 20" f'_c = 5 ksi Grade 60 bars, F_y = 60 ksi

Level	Story Height	Design Ultimate Load (kips)	Reinforcement: Bars	%	Layout	Vertical Arrangement	Actual Factored Load Capacity with e = 4" (kips)
Roof							
	13'	240	4 No. 10	1.27			680
3rd Floor							
	13'	450	4 No. 10	1.27			680
2nd Floor							
	15'	700	8 No. 9	2.00			720
1st Floor							
	5'						
Footing							

Figure 20.25 Design of the interior concrete column for gravity load only.

column forms for cost savings. Column load capacities indicated in Figure 20.25 were obtained from the graphs in Chapter 15.

A general cost-savings factor is the use of relatively low percentages of steel reinforcement. An economical column is therefore one with a minimum percentage, ρ (usually a threshold of 1% of the gross section), of reinforcement. However, other factors often affect design choices for columns, some common ones being the following:

1. Architectural planning of building interiors. Large columns are often difficult to plan around in developing of interior rooms, corridors, stair openings, and so on. Thus, the *smallest* feasible column sizes—obtained with maximum percentages of steel (ρ)—are often desired.
2. Ultimate load response of lightly reinforced columns borders on brittle fracture failure, whereas heavily reinforced columns tend to have a ductile form of ultimate failure. The ductile character is especially desirable for rigid-frame actions in general and particularly for seismic loading conditions.
3. A general rule of practice in rigid-frame design for lateral loadings (wind or earthquakes) is to prefer a form of ultimate response described as *strong column/weak beam failure*. In this example this relates more to the columns in the perimeter bents but may also somewhat condition design choices for the interior columns, since they will take *some* lateral loads when the building as a whole deflects sideways.

Column form may also be an issue that relates to architectural planning or to structural concerns. Round columns work well for some structural actions and may be quite economical for forming, but unless they are totally free standing, they do not fit so well for planning the rest of the building construction. Even square columns of large size may be difficult to plan around in some cases, an example being at the corners of stair wells and elevator shafts. T-shaped or L-shaped columns may be used in special situations.

Large bending moments in proportion to axial compression may also dictate some adjustment of column form or arrangement of reinforcement. When a column becomes essentially beamlike in its action, some of the practical considerations for beam design come into play. In this example these concerns apply to the exterior columns to some degree.

For the intermediate exterior columns there are four actions to consider:

1. The vertical compression due to gravity.
2. Bending moment induced by the interior framing that intersects the wall. These columns are what provides the end resisting moments shown in Figures 20.21 and 20.23.
3. Bending moments in the plane of the wall bent, induced by any unbalanced gravity load conditions (movable live loads) on the spandrels.
4. Bending moments in the plane of the wall bents due to lateral loads.

For the corner columns, the situation is similar to that for the intermediate exterior columns, that is, there is bending on both axes. Gravity loads will produce simultaneous bending on both axes, resulting in a net moment that is diagonal to the column. Lateral loads can cause the same effect since neither wind nor earthquakes will work neatly on the building's major axes, even though this is how design investigation is performed.

Further discussion of the exterior columns is presented in the following considerations for lateral load effects.

Design for Lateral Forces

The major lateral force resisting systems for this structure are as shown in Figure 20.15. In truth, other elements of the construction will also resist lateral distortion of the structure, but by widening the exterior columns in the wall plane and using the very deep spandrel girders, the stiffness of these bents becomes considerable.

Whenever lateral deformation occurs, the stiffer elements will attract the force first. Of course, the stiffest elements may not have the necessary strength and will thus fail structurally, passing the resistance off to other resisting elements. Glass tightly held in flexible window frames, stucco on light wood structural frames, lightweight concrete block walls, or plastered partitions on light metal partition frames may thus be fractured first in lateral movements (as they often are). For the successful design of this building, the detailing of the construction should be carefully done to assure that these events do not occur, in spite of the relative stiffness of the perimeter bents.

In any event, the bents shown in Figure 20.15 will be designed for the entire lateral load. They thus represent the safety assurance for the structure, if not a guarantee against loss of construction.

With the same building profile, the wind loads on this structure will be the same as those determined for the steel structure in Section 20.5. As in the example in that section, the data given in Figure 20.11 are used to determine the horizontal forces on the bracing bent as follows:

$$H_1 = (165.5)(122)/2 = 10{,}096 \text{ lb, say } 10.1 \text{ kips/bent}$$

$$H_2 = (199.5)(122)/2 = 12{,}170 \text{ lb, say } 12.2 \text{ kips/bent}$$

$$H_3 = (210)(122)/2 = 12{,}810 \text{ lb, say } 12.8 \text{ kips/bent}$$

Figure 20.26*a* shows a profile of the north–south bent with these loads applied.

For an approximate analysis consider the individual stories of the bent to behave as shown in Figure 20.26*b*, with the columns developing an inflection point at their midheight points. Because the columns are all deflected the same sideways distance, the shear force in a single column may be assumed to be proportionate to the relative stiffness of the column. If the columns all have the same stiffness, the total load at each story for this bent would simply be divided by 4 to obtain the column shear forces.

Even if the columns are all the same size, however, they may not all have the same resistance to lateral deflection. The end columns in the bent are slightly less restrained at their ends (top and bottom) because they are framed on only one side by a beam. For this approximation, therefore, it is assumed that the relative stiffness of the end columns is one-half that of the intermediate columns. Thus, the shear force in the end columns is one-sixth of the total bent shear force and that in the intermediate column is one-third of the total force. The column shears for each of the three stories is thus as shown in Figure 20.26*c*.

The column shear forces produce bending moments in the columns. With the column inflection points (points of zero moment) assumed to be at midheight, the moment produced by a single shear force is simply the product of the force and half the column height. These column moments must be resisted by the end moments in the rigidly attached beams, and the actions are as shown in Figure 20.27. At each column–beam intersection the sum of the column and beam

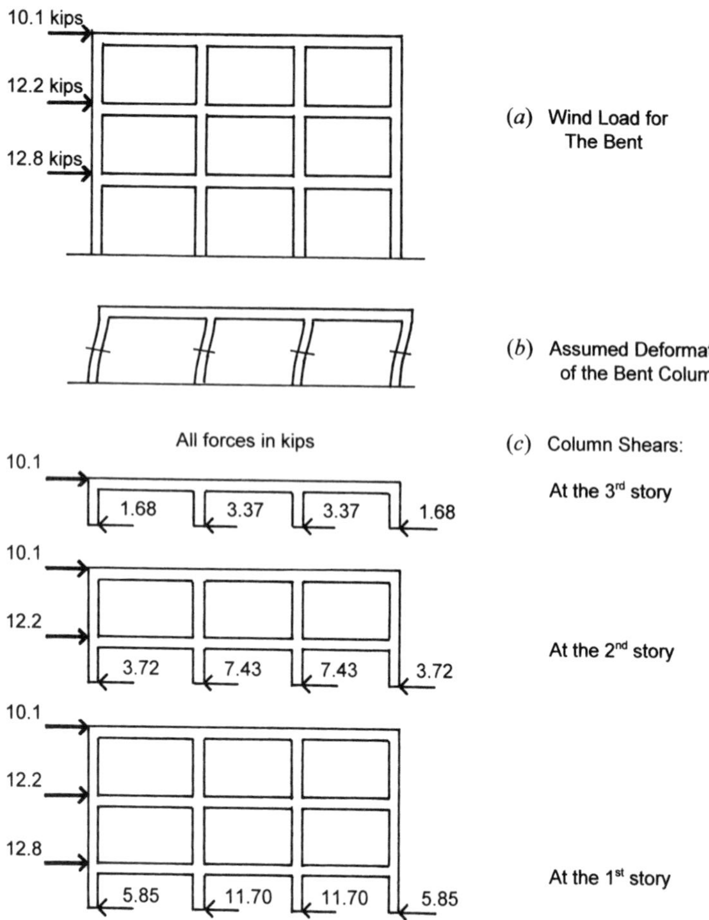

Figure 20.26 Aspects of the lateral load response of the north–south perimeter bents.

moments must be balanced. Thus, the total of the beam moments may be equated to the total of the column moments, and the beam moments may be determined once the column moments are known.

For example, at the second-floor level of the intermediate column, the sum of the column moments from Figure 20.27 is

$$M = 48.3 + 87.8 = 136.1 \text{ kip-ft}$$

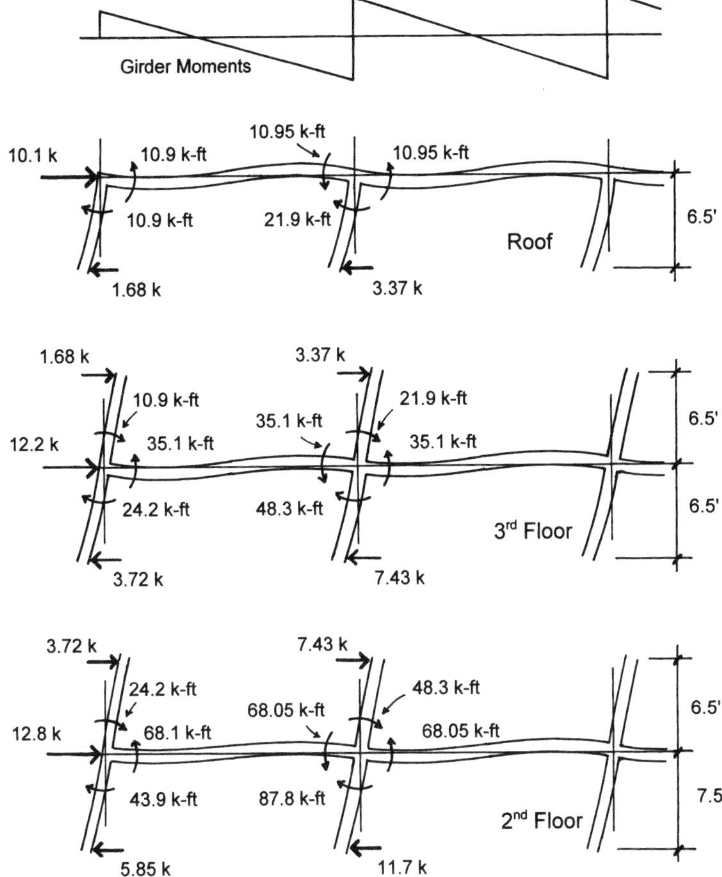

Figure 20.27 Investigation for column and girder bending moments in the north–south bents.

Assuming the two beams framing the column to have equal stiffness at their ends, the beams will share this moment equally, and the end moment in each beam is thus

$$M = \frac{136.1}{2} = 68.05 \text{ kip-ft}$$

as shown in the figure.

The data displayed in Figure 20.27 may now be combined with that obtained from gravity load analyses for a combined load investigation and the final design of the bent members.

Design of the Bent Columns

For the bent columns the axial compression due to gravity must first be combined with any moments induced by gravity for a gravity-only analysis. Then the gravity load actions are combined with the results from the lateral force analysis using the usual adjustments for this combined loading.

Gravity-induced moments for the girders are taken from the girder analysis in Figure 20.24 and are assumed to produce column moments as shown in Figure 20.28. The summary of design conditions for the corner and intermediate columns is given in Figure 20.29. For design, two conditions must be investigated. The first condition is that with the gravity loads only. The second condition adds the lateral load effects to the gravity effects. Different load factors apply for these two conditions. The dual requirements for the columns are given in the bottom two lines of the table in Figure 20.29.

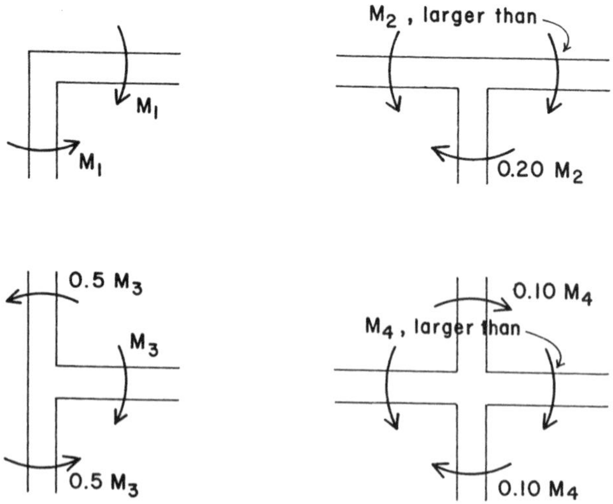

Figure 20.28 Assumptions for approximations of the distribution of bending moments in the bent columns due to gravity loading.

Note: Axial loads in kips, moments in kip-ft, dimensions in inches.

Story	Third	Second	First
Gravity Load Only:			
Axial Live Load	35	67	103
Axial Dead Load	83	154	240
Live Load Moment	32	38	38
Dead Load Moment	128	90	90
Ultimate Gravity Load: (1.2D + 1.6L)			
Axial Load	100+56 = 156	185+107 = 292	288+165 = 453
Moment	154+51 = 205	108+61 = 169	108+61 = 169
e	15.8 in.	7.0 in.	4.5 in.
Combined with Wind:			
Wind Load Moment	21.9	48.3	87.8
Ultimate Moment with Wind (1.2D+1.6W+L)	154+35+32 = 221	108+77+38 = 223	108+141+38 = 287
Ultimate Axial Load with Wind (1.2D + L)	100+35 = 135	185+67 = 252	288+103 = 391
e	19.6 in.	10.6 in.	8.8 in.
Choice of Column from Figure 15.8	?? Design as a beam for M = 221	14 × 20 6 N0. 9	14 × 24 6 No. 10

3rd story: Use 14×24, d = 21 in., $A_s = M_u/\phi f_y jd = (221 \times 12)/0.9[60(0.9 \times 21)] = 2.6$ in.^{2}
Use same as 2nd story, 6 No. 9 bars (3 at each end). Use 14×24 for all stories.

Figure 20.29 Design of the north–south bent columns for combined gravity and lateral loading.

Note that a single column choice from Figure 15.8 is able to fulfill the requirements for all the columns. This is not unusual, as the relationship between load magnitude and moment magnitude changes. A broader range of data for column choices is given in the extensive column design tables in various references.

When bending moment is very high in comparison to the axial load (very large eccentricity), an effective approximate column design can be determined by designing a section simply as a beam with tension reinforcement; the reinforcement is then merely duplicated on both sides of the column.

Design of the Bent Girders

The spandrel girders must be designed for the same two basic load conditions as discussed for the columns. The summary of bending moments for the third-floor spandrel girder is shown in Figure 20.30. Values for the gravity moments are taken from Figure 20.23. Moment induced by wind is that shown in Figure 20.27. It may be noted from the data in Figure 20.30 that the effects of gravity loading prevail and wind loading is not a critical concern for the girder. This would most likely not be the case in lower stories of a much taller building or possibly with a combined loading including major seismic effects.

Figure 20.31 presents a summary of design considerations for the third-floor spandrel girder. The construction assumed here is that shown in Figure 20.21, with the very deep, exposed girder. Some attention should be given to the relative stiffness of the columns and girders, as discussed in Section 3.10. Keep in mind, however, that the girder is almost three times as long as the column and thus may have a considerably stiffer section without causing a disproportionate relationship to occur.

For computation of the required flexural reinforcement, the T-beam effect is ignored and an effective depth of 40 in. is assumed. Required areas of reinforcement may thus be derived as

$$A_s = \frac{M}{\phi f_s jd} = \frac{M \times 12}{0.9(40 \times 0.9 \times 40)} = 0.00926M$$

From Table 13.3, minimum reinforcement is

$$A_s = 0.005bd = 0.005 \times 16 \times 40 = 3.2 \text{ in.}^2, \quad \text{not critical}$$

Values determined for the various critical locations are shown in Figure 20.31. It is reasonable to consider the stacking of bars in two layers in such a deep section, but it is not necessary for the selection of reinforcement shown in the figure.

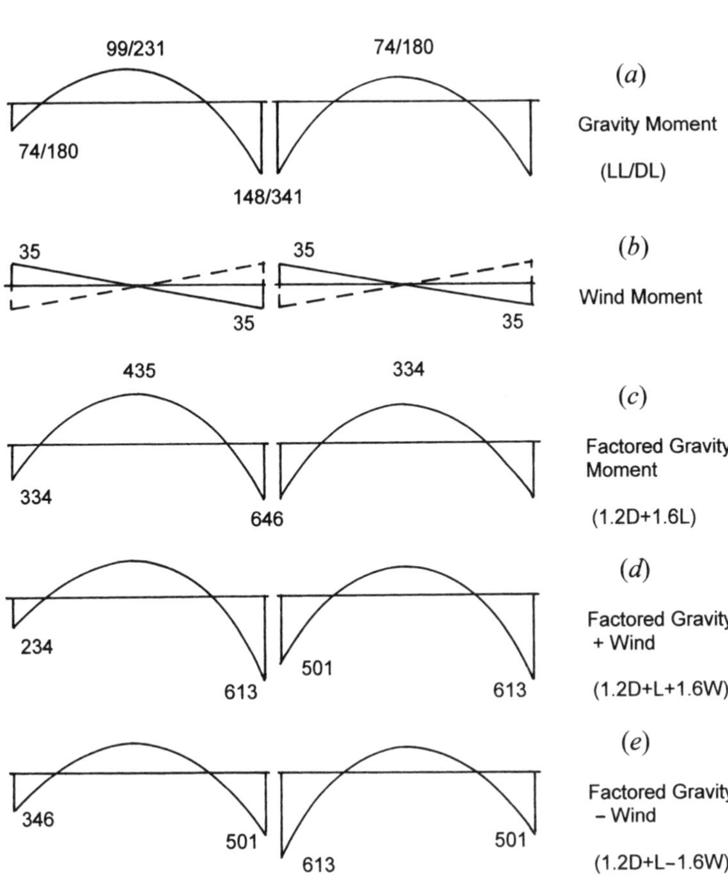

Figure 20.30 Combined gravity and lateral bending moments for the spandrel girders.

The very deep and relatively thin spandrel should be treated somewhat as a wall/slab, and thus the section in Figure 20.31 shows some additional horizontal bars at midheight points. In addition, the stirrups shown should be of a closed form (see Figure 13.14) to serve also as ties, vertical reinforcement for the wall/slab, and (with the extended top) negative moment flexural reinforcement for the adjoining slab. In this situation it would be advisable to use continuous stirrups at

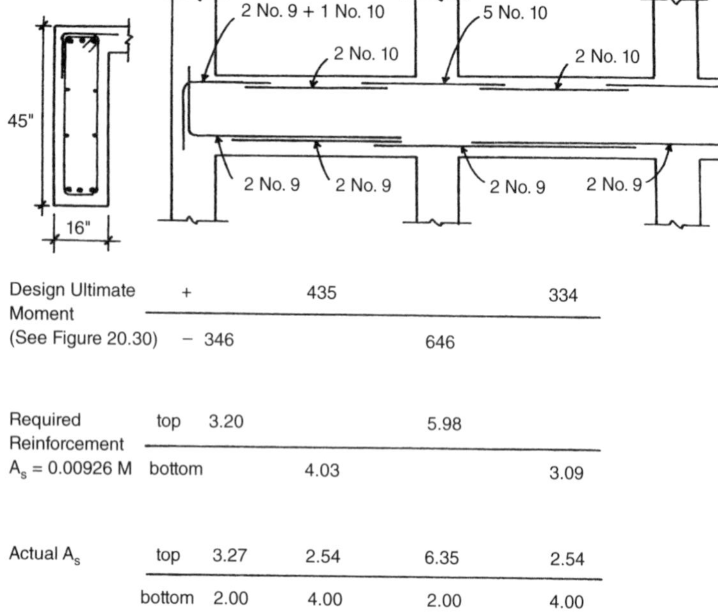

Figure 20.31 Design of the spandrel girder for the combined gravity and lateral loading effects.

a maximum spacing of 18 in. or so for the entire girder span. Closer spacing may be necessary near the supports if the end shear forces require it.

It is also advisable to use some continuous top and bottom reinforcement in spandrels. This relates to some of the following possible considerations:

1. Miscalculation of lateral effects, giving some reserved reversal bending capacity to the girders.
2. A general capability for torsional resistance throughout the beam length (intersecting beams produce this effect).
3. Something there to hold up the continuous stirrups.
4. Some reduction of long-term creep deflection with all sections doubly reinforced. Helps keep load off the window mullions and glazing.

Alternative Floor Structure

Figure 20.32 shows a partial plan and some details for a concrete flat-slab system for the roof and floor structures for Building Three. Features of the system include the following:

1. A general solid 10-in.-thick slab without beams in the major portion of the structure outside the building core.
2. A thickened portion, called a *drop panel*, around the supporting columns, ordinarily extending to one-sixth the span from the

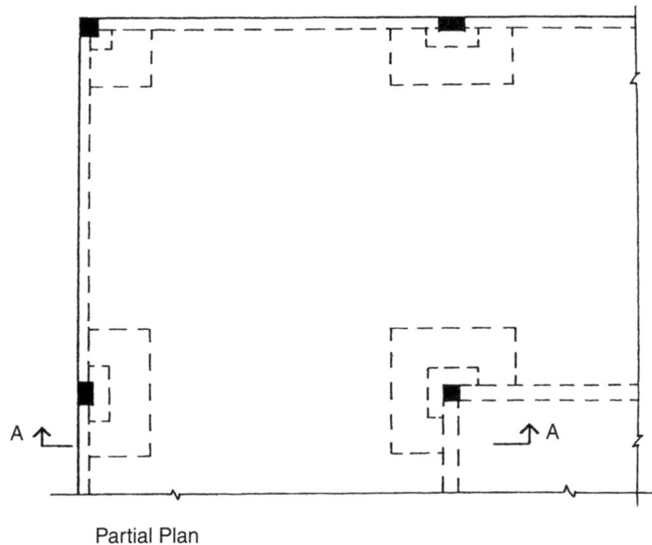

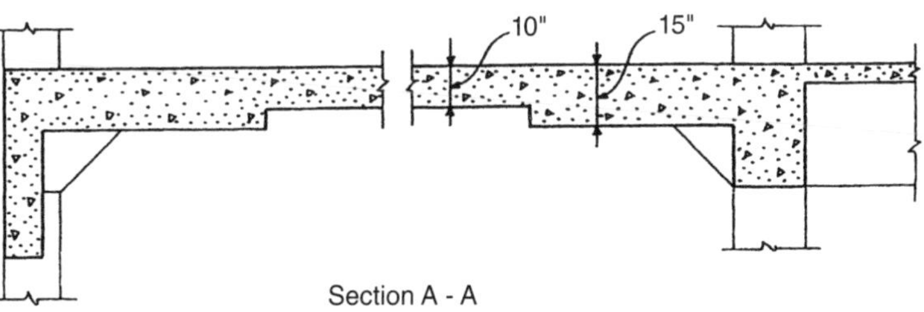

Figure 20.32 Alternative structure for the upper floor using a concrete flat slab with drop panels at columns.

column on all sides. The thickness increase shown is one-half of the general slab thickness.

3. A *column capital*, consisting of a truncated, inverted pyramidal form.
4. Use of the same spandrel beam and the same slab and beam core framing as in the slab-and-beam example.

The flat slab is generally more feasible when live loads are very high and there are a considerable number of continuous bays of the structure in both directions. The system for this building is marginally feasible and might be justified on the basis of some other building design considerations. For example, the story height may be decreased since downward-protruding beams do not interfere with air ducts, sprinkler piping, and so on, in the main portion of the floor. This could mean a building several feet shorter, with resultant savings in curtain walls, columns and partitions, stairs, elevator housing, piping and wiring risers, and so on.

For some occupancies, such as multistory residential, the drop ceiling construction might be eliminated, with all overhead items exposed beneath the considerably simpler form of the concrete structure. This is a real advantage for parking garages and industrial buildings, but not so popular with office buildings.

The flat-slab system is highly indeterminate and a bit complicated for reinforcement. However, it has been used extensively since its development in the early twentieth century, and routine design of common examples is pretty much "canned" for repetition by now.

20.9 DESIGN OF THE FOUNDATIONS

Unless site conditions require the use of a more complex foundation system, it is reasonable to consider the use of simple shallow bearing foundations (footings) for Building Three. Column loads will vary depending on which of the preceding structural schemes is selected. The heaviest loads are likely to occur with the all-concrete structure in Section 20.8.

The most direct solution for concentrated column loads is a square footing, as described in Section 16.3. A range of sizes of these footings is given in Table 16.4. For a free-standing column, the choice is relatively simple, once an acceptable design pressure for the supporting soil is established.

Problems arise when the conditions at the base of a column involve other than a free-standing case for the column. In fact, this is the case for most of the columns in Building Three. Consider the structural plan as indicated in Figure 20.18. All but two of the interior columns are adjacent to construction for the stair towers or the elevator shaft.

For the three-story building, the stair tower may not be a problem, although in some buildings these are built as heavy masonry or concrete towers and are used for part of the lateral bracing system. This might possibly be the case for the structure in Section 20.7.

Assuming that the elevator serves the lowest occupied level in the building, there will be a considerably deep construction below this level to house the elevator pit. If the interior footings are quite large, they may come very close to the elevator pit construction. In this case, the bottoms of these footings would need to be dropped to a level close to that of the bottom of the elevator pit. If the plan layout results in a column right at the edge of the elevator shaft, this is a more complicated problem, as the elevator pit would need to be on top of the column footing.

For the exterior columns at the building edge, there are two special concerns. The first has to do with the necessary support for the exterior building wall, which coincides with the column locations. If there is no basement and the exterior wall is quite light in weight—possibly a metal curtain wall that is supported at each upper level—the exterior columns will likely get their own individual square footings and the wall will get a minor strip footing between the column footings. This scheme is shown in the partial foundation plan in Figure 20.33*a*. Near the columns the light wall will simply be supported by the column footings.

If the wall is very heavy, the solution in Figure 20.33*a* may be less feasible, and it may be reasonable to consider the use of a wide strip footing that supports both the wall and the columns, as shown in Figure 20.33*b*. The ability of the wall to serve as a distributing member for the uniform pressure on the strip footing must be considered, which brings up another concern as well. The other concern has to do with the presence or absence of a basement. If there is no basement, it may be theoretically possible to place footings quite close to the ground surface, with a minimal penetration of the general foundation construction below grade. If there is a basement, there will likely be a reasonably tall concrete wall at the building edge.

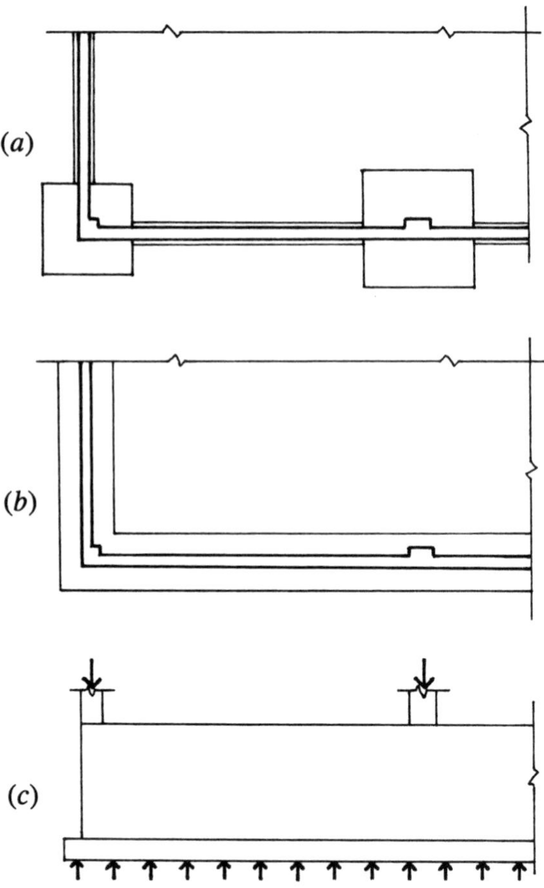

Figure 20.33 Considerations for the foundations: (*a*) partial plan with individual column footings, (*b*) partial plan with continuous wall footing, and (*c*) use of a tall basement wall as a distribution girder for the continuous wall footing.

Regardless of the wall construction above grade, it is reasonable to consider the use of the basement wall as a distributing member for the column loads, as shown in Figure 20.33*c*. Again, this is only feasible for a low-rise building, with relatively modest loads on the exterior columns.

Appendix A

PROPERTIES OF SECTIONS

This appendix deals with various geometric properties of planar (two-dimensional) areas. The areas referred to are the cross-sectional areas of structural members. These geometric properties are used in the analysis of stresses and deformations in the design of the structural members.

A.1 CENTROIDS

The *center of gravity* of a solid is the point at which all of its weight can be considered to be concentrated. Since a planar area has no weight, it has no center of gravity. The point in a planar area that corresponds to the center of gravity of a very thin plate of the same area and shape is called the *centroid* of the area. The centroid is a useful reference for various geometric properties of planar areas.

For example, when a beam is subjected to a bending moment, the materials in the beam above a certain plane in the beam are in compression and the materials below the plane are in tension. This plane is the *neutral stress plane*, also called the neutral surface or the zero stress plane (see Section 3.6). For a cross section of the beam the

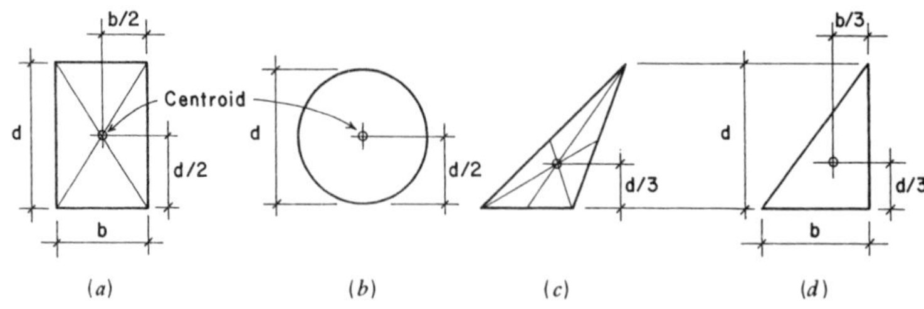

Figure A.1 Centroids of various shapes.

intersection of the neutral stress plane is a line that passes through the centroid of the section and is called the *neutral axis* of the section. The neutral axis is very important for investigation of bending stresses in beams.

The location of the centroid for symmetrical shapes is located on the axis of symmetry for the shape. If the shape is bisymmetrical—that is, it has two axes of symmetry—the centroid is at the intersection of these axes. Consider the rectangular area shown in Figure A.1*a*; obviously, its centroid is at its geometric center and is quite easily determined.

(*Note:* Tables A.3–A.8 and Figure A.11, referred to in the discussion that follows, are located at the end of this appendix.)

For more complex forms, such as those of rolled steel members, the centroid will also be on any axis of symmetry. And, as for the simple rectangle, if there are two axes of symmetry, the centroid is readily located.

For simple geometric shapes, such as those shown in Figure A.1, the location of the centroid is easily established. However, for more complex shapes, the centroid and other properties may have to be determined by computations. One method for achieving this is by use of the *statical moment*, defined as the product of an area times its distance from some reference axis. Use of this method is demonstrated in the following examples.

Example 1. Figure A.2 is a beam cross section that is unsymmetrical with respect to a horizontal axis (such as X–X in the figure). The area is symmetrical about its vertical centroidal axis, but the true location of the centroid requires locating the horizontal centroidal axis. Find the location of the centroid.

Solution: Using the statical moment method, first divide the area into units for which the area and location of the centroid are readily

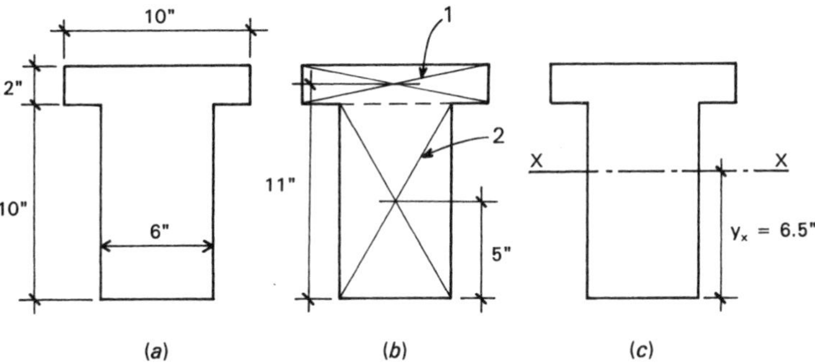

Figure A.2 Reference for Example 1.

determined. The division chosen here is shown in Figure A.2*b* with the two parts labeled 1 and 2. How cross sections are subdivided will not change the location of the centroid.

The second step is to choose a reference axis about which to sum statical moments and from which the location of the centroid is readily measured. A convenient reference axis for this shape is one at either the top or bottom of the shape. With the bottom chosen the distances from the centroids of the parts to this reference axis are shown in Figure A.2*b*.

The computation next proceeds to the determination of the unit areas and their statical moments. This work is summarized in Table A.1, which shows the total area to be 80 in.^{2} and the total statical moment to be 520 in.^{3} Dividing this moment by the total area produces the value of 6.5 in., which is the distance from the reference axis to the centroid of the whole shape, as shown in Figure A.2*c*.

Problems A.1.A–F Find the location of the centroid for the cross-sectional areas shown in Figure A.3. Use the reference axes indicated and compute the distances from the axes to the centroid, designated as *cx* and *cy*, as shown in Figure A.3*b*.

TABLE A.1 Summary of Computations for Centroid: Example 1

Part	Area (in.^{2})	y (in.)	$A \times y$ (in.^{3})
1	$2 \times 10 = 20$	11	220
2	$6 \times 10 = 60$	5	300
Σ	80		520

$$y_x = \frac{520}{80} = 6.5 \text{ in.}$$

A.2 MOMENT OF INERTIA

Consider the area enclosed by the irregular line in Figure A.4*a*. In this area, designated A, a small unit area a is indicated at z distance from the axis marked X–X. If this unit area is multiplied by the square of its distance from the reference axis, the result is the quantity az^2. If all of the units of the area are thus identified and the sum of these products is made, the result is defined as the *second moment* or the *moment of inertia* of the area, designated as I. Thus

$$\sum az^2 = I \quad \text{or specifically} \quad I_{X-X}$$

which is the moment of inertia of the area about the X–X axis.

The moment of inertia is a somewhat abstract item, less able to be visualized than area, weight, or center of gravity. It is, nevertheless, a real geometric property that becomes an essential factor for investigation of stresses and deformations due to bending. Of particular interest is the moment of inertia about a centroidal axis, and—most significantly—about a principal axis for the shape. Figures A.4*b, c, e,* and *f* indicate such axes for various shapes. An inspection of Tables A.3–A.8 will reveal the properties of moment of inertia about the principal axes of the shapes in the tables.

Moment of Inertia of Geometric Figures

Values for moment of inertia can often be obtained from tabulations of structural properties. Occasionally, it is necessary to compute values for a given shape. This may be a simple shape, such as a square, rectangular, circular, or triangular area. For such shapes simple formulas are derived to express the value for the moment of inertia.

Rectangle. Consider the rectangle shown in Figure A.4*c*. Its width is b and its depth is d. The two principal axes are X–X and Y–Y, both passing through the centroid of the area. For this case the moment of inertia with respect to the centroidal axis X–X is

$$I_{X-X} = \frac{bd^3}{12}$$

and the moment of inertia with respect to the Y–Y axis is

$$I_{Y-Y} = \frac{db^3}{12}$$

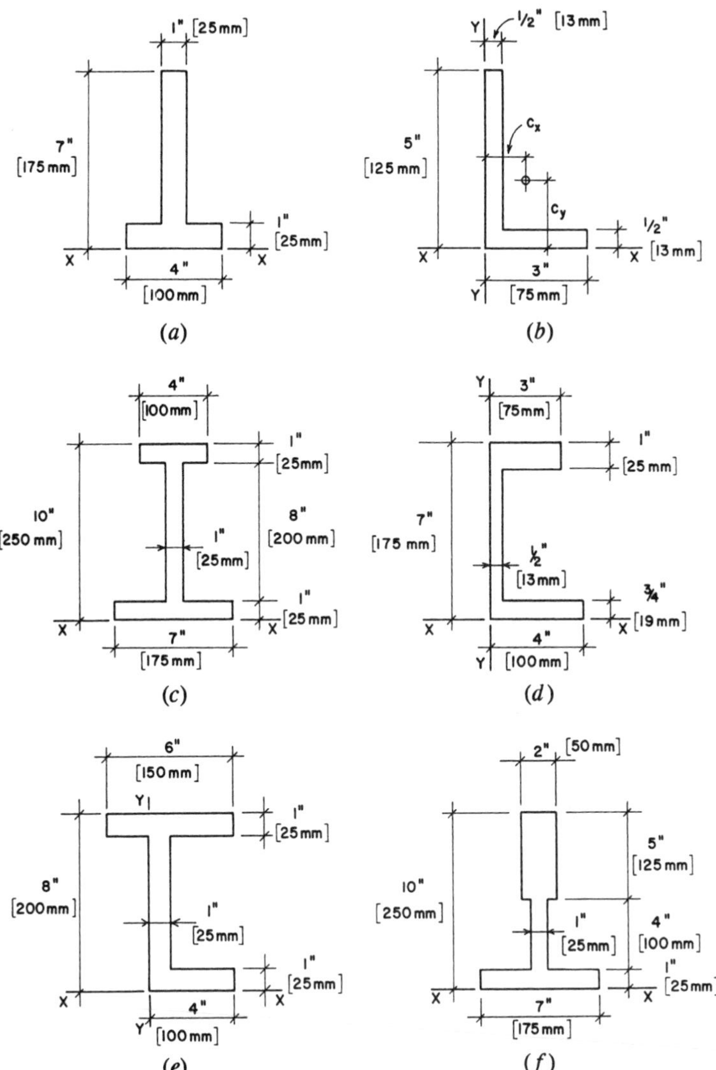

Figure A.3 Reference for Problem A.1.

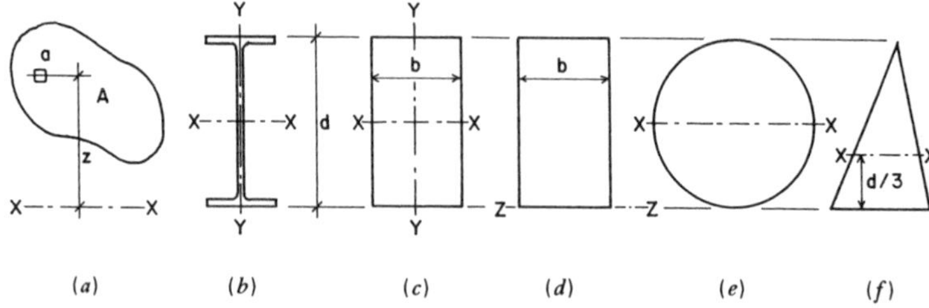

Figure A.4 Consideration of reference axes for the moment of inertia of various shapes of cross sections.

Example 2. Find the value of the moment of inertia for a 6×12-in. wood beam about an axis through its centroid and parallel to the narrow dimension.

Solution: As listed in standard references for wood products, the actual dimensions of the section are 5.5×11.5 in. Then

$$I = \frac{bd^3}{12} = \frac{5.5 \times (11.5)^3}{12} = 697.1 \text{ in.}^4$$

which is in agreement with the value for I_{X-X} in references.

Circle. Figure A.4*e* shows a circular area with diameter d and axis X–X passing through its center. For the circular area the moment of inertia is

$$I = \frac{\pi d^4}{64}$$

Example 3. Compute the moment of inertia of a circular cross section, 10 in. in diameter, about its centroidal axis.

Solution: The moment of inertia is

$$I = \frac{\pi d^4}{64} = \frac{3.1416(10)^4}{64} = 490.9 \text{ in.}^4$$

Triangle. The triangle in Figure A.4*f* has a height h and a base width b. The moment of inertia about a centroidal axis parallel to the base is

$$I = \frac{bh^3}{36}$$

Example 4. If the base of the triangle in Figure A.4*f* is 12 in. wide and the height from the base is 10 in., find the value for the centroidal moment of inertia parallel to the base.

Solution: Using the given values in the formula

$$I = \frac{bh^3}{36} = \frac{12(10)^3}{36} = 333.3 \text{ in.}^4$$

Open and Hollow Shapes. Values of moment of inertia for shapes that are open or hollow may sometimes be computed by a method of subtraction. The following examples demonstrate this process. Note that this is possible only for shapes that are symmetrical.

Example 5. Compute the moment of inertia for the hollow box section shown in Figure A.5*a* about a centroidal axis parallel to the narrow side.

Solution: Find first the moment of inertia of the shape defined by the outer limits of the box:

$$I = \frac{bd^3}{12} = \frac{6(10)^3}{12} = 500 \text{ in.}^4$$

Then find the moment of inertia for the shape defined by the void area:

$$I = \frac{4(8)^3}{12} = 170.7 \text{ in.}^4$$

The value for the hollow section is the difference; thus,

$$I = 500 - 170.7 = 329.3 \text{ in.}^4$$

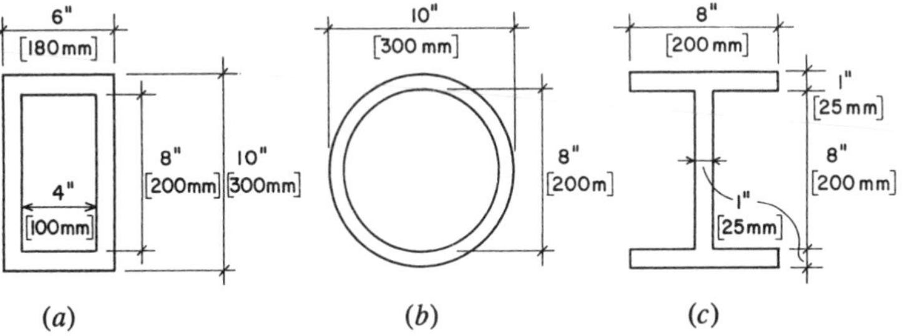

Figure A.5 Reference for Examples 5, 6, and 7.

Example 6. Compute the moment of inertia about the centroidal axis for the pipe section shown in Figure A.5*b*. The thickness of the shell is 1 in.

Solution: As in the preceding example, the two values may be found and subtracted. Or a single computation may be made as follows:

$$I = \frac{\pi}{64}(d_o^4 - d_i^4) = \frac{3.1416}{64}(10^4 - 8^4) = 491 - 201 = 290 \text{ in.}^4$$

Example 7. Referring to Figure A.5*c*, compute the moment of inertia of the I-shaped section about the centroidal axis parallel to the flanges.

Solution: This is essentially similar to the computation for Example 5. The two voids may be combined into a single one that is 7 in. wide. Thus,

$$I = \frac{8(10)^3}{12} - \frac{7(8)^3}{12} = 667 - 299 = 368 \text{ in.}^4$$

Note that this method can only be used when the centroids of the outer shape and the void coincide. For example, it cannot be used to find the moment of inertia for the I shape about its vertical centroidal axis. For this computation the method discussed in the next section must be used.

A.3 TRANSFERRING MOMENTS OF INERTIA

Determination of the moment of inertia of unsymmetrical and complex shapes cannot be done by the simple processes illustrated in the preceding examples. An additional step that must be used is that involving the transfer of moment of inertia about a remote axis. The formula for achieving this transfer is as follows:

$$I = I_o + Az^2$$

In this formula,

I = moment of inertia of the cross section about the required reference axis
I_o = moment of inertia of the cross section about its own centroidal axis, parallel to the reference axis
A = area of the cross section
z = distance between the two parallel axes

These relationships are illustrated in Figure A.6, where *X–X* is the centroidal axis of the area and *Y–Y* is the reference axis for the transferred moment of inertia.

Application of this principle is illustrated in the following examples.

Example 8. Find the moment of inertia of the T-shaped area in Figure A.7 about its horizontal (*X–X*) centroidal axis. (*Note:* The location of the centroid for this section was solved as Example 1 in Section A.1.)

Solution: A necessary first step in these problems is to locate the position of the centroidal axis if the shape is not symmetrical. In this case, the T shape is symmetrical about its vertical axis but not about the horizontal axis. Locating the position of the horizontal axis was the problem solved in Example 1 in Section A.1.

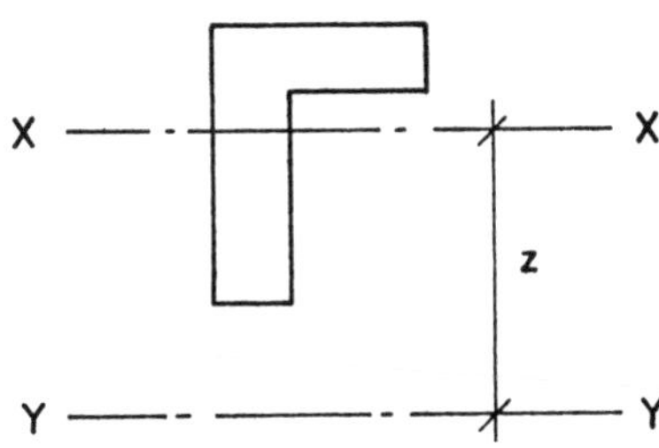

Figure A.6 Transfer of moment of inertia to a parallel axis.

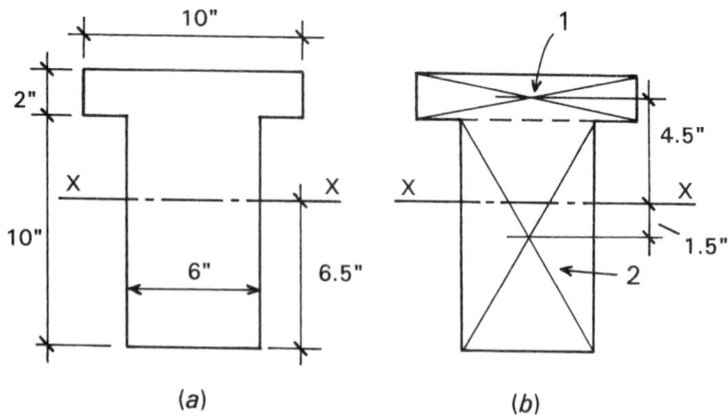

Figure A.7 Reference for Example 8.

The next step is to break the complex shape down into parts for which centroids, areas, and centroidal moments of inertia are readily found. As was done in Example 1, the shape here is divided between the rectangular flange part and the rectangular web part.

The reference axis to be used here is the horizontal centroidal axis. Table A.2 summarizes the process of determining the factors for the parallel axis transfer process. The required value for I about the horizontal centroidal axis is determined to be 1046.7 in.^{4}

A common situation in which this problem must be solved is in the case of structural members that are built up from distinct parts. One such section is that shown in Figure A.8, where a box-shaped cross section is composed by attaching two plates and two rolled channel sections. While this composite section is actually symmetrical about both its principal axes and the locations of these axes are apparent, the values for moment of inertia about both axes must be determined by the parallel axis transfer process. The following example demonstrates the process.

TABLE A.2 Summary of Computations for Moment of Inertia: Example 9

Part	Area (in.^{2})	y (in.)	I_o (in.^{4})	$A \times y^2$ (in.^{4})	I_x (in.^{4})
1	20	4.5	$10(2)^3/12 = 6.7$	$20(4.5)^2 = 405$	411.7
2	60	1.5	$6(10)^3/12 = 500$	$60(1.5)^2 = 135$	635
Σ					1046.7

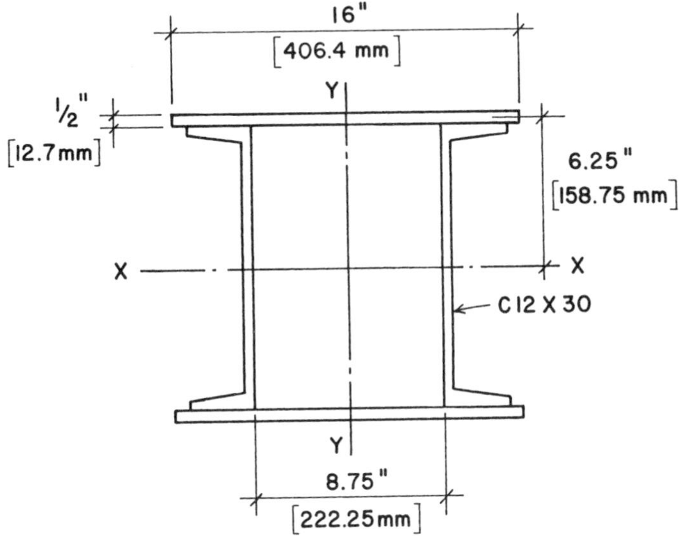

Figure A.8 Reference for Example 9.

Example 9. Compute the moment of inertia about the centroidal *X–X* axis of the built-up section shown in Figure A.8.

Solution: For this situation the two channels are positioned so that their centroids coincide with the reference axis. Thus, the value of I_o for the channels is also their actual moment of inertia about the required reference axis, and their contributions to the required value here is simply 2 times their listed value for moment of inertia about their *X–X* axis, as given in Table A.4: $2(162) = 324$ in.^{4}.

The plates have simple rectangular cross sections, and the centroidal moment of inertia of one plate is thus determined as

$$I_o = \frac{bd^3}{12} = \frac{16 \times (0.5)^3}{12} = 0.1667 \text{ in.}^4$$

The distance between the centroid of the plate and the reference *X–X* axis is 6.25 in., and the area of one plate is 8 in.^{2} The moment of inertia for one plate about the reference axis is thus

$$I_o + Az^2 = 0.1667 + (8)(6.25)^2 = 312.7 \text{ in.}^4$$

and the value for the two plates is twice this, or 625.4 in.^{4}.

Adding the contributions of the parts, the answer is 324 + 625.4 = 949.4 in.^{4}.

Problems A.3.A–F Compute the moments of inertia about the indicated centroidal axes for the cross-sectional shapes in Figure A.9.

Problems A.3.G–I Compute the moments of inertia with respect to the centroidal X–X axes for the built-up sections in Figure A.10. Make use of any appropriate data from the tables of properties for steel shapes.

A.4 MISCELLANEOUS PROPERTIES

Elastic Section Modulus

The term I/c in the formula for flexural stress is called the *section modulus*. Use of the section modulus permits a minor shortcut in the computations for flexural stress or the determination of the bending moment capacity of members. However, the real value of this property is in its measure of relative bending strength of members within the elastic region of the stress–strain curve. As a geometric property, it is a direct index of bending strength for a given member cross section. Members of various cross sections may thus be rank ordered in terms of their bending strength strictly on the basis of their S values. Because of its usefulness, the value of S is listed together with other significant properties in the tabulations for steel and wood members.

For members of standard form (structural lumber and rolled steel shapes), the value of S may be obtained from tables similar to those presented at the end of this appendix. For complex forms not of standard form, the value of S must be computed, which is readily done once the centroidal axes are located and moments of inertia about the centroidal axes are determined.

Example 10. Verify the tabulated value for the section modulus of a 6 × 12 wood beam about the centroidal axis parallel to its narrow side.

Solution: From Table A.8 the actual dimensions of this member are 5.5 × 11.5 in., and the value for the moment of inertia is 697.1 in.^{4}. Then

$$S = \frac{I}{c} = \frac{697.1}{5.75} = 121.235 \text{ in.}^3$$

which agrees with the value in Table A.8.

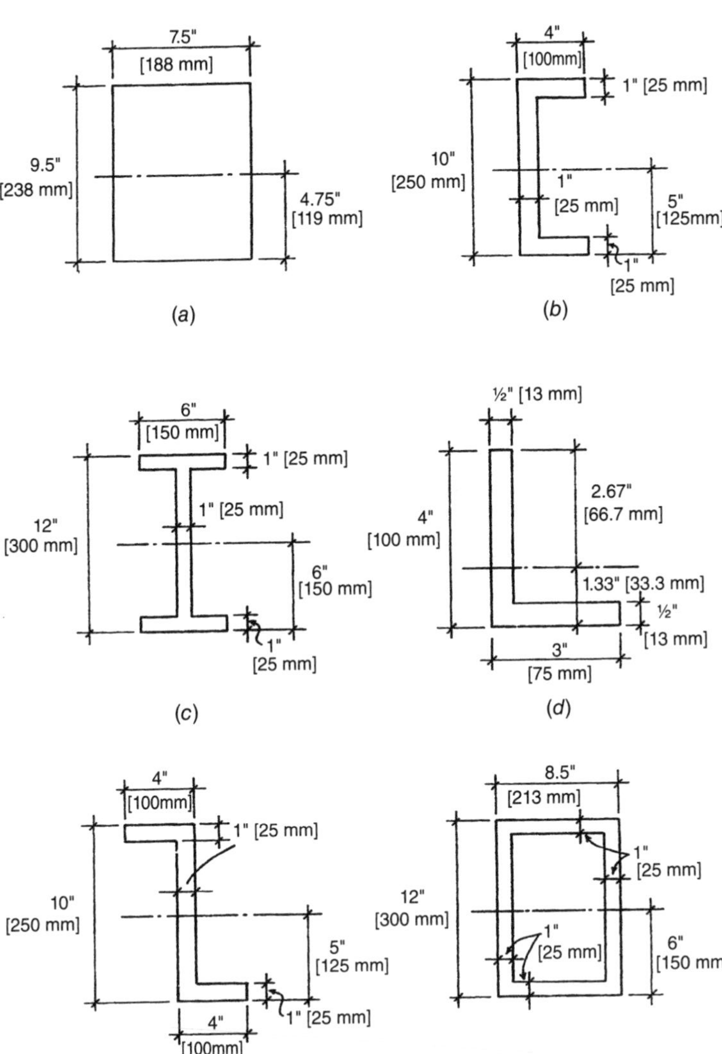

Figure A.9 Reference for Problems A.3.A–F.

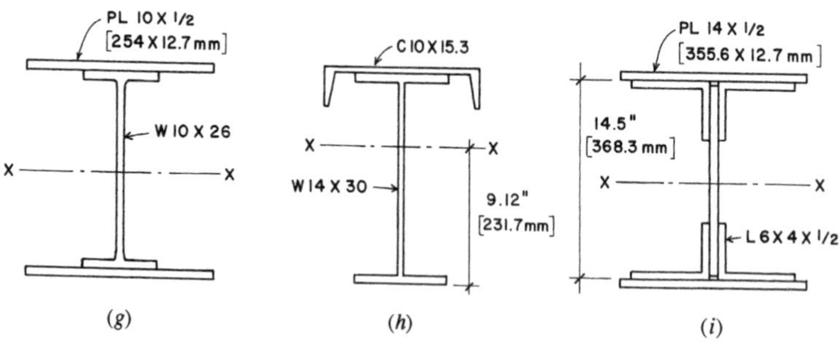

Figure A.10 Reference for Problems A.3.G–I.

Plastic Section Modulus

The plastic section modulus, designated Z, is used in a similar manner to the elastic stress section modulus S. The plastic modulus is used to determine the fully plastic stress moment capacity of a steel beam. Thus,

$$M_p = F_y \times Z$$

The use of the plastic section modulus is discussed in Section 9.4.

Radius of Gyration

For design of slender compression members an important geometric property is the *radius of gyration*, defined as

$$r = \sqrt{\frac{I}{A}}$$

Just as with moment of inertia and section modulus values, the radius of gyration has an orientation to a specific axis in the planar cross section of a member. Thus, if the I used in the formula for r is that with respect to the X–X centroidal axis, then that is the reference for the specific value of r.

A value of r with particular significance is that designated as the *least radius of gyration*. Since this value will be related to the least value of I for the cross section, and since I is an index of the bending stiffness of the member, then the least value for r will indicate the weakest response of the member to bending. This relates specifically to the resistance of slender compression members to buckling.

Buckling is essentially a sideways bending response, and its most likely occurrence will be on the axis identified by the least value of I or r. Use of these relationships is discussed for columns in Parts II and III.

A.5 TABLES OF PROPERTIES OF SECTIONS

Figure A.11 presents formulas for obtaining geometric properties of various simple plane sections. Some of these may be used for single-piece structural members or for the building up of complex members.

Tables A.3–A.8 present the properties of various plane sections. These are sections identified as those of standard industry-produced sections of wood and steel. Standardization means that the shapes and dimensions of the sections are fixed and each specific section is identified in some way.

Structural members may be employed for various purposes, and thus they may be oriented differently for some structural uses. Of note for any plane section are the *principal axes* of the section. These are the two, mutually perpendicular, centroidal axes for which the values will be greatest and least, respectively, for the section; thus, the axes are identified as the major and minor axes. If sections have an axis of symmetry, it will always be a principal axis—either major or minor.

For sections with two perpendicular axes of symmetry (rectangle, H, I, etc.), one axis will be the major axis and the other the minor axis. In the tables of properties the listed values for I, S, and r are all identified as to a specific axis, and the reference axes are identified in a figure for the table.

Other values given in the tables are for significant dimensions, total cross-sectional area, and the weight of a 1-ft-long piece of the member. The weight of wood members is given in the table, assuming an average density for structural softwood of 35 lb/ft^3. The weight of steel members is given for W and channel shapes as part of their designation; thus, a W8 × 67 member weighs 67 lb/ft. For steel angles and pipes the weight is given in the table, as determined from the density of steel at 490 lb/ft^3.

The designation of some members indicates their true dimensions. Thus, a 10-in. channel and a 6-in. angle have true dimensions of 10 and 6 in. For W shapes and pipe, the designated dimensions are nominal, and the true dimensions must be obtained from the tables.

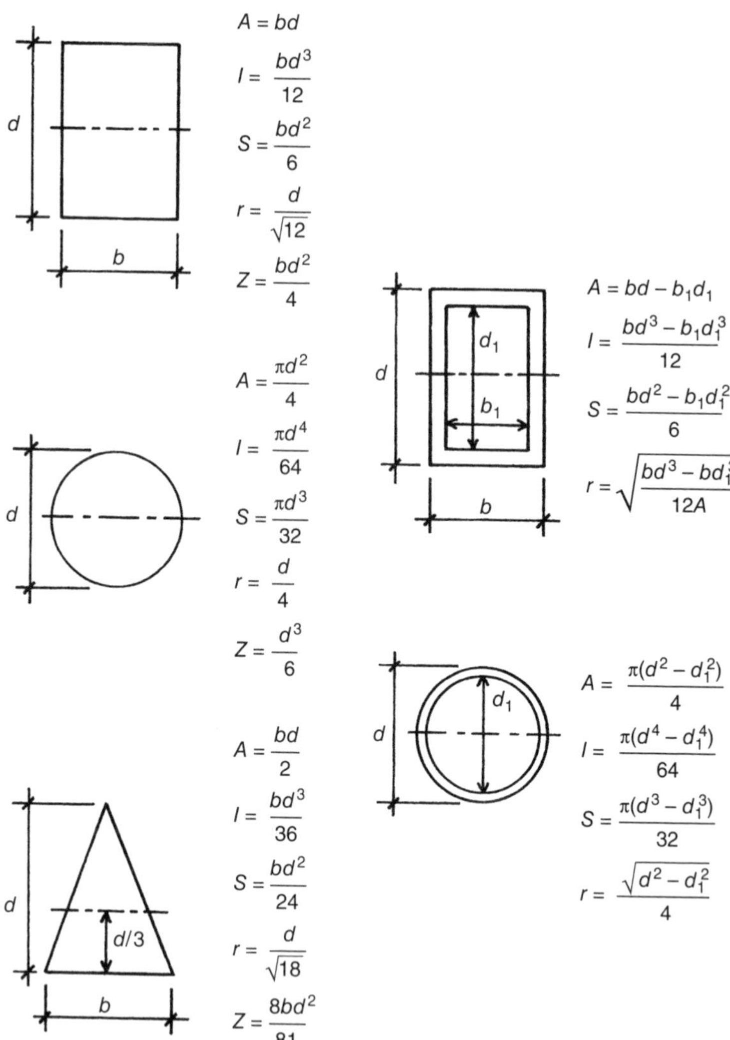

Figure A.11 Properties of various geometric shapes of cross sections: A = area, I = moment of inertia, S = section modulus, and r = radius of gyration.

TABLE A.3 Properties of W Shapes

Shape	Area	Depth	Web Thickness	Flange Width	Flange Thickness		Elastic Properties Axis X–X			Axis Y–Y			Modulus
	A (in.^{2})	d (in.)	t_w (in.)	b_f (in.)	t_f (in.)	k (in.)	I (in.^{4})	S (in.^{3})	r (in.)	I (in.^{4})	S (in.^{3})	r (in.)	Z_x (in.^{3})
W30×116	34.2	30.01	0.565	10.495	0.850	1.625	4930	329	12.0	164	31.3	2.19	378
×108	31.7	29.83	0.545	10.475	0.760	1.562	4470	299	11.9	146	27.9	2.15	346
×99	29.1	29.65	0.520	10.450	0.670	1.437	3990	269	11.7	128	24.5	2.10	312
W27×94	27.7	26.92	0.490	9.990	0.745	1.437	3270	243	10.9	124	24.8	2.12	278
×84	24.8	26.71	0.460	9.960	0.640	1.375	2850	213	10.7	106	21.2	2.07	244
W24×84	24.7	24.10	0.470	9.020	0.770	1.562	2370	196	9.79	94.4	20.9	1.95	224
×76	22.4	23.92	0.440	8.990	0.680	1.437	2100	176	9.69	82.5	18.4	1.92	200
×68	20.1	23.73	0.415	8.965	0.585	1.375	1830	154	9.55	70.4	15.7	1.87	177

(continued)

TABLE A.3 (Continued)

			Web	Flange			Elastic Properties						Modulus
	Area	Depth	Thickness	Width	Thickness		Axis X–X			Axis Y–Y			
Shape	A (in.^{2})	d (in.)	t_w (in.)	b_f (in.)	t_f (in.)	k (in.)	I (in.^{4})	S (in.^{3})	r (in.)	I (in.^{4})	S (in.^{3})	r (in.)	Z_x (in.^{3})
W21×83	24.3	21.43	0.515	8.355	0.835	1.562	1830	171	8.67	81.4	19.5	1.83	196
×73	21.5	21.24	0.455	8.295	0.740	1.500	1600	151	8.64	70.6	17.0	1.81	172
×57	16.7	21.06	0.405	6.555	0.650	1.375	1170	111	8.36	30.6	9.35	1.35	129
×50	14.7	20.83	0.380	6.530	0.535	1.312	984	94.5	8.18	24.9	7.64	1.30	110
W18×86	25.3	18.39	0.480	11.090	0.770	1.437	1530	166	7.77	175	31.6	2.63	186
×76	22.3	18.21	0.425	11.035	0.680	1.375	1330	146	7.73	152	27.6	2.61	163
×60	17.6	18.24	0.415	7.555	0.695	1.375	984	108	7.47	50.1	13.3	1.69	123
×55	16.2	18.11	0.390	7.530	0.630	1.312	890	98.3	7.41	44.9	11.9	1.67	112
×50	14.7	17.99	0.355	7.495	0.570	1.250	800	88.9	7.38	40.1	10.7	1.65	101
×46	13.5	18.06	0.360	6.060	0.605	1.250	712	78.8	7.25	22.5	7.43	1.29	90.7
×40	11.8	17.90	0.315	6.015	0.525	1.187	612	68.4	7.21	19.1	6.35	1.27	78.4
W16×50	14.7	16.26	0.380	7.070	0.630	1.312	659	81.0	6.68	37.2	10.5	1.59	92.0
×45	13.3	16.13	0.345	7.035	0.565	1.250	586	72.7	6.65	32.8	9.34	1.57	82.3
×40	11.8	16.01	0.305	6.995	0.505	1.187	518	64.7	6.63	28.9	8.25	1.57	72.9
×36	10.6	15.86	0.295	6.985	0.430	1.125	448	56.5	6.51	24.5	7.00	1.52	64.0
W14×211	62.0	15.72	0.980	15.800	1.560	2.250	2660	338	6.55	1030	130	4.07	390
×176	51.8	15.22	0.830	15.650	1.310	2.000	2140	281	6.43	838	107	4.02	320
×132	38.8	14.66	0.645	14.725	1.030	1.687	1530	209	6.28	548	74.5	3.76	234
×120	35.3	14.48	0.590	14.670	0.940	1.625	1380	190	6.24	495	67.5	3.74	212
×74	21.8	14.17	0.450	10.070	0.785	1.562	796	112	6.04	134	26.6	2.48	126

×68	20.0	14.04	0.415	10.035	0.720	1.500	723	103	6.01	121	24.2	2.46	115
×48	14.1	13.79	0.340	8.030	0.595	1.375	485	70.3	5.85	51.4	12.8	1.91	78.4
×43	12.6	13.66	0.305	7.995	0.530	1.312	428	62.7	5.82	45.2	11.3	1.89	69.6
×34	10.0	13.98	0.285	6.745	0.455	1.000	340	48.6	5.83	23.3	6.91	1.53	54.6
×30	8.85	13.84	0.270	6.730	0.385	0.937	291	42.0	5.73	19.6	5.82	1.49	47.3
W12×136	39.9	13.41	0.790	12.400	1.250	1.937	1240	186	5.58	398	64.2	3.16	214
×120	35.3	13.12	0.710	12.320	1.105	1.812	1070	163	5.51	345	56.0	3.13	186
×72	21.1	12.25	0.430	12.040	0.670	1.375	597	97.4	5.31	195	32.4	3.04	108
×65	19.1	12.12	0.390	12.000	0.605	1.312	533	87.9	5.28	174	29.1	3.02	96.8
×53	15.6	12.06	0.345	9.995	0.575	1.250	425	70.6	5.23	95.8	19.2	2.48	77.9
×45	13.2	12.06	0.335	8.045	0.575	1.250	350	58.1	5.15	50.0	12.4	1.94	64.7
×40	11.8	11.94	0.295	8.005	0.515	1.250	310	51.9	5.13	44.1	11.0	1.93	57.5
×30	8.79	12.34	0.260	6.520	0.440	0.937	238	38.6	5.21	20.3	6.24	1.52	43.1
×26	7.65	12.22	0.230	6.490	0.380	0.875	204	33.4	5.17	17.3	5.34	1.51	37.2
W10×88	25.9	10.84	0.605	10.265	0.990	1.625	534	98.5	4.54	179	34.8	2.63	113
×77	22.6	10.60	0.530	10.190	0.870	1.500	455	85.9	4.49	154	30.1	2.60	97.6
×49	14.4	9.98	0.340	10.000	0.560	1.312	272	54.6	4.35	93.4	18.7	2.54	60.4
×39	11.5	9.92	0.315	7.985	0.530	1.125	209	42.1	4.27	45.0	11.3	1.98	46.8
×33	9.71	9.73	0.290	7.960	0.435	1.062	170	35.0	4.19	36.6	9.20	1.94	38.8
×19	5.62	10.24	0.250	4.020	0.395	0.812	96.3	18.8	4.14	4.29	2.14	0.874	21.6
×17	4.99	10.11	0.240	4.010	0.330	0.750	81.9	16.2	4.05	3.56	1.78	0.844	18.7

Source: Adapted from data in the *Steel Construction Manual* (Ref. 5), with permission of the publishers, American Institute of Steel Construction. This table is a sample from an extensive set of tables in the reference document.

TABLE A.4 Properties of American Standard Channels

Shape	Area A (in.^{2})	Depth d (in.)	Web Thickness t_w (in.)	Flange Width b_f (in.)	Flange Thickness t_f (in.)	k (in.)	Elastic Properties, Axis X–X I (in.^{4})	Axis X–X S (in.^{3})	Axis X–X r (in.)	Axis Y–Y I (in.^{4})	Axis Y–Y S (in.^{3})	Axis Y–Y r (in.)	x^a (in.)	$e_o^{\,b}$ (in.)
C15 × 50	14.7	15.0	0.716	3.716	0.650	1.44	404	53.8	5.24	11.0	3.78	0.867	0.798	0.583
× 40	11.8	15.0	0.520	3.520	0.650	1.44	349	46.5	5.44	9.23	3.37	0.886	0.777	0.767
× 33.9	9.96	15.0	0.400	3.400	0.650	1.44	315	42.0	5.62	8.13	3.11	0.904	0.787	0.896
C12 × 30	8.82	12.0	0.510	3.170	0.501	1.13	162	27.0	4.29	5.14	2.06	0.763	0.674	0.618
× 25	7.35	12.0	0.387	3.047	0.501	1.13	144	24.1	4.43	4.47	1.88	0.780	0.674	0.746
× 20.7	6.09	12.0	0.282	2.942	0.501	1.13	129	21.5	4.61	3.88	1.73	0.799	0.698	0.870

C10 × 30	8.82	10.0	0.673	3.033	0.436	1.00	103	20.7	3.42	3.94	1.65	0.669	0.649	0.369
× 25	7.35	10.0	0.526	2.886	0.436	1.00	91.2	18.2	3.52	3.36	1.48	0.676	0.617	0.494
× 20	5.88	10.0	0.379	2.739	0.436	1.00	78.9	15.8	3.66	2.81	1.32	0.692	0.606	0.637
× 15.3	4.49	10.0	0.240	2.600	0.436	1.00	67.4	13.5	3.87	2.28	1.16	0.713	0.634	0.796
C9 × 20	5.88	9.0	0.448	2.648	0.413	0.94	60.9	13.5	3.22	2.42	1.17	0.642	0.583	0.515
× 15	4.41	9.0	0.285	2.485	0.413	0.94	51.0	11.3	3.40	1.93	1.01	0.661	0.586	0.682
× 13.4	3.94	9.0	0.233	2.433	0.413	0.94	47.9	10.6	3.48	1.76	0.962	0.669	0.601	0.743
C8 × 18.75	5.51	8.0	0.487	2.527	0.390	0.94	44.0	11.0	2.82	1.98	1.01	0.599	0.565	0.431
× 13.75	4.04	8.0	0.303	2.343	0.390	0.94	36.1	9.03	2.99	1.53	0.854	0.615	0.553	0.604
× 11.5	3.38	8.0	0.220	2.260	0.390	0.94	32.6	8.14	3.11	1.32	0.781	0.625	0.571	0.697
C7 × 14.75	4.33	7.0	0.419	2.299	0.366	0.88	27.2	7.78	2.51	1.38	0.779	0.564	0.532	0.441
× 12.25	3.60	7.0	0.314	2.194	0.366	0.88	24.2	6.93	2.60	1.17	0.703	0.571	0.525	0.538
× 9.8	2.87	7.0	0.210	2.090	0.366	0.88	21.3	6.08	2.72	0.968	0.625	0.581	0.540	0.647
C6 × 13	3.83	6.0	0.437	2.157	0.343	0.81	17.4	5.80	2.13	1.05	0.642	0.525	0.514	0.380
× 10.5	3.09	6.0	0.314	2.034	0.343	0.81	15.2	5.06	2.22	0.866	0.564	0.529	0.499	0.486
× 8.2	2.40	6.0	0.200	1.920	0.343	0.81	13.1	4.38	2.34	0.693	0.492	0.537	0.511	0.599

[a]Distance to centroid of section.

[b]Distance to shear center of section.

Source: Adapted from data in the *Steel Construction Manual* (Ref. 5), with permission of the publishers, American Institute of Steel Construction. This table is a sample from an extensive set of tables in the reference document.

TABLE A.5 Properties of Single-Angle Shapes

Size and Thickness (in.)	k (in.)	Weight per ft (lb)	Area A (in.^{2})	Axis X–X I (in.^{4})	S (in.^{3})	r (in.)	y (in.)	Axis Y–Y I (in.^{4})	S (in.^{3})	r (in.)	x (in.)	Axis Z–Z r (in.)	Tan a
$8\times8\times1\frac{1}{8}$	1.75	56.9	16.7	98.0	17.5	2.42	2.41	98.0	17.5	2.42	2.41	1.56	1.000
$\times1$	1.62	51.0	15.0	89.0	15.8	2.44	2.37	89.0	15.8	2.44	2.37	1.56	1.000
$8\times6\times\frac{3}{4}$	1.25	33.8	9.94	63.4	11.7	2.53	2.56	30.7	6.92	1.76	1.56	1.29	0.551
$\times\frac{1}{2}$	1.00	23.0	6.75	44.3	8.02	2.56	2.47	21.7	4.79	1.79	1.47	1.30	0.558
$6\times6\times\frac{5}{8}$	1.12	24.2	7.11	24.2	5.66	1.84	1.73	24.2	5.66	1.84	1.73	1.18	1.000
$\times\frac{1}{2}$	1.00	19.6	5.75	19.9	4.61	1.86	1.68	19.9	4.61	1.86	1.68	1.18	1.000
$6\times4\times\frac{5}{8}$	1.12	20.0	5.86	21.1	5.31	1.90	2.03	7.52	2.54	1.13	1.03	0.864	0.435
$\times\frac{1}{2}$	1.00	16.2	4.75	17.4	4.33	1.91	1.99	6.27	2.08	1.15	0.987	0.870	0.440
$\times\frac{3}{8}$	0.87	12.3	3.61	13.5	3.32	1.93	1.94	4.90	1.60	1.17	0.941	0.877	0.446

5 × 3½ × ½	1.00	13.6	4.00	9.99	2.99	1.58	1.66	4.05	1.56	1.01	0.906	0.755	0.479
× 3/8	0.87	10.4	3.05	7.78	2.29	1.60	1.61	3.18	1.21	1.02	0.861	0.762	0.486
5 × 3 × ½	1.00	12.8	3.75	9.45	2.91	1.59	1.75	2.58	1.15	0.829	0.750	0.648	0.357
× 3/8	0.87	9.8	2.86	7.37	2.24	1.61	1.70	2.04	0.888	0.845	0.704	0.654	0.364
4 × 4 × ½	0.87	12.8	3.75	5.56	1.97	1.22	1.18	5.56	1.97	1.22	1.18	0.782	1.000
× 3/8	0.75	9.8	2.86	4.36	1.52	1.23	1.14	4.36	1.52	1.23	1.14	0.788	1.000
4 × 3 × ½	0.94	11.1	3.25	5.05	1.89	1.25	1.33	2.42	1.12	0.864	0.827	0.639	0.543
× 3/8	0.81	8.5	2.48	3.96	1.46	1.26	1.28	1.92	0.866	0.879	0.782	0.644	0.551
× 5/16	0.75	7.2	2.09	3.38	1.23	1.27	1.26	1.65	0.734	0.887	0.759	0.647	0.554
3½ × 3½ × 3/8	0.75	8.5	2.48	2.87	1.15	1.07	1.01	2.87	1.15	1.07	1.01	0.687	1.000
× 5/16	0.69	7.2	2.09	2.45	0.976	1.08	0.990	2.45	0.976	1.08	0.990	0.690	1.000
× 5/16	0.75	6.1	1.78	2.19	0.927	1.11	1.14	0.939	0.504	0.727	0.637	0.540	0.501
3 × 3 × 3/8	0.69	7.2	2.11	1.76	0.833	0.913	0.888	1.76	0.833	0.913	0.888	0.587	1.000
× 5/16	0.62	6.1	1.78	1.51	0.707	0.922	0.865	1.51	0.707	0.922	0.865	0.589	1.000
3 × 2½ × 3/8	0.75	6.6	1.92	1.66	0.810	0.928	0.956	1.04	0.581	0.736	0.706	0.522	0.676
× 5/16	0.69	5.6	1.62	1.42	0.688	0.937	0.933	0.898	0.494	0.744	0.683	0.525	0.680
3 × 2 × 3/8	0.69	5.9	1.73	1.53	0.781	0.940	1.04	0.543	0.371	0.559	0.539	0.430	0.428
× 5/16	0.62	5.0	1.46	1.32	0.664	0.948	1.02	0.470	0.317	0.567	0.516	0.432	0.435
2½ × 2½ × 3/8	0.69	5.9	1.73	0.984	0.566	0.753	0.762	0.984	0.566	0.753	0.762	0.487	1.000
× 5/16	0.62	5.0	1.46	0.849	0.482	0.761	0.740	0.849	0.482	0.761	0.740	0.489	1.000
2½ × 2 × 3/8	0.69	5.3	1.55	0.912	0.547	0.768	0.831	0.514	0.363	0.577	0.581	0.420	0.614
× 5/16	0.62	4.5	1.31	0.788	0.466	0.776	0.809	0.446	0.310	0.584	0.559	0.422	0.620

Source: Adapted from data in the *Steel Construction Manual* (Ref. 5), with permission of the publishers, American Institute of Steel Construction. This table is a sample from an extensive set of tables in the reference document.

TABLE A.6 Properties of Double-Angle Shapes with Long Legs Back to Back

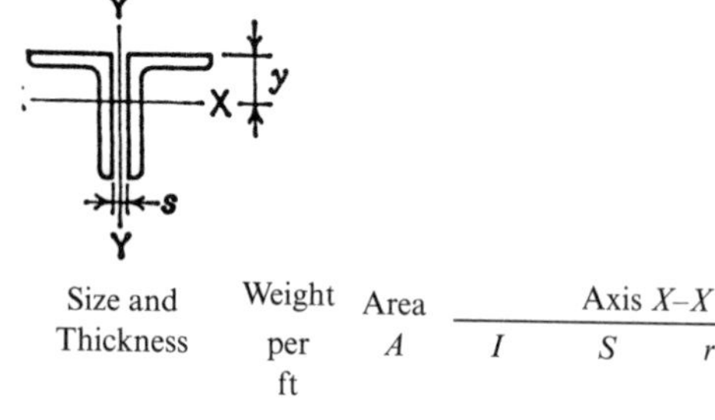

Size and Thickness (in.)	Weight per ft (lb)	Area A (in.^{2})	Axis X–X I (in.^{4})	S (in.^{3})	r (in.)	y (in.)	Axis Y–Y Radii of Gyration Back to Back of Angles, in. 0	3/8	3/4
8×6×1	88.4	26.0	161.0	30.2	2.49	2.65	2.39	2.52	2.66
×3/4	67.6	19.9	126.0	23.3	2.53	2.56	2.35	2.48	2.62
×1/2	46.0	13.5	88.6	16.0	2.56	2.47	2.32	2.44	2.57
6×4×3/4	47.2	13.9	49.0	12.5	1.88	2.08	1.55	1.69	1.83
×1/2	32.4	9.50	34.8	8.67	1.91	1.99	1.51	1.64	1.78
×3/8	24.6	7.22	26.9	6.64	1.93	1.94	1.50	1.62	1.76
5×3 1/2×1/2	27.2	8.00	20.0	5.97	1.58	1.66	1.35	1.49	1.63
×3/8	20.8	6.09	15.6	4.59	1.60	1.61	1.34	1.46	1.60
5×3×1/2	25.6	7.50	18.9	5.82	1.59	1.75	1.12	1.25	1.40
×3/8	19.6	5.72	14.7	4.47	1.61	1.70	1.10	1.23	1.37
×5/16	16.4	4.80	12.5	3.77	1.61	1.68	1.09	1.22	1.36
4×3×1/2	22.2	6.50	10.1	3.78	1.25	1.33	1.20	1.33	1.48
×3/8	17.0	4.97	7.93	2.92	1.26	1.28	1.18	1.31	1.45
×5/16	14.4	4.18	6.76	2.47	1.27	1.26	1.17	1.30	1.44
3 1/2×2 1/2×3/8	14.4	4.22	5.12	2.19	1.10	1.16	0.976	1.11	1.26
×5/16	12.2	3.55	4.38	1.85	1.11	1.14	0.966	1.10	1.25
×1/4	9.8	2.88	3.60	1.51	1.12	1.11	0.958	1.09	1.23
3×2×3/8	11.8	3.47	3.06	1.56	0.940	1.04	0.777	0.917	1.07
×5/16	10.0	2.93	2.63	1.33	0.948	1.02	0.767	0.903	1.06
×1/4	8.2	2.38	2.17	1.08	0.957	0.993	0.757	0.891	1.04
2 1/2×2×3/8	10.6	3.09	1.82	1.09	0.768	0.831	0.819	0.961	1.12
×5/16	9.0	2.62	1.58	0.932	0.776	0.809	0.809	0.948	1.10
×1/4	7.2	2.13	1.31	0.763	0.784	0.787	0.799	0.935	1.09

Source: Adapted from data in the *Steel Construction Manual* (Ref. 5), with permission of the publishers, American Institute of Steel Construction. This table is a sample from an extensive set of tables in the reference document.

TABLE A.7 Properties of Standard Weight Steel Pipe

Dimensions					Properties			
Nominal Diameter (in.)	Outside Diameter (in.)	Inside Diameter (in.)	Wall Thickness (in.)	Weight per ft (lb)	A (in.^{2})	I (in.^{4})	S (in.^{3})	r (in.)
3	3.500	3.068	0.216	7.58	2.23	3.02	1.72	1.16
3½	4.000	3.548	0.226	9.11	2.68	4.79	2.39	1.34
4	4.500	4.026	0.237	10.79	3.17	7.23	3.21	1.51
5	5.563	5.047	0.258	14.62	4.30	15.2	5.45	1.88
6	6.625	6.065	0.280	18.97	5.58	28.1	8.50	2.25
8	8.625	7.981	0.322	28.55	8.40	72.5	16.8	2.94
10	10.750	10.020	0.365	40.48	11.9	161	29.9	3.67
12	12.750	12.000	0.375	49.56	14.6	279	43.8	4.38

Source: Adapted from data in the *Steel Construction Manual* (Ref. 5), with permission of the publishers, American Institute of Steel Construction. This table is a sample from an extensive set of tables in the reference document.

TABLE A.8 Properties of Structural Lumber

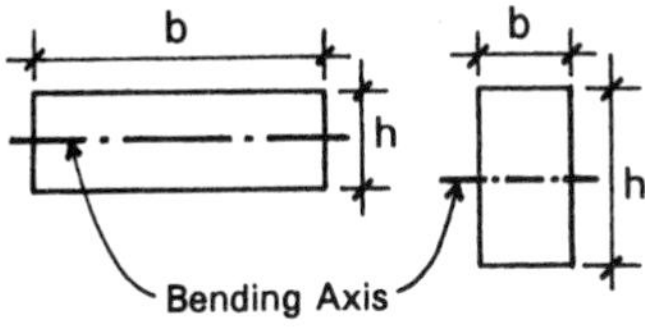

Dimensions (in.)			X–X Axis		Y–Y Axis		Weight at 35 lb/ft^3 Density
		Area	Section Modulus	Moment of Inertia	Section Modulus	Moment of Inertia	
Nominal $b \times h$	Actual $b \times h$	A (in.^{2})	S (in.^{3})	I (in.^{4})	S (in.^{3})	I (in.^{4})	(lb/ft)
2×3	1.5×2.5	3.75	1.563	1.953	0.938	0.703	0.911
2×4	1.5×3.5	5.25	3.063	5.359	1.313	0.984	1.276
2×6	1.5×5.5	8.25	7.563	20.80	2.063	1.547	2.005
2×8	1.5×7.25	10.88	13.14	47.63	2.719	2.039	2.643
2×10	1.5×9.25	13.88	21.39	98.93	3.469	2.602	3.372
2×12	1.5×11.25	16.88	31.64	178.0	4.219	3.164	4.102
2×14	1.5×13.25	19.88	43.89	290.8	4.969	3.727	4.831
3×4	2.5×3.5	8.75	5.104	8.932	3.646	4.557	2.127
3×6	2.5×5.5	13.75	12.60	34.66	5.729	7.161	3.342
3×8	2.5×7.25	18.13	21.90	79.39	7.552	9.440	4.405
3×10	2.5×9.25	23.13	35.65	164.9	9.635	12.04	5.621
3×12	2.5×11.25	28.13	52.73	296.6	11.72	14.65	6.836

(*continued*)

TABLE A.8 (*Continued*)

Dimensions (in.)			X–X Axis		Y–Y Axis		Weight at 35 lb/ft³ Density
		Area	Section Modulus	Moment of Inertia	Section Modulus	Moment of Inertia	
Nominal $b \times h$	Actual $b \times h$	A (in.^{2})	S (in.^{3})	I (in.^{4})	S (in.^{3})	I (in.^{4})	(lb/ft)
3×14	2.5×13.25	33.13	73.15	484.6	13.80	17.25	8.051
3×16	2.5×15.25	38.13	96.90	738.9	15.89	19.86	9.266
4×4	3.5×3.5	12.25	7.146	12.51	7.146	12.51	2.977
4×6	3.5×5.5	19.25	17.65	48.53	11.23	19.65	4.679
4×8	3.5×7.25	25.38	30.66	111.1	14.80	25.9	6.168
4×10	3.5×9.25	32.38	49.91	230.8	18.89	33.05	7.869
4×12	3.5×11.25	39.38	73.83	415.3	22.97	40.20	9.570
4×14	3.5×13.25	46.38	102.4	678.5	27.05	47.34	11.27
4×16	3.5×15.25	53.38	135.7	1034	31.14	54.49	12.97
5×5	4.5×4.5	20.25	15.19	34.17	15.19	34.17	4.922
6×6	5.5×5.5	30.25	27.73	76.26	27.73	76.26	7.352
6×8	5.5×7.5	41.25	51.56	193.4	37.81	104.0	10.03
6×10	5.5×9.5	52.25	82.73	393.0	47.90	131.7	12.70
6×12	5.5×11.5	63.25	121.2	697.1	57.98	159.4	15.37
6×14	5.5×13.5	74.25	167.1	1128	68.06	187.2	18.05
6×16	5.5×15.5	85.25	220.2	1707	78.15	214.9	20.72
6×18	5.5×17.5	96.25	280.7	2456	88.23	242.6	23.39
6×20	5.5×19.5	107.3	348.6	3398	98.31	270.4	26.07
6×22	5.5×21.5	118.3	423.7	4555	108.4	298.1	28.74
6×24	5.5×23.5	129.3	506.2	5948	118.5	325.8	31.41
8×8	7.5×7.5	56.25	70.31	263.7	70.31	263.7	13.67
8×10	7.5×9.5	71.25	112.8	535.9	89.06	334.0	17.32
8×12	7.5×11.5	86.25	165.3	950.5	107.8	404.3	20.96
8×14	7.5×13.5	101.3	227.8	1538	126.6	474.6	24.61
8×16	7.5×15.5	116.3	300.3	2327	145.3	544.9	28.26
8×18	7.5×17.5	131.3	382.8	3350	164.1	615.2	31.90
8×20	7.5×19.5	146.3	475.3	4634	182.8	685.5	35.55
8×22	7.5×21.5	161.3	577.8	6211	201.6	755.9	39.19
8×24	7.5×23.5	176.3	690.3	8111	220.3	826.2	42.84
10×10	9.5×9.5	90.25	142.9	678.8	142.9	678.8	21.94
10×12	9.5×11.5	109.3	209.4	1204	173.0	821.7	26.55
10×14	9.5×13.5	128.3	288.6	1948	203.1	964.5	31.17
10×16	9.5×15.5	147.3	380.4	2948	233.1	1107	35.79
10×18	9.5×17.5	166.3	484.9	4243	263.2	1250	40.41
10×20	9.5×19.5	185.3	602.1	5870	293.3	1393	45.03
10×22	9.5×21.5	204.3	731.9	7868	323.4	1536	49.64
10×24	9.5×23.5	223.3	874.4	10270	353.5	1679	54.26

TABLE A.8 (*Continued*)

Dimensions (in.)		Area	X–X Axis		Y–Y Axis		Weight at 35 lb/ft^3 Density
			Section Modulus	Moment of Inertia	Section Modulus	Moment of Inertia	
Nominal $b \times h$	Actual $b \times h$	A (in.^{2})	S (in.^{3})	I (in.^{4})	S (in.^{3})	I (in.^{4})	(lb/ft)
12×12	11.5×11.5	132.3	253.5	1458	253.5	1458	32.14
12×14	11.5×13.5	155.3	349.3	2358	297.6	1711	37.73
12×16	11.5×15.5	178.3	460.5	3569	341.6	1964	43.32
12×18	11.5×17.5	201.3	587.0	5136	385.7	2218	48.91
12×20	11.5×19.5	224.3	728.8	7106	429.8	2471	54.51
12×22	11.5×21.5	247.3	886.0	9524	473.9	2725	60.10
12×24	11.5×23.5	270.3	1058	12440	518.0	2978	65.69
14×14	13.5×13.5	182.3	410.1	2768	410.1	2768	44.30
14×16	13.5×15.5	209.3	540.6	4189	470.8	3178	50.86
14×18	13.5×17.5	236.3	689.1	6029	531.6	3588	57.42
14×20	13.5×19.5	263.3	855.6	8342	592.3	3998	63.98
14×22	13.5×21.5	290.3	1040	11180	653.1	4408	70.55
14×24	13.5×23.5	317.3	1243	14600	713.8	4818	77.11
16×16	15.5×15.5	240.3	620.6	4810	620.6	4810	58.39
16×18	15.5×17.5	271.3	791.1	6923	700.7	5431	65.93
16×20	15.5×19.5	302.3	982.3	9578	780.8	6051	73.46
16×22	15.5×21.5	333.3	1194	12840	860.9	6672	81.00
16×24	15.5×23.5	364.3	1427	16760	941.0	7293	88.53
18×18	17.5×17.5	306.3	893.2	7816	893.2	7816	74.44
18×20	17.5×19.5	341.3	1109	10810	995.3	8709	82.94
18×22	17.5×21.5	376.3	1348	14490	1097	9602	91.45
18×24	17.5×23.5	411.3	1611	18930	1199	10500	99.96
20×20	19.5×19.5	380.3	1236	12050	1236	12050	92.42
20×22	19.5×21.5	419.3	1502	16150	1363	13280	101.9
20×24	19.5×23.5	458.3	1795	21090	1489	14520	111.4
22×22	21.5×21.5	462.3	1656	17810	1656	17810	112.4
22×24	21.5×23.5	505.3	1979	23250	1810	19460	122.8
24×24	23.5×23.5	552.3	2163	25420	2163	25420	134.2

Source: Compiled from data in the *National Design Specification for Wood Construction* (Ref. 3), with permission of the publisher, American Forest and Paper Association.

GLOSSARY

The material presented in this glossary constitutes a brief dictionary of words and terms frequently encountered in discussions of the design of building structures. Many of the words and terms have reasonably well-established meanings; in those cases we have tried to be consistent with the accepted usage. In some cases, however, words and terms are given different meanings by different authors or by groups that work in different fields—in which case the definition here is that used for the work in this book.

Some words and terms are commonly misused with regard to their precise meaning, an example being *unreinforced,* which would imply something from which reinforcing has been removed, whereas it is commonly used to refer to something that was never reinforced in the first place. Where such is the case, we have given the commonly used meaning here.

To be clear in its requirements, a legal document such as a building code often defines some words and terms. Care should be exercised when reading such documents to be aware of these precise meanings.

For fuller explanation of some of the words and terms here, as well as many not given here, the reader should use the Index to find the related discussion in the text.

Adequate Just enough; sufficient. Indicates a quality of bracketed acceptability—on the one hand, not insufficient, but on the other hand, not superlative or excessive.

Aggregate Inert, loose material that makes up the largest part (typically two-thirds to three-fourths) of the bulk of concrete; what the water and cement paste holds together; ordinarily consists of stone—ranging in size from medium fine sand to coarse gravel.

Allowable Stress See *Stress*.

Allowable Strength Design (ASD) Structural design method that is a strength design method similar to LRFD using unfactored loading (service loading). This method is currently used only in steel design.

Allowable Stress Design (ASD) Structural design method that employs limits based on allowable stresses and responses to service (actual usage) load conditions.

Analysis Separation into constituent parts. In engineering, the investigative determination of the detail aspects of a particular phenomenon. May be qualitative—meaning a general evaluation of the nature of the phenomenon—or quantitative—meaning the numerical determination of the magnitude of the phenomenon. See also *Synthesis*.

Anchorage Attachment for resistance to movement; usually opposing uplift, overturn, sliding, or horizontal separation. *Tiedown*, or *holddown*, refers to anchorage against uplift or overturn. *Positive anchorage* refers to fastening that does not easily loosen.

Beam A structural element that sustains transverse loading and develops internal forces of bending and shear in resisting loads. Also called a *girder* if very large, a *joist* if small or in closely spaced sets, a *rafter* if used for a roof, and a *header* or *lintel* if used over an opening in a wall.

Bearing Foundation Foundation that transfers loads to soil by direct vertical contact pressure (bearing). Usually refers to a *shallow bearing foundation*, which is placed directly beneath the lowest part of the building and not very far from the ground surface. See also *Footing*.

Bending Turning action that causes change in the curvature of linear elements; characterized by the development of opposed internal stresses of compression and tension. See also *Moment*.

Bent A planar framework, or some defined portion of one, that is intended for resistance to both horizontal and vertical loads in the plane of the frame.

Box System A lateral bracing system in which horizontal loads are resisted not by a column-and-beam system but rather by planar elements (shear walls and horizontal diaphragms) or braced frames (trusses).

Braced Frame Building code term for a trussed frame used for lateral bracing.

Bracing The general term used for elements that provide support against sideways movements due to lateral loads or to the buckling of slender elements.

Brittle Fracture Sudden failure, usually due to tension or shear; the usual failure of brittle materials, such as glass, plaster, and concrete.

Buckling Collapse, in the form of sudden sideways deflection or of torsional rotation (twisting).

Building Code Legal document for regulation of building form, features, and construction. Model codes are developed by recommending organizations; real codes are enacted as ordinances by some governmental unit (city, county, state).

Built-Up Member Structural member assembled from two or more parts in a manner that results in the combined parts working as a single unit.

Calculation Ordered, rational determination, usually by mathematical computations.

Centroid The geometric center of an object, usually analogous to the center of gravity. The point at which the entire mass of the object may be considered to be concentrated when considering moment of the mass.

Cold-Formed Element Structural element produced from sheet steel by bending, rolling, or stamping without heating of the steel.

Composite Panel Structural panel with wood veneer faces and a fiberboard core. In thick panels there is also a center wood veneer.

Compression Force action that tends to press adjacent particles of a material together and to cause shortening of objects in the direction of the compressive force.

Concrete Masonry Unit (CMU) Precast concrete unit; or good old concrete block.

Connection The union or joining of two or more distinct elements. In a structural assemblage, a connection device itself becomes an entity, with interactions of the connected elements visualized in terms of their actions on the connecting device.

Continuity The character of continuous, monolithic structural elements, wherein actions of adjacent elements are influenced by their continuous nature, such as with multistory columns, multispan beams, and multielement rigid frames.

Core Bracing Concentration of the vertical elements of a lateral bracing system at a central location in the building; usually at the location of elevators, stairs, and vertical service elements.

Creep Plastic deformation at constant stress levels that occurs over time (basically under dead load); a common effect in structures of concrete.

Curtain Wall An exterior wall of a building that is supported entirely by the building structure, rather than being self-supporting or a bearing wall.

Dead Load See *Load.*

Deflection Lateral movement of a structure under loading, such as the sag of a beam.

Determinate Structure A structure having the exact sufficiency for stability and therefore being subject to investigation by consideration of the equilibrium of simple static forces alone. See also *Indeterminate Structure.*

Diaphragm A planar element (wall, floor deck, etc.) used to resist forces in its own plane by shear action. See also *Horizontal Diaphragm* and *Shear Wall.*

Doubly Reinforced Beam A concrete beam with both tension and compression reinforcement.

Ductility Stress–strain (load–deformation) behavior that results from the plastic yielding of materials. To be significant—qualifying a material as ductile—the plastic yield before failure should be several times the elastic strain up to the point of plastic yield.

Elastic Behavior Used to describe two aspects of stress–strain behavior. The first is a constant stress–strain proportionality, or constant modulus of elasticity, as represented by a straight-line form of the stress–strain graph. The second is the stress-level limit within which all strain is recoverable; that is, there is no permanent deformation. The latter phenomenon may occur even though the stress–strain relationship is nonlinear (e.g., as it is for wood).

Engineered Wood General term for products produced from wood other than single pieces of sawn wood.

Equilibrium A balanced state or condition, usually used to describe a situation in which opposed force effects neutralize each other to produce a net effect of zero.

Factored Load Service load multiplied by a factor to produce an adjusted load for strength design.

Field Assemblage Construction work performed at the construction site (the field). Refers mostly to production and erection of steel frames.

Flexible See *Stiffness*.

Footing A shallow bearing-type foundation element consisting of a concrete pad cast directly into an excavation.

Free-Standing Wall See *Wall*.

Function Capability; intended use.

Grade 1. The level of the ground surface. 2. Rated quality (capability, capacity, refinement, etc.) of material.

Grade Beam A foundation element at or near the finished ground level that acts as a footing, a tie, or a spanning element.

Grain 1. A discrete particle of material that constitutes a loose material, such as soil. 2. The fibrous orientation of wood.

Grout Lean concrete (predominantly water, cement, and sand) used as a filler in the voids of masonry units, under steel bearing plates, and so on.

Header A beam at the edge of an opening in a roof or floor or at the top of an opening in a wall.

Horizontal Diaphragm Usually a roof or floor deck used as part of a lateral bracing system. See *Diaphragm*.

Hot Rolling Industrial process in which an ingot (lump) of steel is heated to the softening point and then squeezed between rollers repeatedly to produce a linear element with a constant cross section.

Indeterminate Structure In general, any structure whose load-resisting behavior cannot be determined by simple consideration of static equilibrium.

Inelastic See *Stress–Strain Behavior*.

Joist See *Beam*.

Kern Limit Limiting dimension for the eccentricity (off-center condition) of a compression force if tension stress is to be avoided.

Lateral Sideways. Used to describe something that is perpendicular to a major axis or direction. With respect to the vertical direction of gravity forces, primary effects of wind, earthquakes, and horizontal soil pressures are called *lateral effects*. Horizontal buckling of beams is called *lateral buckling*.

Lateral Unsupported Length For a linear structural element (beam, column), the distance between points of assured lateral bracing.

Live Load See *Load*.

Load The active force (or combination of forces) exerted on a structure. *Dead Load* is permanent gravity load, including the weight of the structure itself. *Live Load* is literally any load that is not permanent, although the term is ordinarily applied to distributed surface loads on roofs and floors. *Service Load* is that to which the structure is expected to be subjected. *Factored Load* is the service load modified by amplification factors for use in strength design.

Load and Resistance Factor Design (LRFD) See *Strength Design*.

Member One of the distinct elements of an assemblage.

Moment Action tending to produce turning or rotation. Product of a force times a distance (lever arm); yields a measurement unit of force times distance—for example, foot-pounds, kilonewton-meters, and so on. Bending moment causes curvature of linear elements; torsional moment causes twisting rotation.

Moment of Inertia The second moment of an area about a fixed line (axis) in the plane of the area. A purely mathematical property, not subject to direct physical measurement. Has significance in that it can be quantified for any geometric shape and is a measurement of certain structural responses, such as deflection of beams.

Net Section Cross-sectional area of a structural member reduced by holes, notches, and so on. Most significant in determination of tension response.

Normal 1. The ordinary, usual, unmodified state of something. 2. Perpendicular, such as pressure on a surface.

Open-Web Joist A light steel truss, usually with parallel chords, commonly used in closely spaced sets—as with wood floor joists. A manufactured product.

Optimal Best; most satisfying. The best solution to a set of criteria is the optimal one. When the criteria have opposed values, there may be no single optimal solution, except by the superiority of a single criterion—such as the lightest, the strongest, the cheapest, and so on.

Overturn Rotational effect consisting of toppling or tipping over; an effect of lateral loads on vertical elements.

***P*–delta Effect** Secondary bending effect on vertical members of a frame, induced by the vertical loads acting on the laterally displaced (deflected) members.

Pedestal A short pier or upright compression member. A column qualified by a ratio of unsupported (unbraced) height to least lateral dimension of 3 or less.

Perimeter Bracing Vertical elements of a lateral bracing system located at the building perimeter.

Plain Concrete Concrete cast without reinforcement or prestressing.

Plastic In structural investigation, the type of stress–strain response that occurs in ductile behavior, beyond the yield stress point; usually results in permanent deformation.

Plastic Hinge Rotational effect that occurs in steel members when the entire cross section is subjected to yield stress.

Plastic Moment Resisting moment produced at the point of development of a plastic hinge.

Poured-in-Place Concrete Concrete cast where it is intended to stay; also called *site cast*.

Precast Concrete Concrete members cast at a location other than that at which they are to be used.

Principal Axes The set of mutually perpendicular axes through the centroid of an area, about which the moments of inertia will be maximum and minimum. Called individually the *major axis* and the *minor axis*.

Radius of Gyration A defined mathematical property: the square root of the product of the moment of inertia divided by the area of a section.

Reaction Response. In structural investigation, the response of the structure to the loads or the response of the supports to the loaded structure. Mostly used to describe the latter.

Reference Design Values Values for allowable stress and modulus of elasticity for wood with no modification for usage conditions.

Reinforce To strengthen, usually by adding something.

Resistance Factor Reduction factor for adjustment of the ultimate resistance of a structural element to a force action—bending, compression, shear, and so on.

Restoring Moment Resistance to overturn due to the weight of the laterally loaded element.

Rigid Bent See *Rigid Frame*.

Rigid Frame Common term for a framework in which members are connected by joints that are capable of transmitting bending moments to the ends of the members. The term *rigid* derives not so much from the character of the frame as from that of the rigid joints. Now more accurately described as a *moment-resisting space frame*—a mouthful but more accurate.

Rigidity Degree of resistance to deformation; highly resistive elements are *stiff* or *rigid*, elements with low resistance are *flexible*.

Rolled Shape Linear steel member with a cross section produced by *hot rolling*.

Safety Relative unlikelihood of failure; absence of danger. The *safety factor* is the ratio of the structure's ultimate resistance to the actual demand (service load) on the structure.

Section The two-dimensional area or profile obtained by passing a plane through a form. *Cross section* usually implies a section at right angles to another section or to the linear axis of an object (such as a vertical cross section of a horizontal beam).

Sense See *Sign*.

Service Conditions Situations arising from the usage of a structure. See also *Load.*

Service Load See *Load.*

Shear A force effect that is lateral (perpendicular) to a structure, or one that involves a slipping effect, as opposed to a push–pull effect on a cross section.

Shear Wall A vertical diaphragm; acts as a bracing element for horizontal force (shear) by developing shear stress in the plane of the wall.

Shop Assemblage Refers to construction work performed at a production facility (the shop), as opposed to work done at the construction site (the field). Refers mostly to production and erection of steel frames.

Sign Algebraic notation of sense—positive, negative, or neutral. Relates to direction of forces—if up is positive, down is negative; to stress—if tension is positive, compression is negative; to rotation—if clockwise is positive, if counterclockwise is negative.

Site Cast See *Poured-in-Place Concrete.*

Slab A horizontal, planar element of concrete. Occurs as a roof or floor deck in a framed structure (called a *supported slab*) or as a pavement poured directly on the ground surface (called a *slab on grade*).

Slenderness Relative thinness; a measurement of resistance to buckling.

Stability Refers to the inherent capability of a structure to develop force resistance as a result of its form, orientation, articulation of its parts, type of connections, methods of support, and so on. Is not related to quantified strength or stiffness, except when actions involve buckling of slender elements.

Stiffness See *Rigidity.*

Strain Deformation resulting from stress; measured as a percentage change and is thus dimensionless.

Strength Capacity to resist force.

Strength Design One of two fundamental design methods for assuring a margin of structural safety. *Strength design*, also called *ultimate strength design*, is performed by using a design ultimate load (a magnification of the service load) and comparing it to the ultimate resistance of the structure. When strength design is performed with both factored loads and factored resistances, it is called *load and resistance factor design* (LRFD). When strength design is performed with unfactored loads and factored resistances, it is called *allowable strength design* (ASD).

Stress The mechanism of force within a material of a structure, visualized as a pressure effect (tension or compression) or a shear effect on the surface of a unit of material and quantified as force per unit area. *Allowable stress* is a limit established for design by stress methods; *ultimate stress* is that developed at a failure condition.

Stress Design One of two fundamental design methods for assuring a margin of structural safety. *Allowable stress design* (ASD) is performed by analyzing stresses produced by *service loads* and comparing them to established limits.

Stress–Strain Behavior The relation of stress to strain in a material or structure; usually visualized by a stress–strain graph covering the range from no load to failure. Various aspects of the form of the graph define particular behavior characteristics of the material. A straight line indicates an *elastic* relationship; a curve indicates *inelastic* behavior. A sudden bend in the graph usually indicates a plastic strain or *yield* that results in some permanent deformation. The slope of the graph (if straight), or of a tangent to the curve, indicates the relative stiffness of the material; measured by the tangent of the angle (stress–strain) and called the *modulus of elasticity*.

Structure That which gives form to something and works to resist changes in the form due to the actions of various forces.

Stud One of a set of closely spaced columns used to produce a framed wall.

Synthesis The process of combining a set of components into a whole; opposite of analysis.

System A set of interrelated elements; an ordered assemblage; an organized procedure or method.

Tension Force action that tends to separate adjacent particles of a material or pull elements apart. Produces straightening effects and elongation.

Tiedown See *Anchorage*.

Torsion Rotational (moment) effect involving twisting in a plane perpendicular to the linear axis of an element.

Truss A framework of linear elements that achieves stability through triangular formations of the elements.

Unreinforced Grammatically incorrect but commonly used term referring to concrete or masonry structures without reinforcement. Unreinforced concrete is also called *plain concrete*.

Uplift Net upward (lifting) force effect; may be due to wind, overturning moment, or an upward seismic acceleration.

Vector A mathematical quantity having direction as well as magnitude and sense (sign). Comparison is made to *scalar* quantities having only magnitude and sense, such as time and temperature. A vector may be represented by an arrow with its length proportional to the magnitude, the angle of its line indicating the direction, and the arrowhead representing the sign.

Vertical Diaphragm See *Shear Wall*.

Wall A vertical, usually planar, building element. *Foundation walls* are those partly or totally below ground. *Bearing walls* are used to carry vertical loads. *Shear walls* are used as bracing elements for horizontal forces in the plane of the wall. *Free-standing walls* are walls whose tops are not laterally braced. *Retaining walls* resist horizontal soil pressures perpendicular to the wall plane. *Curtain walls* are nonstructural exterior walls. *Partition walls* are nonstructural interior walls.

Wet Concrete Freshly mixed concrete before hardening.

Yield See *Stress–Strain Behavior*.

REFERENCES

1. *Minimum Design Loads for Buildings and Other Structures*, ASCE/SEI 7-10, American Society of Civil Engineers, Reston, VA, 2013.
2. J. Ambrose and D. Vergun, *Simplified Building Design for Wind and Earthquake Forces*, 3rd ed., Wiley, Hoboken, NJ, 1995.
3. *National Design Specification for Wood Construction*, American Wood Council, Leesburg, VA, 2014.
4. *International Building Code*, International Code Council, Inc., Country Club Hills, IL, 2011.
5. *Steel Construction Manual*, 14th ed., American Institute of Steel Construction, Chicago, IL, 2011.
6. *Standard Specifications, Load Tables, and Weight Tables for Steel Joists and Joist Girders*, Steel Joist Institute, Myrtle Beach, SC, 2010.
7. *Guide for Specifying Steel Joists with LRFD*, Steel Joist Institute, Myrtle Beach, SC, 2000.
8. C. M. Uang, S. W. Wattar, and K. M. Leet, Proposed Revision of the Equivalent Axial Load Method for LRFD Steel and Composite Beam—Column Design, *Engineering Journal*, AISC, Fall 1990.
9. *Steel Deck Institute Design Manual for Composite Decks, Form Decks, and Roof Decks*, Steel Deck Institute, St. Louis, MO, 2013.

10. *Building Code Requirements for Structural Concrete*, ACI 318-14, American Concrete Institute, Detroit, MI, 2014.

11. J. Ambrose, *Simplified Design of Building Foundations*, 2nd ed., Wiley, New York, 1988.

12. *Introduction to Lateral Design*, APA—The Engineered Wood Association, Tacoma, WA, 2003.

QUICK REFERENCE TO USEFUL DATA

Index

C

D

M

N

R

S

Y

Z